Nonconvex Optimization and Its Applications

VOLUME 85

GLOBAL OPTIMIZATION

Scientific and Engineering Case Studies

GLOBAL OPTIMIZATION

Scientific and Engineering Case Studies

Edited by

JÁNOS D. PINTÉR
Pintér Consulting Services Inc., Halifax, Nova Scotia, Canada

 Springer

Library of Congress Control Number: 2005937082

ISBN-10: 0-387-30408-8   e-ISBN: 0-387-30927-6
ISBN-13: 978-0387-30408-3

Printed on acid-free paper.

AMS Subject Classifications: 90C26, 65K99, 90C90, 68Uxx, 78A99

Printed in the United States of America.

9 8 7 6 5 4 3 2 1

springer.com

*"Far and away the best prize that life offers is
the chance to work hard at work worth doing."*

Theodore Roosevelt (1858–1919)

*This volume is dedicated to people around the world
who need our work – research, models, and techniques –
to help them live, and to live better.*

Contents

Preface

Man-made systems and controlled procedures can often be described, at least to a postulated "reasonable degree of accuracy", by continuous linear functions. For prominent instances of such descriptions, one may think of production and distribution systems and their basic quantitative models known from the operations research, management science, and industrial engineering literature. For illustrative purposes, we refer to the *50th Anniversary Issue of Operations Research* (2002), and to the many topical entries of Greenberg's (2005) *Mathematical Programming Glossary* (on the web), and the *Handbook of Applied Optimization* edited by Pardalos and Resende (2002).

If we attempt to analyze natural – physical, chemical, geological, biological, environmental – systems and their governing processes, then nonlinear functions will play an essential role in their quantitative description. Of course, there are also man-made systems that exhibit pronounced nonlinearities as illustrated by various scientific, engineering, econometric and financial studies. From the extensive related literature, consult e.g. Aris (1999), Bartholomew-Biggs (2005), Bracken and McCormick (1968), Corliss and Kearfott (1999), Diwekar (2003), Edgar, Himmelblau and Lasdon (2001), Floudas and Pardalos (2000), Gershenfeld (1999), Grossmann (1996), Hansen and Jørgensen (1991), Hendrix (1998), Kampas and Pintér (2006), Lopez (2005), Mistakidis and Stavroulakis (1997), Murray (1983), Papalambros and Wilde (2000), Pardalos, Shalloway, and Xue (1996), Pintér (1996a, 2001, 2006), Schittkowski (2002), Sethi and Thompson (2000), Tawarmalani and Sahinidis (2002), Wilson, Turcotte, and Halpern (2002), Zabinsky (2003).

Prescriptive (control, management, optimization) models based on a nonlinear systems description often may – or provably do – possess multiple local optima. The objective of global optimization (GO) is to find the "best possible" solution of multiextremal problems. Formally, the prototype continuous global optimization problem (GOP) can be stated as

(1) $\min f(x)$

subject to

(2) $x \in D = \{ l \leq x \leq u; \ f_j(x) \leq 0 \ j=1,...,J \} \subset R^n$.

The relations (1)-(2) describe a very general optimization model type defined by the following key ingredients:

- x real-valued n-vector that describes the decision alternatives
- $f(x)$ continuous objective function; $f_0(x) := f(x)$

- D non-empty set of feasible decisions
- $f_j(x)$ continuous constraint functions, for $j=1,...,J$
- l, u explicit, finite (component-wise) bounds of x.

Applying these basic analytical assumptions, it is easy to verify (by the extreme value theorem of classical analysis) that the optimal solution set of the GOP is non-empty. We shall denote the set of globally optimal solutions by X^*.

The solution of the GOP theoretically requires the determination of the set X^*, or at least an exact global solution $x^* \in X^*$ and the corresponding optimum value $f^* = f(x^*)$. In practice, this is often not possible (not only in the context of global optimization, but across the various classes of continuous optimization problems). Therefore our standard numerical objective is the sufficiently precise approximation of X^* or an x^*, and of f^*, based on a finite number of algorithmically chosen search steps. This objective requires the generation of a sample point sequence $\{x_k\}$ and corresponding model function evaluations $\{f_j(x_k)$ for $j=0,1,...,J\}$. The function evaluations optionally may include also the calculation or estimation of higher-order or other – local or global – information.

To illustrate the potential difficulty of the general GOP, consider the following (merely two-dimensional, and only box-constrained) model instance. (For simplicity, the variables are denoted by x and y, to avoid the use of indices.)

(3) $\min f(x,y)$ $f(x,y) := (sin(xy) + sin(3y-5x) + sin(x^2-4y) - 2)^2$

subject to

(4) $-3 \leq x \leq 3$ $-2 \leq y \leq 5$.

The surface and contour plots of the function (3) in the region (4) are shown by Figures 1 and 2.

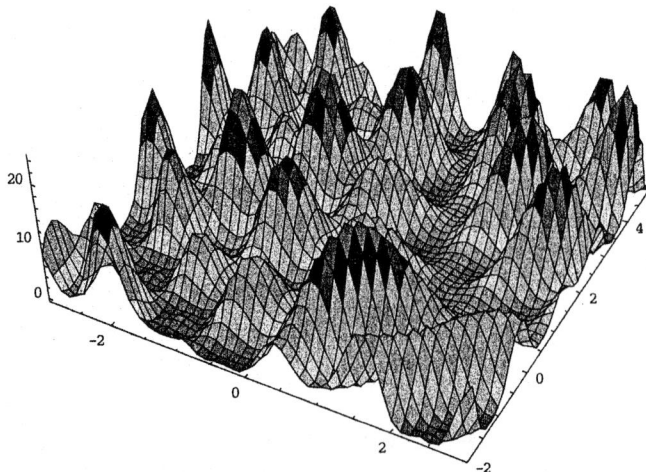

Figure 1. Surface plot of the objective function in the GOP (3)-(4).

These figures illustrate two important facts.

1) Global optimization models can be difficult (perhaps immensely difficult), even in very low-dimensions.
2) The classical repertoire of (local) numerical optimization is not suitable on its own to handle this problem and similar multi-extremal models. Indeed, depending on the starting point of any given local optimization method, it can easily get "trapped" in one of the many valleys (regions of attraction) of the objective function surface.

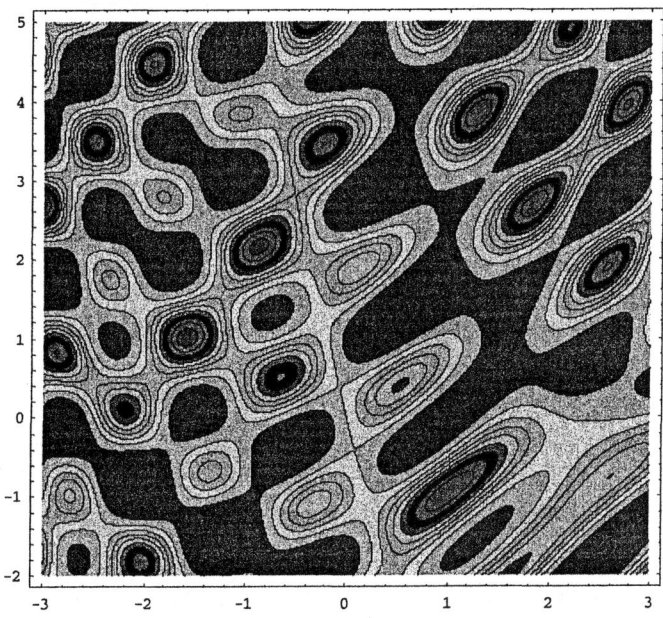

Figure 2. Contour plot of the objective function in the GOP (3)-(4).

As a side note for the interested reader, several numerical global solutions of the GOP (3)-(4) are listed below:

x^*	y^*	$f(x^*,y^*)$
−0.4891396742	−0.2256392574	$1.6537752573033925 \cdot 10^{-16}$
−1.1840237053	2.9661319066	$1.3696484154236246 \cdot 10^{-15}$
−0.4129962319	−0.3441255592	$4.3970820983759176 \cdot 10^{-17}$
−2.8343606253	−0.0078477508	$1.090931979062312 \cdot 10^{-15}$

In the case of multiple global optima, it may be of interest to select a specific solution such as the one that has a minimal Euclidean norm. The corresponding solution is

$$x^* \approx -0.4148624957, \ y^* \approx -0.254331095, \ \|(x^*, y^*)\| \approx 0.4866160665476753.$$

Each of these approximate global solutions have been generated using the MathOptimizer Professional software with an external Lipschitz Global Optimizer (LGO) solver link (Pintér, 1996a; Pintér, 1996...2005; Pintér and Kampas, 2003). The solution times are less than 0.3 second in each case, on a desktop PC equipped with an Intel Pentium 4 1.6 GHz processor.

Proceeding further, let us remark next that even relatively "simple" specific instances of the GOP may have an exponentially increasing number of global and local solutions. One can imagine a 10- or 100-variable direct extension of the model (3)-(4) in which selected two-dimensional search subspaces would show similar complexity to Figures 1 and 2. Since there are no universally applicable analytical criteria to verify global optimality, the required algorithmic search effort could grow exponentially in terms of the model size, since the algorithm should have to "visit" around the entire search region in sufficient detail. The model size of the GOP is characterized here simply by n and m, without further consideration given to the specific form of the model constraint functions. These functions could also be complicated: for a visual example, one can think of a collection of disjoint subsets ("islands of an archipelago") that together form D.

The general GO paradigm expressed by the model statement (1)-(2) is in contrast to traditional optimization methods. The latter – as a rule – will find only local optima of the GOP, based on a user-supplied initial (local) guess of the solution. Global optimization encompasses and extends local nonlinear optimization. This is certainly valid in a formal theoretical sense, but is pertinent also in numerical practice since GO strategies eventually need to have the convergence guarantee and precision of local search methodology, at least with respect to the global solution(s).

The field of global optimization has been gaining increasing attention in recent decades, and it has reached a certain level of maturity. The number of textbooks focused on GO is in the hundreds worldwide. The book series titled *Nonconvex Optimization and Its Applications* in itself includes nearly one hundred volumes, as of 2005. From this series, consult e.g. the introductory volume by Horst, Pardalos, and Thoai (1995), the two *Handbooks* edited by Horst and Pardalos (1995) and by Pardalos and Romeijn (2002), or the volumes by Kearfott (1996), Pintér (1996a), Tawarmalani and Sahinidis (2002), and Zabinsky (2003). The recent book chapter by Neumaier (2004) also provides a detailed overview of rigorous deterministic GO approaches.

Algorithmic advances – together with readily accessible and relatively inexpensive computational power – have led to a growing range of global optimization software implementations. This development has been greatly facilitated by significant progress in the areas of core professional (mainly C and

Fortran) compilers, spreadsheet-based modeling, algebraic modeling languages (with a focus on optimization), and integrated scientific-technical computing systems. Without going into details on any of these software systems, please consult e.g. the following references and the software products discussed therein.

- Compiler platforms: Lahey Computer Systems (2003), Microsoft (2005)
- Spreadsheet-based modeling: Frontline Systems (2005), Bertsimas and Freund (2000), Winston and Albright (1997)
- Algebraic modeling languages:
    AIMMS (Paragon Decision Technology, 2005)
    AMPL (Fourer, Gay, and Kernighan, 1993)
    GAMS (Brooke, Kendrick, and Meeraus, 1988)
    LINDO Solver Suite (Schrage, 2001; LINDO Systems, 2005)
    LPL (Virtual Optima, 2005)
    MPL (Maximal Software, 2005)
    TOMLAB for MATLAB (TOMLAB Optimization, 2004)
- Integrated scientific-technical computing systems:
    Maple (Lopez (2005), Maplesoft (2005), Parlar (2000), Wright (2002))
    MATLAB (Moler (2004), The MathWorks (2005), Venkataraman (2002), Wilson, Turcotte, and Halpern (2002))
    Mathematica (Maeder (2000), Trott (2004, 2005), Wolfram (2004), Wolfram Research (2005))

With respect to modeling environments and GO software implementations, see also the edited volumes (Kallrath, 2004), (Liberti and Maculan, 2005), and (Pintér, 2005). The websites maintained by Fourer (2005), Mittelmann and Spellucci (2005), Neumaier (2005), and by the Optimization Technology Center (2005) also offer valuable topical information.

While most GO software products a decade ago have been arguably more "academic" than "professional" (Pintér, 1996b), today a growing number of companies offer professionally developed and maintained GO software, often as a solver component or option of modeling language and systems. Global optimization is also becoming part of the "mainstream" operations research curriculum: for instance, the prominent textbook by Hillier and Lieberman (2005) now offers also GO demo software (the model size-limited MPL/LGO implementation) as part of its electronic supplement.

The present volume illustrates the applicability of global optimization strategies and software to a broadening range of practically important issues. The emphasis is on real-world applications – including also open problems – that *apparently need genuine GO methodology. The contributed chapters cover* applications from the following areas:

- agroecosystem management
- assembly line design
- bioinformatics
- biophysics

- "black box" systems optimization
- chemical process optimization
- chemical product design
- composite structure design
- computational modeling of atomic and molecular structures
- controller design for induction motors
- electrical engineering design
- feeding strategies in animal husbandry
- inverse position problem in advanced kinematics
- laser design
- learning in neural nets
- mechanical engineering design
- numerical solution of equations
- radiotherapy planning
- robot design
- satellite data analysis
- water resources systems
- wireless communication networks

These applications can be broadly classified as belonging to the areas of natural sciences (agriculture, biology, computational chemistry, environment) and engineering (design and process optimization), while mathematical modeling, optimization and computer science are the unifying concepts. Let us remark here that in addition to engineering and scientific applications (represented by the studies of this volume) important areas of GO applications emerge also in econometrics and finance.

Due to the large variety of model types encompassed by the general GO paradigm (1)-(2), there is no "universally best" global optimization strategy or software that will handle all GO models with theoretical rigor and competitive efficiency, within the framework of a prefixed amount of resources (time or model function evaluations). This is true even if the size of models (variables, functions) is a priori limited: recall Figures 1 and 2, which illustrate the potential difficulty of merely two-dimensional, box-constrained models. Of course, this does not mean that models of practical relevance can not be tackled successfully; however, one may have to (in fact, typically should) rely on modeling insight, intelligent – and perhaps model-dependent – combinations and/or adaptations of GO approaches. As a general guideline, even a rudimentary global scope search can lead to better solutions than the most sophisticated local search method started from the "wrong valley"... (Recall again Figures 1 and 2.)

The solution strategies discussed in this volume illustrate the above points by encompassing a range of practically viable methods. The contributing authors have made an honest effort to illustrate not only the successes but also the difficulties and the current limitations of practical global optimization.

Specifically, the chapters discuss both rigorous (theoretically globally convergent) and heuristic GO approaches such as

- adaptive random search
- branch-and-bound strategies
- evolutionary search approaches
- flexible simplex search heuristics
- genetic algorithms
- hybrid (stochastic-deterministic) approaches
- mixed integer nonlinear programming methods
- neural networks
- simulated annealing
- stochastic simulation
- "traditional" local nonlinear optimization

The methods listed above can be broadly categorized as follows:

- theoretically rigorous with a deterministic guarantee of global convergence (branch-and-bound, "exact" mixed integer GO, and other approches)
- theoretically rigorous with a probabilistic guarantee of global convergence (adaptive random search, properly designed combined stochastic-deterministic approaches, properly implemented simulated annealing, and others)
- "obvious" extensions of traditional local search methodology (such as a limited globalized search effort combined with local optimization)
- heuristic direct search methods (e.g., flexible simplex search)
- metaheuristics (evolutionary search, genetic algorithms, neural networks)

Although the last three GO approaches do not have generally valid (provable) theoretical convergence properties in the continuous GO context, such heuristic methods still can be very useful in practice. Note furthermore that these heuristic methods can be adapted to guarantee at least stochastic global convergence. In addition to the already mentioned references on rigorous deterministic or stochastic global optimization, a few useful references on heuristic methods are Glover and Laguna (1997), Goldberg (1989), Michalewicz (1996), Osman and Kelly (1996), Rothlauf (2002), Rudolph (1997), Voss, Martello, Osman, and Roucairol (1999). It is worthwile pointing out that although so far most heuristic approaches have been designed to solve combinatorial (discrete) optimization models, such methods can be adapted to tackle also continuous models.

Let us also remark that the GO literature offers a growing number of sufficiently detailed comparative numerical studies which shed light on the applicability of the most prominent methods and software to models of realistic complexity, in addition to "standard" GO test models that have been used for a few decades. Consult e.g. the test model library compiled by Floudas, Pardalos, Adjiman, Esposito, Gumus, Harding, Klepeis, Meyer, and Schweiger (1999), as well as topical expositions by Pintér (2002, 2003), Ali, Khompatraporn, and Zabinsky (2005), Khompatraporn, Pintér, and Zabinsky (2005). The websites by Fourer (2005), Mittelmann and Spellucci (2005), Neumaier (2005), and the

Optimization Technology Center (2005) also offer useful information regarding this point.

The individual chapters of the present volume have been written with the objective of addressing both experts and non-experts in the specific application area discussed. Therefore the authors attempted to follow a "tutorial" style, providing sufficient background to the key issues, model formulation and solution approaches presented. Most chapters are strongly application-oriented, in accordance with the overall objectives of this book.

We trust that our work will be of interest to researchers and practitioners in academia, research and consulting organizations, and industry. The book presents GO challenges and real-world case studies in sufficient detail, to enable graduate level classroom discussions and independent studies. The book can also be used in the framework of a practically motivated seminar or lecture series on nonlinear modeling and optimization.

The contributing authors and myself welcome your comments and suggestions related to this volume. Thanks for your attention, and enjoy the book!

János D. Pintér
PCS Inc.
Halifax, NS, Canada

References to Preface

This illustrative list of references includes books, software systems, and websites that are (mostly or at least partially) related to the practice of nonlinear systems modeling and global optimization, including software and applications. Many of the websites listed below offer extensive further information of relevance.

Ali, M.M., Khompatraporn, Ch., and Zabinsky, Z.B. (2005) A numerical evaluation of several stochastic algorithms on selected continuous global optimization test problems. *Journal of Global Optimization* 31, 635-672.

Aris, R. (1999) *Mathematical Modeling: A Chemical Engineer's Perspective.* Academic Press, San Diego, CA..

Bartholomew-Biggs, M. (2005) *Nonlinear Optimization with Financial Applications.* Kluwer Academic Publishers, Dordrecht.

Bertsimas, D. and Freund, R.M. (2000) *Data, Models and Decisions: The Fundamentals of Management Science.* South-Western College Publishing, Cincinnati, OH.

Bracken, J. and McCormick, G.P. (1968) *Selected Applications of Nonlinear Programming.* Wiley, New York.

Brooke, A., Kendrick, D. and Meeraus, A. (1988) *GAMS: A User's Guide.* The Scientific Press, Redwood City, CA. Revised versions are available from the GAMS Development Corporation, Washington, DC; http://www.gams.com.

Corliss, G.F. and Kearfott, R.B. (1999) Rigorous global search: industrial applications. In: Csendes, T., ed. *Developments in Reliable Computing,* pp. 1-16. Kluwer Academic Publishers, Dordrecht.

Diwekar, U. (2003) *Introduction to Applied Optimization.* Kluwer Academic Publishers, Dordrecht.

Edgar, T.F., Himmelblau, D.M., and Lasdon, L.S. (2001) *Optimization of Chemical Processes.* (2nd Edn.) McGraw-Hill, New York.

Floudas, C.A., Pardalos, P.M., Adjiman, C., Esposito, W.R., Gumus, Z.H., Harding, S.T., Klepeis, J.L., Meyer, C.A., and Schweiger, C.A. (1999) *Handbook of Test Problems in Local and Global Optimization.* Kluwer Academic Publishers, Dordrecht.

Floudas, C.A. and Pardalos, P.M., Eds. (2000) *Optimization in Computational Chemistry and Molecular Biology: Local and Global Approaches.* Kluwer Academic Publishers, Dordrecht.

Fourer, R., Gay, D.M., and Kernighan, B.W. (1993) *AMPL: A Modeling Language for Mathematical Programming.* The Scientific Press, Redwood City, CA. For the latest edition, see http://www.ampl.com.

Fourer, R. (2005) *Nonlinear Programming Frequently Asked Questions.* Optimization Technology Center of Northwestern University and Argonne National Laboratory, Evanston, IL and Argonne, IL. http://www-unix.mcs.anl.gov/otc/Guide/faq/nonlinear-programming-faq.html.

Frontline Systems (2005) *Premium Solver Platform – Solver Engines. User Guide.* Frontline Systems, Inc., Incline Village, NV. http://www.solver.com.

Gershenfeld, N. (1999) *The Nature of Mathematical Modeling.* Cambridge University Press, Cambridge.

Greenberg, H.J. (2005) *Mathematical Programming Glossary.* http://carbon. cudenver.edu/~hgreenbe/glossary.

Glover, F. and Laguna, M. (1997) *Tabu Search.* Kluwer Academic Publishers, Dordrecht.

Goldberg, D.E. (1989) *Genetic Algorithms in Search, Optimization, and Machine Learning.* Addison-Wesley, Reading, MA.

Grossmann, I.E., Ed. (1996) *Global Optimization in Engineering Design.* Kluwer Academic Publishers, Boston / Dordrecht / London.

Hansen, P.E. and Jørgensen, S.E., Eds. (1991) *Introduction to Environmental Management.* Elsevier, Amsterdam.

Hendrix, E.M.T. (1998) *Global Optimization at Work.* Ph.D. Thesis, University of Wageningen, Netherlands.

Hillier, F.J. and Lieberman, G.J. (2005) *Introduction to Operations Research.* (8th Edition.) McGraw-Hill, New York.

Horst, R. Pardalos, P.M., and Thoai, N.V. (1995) *Introduction to Global Optimization.* Kluwer Academic Publishers, Dordrecht.

Horst, R. and Pardalos, P.M., Eds. (1995) *Handbook of Global Optimization. Volume 1.* Kluwer Academic Publishers, Dordrecht.

Kallrath, J., Ed. (2004) *Modeling Languages in Mathematical Optimization.* Kluwer Academic Publishers, Dordrecht.

Kampas, F.J. and Pintér, J.D. (2006) *Advanced Optimization: Scientific, Engineering, and Economic Applications with Mathematica Examples.* Elsevier, Amsterdam. (To appear.)

Kearfott, R.B. (1996) *Rigorous Global Search: Continuous Problems.* Kluwer Academic Publishers, Dordrecht.

Khompatraporn, Ch., Pintér, J.D., and Zabinsky, Z.B. (2005) Comparative assessment of algorithms and software for global optimization. *Journal of Global Optimization* 31, 613-633.

Lahey Computer Systems (2003) *Fortran 95 User's Guide.* Lahey Computer Systems, Inc., Incline Village, NV. http://www.lahey.com.

Liberti, L., and Maculan, N., Eds. (2005) *Global Optimization: From Theory to Implementation.* Springer Science + Business Media, New York, 2005.

LINDO Systems (2005) *Solver Suite.* LINDO Systems, Inc., Chicago, IL. http://www.lindo.com.

Lopez, R.J. (2005) *Advanced Engineering Mathematics with Maple.* (Interactive electronic edition.) Waterloo Maple Inc., Waterloo, ON.

Maeder, R.E. (2000) *Computer Science with Mathematica.* Cambridge University Press, Cambridge, UK.

Maplesoft (2005) *Maple.* Maplesoft, Inc., Waterloo, ON. http://www. maplesoft.com.

Maximal Software (2005) *MPL Modeling System.* Maximal Software, Inc. Arlington, VA. http://www.maximal-usa.com.

Michalewicz, Z. (1996) *Genetic Algorithms + Data Structures = Evolution Programs.* (3rd Edition.) Springer, New York.

Microsoft (2005) *Microsoft Visual Studio Developer Center.* Microsoft Corporation, Redmond, WA. http://www.microsoft.com/vstudio/.

Mistakidis, E.S. and Stavroulakis, G.E. (1997) *Nonconvex Optimization. Algorithms, Heuristics and Engineering Applications of the F.E.M.* Kluwer Academic Publishers, Dordrecht.

Mittelmann, H.D. and Spellucci, P. (2005) *Decision Tree for Optimization Software.* http://plato.la.asu.edu/guide.html.

Moler, C. (2004) *Numerical Computing with MATLAB.* SIAM, Philadelphia.

Murray, J.D. (1983) *Mathematical Biology.* Springer, Berlin.

Neumaier, A. (2004) Complete search in continuous global optimization and constraint satisfaction. In: Iserles, A. (Ed.) *Acta Numerica 2004*, pp. 1-99. Cambridge University Press, Cambridge.

Neumaier, A. (2005) *Global Optimization.* http://www.mat.univie.ac.at/~neum/ glopt. html.

Nonconvex Optimization and Its Applications. A book series of Kluwer Academic Publishers, Dordrecht, published since 1995; now published by Springer Science + Business Media, New York.

Operations Research (2002) *50th Anniversary Issue* (includes articles written by invited contributors). INFORMS, Linthicum, MD. For further information, visit http://www.informs.org.

Optimization Technology Center (2005) *NEOS Guide to Optimization Software.* http://www-fp.mcs.anl.gov/otc/Guide/SoftwareGuide/.

Osman, I.H. and Kelly, J.P., Eds. (1996) *Meta-Heuristics: Theory and Applications.* Kluwer Academic Publishers, Dordrecht.

Papalambros, P.Y. and Wilde, D.J. (2000) *Principles of Optimal Design.* Cambridge University Press, Cambridge.

Paragon Decision Technology (2005) AIMMS. Paragon Decision Technology B.V., Haarlem, The Netherlands. http://www.aimms.com.

Pardalos, P.M. and Resende, M.G.C., Eds. (2002) *Handbook of Applied Optimization.* Oxford University Press, Oxford.

Pardalos, P.M. and Romeijn, H.E., Eds. (2002) *Handbook of Global Optimization. Volume 2.* Kluwer Academic Publishers, Dordrecht.

Pardalos, P.M., Shalloway, D. and Xue, G., Eds. (1996) *Global Minimization of Nonconvex Energy Functions: Molecular Conformation and Protein Folding.* DIMACS Series, Vol. 23, American Mathematical Society, Providence, RI.

Parlar, M. (2000) *Interactive Operations Research with Maple: Methods and Models.* Birkhäuser, Boston.

Pintér, J.D. (1996a) *Global Optimization in Action.* Kluwer Academic Publishers, Dordrecht.

Pintér, J.D. (1996b) Continuous global optimization software: A brief review. *Optima* 52, 1-8. See also http://plato.la.asu.edu/gom.html.

Pintér, J.D. (1996...2005) *LGO—A Model Development System for Continuous Global Optimization. User's Guide*. Pintér Consulting Services, Inc., Halifax, NS. http://www.pinterconsulting.com.

Pintér, J.D. (2001) *Computational Global Optimization in Nonlinear Systems: An Interactive Tutorial*. Lionheart Publishing, Atlanta, GA.

Pintér, J.D. (2002) Global optimization: software, test problems, and applications. In: Pardalos, P. M. and Romeijn, H. E., Eds., *Handbook of Global Optimization, Vol. 2*, pp. 515-569. Kluwer Academic Publishers, Dordrecht.

Pintér, J.D. (2003) GAMS/LGO nonlinear solver suite: key features, usage, and numerical performance. (Submitted for publication.) GAMS Solver Documentation Pages http://www.gams.com/solvers/solvers.htm#LGO.

Pintér, J.D. (2005) Nonlinear optimization in modeling environments: software implementations for compilers, spreadsheets, modeling languages, and integrated computing systems. In: Jeyakumar, V. and Rubinov, A.M., Eds., *Continuous Optimization: Current Trends and Applications*, pp. 147-173. Springer Science + Business Media, New York.

Pintér, J.D. (2006) *Applied Nonlinear Optimization in Modeling Environments*. CRC Press, Boca Raton, FL. (To appear.)

Pintér, J.D. and Kampas, F.J. (2003) *MathOptimizer Professional – An Advanced Modeling and Optimization System for Mathematica Users with an External Solver Link: User Guide*. Pintér Consulting Services, Inc., Halifax, NS. http://www.pinterconsulting.com.

Rothlauf, F. (2002) *Representations for Genetic and Evolutionary Algorithms*. Physyca-Verlag / Springer, Heidelberg.

Rudolph, G. (1997) *Convergence Properties of Evolutionary Algorithms*. Verlag Dr. Kovac, Hamburg.

Schittkowski, K. (2002) *Numerical Data Fitting in Dynamical Systems*. Kluwer Academic Publishers, Dordrecht.

Schrage, L. (2001) *Optimization Modeling with LINGO*. LINDO Systems, Inc., Chicago, IL.

Sethi, S.P. and Thompson, G.L. (2000) *Optimal Control Theory. Applications to Management Science and Economics*. (2nd Edition.) Kluwer Academic Publishers, Dordrecht.

Tawarmalani, M. and Sahinidis, N.V. (2002) *Convexification and Global Optimization in Continuous and Mixed-integer Nonlinear Programming*. Kluwer Academic Publishers, Dordrecht.

The MathWorks (2005) *MATLAB*. The MathWorks, Inc., Natick, MA. http://www.mathworks.com.

TOMLAB Optimization (2004) *TOMLAB*. TOMLAB Optimization AB, Västerås, Sweden. http://www.tomlab.biz.

Trott, M. (2004, 2005) *The Mathematica GuideBook, Volumes 1-4*. Springer Science + Business Media, New York.

Venkatamaran, P. (2002) *Applied Optimization with MATLAB Programming*. Wiley, New York.

Virtual Optima (2005) *LPL: A Mathematical Modeling Language*. (Author: T. Hürlimann.) Virtual Optima, Inc., Fribourg, Switzerland. http://www.virtual-optima.com.

Voss, S., Martello, S., Osman, I.H., and Roucairol, C., Eds. (1999) *Meta-Heuristics: Advances and Trends in Local Search Paradigms for Optimization*. Kluwer Academic Publishers, Dordrecht.

Wilson, H.B., Turcotte, L.H., and Halpern, D. (2002) *Advanced Mathematics and Mechanics Applications Using MATLAB*. (3rd Edition.) CRC Press, Boca Raton, FL.

Winston, W.L. and Albright, S.C. (2001) *Practical Management Science: Spreadsheet Modeling and Applications*. (2nd Edition.) Duxbury Press, Belmont, CA.

Wolfram, S. (2004) *The Mathematica Book*. Wolfram Media and Cambridge University Press. (5th Edition.)

Wolfram Research (2005) *Mathematica*. Wolfram Research, Inc., Champaign, IL. http://www.wolfram.com.

Wright, F. (2002) *Computing with Maple*. CRC Press, Boca Raton, FL.

Zabinsky, Z.B. (2003) *Stochastic Adaptive Search for Global Optimization*. Kluwer Academic Publishers, Boston / Dordrecht / London.

Acknowledgments

First of all, I would like to thank all contributing authors for their work, patience, and support that made this edited book project possible.

I also thank all reviewers for providing constructive comments that helped to improve both contents and style of the contributions. The following colleagues served as reviewers of one or several chapters: Luke Achenie, Julio Banga, George Corliss, Jon Doye, Bernd Hartke, Glenn Isenor, Andrzej Osyczka, Brahim Rekiek, Graham Wood, Quan Zhu, and Francesco Zirilli, in addition to the editor.

Baker Kearfott and Arnold Neumaier served as reviewers for the entire edited volume: thanks for your constructive comments.

I wish to thank Springer, in particular John Martindale and Robert Saley, for supporting and managing this project.

Last but very far from least, I thank Susan for her love and quiet, but ever-present, support.

JDP

Chapter 1

A GLOBAL OPTIMIZATION STRATEGY AND ITS USE IN SOLVENT DESIGN

L. E. K. Achenie, G. M. Ostrovsky and M. Sinha
University of Connecticut

Abstract: Solvent design can be modeled as a mixed integer nonlinear programming problem (MINLP) in which discrete variables denote the presence or absence of molecular structural entities and to what extent they occur in the pure component compound or mixture. On the other hand, continuous variables denote process variables such as temperature and flow rates. In the MINLP model the number of discrete variables can range from several tens to several hundreds. Therefore the use of the standard branch-and-bound method for solving the problem can be computationally intensive since all the variables (discrete and or continuous) must be used as branching variables. To overcome this problem, we have proposed a new strategy in which branching is done using branching functions instead of all the search variables. This approach results in a decrease in the number of branching variables. During branch and bound, the bounding operation is performed in the search variables space, while the branching operation is performed in a reduced dimension space defined by the branching (or splitting) functions. The branching functions are determined from the special tree function representation of both the objective function and constraints. The suggested MINLP solution approach is demonstrated on a solvent design application.

Key words: Solvent design, MINLP, branching function, special tree function, branch and bound.

1.   INTRODUCTION

Chemical product design addresses the design of single component chemical compounds and/or mixtures (blends) of compounds with pre-specified thermo-physical properties. In recent years, the traditional wet chemistry based chemical product design is being supplemented with computer-aided approaches, namely computer-aided molecular design (CAMD). The CAMD problem can often be posed as a mathematical program in which a number of binary and continuous variables define the search space (Duvedi and Achenie, 1996; Churi and Achenie, 1996; Maranas, 1997; Odele and Machietto, 1993; Pistikopoulos and Stefanis, 1998). A binary variable is an integer variable that can have one of two possible values, for example 0 and 1. This chapter discusses a globally optimal branch and bound approach to solving the resulting mathematical program. The approach is more fully discussed in a similar chapter in Sinha et al., 2002.

2.   PROBLEM DEFINITION

A typical molecular design problem may be modeled as a single objective minimization or maximization subject to structural and performance constraints. Thus a CAMD problem for single component molecular design in which thermo-physical property matching is sought may be modeled as

$$\min_{x,v,\theta} f(x,v,\theta) \tag{1}$$

$$\varphi_j(x,v,\theta) \le 0, \qquad j = 1,...,m_1 \tag{2}$$

$$h_i(x,v,\theta) = 0, \qquad i = 1,...,m_2 \tag{3}$$

where v is a vector of binary variables that define the molecular structure, x is a vector of continuous variable such as process variables (pressure, temperature, etc.) and θ is a vector of group contribution parameters. Note that additional binary variables may be included in v to indicate additional constraints on the kind of molecular structures that can be generated. $f(x,v,\theta)$ is the performance objective function (for example some undesirable property such as a compound's ozone depletion potential). The group contribution model is a structure-property correlation that has found wide use in the chemical process industry.

The constraints involve (a) structural feasibility, (b) physical property targets, and (c) process constraints. The constraints associated with structural feasibility are usually linear. Physical property targets often have the form $p_k^L \leq p_k(x,v,\theta) \leq p_k^U$. If $p_k(x,v,\theta)$ is modeled using group contribution, then it may have the form

$$p_k = \sum_j n_j \theta_j^1 / \sum_j n_j \theta_j^2 \ .$$

Here θ_j^1 and θ_j^2 are elements in θ and n_j is the number of θ_j^1 or θ_j^2 present in the molecule. Transformation of such constraints into a linear form is straightforward. The function $p_k(x,v,\theta)$ can also have the form

$$p_k = f_{NL}^1 \left(\sum_j n_j \theta_j^a \right) \bigg/ f_{NL}^2 \left(\sum_j n_j \theta_j^b \right)$$

where f_{NL}^1 and f_{NL}^2 are nonlinear functions; in addition θ_j^a and θ_j^b are parameters. Property constraints, which employ the given form, include solubility parameter based models often used in solvent design. It is not always possible to reformulate these constraints into linear or convex forms.

The nonlinear mathematical programming model for the CAMD problem (PMD) has the following features: (a) it is a nonconvex mixed integer nonlinear problem (MINLP) problem involving a large number of binary variables, (b) the number of linear constrains is larger than the number of nonlinear constraints, and (c) most of the components of the design vector (u) participate in the nonlinear terms. Previous attempts using global optimization are either geared to small size problems or use soft computing approaches (such as simulated annealing and genetic algorithms). The approach discussed here is based on the branch and bound (BB) algorithm. The basic BB algorithm may encounter a large number of branching variables for product design problems. To address this, the branch-and-bound global optimization algorithm presented here exploits the problem structure and allows significant reduction in branching expressions. A discussion of the algorithm is based on the papers (Sinha, Achenie and Ostrovksy, 1999) and (Ostrovksy, Achenie and Sinha, 2000).

In group contribution based computer aided single component product design, solvents are formed from certain combinations of a set of structural groups. The pre-specified set of m structural groups is called the *basis set*. The size and composition of the basis set depends on the intended application, the availability of accurate property prediction models and the computational resources available. First, we define a set of variables based on an initial set of structural groups as

$$
u_{ik} = \begin{cases} 1 & \text{if the } i\text{-th group in the molecule is the } k\text{-th structural} \\ & \text{group in the basis-set} \\ 0 & \text{otherwise} \end{cases} \qquad \textit{Churi - Achenie model}
$$

$$
u_i = \begin{cases} 1 & \text{if the i-th structural group in the basis-set} \\ & \text{is in the molecule} \\ 0 & \text{otherwise} \end{cases} \qquad \textit{Odele - Machietto model}
$$

$$(4)$$

Odele and Machietto (1993) proposed a formulation that ensured that the valence of each structural group was satisfied. This formulation only accounts for the presence and absence of structural groups in the molecule. However, it does not consider the information that determines how the groups are connected to each other in the molecule. To overcome this limitation, Churi and Achenie (1996) proposed a model that gives complete information with regard to how the groups are connected to each other. Presently there is no known group contribution method that takes advantage of the connectivity information of the Churi-Achenie model. In the latter model, the following variables were introduced

$$
z_{ijp} = \begin{cases} 1 & \text{if the } i\text{-th group's } j\text{-th site is attached to the } p^{\text{th}} \text{ group} \\ 0 & \text{otherwise} \end{cases}
$$

$$
w_i = \begin{cases} 1 & \text{if the } i\text{-th group in a molecule does not have a group attached} \\ 0 & \text{otherwise} \end{cases}
$$

$$(5)$$

For single component solvents structural constraints are imposed for (a) limiting the number of structural groups in a molecule; (b) ensuring that the number of bonds attached to a group equals the valence of the group; and (c) ensuring that each group in a molecule is attached to at least one other group. The formulation is effective in specifying whether the molecule is acyclic or cyclic. Moreover the maximum number of cycles can also be controlled. This representation is also effective in distinguishing between isomers. If the chemical process is not accounted for, then the pure component molecular design problem involves only binary variables. The maximum number of groups in a molecule is n_{max}; the number of groups in the basis set is m with the maximum valence of s_{max}. In this case the search dimension is then given by $n_{max} \times m + n_{max} \times s_{max} \times n_{max} + n_{max}$. Here the number of binary variables is equal to the sum of the dimensions of u, z and w, respectively (assuming the Churi-Achenie model is used). The number of linear structural constraints employed are $n_{max}^2 + n_{max} \times m + 3n_{max} + s_{max} + 1$. For example, a CAMD problem with $n_{max} = 5$, $m = 10$, and $s_{max} = 2$ results in 93 linear constraints. The number of nonlinear constraints is generally small compared to the number of linear constraints. Let all the binary variables in the problem be assembled in the vector v (q-dimensional). If the Odele-Machietto model is employed then $v \equiv u$; on the other hand if the Churi-Achenie model is employed then $v \equiv [u, z, w]$. Then the solvent design problem (see Eq. (1), (2), (3)) can be expressed compactly as a mixed integer nonlinear program in the general form

$$\textbf{P:} f = \min_{x, v \in D} f(x, v) \tag{6}$$

such that

$$D = \{x, v : c \le x \le d, \varphi_i(x, v) \le 0,$$
$$i = 1, \dots m,\ h(x, v) = 0, x \in X \subseteq \Re^n,\ v \in \{0,1\}^q\}$$

3.   PROPOSED SOLUTION METHOD

3.1   Branch-and-Bound Preliminaries

The branch and bound (BB) method (Horst and Tuy, 1990) has been used for solving several problems in chemical engineering (Ostrovsky *et. al.*, 1990, Friedler *et. al.*, 1998, Quesada and Grossmann, 1995, Ryoo and Sahinidis, 1996, Maranas and Floudas 1997, Adjiman *et. al.*, 1998). The generic BB method looks for a minimum of the objective function $f(x,v)$ by partitioning the region D into subregions D_i with respect to the search variables. At each iteration, a subregion D_i is further partitioned into D_{ip} and D_{iq} ($D_i = D_{ip} \cup D_{iq}$). The generic BB method consists of the following:

(i)   An algorithm for estimating a lower bound (LB) μ_i on the objective function $f(x,v)$ in any subregion $D_i \in D$ such that
$$\mu_i \leq f(x,v) \quad \forall x, v \in D_i$$

(ii)   An algorithm for estimating an upper bound (UB) η_j on $f(x,v)$ in any $D_i \in D$ such that $\eta_j \geq f(x,v) \quad \forall x, v \in D_j$

(iii)   An algorithm for partitioning D_i

Designate the set of subregions at the k-th iteration of the BB method as $L^{(k)} = (D_i, i = 1,..., N_k)$. Let $I^{(k)}$ be the index set of the subregions belonging to $L^{(k)}$. Then the algorithm for the BB method is as follows

Step 1: Set *k=1*. Give an initial set $L^{(0)}$ of the subregions D_i *(i=1,...,N₀,* usually *N₀=1)*.

Step 2: Calculate an LB for each $D_i \in L^{(k)}$

Step 3: Determine the subregion with the least LB. Let it be the l_m-th region then

$$\mu_{l_m} = \min_{l \in I^{(k)}} \mu_l \tag{7}$$

Step 4: Split D_{l_m} into two subregions D_p and D_q ($D_{lm} = D_p \cup D_q$) such that

$$D_p = \{x : x \in D_{lm}, x_s \leq c_s\}, D_q = \{x : x \in D_{lm}, x_s \geq c_s\}$$

The variable, x_s, is the *branching variable* and c_s is the *branching point*.

Step 5: Determine LB and UB for subregions p and q.

Step 6: Determine the least upper bound $\eta^{(k)}$ at the k-th iteration.

$$\eta^{(k)} = \min(\eta^{k-1}, \eta_p, \eta_q) \qquad (8)$$

For the first iteration $\eta^{(0)} = \infty$

Step 7: If $\eta^{(k)} - \mu_{l_m} \leq \varepsilon$ then STOP.

Step 8: If

$$\mu_j > \eta^k \qquad (9)$$

is met for $j = p$ or $j = q$ then the corresponding subregion is eliminated from consideration.

Step 9: Form a new set $L^{(k)}$ of the remaining subregions as follows

$$L^{(k)} = (D_1, ..., D_{l_m-1}, D_{l_m+1}, ..., D_{N_{k-1}}, \overline{L})$$

where

$$\overline{L} = \begin{cases} D_p, D_q & \text{if } \mu_j \leq \eta^{(k)} \quad j = p, q \\ D_q, & \text{if } \mu_q < \eta^{(k)} < \mu_p \\ D_p, & \text{if } \mu_q < \eta^{(k)} < \mu_p \end{cases}$$

Step 10: Set $k=k+1$, and go to Step 3

Each BB method needs to develop algorithms for partitioning and for estimating lower and upper bounds. Thus we describe algorithms we have developed for estimating lower and upper bounds for the mixed integer nonlinear program arising from our formulation of the computer aided molecular design problem. Let us consider the partitioning algorithm. At each iteration in a standard BB method, the "optimal" subregion D_{l_m} is partitioned into two subregions D_p and D_q using the constraints $x_i \le x_i^*$ and $x_i \ge x_i^*$ or $v_i \le v_i^*$ and $v_i \ge v_i^*$ as follows

$$D_p = \{x : x \in D_{lm}, x_s \le c_s\}, D_q = \{x : x \in D_{lm}, x_s \ge c_s\}$$

The variable, x_s, is the *branching variable* and c_s is the *branching point*. Different BB methods have different ways of selecting these. Thus in this case $n+q$ branching variables are used. In a realistic product design problem, the number of branching variables can be several hundred. It is known that the number of branching nodes grows exponentially. To alleviate this problem, we will use the following new partition algorithm. Instead of branching on the variables (x,v), we will use appropriate functions $\psi_j(x,v)$, $j = 1,...p$ of the search variables *for* branching. Subsequently, D_i will be determined by the set of inequalities

$$a_j^i \le \psi_j(x,v) \le b_j^i, j = 1,...p,$$

where the lower and upper bounds a_j^i and b_j^i are the dimensions of the multidimensional box (subregion) D_i are determined by the branch-and-bound strategy. Thus D_i has the form

$$D_i = \{x,v : x,v \in D; a_j^i \le \psi_j(x,v) \le b_j^i, j = 1,...p, \}$$

Problem **P** for subregion D_i is written as

$$P_i^L: \quad f_i = \min_{x,v \in D_i} f(x,v) \qquad\qquad (10)$$

A direct solution of the above problem is very difficult. Instead, the approach to be described finds the solution indirectly by successively estimating lower and upper bounds for the performance objective function f_i. In the limit, these bounds should collapse into one to give a solution to the above problem. Thus it is appropriate to discuss how these bounds are obtained.

3.2   Lower Bound Algorithm

A lower bound f_i^L for f_i on D_i is obtained by solving the following problem

$$\mathbf{P_i^L}: f_i^L = \min_{x,v \in \tilde{D}_i} L[f(x,v); \tilde{D}_i]$$

where

$$\tilde{D}_i = \{ x,v : L[\varphi_k ; D_i] \leq 0 \ \ k=1,...,m;$$
$$L[\psi_j ; D_i] \leq b_j^i; L[-\psi_j \ D_i] \leq -a_j^i;$$
$$v \in \{0,1\}^q \}$$

and $L[g(x,v); D_i]$ is a convex underestimator for the generic function $g(x,v)$. Then it is easy to verify that $D_i \subset \tilde{D}_i$.

Some alternatives for estimating lower bounds are: (a) The use of linear or convex nonlinear underestimators; (b) Enforcing the integrality of all the binary variables v at each iteration (Pantelides, 1996); (c) The variables v are considered as continuous variables such that $0 \leq v \leq 1$. In the latter, the variables become binary only at termination of the algorithm. We will construct linear underestimators and we will enforce integrality of v at each iteration as in (b). The resulting problem (PiL) is a mixed integer linear program (MILP).

3.3   Upper Bound Algorithm

The upper bound f_i^U for f_i on D_i can be found by computing $f_i^U = f(\bar{x},v)$, where $[\bar{x},v]$ is a feasible point for problem (10). The latter can be obtained by solving

$$\gamma^* = \min_{x,v,\gamma} \gamma$$

$$\overline{\varphi}_j(x,v) \le \gamma, j = 1...(m+2p)$$

(11)

where

$$\overline{\varphi}_j = \begin{cases} \varphi_j, j = 1...m \\ -\psi_{j-m} + a^i_{j-m}, j = (m+1)...(m+p) \\ -b^i_{j-(m+p)} + \psi_{j-(m+p)} j = (m+1+p)....(m+2p) \end{cases}$$

This is a nonconvex problem and therefore computationally intensive to solve at each iteration. To circumvent this, we obtain an upper estimate γ of the value γ^* by solving the problem

$$\mathbf{P^U}: \quad \overline{\gamma} = \min_{x,v,\gamma} \gamma$$

$$U[\overline{\varphi}_j(x,v); D_i] \le \gamma, j = 1...(m+2p)$$

where $U[\overline{\varphi}_j(x,v); D_i]$ is a linear overestimator of $\overline{\varphi}_j(x,v)$ on D_i such that

$$U[\overline{\varphi}_j(x,v); D_i] \ge \overline{\varphi}_j(x,v); \forall (x,v) \in D_i.$$

Let $\overline{D}_i = \{x,u : U[\overline{\varphi}_j(x,v); D_i] \le 0\}$, then $\overline{D}_i \subset D_i$ and $\overline{\gamma} \ge \gamma *$. Problem $\mathbf{P^U}$ is an MILP. It should be noted that we could terminate the solution to $\mathbf{P^U}$ whenever $\overline{\gamma} \le 0$.

During evaluation of the lower and upper bounds for subregion D_i, the following situations may arise at the k-th step of the branch-and-bound algorithm: (i) $\tilde{D}_i \ne \varnothing, \overline{\gamma} < 0$, (ii) $\tilde{D}_i \ne \varnothing, \overline{\gamma} > 0$, and (iii) $\tilde{D}_i = \varnothing$. In (i), we can calculate both the lower and upper bounds, while in (ii) we can only calculate the lower bound since we cannot ensure that the point obtained by solving problem P^U will be feasible for the problem (3). Finally in (iii), $\tilde{D}_i$ does not contain solution points and consequently it can be excluded from consideration. The branching point $\psi^* = \psi(x^*, u^*)$ is determined at the solution point of the lower bound problem $\mathbf{P^L}$.

3.4   Linear Estimators and Branching Functions

The main challenge in a BB based method is the construction of underestimators and overestimators. McCormick (1976) suggested the factorable programming technique for constructing convex underestimator for a function represented in factorable form. Sherali and Alameddine (1992) suggested a general approach for constructing underestimators for arbitrary polynomial functions. A method for construction of underestimators for more general functions is proposed in the α-BB global optimization method (Adjiman *et al*, 1998). The dimension of the lower bound problem, in all the above approaches, can be much larger than the dimension of the original problem. Here we present an alternative approach in which the lower bound problem has dimension not greater than the dimension of the original problem.

Let us consider a class of functions φ_i that can be represented as a *tree graph* (Fig. 1). Denote the root node of the graph as A_1^N. The set of nodes A_j^{N-k}, which are k branches apart from the root node, are at the N-k^{th} level. Let the k-th level of the tree graph has p_k nodes. Each node A_j^{N-k} has q_j^{N-k} descendants. Assign a differentiable function $\varphi_j^{(N-k)}$ of many variables and $q_j^{(N-k)}$ continuously differentiable functions $f_{ji}^{(N-k)}(y)$ of one variable y to each node A_j^{N-k} of the $(N$-$k)$- th level. $(k=1,...,N$-$1)$. The original function φ_i corresponds to the root node. Thus the following relations hold

$$\varphi_i^{(N-k)} = \sum_{j \in Q^{N-k}} c_{ji}^{(N-k)} f_{ji}^{(N-k)}(\varphi_j^{(N-k-1)}) \tag{12}$$

Here $Q_j^{(N-k)}$ is the set of descendant nodes of $A_j^{(N-k)}$. The variable x_i corresponds to a leaf node. Without loss of generality, we will assume that the leaf nodes are associated with the first level. Otherwise we employ identical transformations to relate the variable x_i *to* the first level. Suppose for example the variable x_i is associated with the second level. Then we can introduce the transformation $\varphi_i^{(2)} = x_i$. In so doing we have related x_i to the first level as well.

A function $f(x)$ is defined as a special tree function (STF) if each node of the computational graph corresponding to it is characterized by relation (see Eq. (10)) . Thus the STF is a superposition of univariate concave or convex functions connected by simple arithmetic operations, namely addition, subtraction, multiplication on some constant coefficient and operations corresponding to univariate functions $f_{ji}^{(N-k)}(y)$ in intermediate nodes. There exists different ways for transformation of a tree function into an STF.

The simplest way consists in the use of the following transformation for removing the multiplication operation.

$$f(x)g(x) = \tfrac{1}{4}(f(x)+g(x))^2 - \tfrac{1}{4}(f(x)-g(x))^2$$

We propose a strategy for constructing a linear underestimator for the function $\varphi_1^{(N)}$ corresponding to the root node A_1^N. Note that $\varphi_1^{(N)}$ is a complex multilevel function of the variables $x_1,...,x_n$ at the first level. We will assume that all the coefficients $c_{ji}^{(N-k)}$ are positive. If a coefficient $c_{ji}^{(N-k)}$ is negative we can introduce new notations $\bar{c}_{ji}^{(N-k)} = -c_{ji}^{(N-k)}$ and $\bar{f}_{ji}^{(N-k)} = -f_{ji}^{(N-k)}$, and replace $c_{ji}^{(N-k)} f_{ji}^{(N-k)}$ by $\bar{c}_{ji}^{(N-k)} \bar{f}_{ji}^{(N-k)}$. Here $\bar{c}_{ji}^{(N-k)} > 0$. Let

$$\varphi_j^{(N-k)} \in S_j^{(N-k-1)}$$

where

$$S_i^{(N-k-1)} = \{\varphi_j^{(N-k-1)} : \overline{\varphi}_j^{(N-k-1)} \le \varphi_j^{(N-k-1)} \le \overline{\overline{\varphi}}_j^{(N-k-1)}\} \tag{13}$$

If we know the bounds for the variables x_i at the first level, we can estimate bounds for all functions φ_l (at all levels) by using interval arithmetic (Moore, 1966). A linear underestimator of the function $\varphi_i^{(N-k)}$ in the region $S_i^{(N-k)}$ with respect to the functions $\varphi_j^{(N-k-1)}$ $j \in Q_i^{N-k}$ will be designated as $L[\varphi_i^{(N-k)}; S_i^{(N-k)}]$.

One can find a linear relation between $L[\varphi_i^{(N-k)}; S_i^{(N-k)}]$ and the linear underestimators of $\varphi_i^{(N-k-1)}$ at the descendant nodes $A_j^{(N-k-1)}$ as

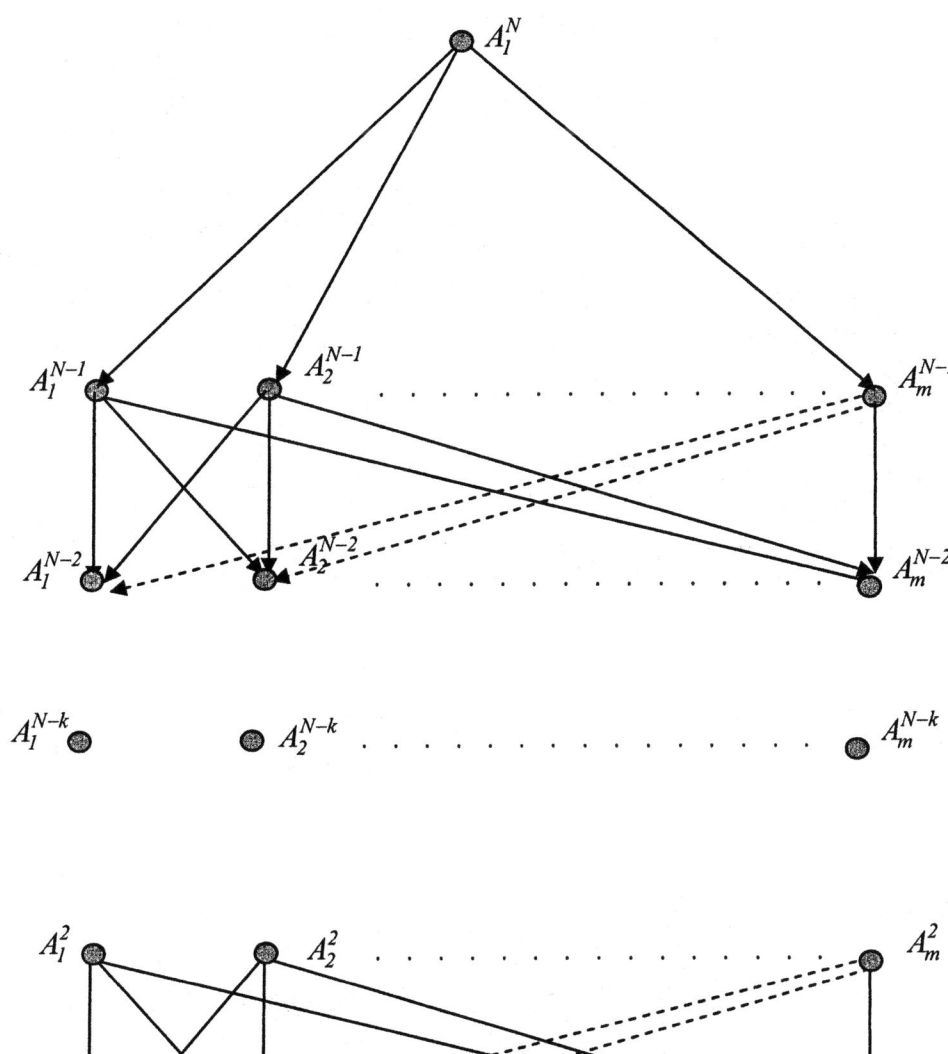

Figure 1. A multilevel representation of a tree

$$\overline{\varphi}_i^{(N-k)} = \sum_{j \in Q_i^{N-k}} c_{ji}^{(N-k)} L[f_{ji}^{(N-k)}(\varphi_j^{(N-k-1)}); S_{ji}^{(N-k-1)}]$$
$$\leq \sum_{j \in Q_i^{N-k}} c_{ji}^{(N-k)} f_{ji}^{(N-k)}(\varphi_j^{(N-k-1)}) \tag{14}$$

Now we will construct a linear underestimator for the function $f_{ji}^{(N-k)}(\varphi_j^{(N-k-1)})$ at the *(N-k)*-th level with respect to $\varphi_j^{(N-k-1)}$. Let the latter satisfy the Eq. (13). To simplify the notation for subsequent developments, let $y = \varphi_j^{(N-k-1)}$ and consider the function *f(y)* in the region $S_y = \{y : \underline{y} \leq y \leq \overline{y}\}$. If *f(y)* is concave then in S_y the linear underestimator has the form

$$L[f(y); S_y] = f(\underline{y}) + \frac{[f(\overline{y}) - f(\underline{y})]}{\overline{y} - \underline{y}}(y - \underline{y}) \tag{15}$$

If instead *f(y)* is convex then a linear underestimator is given by the tangent to *f(y)* at the point

$$y_m = \frac{(\underline{y} + \overline{y})}{2}.$$

In this case the underestimator is given by the following formula

$$L[f(y); S_y] = f'(y_m)(y - y_m) + f(y_m) \tag{16}$$

Here $f'(y_m)$ is the derivative of the function *f(y)* at the point y_m .Substituting the expressions for linear underestimators of $f_{ji}^{(N-k)}$ in $\overline{\varphi}_i^{(N-k)}$ we obtain

$$L[\varphi_i^{(N-k)}; S_i^{(N-k)}] = \sum_{j \in Q_i^{N-k}} d_j \varphi_j^{(N-k-1)} \tag{17}$$

Again we will assume that $d_j > 0$; otherwise we can employ the transformation discussed earlier. Hence we finally obtain the following expression for the linear underestimator of $\varphi_i^{(N-k)}$ as,

$$L[\varphi_i^{(N-k)}; S_i^{(N-k)}] = \sum_{j \in Q_i^{N-k}} d_j L[\varphi_j^{(N-k-1)}; S_j^{(N-k-1)}] \tag{18}$$

At the *(N-k-1)*-th level, we need to know the sign of d_j which is determined at the upper level, *(N-k)*. Therefore, starting from the *N*-th level and moving down to the 2-nd level, we obtain all relations as expressed in Eq. (18) for $k=0,1...,N-1$. A linear underestimator for the function $\varphi_i^{(N)}$ can be represented as a linear function of the variables x_i (associated with the first level) as follows

$$L[\varphi_1^{(N)};S_1^{(N)}] = \sum_{j=1}^{N} c_j x_j \qquad (19)$$

From the above consideration the following algorithm for construction of a linear underestimator for a tree function follows.

Summarizing, the construction of linear underestimator involves:

1. A bottom to top sweep to obtain all bounds
$[\underline{\varphi}_i^k, \overline{\varphi}_i]$ $\forall$ $k = 1,..., N$ and $i = 1,..., p_k$)
2. A top to bottom sweep to obtain the relations (Eq. (18)) for all levels
3. A bottom to top sweep to obtain $L[\varphi_i^{(N-k)};S_i^{(N-k)}]$ as linear functions of x, and u.

We will refer to this method as the sweep method. A similar procedure can be used for construction of linear overestimators. It is important that the underestimator is a linear function of the variables x and v. We note the following. The dimension of the lower bound problem $\mathbf{P^L}$ is the same as dimension of the original problem $\mathbf{P}$.

3.5   Selection of Branching Function

In a conventional BB method, the branching variables are the search variables x_i. However, the larger dimensionality of x_i $(i = 1,...,n)$ can result in a rapid growth in the number of branches in the BB tree. To address this problem, we consider an alternative selection of the branching expressions: *we employ the arguments* $\varphi_j^{(N-k-1)}$ *of all the functions* $f_{ji}^{(N-k)}$ *as branching variables.* Branching on $\varphi_j^{(N-k-1)}$ will decrease the intervals described by (13). Therefore, a tighter linear underestimators of $f_{ji}^{(N-k)}$ will be obtained since $\max (f_{ji}^{(N-k)} - L[\varphi_i^{(N-k)};S_i^{(N-k)}]$) will tend to zero as the size of $c_{ji}^{(N-k)}$ strives to zero. Only independent functions φ_i can be used as branching functions. The suggested approach to selection of branching expressions will be advantageous if the number of independent functions from the functions $\varphi_j^{(N-k-1)}$ is less than the number of variables x_i $(i = 1,...,n)$. In our formulation of the molecular design problem, this is indeed the case.

4.   SOLVENTS BACKGROUND

Solvents are extensively used as a major component of ink in the printing industry. The function of solvents in ink is to act as a vehicle for polymeric resins, pigments and dyes. The ink solvent also assists in wetting and dispersion of dyes and pigments. In letterpress and offset lithographic printing processes, the ink is carried to the plate by means of a train of rubber rollers commonly called "blankets" as shown Fig 1. Thus a thin film of ink is distributed over a large surface area on the blankets. These ink solvents are volatile and evaporate to leave behind the pigments and resins on the blanket surface. Cleaning is required whenever the residue build-up affects the print quality and between print jobs. Paper fibers, ink residue, paper coating and dried ink, are types of material that must be removed from the rubber blankets.

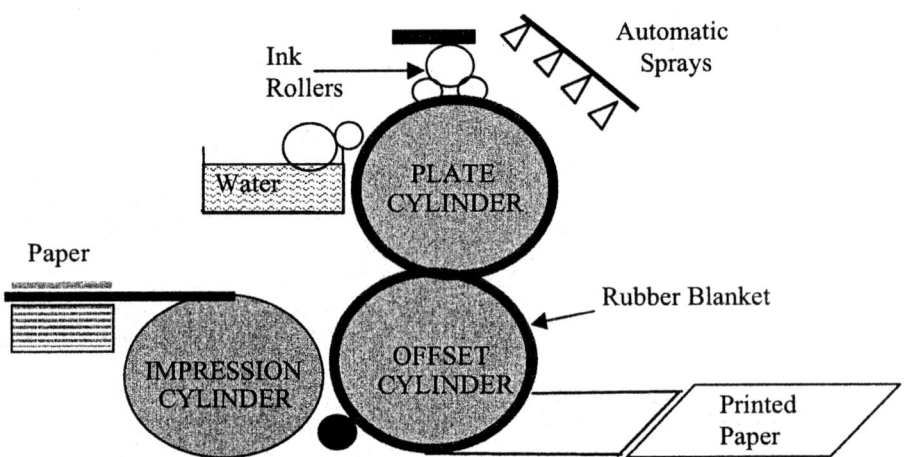

Figure 2. Schematic of Lithographic Printing

One of the most used solvents in lithographic printing is the "blanket wash" which is specially formulated to clean ink and other residue from rubber blankets. Blanket cleaning is accomplished automatically or manually. In an automatic blanket wash process, as shown in Fig. 1, the blanket wash is jet sprayed onto the blanket. Therefore a large amount of the wash is lost by evaporation even before it makes contact with the blanket.

Blanket wash solvents are mostly solvent mixtures as opposed to single component solvents. As such, next to solvent performance, one of the most pressing concerns of the printing industry with regard to the environment is the volatile organic component (VOC) level of solvents.   At present the VOC levels of solvents used in the printing industry are unusually high, well over 80% and far beyond the industry target of 30%.   For example, a commonly used blanket wash, "VM&P naphtha" has a 100% VOC content (United States Environmental Protection Agency, 1997a).

Blanket washes and solvents for "rag and bucket" operations are chosen based on their performance and their impact on the environment, health and safety. There is a wide variation in the performance attributes of cleaning solvents by different vendors.   To enhance the cleaning operation, companies sometimes mix solvents from different vendors. However, this trial and error approach is costly and may not necessarily yield the solvent mixture with the desired performance attributes. In addition, the solvent for a cleaning operation may not meet safety, health and environmental restrictions.

Another important issue is minimizing the effect of a solvent on the surface characteristics of the rubber blanket by inducing swelling. Swelling severely affects the print quality in lithographic processes. Thus, there is a need to account for this in blanket wash design.

The goal of this case study is to design globally optimal solvents to be used for cleaning in lithographic. These solvents should (i) have a minimal drying time, (ii) dissolve residue ink, (iii) not swell the blanket, and (iv) be environmentally benign.   Drying time is correlated with the heat of vaporization of the solvent.   The ink residue is assumed to consist of phenolic resins.

5.   PROBLEM DEFINITION

The problem as posed can be modeled as a multicriteria optimization problem. However, in the printing industry, there are rather loose and minimal requirements on these attributes. Therefore these attributes are regarded as constraints with given targets (similar to goal programming, Tamiz, 1996). A straightforward approach to modeling the problem as a special kind of multicriteria problem is to consider a lumped objective in which the different criteria appear as terms with appropriate weights. However this approach forces the solvent formulation engineer to think of appropriate weights (usually of no physical meaning) to employ, a rather non-trivial task. A more meaningful and rigorous approach is to consider the problem as a multi-level optimization problem. The latter is rather difficult to solve and has usually been restricted to bi-level optimization problems in which the decision variables are continuous.

We reiterate that the goal of this case study is to design optimal solvents to be used as cleaning agents in the printing industry. These solvents should (i) have a minimal drying time, (ii) dissolve residue ink, (iii) not swell the blanket, and (iv) be environmentally benign. Drying time is correlated with the heat of vaporization of the solvent.

The ink residue is assumed to consist of phenolic resins. Solvents that can effectively dissolve the ink residue obey the solute-solvent interaction

$$R^{ij} = 4(\delta_D - \delta_D^*)^2 + (\delta_P - \delta_P^*)^2 + (\delta_H - \delta_H^*)^2 \leq (R^*)^2$$

where $\delta_d^* = 23.3$, $\delta_P^* = 6.6$, $\delta_D = 8.3$ and $(R^*)^2 = 19.8$, and δ_d, δ_P, δ_D are determined from a model, for example a group contribution model (see Table 1).

The heat of vaporization, boiling point and melting point solvent properties are calculated using the Constantinou and Gani (1994) method. The fragment-based method is used to calculate K_{ow} (Lyman et. al., 1981). The group contribution parameters for solubility parameter calculation are based on van Krevlen and Hoftyzer's method (Barton, 1985). The models and their reference are summarized in Table 1.

Table 1. Property Prediction Models for CAMD_1 and CAMD_2

Property	Reference
Solubility Parameter	Barton, 1985
Boiling Point	Constantinou and Gani, 1994
Melting Point	Constantinou and Gani, 1994
Heat of Vaporization	Constantinou and Gani, 1994
Partition Coefficient (log K_{OW})	Lyman et al., 1981

We note that the nonlinear property prediction constraints (φ_j in $\mathbf{P_{MD}}$) do not employ the z_{ijp} and w_i variables from the Churi-Achenie octet rule implementation (see Chapter 3). Thus the problem is nonlinear with respect to only the u_{ik} variables. In the property prediction models, the nonlinearities are present in all the u_{ik} variables. The estimators for the case study are constructed in the appendix. These estimators are then used in the proposed branch and bound technique to solve the problem.

5.1   Case Study 1

In this case study, structural feasibility constraints are employed to ensure feasible molecular structures. For simplicity introduce the notation

$$\psi_1 = \psi_H, \psi_2 = \psi_P, \psi_3 = \psi_V, \psi_4 = \psi_D$$

The resulting molecular design formulation is shown below.

CAMD_1:

$$\min \sum_i \sum_j u_{ij} (\Delta H_V)_j \tag{20}$$

$$\sum_i \sum_j u_{ij} \le n_{max} \tag{21}$$

$$\sum_i \sum_j u_{ij} (2 - v_j) = 2 \tag{22}$$

$$\exp((\sum_i \sum_j u_{ij} (T_b)_j) / 204.4) \ge 323 \tag{23}$$

$$\exp((\sum_i \sum_j u_{ij} (T_m)_j) / 102.425) \le 223 \tag{24}$$

$$\sum_i \sum_j u_{ij} (\chi^o)_j + \sum_i \sum_j u_{ij} (\chi^1)_j \le 4.0 \tag{25}$$

$$4(\delta_D - 23.3)^2 + (\delta_P - 6.6)^2 + (\delta_H - 8.3)^2 \le (19.8)^2 \tag{26}$$

$$\psi_D - 6.3\psi_V \geq 0 \tag{27}$$

$$\bar{\psi}_i \leq \psi_i \leq \bar{\psi}_i, \quad i = 1,2,3,4 \tag{28}$$

To solve **CAMD_1**: we proceed as follows

Step 1:

(a) We choose as basis set twelve groups, namely CH_3-, CH_2-, Ar-, -Ar-, -OH, CH_3CO-, $-CH_2CO$-, $-COOH$, CH_3COO-, $-CH_2COO$-, $-CH_3O$, and -CH_2O-.

(b) The design variables are given by the structural variables u_{ij}, which determine whether a particular structural group is present in the molecule.

Step 2: The performance objective is given by the double summation in Eq. (1), which gives the heat of vaporization of the compound.

Step 3: Constraints are employed in order to ensure that the last seven groups in the basis set are not allowed to occur more than twice in a compound as follows

$$\sum_{j=5,\dots,12} u_{ij} \leq 2.$$

The constraint $\delta_P \geq 6.3$, will ensure minimal blanket swelling. The environmental impact of solvents is accounted for by requiring that the maximum value of the partition coefficient (log K_{ow}) be 4.0. To ensure that the solvent is a liquid at ambient temperature, the limits on boiling point (T_b) and melting point (T_m) are imposed. The constraints are Eqs. (23) through (28). Eqs. (23) to (26) are the property target constraints on blanket swelling, and Eq. (27) are constraints imposed by the branching functions. Eq. (28) are simple bounds on the branching functions.

Step 4: Decide whether to use the Odele-Machietto or the Churi-Achenie Octet Rule Model. Here we employ the much simpler (although restrictive) Odele-Machietto model for acyclic compounds where v_j is the valence of jth

structural group. The model is given in Eq. (22). We also include the molecular structural constraints (Eqs. (21) and (22)).

Step 5: Using information from previous steps, assemble the mathematical program, i.e. the performance objective, constraints, design variables and the Octet Rule Model. Eqs. (20) through (28) make up the mathematical program.

Step 6: Construct linear estimators of the performance objective and the constraints.

Step 7: Enter an iterative loop using the branch and bound (BB) procedure in Section 3.3.1. There are two nonconvex constraints. The splitting functions employed are ψ_D, ψ_P, ψ_H and ψ_V. The MILP solver used is a public domain code lp_solve by Hartmut Schwab available at (ftp.es.ele.tue.nl/pub/lp_solve). This solver uses the simplex algorithm. lp_solve uses a rather simple depth first strategy. Identify the optimal molecule using information from the solution.

Three different runs were investigated for case study 1. The three runs correspond to n_{max} of 3, 4, 5, 6, 7, and 10 (**CAMD_1a, CAMD_1b, CAMD_1c, CAMD_1d, CAMD_1e,** and **CAMD_1f,** respectively). The corresponding problem dimensions are 36, 48, 60, 72, 84 and 120. For all cases the number of constraints are 15. The termination criterion used is an absolute tolerance of 10^{-3}. The results are shown in Table 2.

Problem **CAMD_1a** has a very limited search space. A feasible solution was found in the first iteration in the branch-and-bound algorithm. In **CAMD_1c**, the algorithm took 31 iterations and 351.4 seconds on a 333-MHz DELL Pentium II personal computer. The maximum number of sub-regions constructed is 16. The globally optimal solution corresponded to methyl-ethyl ketone (MEK or CH_3-CH_2-CO-CH_3) with objective function 35.471 kJ/ mole. This compound was found at the 10th iteration with a valid upper bound of 35.471 and a lower bound of 33.99. Since the difference between the upper and lower bound was more than the tolerance, the algorithm continued executing. The algorithm finally converged to MEK as the global solution after 21 more iterations.   The two other feasible compounds found were propanol (CH_3-CH_2CH_2-OH) and diethyl-ketone (CH_3-CH_2-CO-CH_2-CH_3). The objective function values for propanol and diethyl-ketone were 44.77 kJ/mole and 40.12kJ/mole, respectively.

We note that at any iteration, the solution of the relaxed MILP problem is a structurally feasible compound since all the structural constraints are linear. During the execution of the algorithm, fifteen different compounds were found. Of these, two other compounds satisfied the specified or performance constraints. For case **CAMD_1e**, the number of iterations is 46 and 3 compounds are designed. The maximum number of subregions created is 21. In **CAMD_1f**, the number of iterations is 67. The maximum number of subregions created is 21. Even though the number of iterations does not grow very much, the CPU time increases. This is because the CPU time associated with each LP solution increases significantly when the number of variables increases. Another desirable property of this algorithm is that a very small number of subregions are created.

For the three cases, the number of subregions created are 16, 21 and 21, respectively. Thus the algorithm is very efficient in terms of storage requirements. It should be noted that as the dimension of the problem increases from 60 to 120, the number of iterations only increases from 31 to 67. This is perhaps the consequence of the fact that the number of branching variables, namely 4, is the same in all the cases.

Table 2. Application of Reduced Space BB algorithm to CAMD_1

Case	n_{max}	Var	Constr	Iter	CPU (min)	Max # of subregions
CAMD_1a	3	36	15	1	0.045	1
CAMD_1b	4	48	15	18	0.86	12
CAMD_1c	5	60	15	31	5.85	16
CAMD_1d	6	72	15	42	17.21	20
CAMD_1e	7	84	15	46	48.45	21
CAMD_1f	10	120	15	67	713.5	21

Recall that in all the example problems above, although the number of variables u_{ij} increased from 60 to 120, the number of branching functions is unchanged at 4. In contrast, if we employ the standard full space BB algorithm, we will need to perform branching with respect to all the variables u_i. Here, the number of branching variables ranges from 60 to 120.

5.2   Case study 2

In this case, the same formulation is solved with the Churi-Achenie model (see Step 4 above). The connectivity variables z and w are employed in the structural representation as described in Section 2. The second constraint in **CAMD_1** is replaced by the following set of structural constraints. This leads to a large increase in the number of linear structural constraints.

$$\sum_{p=1}^{m} \sum_{j=1}^{S_{max}} z_{ijp} = \sum_{k=1}^{m} u_{ik} v_k \qquad i = 1...n_{max}$$

(29)

$$\sum_{p=1}^{i-1} \sum_{j=1}^{s\,max} z_{ijp} > -w_i \qquad i = 2....n_{max}$$

(30)

$$\sum_{i=1}^{nmax} \sum_{k=1}^{m} u_{ik} + \sum_{i=1}^{n\,max} w_i = n_{max}$$

(31)

$$w_1 = 0$$

(32)

$$w_i \leq w_{i+1} \quad i=1...(n_{max}-1)$$

(33)

$$\sum_{j=w+1}^{S_{max}} \sum_{p=1}^{n_{max}} z_{ijp} + M u_{ik} \leq M \qquad i = 1...n\,max, k = 1....m$$

(34)

$$\sum_{j=1}^{S\,max} z_{ijp} = \sum_{j=1}^{S\,max} z_{pji} \qquad i = 1...(n_{max} - 1), p = (i+1)...n_{max}$$

(35)

$$\sum_{p=1}^{n} z_{ijp} \leq 1 \qquad i = 1...n_{max}, j = 1...s_{max}$$

(36)

$$\sum_{k=1}^{m} u_{ik} - \sum_{k=1}^{m} u_{i-1,k} \leq 0 \qquad i = 2...n_{max}$$

(37)

This formulation is solved for n_{max} equal to 3, 4 and 5 (**CAMD_2a, CAMD_2b**, and **CAMD_2c**). The numbers of search variables are 57, 84 and 115 respectively. The corresponding numbers of constraints are 67, 84 and 113. Note that the formulation is nonlinear with respect to only the u_{ik} variables. The results are summarized in Table 3.

The number of variables that participate in the nonlinear term is the dimension of u_{ik} variables.   The remaining variables determine the connectivity information and appear only in the linear terms in **CAMD_2**. The dimensions of the vector of variables u_{ik} in the three runs are 36, 48 and 60 (**CAMD_2a, CAMD_2b** and **CAMD_2c**).

Table 3. Application of Reduced-Space BB algorithm to problem **CAMD_2**

Case	n_{max}	# of Vars	# of Const	# of Iters	CPU (min)	Max # of subregions
CAMD_2a	3	57	67	1	0.1	1
CAMD_2b	4	89	89	18	3.36	9
CAMD_2c	5	115	113	22	14.5	11

CAMD_2a corresponds to a problem with a reduced search space restricted by $n_{max}= 3$. For this case the global optimal solution was found in only one iteration.   When the search space was increased to 89 and 115 variables, the number of iterations also increased to 18 and 22.   For the run **CAMD_2c** one of the feasible compound found in an intermediate step is - CH_2O-CH_2COO-CH_2O-CH_2-, a cyclic compound. The structural constraints used in case study 2 allow design of cyclic compounds. The constraints in case study 1 are restricted to only acyclic compounds.

For about the same number of variables, the number of iterations in case study 2 (**CAMD_2**) is relatively smaller than case study 1.   In addition, the maximum number of nodes generated in case study 2 is much smaller that in case study 1. This can be attributed to the fact that in **CAMD_2** the number of variables appearing in nonlinear term is much smaller compared to problems of similar dimension in **CAMD_1**.

5.3   Case Study CAMD_3

In this case study, solvents are designed with entirely different criteria. Here the most desirable attribute of the solvent is recoverability. That is, after the blanket wash operation is performed the solvents compounds that evaporate are recovered by a solvent recovery system. This case study attempts to find a solvent compound that will be least expensive to recover.   Many competing solvent recovery techniques can be applied, namely condensation, gas adsorption and gas absorption. Here the recovery system is restricted to the condensation.

A typical condensation recovery system consists of a compressor that takes in the printing solvent-laden exhaust gases (from the ventilation system) and compresses them to a higher pressure. These high-pressure gases are passed through a condenser that cools this stream. Next it is flashed to recover the solvent. The details of the recovery operation have been discussed elsewhere (Sinha, 1999). Here the objective is to find the solvent compound that will have minimal total annualized cost (TAC) associated with recovery. Here we will use as branching functions (except for the functions in (6.4)) the following functions

$$\psi_5 = H_{Vo} + \sum_i \sum_j u_{ij} H_{Vj}$$

$$\psi_6 = V_o + \sum_i \sum_j u_{ij} T_{bj}$$

$$\psi_7 = P_{comp}$$

$$\psi_8 = T_{cond}$$

The **CAMD_3** case study with recovery considerations is:

$$\min_{U_{ij}} \text{TAC} = 85675*(P_{comp}^{0.284} - 1) + 99.03(298P_{comp}^{0.284} - T_{cond}) +$$

$$9.69 \times 10^{-3} \left(\frac{\Delta H_V}{V_M} \right)$$

$$\sum_i \sum_j u_{ij} \leq 4 \tag{38}$$

$$\sum_i \sum_j u_{ij}(2 - v_j) = 2 \tag{39}$$

$$\log_{10}(V_m) - \log_{10}(P'_{comp}) - 2.7(T_B / T_{cond})^{1.7} \leq -11.47 \tag{40}$$

$$\exp((\sum_i \sum_j u_{ij} t_{mj})/102.425) \le 223 \tag{41}$$

$$500 \le T_{bo} + \sum_i \sum_j u_{ij} T_{bj} \le 700 \tag{42}$$

$$20 \le H_{Vo} + \sum_i \sum_j u_{ij} H_{Vj} \le 80 \tag{43}$$

$$\sum_i \sum_j u_{ij} (\chi^o)_j + \sum_i \sum_j u_{ij} (\chi^1)_j \le 4.0 \tag{44}$$

$$4(\delta_D - 23.3)^2 + (\delta_P - 6.6)^2 + (\delta_H - 8.3)^2 \le (19.8)^2 \tag{45}$$

$$\psi_D - 6.3\psi_V \ge 0 \tag{46}$$

$$\bar{\psi}_i \le \psi_i \le \bar{\psi}_i, \quad i = 1,...,8 \tag{47}$$

Here Eq. (39) is Odele's octet rule implementation, Eqs. (40) to (46) are recovery, melting point, boiling point, heat of vaporization, octanol-water partition, solvent power and swelling constraints, respectively, Eq. (47) represent constraints imposed by the branching functions.

The following modified basis with 15 groups is used in this study: [CH3-, CH$_2$-, -OH, CH$_3$CO-, -CH$_2$CO-, -COOH, CH$_3$COO-, -CH$_2$COO-, -CH$_3$O, -CH$_2$O-, CH2=CH-, -CH=CH-, -CH$_2$NH$_2$, =CHNH$_2$, CH$_3$NH-]. The aromatic groups are removed and some groups with nitrogen are added to include amine or other compounds with nitrogen.

There is a total of 8 splitting functions. The last 4 splitting functions are used for construction of linear underestimators for the objective function and underestimator and overestimator for the recovery constraint. The construction of estimators is discussed in the attached appendix. This case study has 60 variables and 3 nonlinear constraints. Moreover, the objective function is nonlinear. The condenser temperature however can range between 150 °K and 298 °K. This results in poor scaling and causes difficulty during optimization. To overcome this we have scaled the condenser temperature between 0.1 and 0.9 such that $T' = 185T_{cond} + 131.5$ where T' is the scaled condenser temperature.

The globally optimal compound designed is a diester with the structure $CH_3-(CH_2COO)_2-CH_2NH_2$. The recovery cost associated with this compound is \$25,981. The corresponding compressor pressure is 2 atm and the condenser temperature is 288.75°K. The algorithm took 56 iterations and a CPU time of 41.6 seconds. At termination, the number of nodes (i.e. subregions) is 20.

The above problem was solved again with local optimization software *DICOPT* in GAMS (Brooke, 1996). The optimal compound found by *DICOPT* is $HO-CH_2COO-CH_3NH$, an ester. The objective function associated with this compound is 106,327, the compressor pressure is 10 atms and condenser temperature is 298 °K. We note that the extra effort associated with the global optimization is justified and results in almost 4 times reduction in the recovery cost.

6.   DISCUSSION AND CONCLUSIONS

The molecular design problem is reduced to solving an MINLP problem in which the number of binary variables u_{ij} can range from several tens to several hundreds. The use of the standard branch-and-bound method for solving the problem can be computationally intensive since all the variables u_{ij} must be used as branching variables. To overcome this problem, we have proposed a new strategy. The main idea of the method consists in that we do branching using branching functions instead of all the search variables. This approach results in a decrease in the number of branching variables in our molecular design framework. For example, in case study 1, a problem with 120 nonlinear variables is solved with just 4 splitting variables. This is also demonstrated in the case studies. The maximum number of nodes stored in memory during the search is 21 for **CAMD_1e** and **CAMD_1f** and 20 for CAMD_3.

In other words, during branch and bound, the bounding operation is performed the search variables space, while the branching operation is performed in a reduced dimension space defined by the branching (or splitting) functions.

The branching functions are determined from the special tree function representation of both the objective function and constraints. In order to construct the corresponding linear underestimators, we employed the sweep method we developed in our research (Sinha, Achenie and Ostrovksy, 1999) and (Ostrovsky, Achenie and Sinha, 2000).

The proposed algorithm scales well. Specifically, as the problem size increases, the computational effort increases almost linearly. We anticipate that this linear behavior will be exhibited also in large molecular design models.

7.   APPENDIX: CONSTRUCTION OF ESTIMATORS

One very important property for solvent is its ability to dissolve the solute. A solute-solvent interaction is often characterized by the Hansen solubility parameter δ_T (Archer 1996). This parameter is characterized by the three intermolecular interactions, namely hydrogen bonding interaction (δ_H), polar interactions (δ_P) and nonpolar (dispersive) interaction (δ_D) (Hansen, 1971). The mathematical expression for the solvent selection criterion based on the Hansen solubility parameter is

$$R^{ij} = 4(\delta_D - \delta_D^*)^2 + (\delta_P - \delta_P^*)^2 + (\delta_H - \delta_H^*)^2 \leq (R^*)^2 \qquad (48)$$

$$\delta_D = \frac{\psi_D}{\psi_V}, \delta_P = \frac{\sqrt{\psi_P}}{\psi_V}, \delta_H = \frac{\sqrt{\psi_H}}{\psi_V} \qquad (49)$$

where

$$\psi_D = \sum_i \sum_j u_{ij} * F_{Dj}; \psi_P = \sum_i \sum_j u_{ij} * (F_{Pj})^2;$$
$$\psi_H = \sum_i \sum_j u_{ij} * (-U_{Hj}); \psi_V = V_o + \sum_i \sum_j u_{ij} V_j \qquad (50)$$

Here F_{Di}, F_{Pi} and U_{Hi} are the group contribution parameters associated with the dispersion, polar and hydrogen bonding solubility parameters respectively (Barton 1985). Substituting Eq. (30) into Eq. (39) we obtain

$$4\left(\psi_D - \delta_D^* \ \psi_V\right)^2 + \left(\sqrt{\psi_P} - \delta_P^*\psi_V\right)^2 +$$

$$\left(\sqrt{\psi_H\psi_V} - \delta_H^*\psi_V\right)^2 - (R^*\psi_V)^2 \le 0 \tag{51}$$

Note that the nonconvex Hansen solubility design criteria make the solvent design problem multiextremal. R* is the interaction radius associated with the solute. The distance between the solute and solvent solubility parameters is R^{ij} and can be computed as shown in Eq. (39). V_j is the molar volume of the solvent. We will now construct linear underestimators for this important constraint. Eq. (32) is made up of four separable terms. The first and the fourth terms are squares of linear equations. The second and the third terms are relatively more complicated. Using Eq. (30) one can obtain the STF representation of the third term in the form

$$\varphi_1^5 = (\varphi_1^4)^2 \ , \ \varphi_1^4 = \frac{1}{16}\left[\left(\varphi_1^3\right)^2 - \left(\varphi_2^3\right)^2 - 4\delta_H^*\varphi_2^2\right], \quad \varphi_1^3 = \sqrt{\varphi_1^2} + \sqrt{\varphi_2^2} \ ,$$

$$\varphi_1^3 = \sqrt{\varphi_1^2} - \sqrt{\varphi_2^2} \ , \ \varphi_1^2 = \psi_H \ \text{and} \ \varphi_2^2 = \psi_V$$

Here the functions $\varphi_1^5, \varphi_1^4, \ \varphi_1^3, \varphi_2^3, \varphi_1^2, \varphi_2^2$ corresponds to the fifth, third and second levels respectively.

The first level functions are the variables u_{ik}. The branching set branching functions contain $\varphi_1^4, \varphi_1^3, \varphi_2^3, \psi_H, \psi_P, \psi_V, \psi_D$. It is easy to see that the first three functions are expressed as functions of the variables $\psi_H, \psi_P, \psi_V, \psi_D$. Therefore the functions $\psi_H, \psi_P, \psi_V, \psi_D$ can be used as branching functions. Let us consider for illustration construction of an underestimator for the third term in Eq. (32). First we need to find bounds for $\varphi_1^4, \ \varphi_1^3, \varphi_2^3, \varphi_1^2, \varphi_2^2$. Since we know the bounds for u_{ik} we can determine the bounds for φ_1^2, φ_2^2. The ranges of the functions φ_1^2, φ_2^2 are used to construct the ranges of φ_1^3 and φ_2^3 at the third level. These are then used to construct the range of φ_1^4, at the fourth level. Thus the bounds are estimated in a bottom up sweep.

The linear underestimators are constructed in a reverse sweep that starts at the fifth level and goes down. First, a linear underestimator of φ_1^5 is constructed in terms of φ_1^4 as follows $L[\varphi_1^5; S^4] = \mu_1(\varphi_1^4) + \mu_2$. Here S^4

$= (\overline{\varphi_1^4}, \overline{\overline{\varphi_1^4}})$. The sign of μ_l, which depends on the interval $(\overline{\varphi_1^4}, \overline{\overline{\varphi_1^4}})$, determines whether the function $\mu_1(\varphi_1^4)$ is convex or concave with respect to variables φ_1^3 and φ_2^3. The underestimator now has the following form

$$L[\varphi_1^5; S^3] = (\mu_3 - \mu_4)(\sqrt{\psi_H}) + (\mu_3 + \mu_4)(\sqrt{\psi_V}) + \mu_5(\psi_V) + \mu_6 .$$

The signs of $(\mu_3 - \mu_4)$ and $(\mu_3 + \mu_4)$ determine if the corresponding functions are concave or convex. Subsequently, the underestimator is constructed with respect to ψ_H and ψ_V. After rearranging the terms, the linear underestimator is represented as $L[\varphi_1^5, S^2] = \mu_8 \psi_H + \mu_9(\psi_V) + \mu_{10} .$

We reiterate that the subregion is not in terms of search variable u_{ik}, but rather in terms of functions of u_{ik}. Based on the region S the coefficients μ_l, $(\mu_3 - \mu_4)$ and $(\mu_3 + \mu_4)$ are calculated and a decision about construction of the underestimator is made at two levels (Ostrovksy, Achenie and Sinha, 2000). This makes the algebraic structure of the underestimator adaptive.

8.   REFERENCES

Adjiman, C. S., Dallwig, S., Floudas, C. A., and Neumair, A. (1998). A global Optimization method, alpha-BB, for general twice-differentiable NLPs --I. Theoretic Advances. *Computers and Chemical Engineering*, 22(9), 1137-1158.

Archer, W. L. (1996). *Industrial Solvent Handbook*, Marcel Dekker Inc.

Barton, A. F. (1985). *CRC Handbook of Solubility Parameters and Other Cohesion Parameters*, CRC Press, Inc., Boca Raton, Florida.

Brooke, A. (1996) GAMS - A User's Guide, Scientific Press, San Francisco, CA

Churi, N., and Achenie, L. E. K. (1996). Novel Mathematical Programming Model for Computer Aided Molecular Design. *Industrial and Engineering Chemistry Research*, 35(10), 3788-3794.

Constantinou, L., and Gani, R. (1994). New Group Contribution Method for Estimating Properties of Pure Compounds. *AIChE Journal*, 40, 1697-1710.

Duvedi, A. P., and Achenie, L. E. K. (1996). Designing Environmentally Safe Refrigerants Using Mathematical Programming. *Chemical Engineering Science*, 51, 3727-3739.

Friedler, F., Fan, L. T., Kalotai, L., and Dallos, A. (1998). A combinatorial approach for generating candidate compounds with desired properties based on group contribution. *Computers and Chemical Engineering*, 22(6), 809-817.

Hansen, C. M., and Beerbower, A. (1971). Solubility Parameters. Kirk-Othmer Encyclopedia of Chemical Technology, A. Standen, ed., Interscience, New York.

Horst, R., and Tuy, H. (1990). *Global Optimization: Deterministic Approaches*, Springer-Verlag, Heidelberg.

Lyman, W. J., Reehl, W. F., and Rosenblatt, D. H. (1981). *Handbook of Chemical Property Estimation Methods*, McGraw-Hill Book Company.

Maranas, C. D. (1997). Optimal Molecular Design under Property Prediction Uncertainty. *AIChE Journal*, 43(5), 1250-1263.

McCormick, G. P. (1976). Computability of global solutions to factorable nonconvex programs. Part I -- convex underestimating problems. *Math. Program.*, 10, 147-175.

Moore, R. E. (1966). *Interval Analysis*, Prentice-Hall, Englewood Cliffs, New Jersey.

Odele, O., and Machietto, S. (1993). Computer Aided Molecular Design: A Novel Method for Optimal Solvent Selection. *Fluid Phase Equilibria*, 82, 47-54.

Ostrovsky, G. M., Ostrovsky, M. G., and Mikhailow, G. W. (1990). Discrete Optimization of chemical processes. *Computers and Chemical Engineering*, 14(1), 111.

Ostrovsky, G., Achenie, L. E. K., and Sinha, M. "A Reduced Dimension Branch-and-Bound Algorithm for Molecular Design," *(to appear in Journal of Global Optimization, circa 2000)*

Pantelides, (1996). *Global Optimization of General Process Models*. In I.E. Grossmann , ed. *Global Optimization in Engineering Design*, Kluwer Academic Publishers.

Pistikopoulos, E. N., and Stefanis, S. K. (1998). Optimal solvent design for environmental impact minimization. *Computers and Chemical Engineering*, 22(6), 717-733.

Quesada, I., and Grossmann, I. E. (1995). A Global Optimization Algorithm for Linear Fractional and Bilinear Programs. *Journal of Global Optimization*, 6, 39-76.

Ryoo, H. S., and Sahinidis, N. V. (1996). A Branch-and-Reduce Approach to Global Optimization. *Journal of Global Optimization*, 8, 107-138.

Sherali, H. D., and Alameddine, A. (1992). A new reformulation-linearization technique for bilinear programming problems. *Journal of Global Optimization*, 2, 379-410.

Sinha, M. A Systems Engineering Framework for Solvent Design. *Ph.D. Thesis,* University of Connecticut, 1999.

Sinha, M., Achenie, L. E. K. and Ostrovsky, G. M. "Design of Environmentally Benign Solvents via Global Optimization," Comp. Chem Eng. **23,** 1381-1394, 1999.

Sinha, M. Achenie, L. E. K., and Ostrovsky, G., *In* Computer Aided Molecular Design: Theory and Practice, Editors: Luke E. K. Achenie, Rafiqul Gani, & V. Venkatasubramanian, *Elsevier Publishers*, 2002, isbn 0-444-51283-7).

Tamiz, M. (1996). Multi-Objective Programming and Goal Programming Theories and Applications, Springer, York.

Vaidyanathan, R., and El-Halwagi, M. (1994). Computer-Aided Design of High Performance Polymers. *J. Elastom Plasti.*, 26(3), 277.

Venkatasubramanium, V., and Chan, K. (1989). A neural network methodology for process fault diognosis. *AIChE Journal*, 35, 1993.

Chapter 2

FEEDING STRATEGIES FOR MAXIMISING GROSS MARGIN IN PIG PRODUCTION

David L. J. Alexander
Institute of Information Sciences and Technology
Massey University
New Zealand
D.Alexander@massey.ac.nz

Patrick C. H. Morel
Institute of Food, Nutrition and Human Health
Massey University
New Zealand
P.C.Morel@massey.ac.nz

Graham R. Wood
Department of Statistics
Macquarie University
Australia
GWood@efs.mq.edu.au

Abstract   Nonlinear optimisation and a pig growth model are combined with the traditional use of linear programming to maximise gross margin per pig place per year for the pig producer. Emphasis in this paper is on description of the problem and analysis of the objective function.

Keywords: genetic algorithm, growth model, linear programming, Monte Carlo, Nelder-Mead algorithm, simulated annealing

1.   Introduction

Production of pig meat worldwide exceeds that of any other meat. Pigs are generally fed in a controlled environment, with least cost diets determined using linear programming. Of greater importance to the pig producer than least cost diets, however, is the maximisation of gross margin (per pig place or per pig place per year). Two relatively recent developments, pig growth models and efficient methods for optimisation of nonlinear functions of high dimension, in conjunction with the traditional use of linear programming, now make gross margin maximisation possible. The method of solution involves a synthesis of optimisation techniques, linear programming being employed as a sub-routine within the framework of a wider search algorithm. The purpose of this paper is to describe this problem, together with the solution methodology and associated practical outcomes.

The format of the paper is as follows. In the next section the optimisation problem is formulated, the domain described and the objective function detailed. Successful solution approaches are described in Section 3. The nature of the objective function is discussed in Section 4. A summary then concludes the paper.

2.   The problem

Pig farmers growing pigs from weaning until slaughter wish to maximise gross margin, namely

Gross Return – Total Feed Costs – Weaner Cost,

per pig place per year. The main area over which the pig producer has control is the feeding of the pig. A single feeding regimen can be summarised, as described in DeLange, 1995, by three parameters, d, r and p. A *feeding strategy* F is then a finite sequence of (d, r, p) triples, with each triple describing the diet for a fixed period, say one week. Here

d = digestible energy density, in MegaJoules per kilogram

r = minimum lysine to digestible energy ratio, in grams per
        MegaJoule

p = proportion of the *ad libitum* daily digestible energy intake

The *ad libitum* digestible energy intake for the pig at any given liveweight is determined on farm using monitor pens of pigs. Feed supplied, feed wasted and pig liveweight are recorded in order to determine the voluntary daily feed intake of the pig.

For the purposes of this exposition we work with a feeding strategy of ten (d, r, p) triples, each fed for one week, so the general point in the domain has the form

$$F = (d_1, r_1, p_1; \ d_2, r_2, p_2; \ \ldots; \ d_{10}, r_{10}, p_{10})$$

For the particular problem examined here, relevant to the New Zealand situation, a growth period of ten weeks is sufficient to include the most profitable solutions.

The energy content of the ingredients in the diet requires that d be at least 12 MJ/kg and no more than 16 MJ/kg. A sensible range for r is from 0.4 to 1.0 grams of lysine per MegaJoule; values outside this range are feasible, but this range has been found wide enough to encompass the most profitable diets and is in accordance with the opinion of pig nutritionists. Finally, p is allowed to vary between 0.5 and 1.0; nutritionists agree that pigs require at least 50% of their maximum voluntary digestible energy intake. In summary, the general point in the domain lies in the hypercuboid in $\mathbf{R}^{30}$ given by $([12, 16] \times [0.4, 1] \times [0.5, 1))^{10}$.

Thus the objective function of interest is $g(F)$, the gross margin per pig place per year associated with feeding strategy F. This is calculated in two steps:

1 Calculation of $g(F, x)$, the gross margin per pig place per year when feeding strategy F is administered for $x \leq 70$ days, and then

2 Calculation of $g(F) = \max_x g(F, x)$

We now provide more detail on each of these steps.

In the first step, the gross margin per pig place per year is calculated for a given feeding strategy F and number of days $x \leq 70$ for which it is fed, as detailed in Figure 1.1. This calculation successively relies on a linear program, the pig growth model and the price schedule.

A sample linear program is illustrated in Table 1.1. Using this we minimise the diet cost per kilogram, subject to the d and r values in the (d, r, p) triple. Parameter d is incorporated in an equality constraint for the digestible energy total. Together with parameter r it also dictates the minimum lysine level in the diet. The other amino acids are constrained to be present in at least certain proportions to lysine, to ensure ideal protein balance. Parameter p and the *ad libitum* digestible energy intake curve then determine the total feed intake and so the minimum daily feed cost.

The pig growth model used, from DeLange, 1995, requires three parameters particular to the pig genotype:

$$
\begin{aligned}
P_0 \;&=\; \text{initial mass of protein in the weaner pig} \\
Pd_{\max} \;&=\; \text{maximum daily protein deposition} \\
\min LP \;&=\; \text{minimum allowable lipid to protein ratio}
\end{aligned}
$$

Given a feeding regimen and initial chemical body composition the pig growth model is then capable of "growing" a pig for x days, outputting backfat thickness and carcass weight.

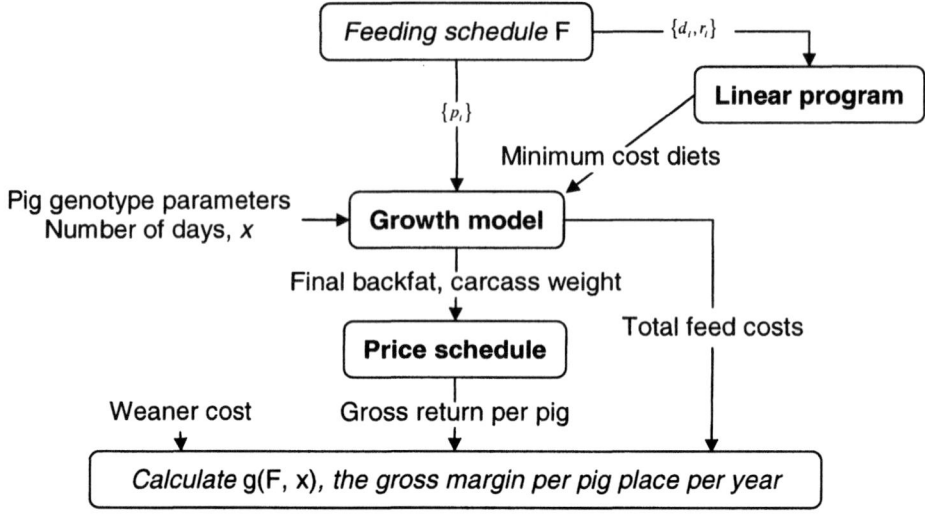

Figure 1.1. The gross margin per pig place per year when feeding strategy F is used for x days is calculated as shown in the flow chart. A linear program finds the minimum cost diets for the given feeding strategy. The pig growth model and price schedule are then used to find the market return per pig, from which the gross margin per pig place per year $g(F, x)$ may be calculated.

Price schedules for pigs at slaughter typically depend on backfat thickness and carcass weight. A 1998 New Zealand schedule is reproduced in Table 1.2. For pigs in the best categories (backfat thickness from 6mm to 9mm and carcass weight below 55kg) the producer receives NZ\$3.10/kg.

In the second step, $g(F)$ is easily found as the maximum in the array

$$(g(F, 1), \; g(F, 2), \; \ldots, \; g(F, 70))$$

Figure 1.2 summarises the problem: a linear program and a pig growth model allow us to evaluate the objective function, which is then maximised over a high dimensional hypercuboid using a nonlinear program.

3.   Solution algorithms

For this problem the objective function is of the "black box" variety; it is not known analytically. Many objective function evaluations are thus required by the solution algorithms, involving substantial computation. The domain is also of moderately high dimension. Simulated annealing (Metropolis et al., 1953) and genetic algorithms (Holland, 1975) are both successful stochastic approaches to the problem, while a satisfactory de-

Table 1.1. The linear program providing the least cost diet.

	Barley		Soybean		$\cdots$		Meat and Bone		
minimise:	$0.25B$	$+$	$0.715S$	$+$	$\cdots$	$+$	$0.5M$		
Digestible energy	$13.2B$	$+$	$15.86S$	$+$	$\cdots$	$+$	$12.5M$	$=$	d
Lysine lower bound	$3.19B$	$+$	$27.33S$	$+$	$\cdots$	$+$	$16.19M$	$\geq$	dr
Balanced	$2.85B$	$+$	$10.74S$	$+$	$\cdots$	$+$	$20.3M$	$\geq$	$0.62dr$
amino	$1.2B$	$+$	$4.68S$	$+$	$\cdots$	$+$	$1.09M$	$\geq$	$0.19dr$
acid	$2.42B$	$+$	$11.48S$	$+$	$\cdots$	$+$	$9.95M$	$\geq$	$0.32dr$
lower	$8.34B$	$+$	$36.75S$	$+$	$\cdots$	$+$	$18.91M$	$\geq$	$0.95dr$
bounds	$3B$	$+$	$16.22S$	$+$	$\cdots$	$+$	$9.54M$	$\geq$	$0.67dr$
	$6.85B$	$+$	$33.25S$	$+$	$\cdots$	$+$	$21.5M$	$\geq$	dr
					$\vdots$				
Mineral	$0.5B$	$+$	$3S$	$+$	$\cdots$	$+$	$105M$	$\geq$	8
bounds	$0.5B$	$+$	$3S$	$+$	$\cdots$	$+$	$105M$	$\leq$	13
	$3.1B$	$+$	$2.6S$	$+$	$\cdots$	$+$	$52M$	$\geq$	7
	$3.1B$	$+$	$2.6S$	$+$	$\cdots$	$+$	$52M$	$\leq$	11
	$0.2B$	$+$	$0.1S$	$+$	$\cdots$	$+$	$5.5M$	$\geq$	10
	$0.2B$	$+$	$0.1S$	$+$	$\cdots$	$+$	$5.5M$	$\leq$	20
					$\vdots$				
Ingredient	$-0.4B$	$+$	$0.6S$	$-$	$\cdots$	$-$	$0.4M$	$\leq$	0
upper	$-0.1B$	$-$	$0.1S$	$-$	$\cdots$	$-$	$0.1M$	$\leq$	0
bounds	$-0.3B$	$-$	$0.3S$	$-$	$\cdots$	$-$	$0.3M$	$\leq$	0
	$-0.4B$	$-$	$0.4S$	$-$	$\cdots$	$-$	$0.4M$	$\leq$	0
	$-0.4B$	$-$	$0.4S$	$-$	$\cdots$	$-$	$0.4M$	$\leq$	0
					$\vdots$				
Diet mass	B	$+$	S	$+$	$\cdots$	$+$	M	$=$	1

Table 1.2. A New Zealand price schedule giving prices in cents per kg for pigs at slaughter, as at 24 August 1998. A levy of \$9.40 per carcass is deducted.

Fat (mm)	Carcass Weight (kg)									
	35.0 and under	35.1 to 40.0	40.1 to 45.0	45.1 to 50.0	50.1 to 55.0	55.1 to 60.0	60.1 to 65.0	65.1 to 70.0	70.1 to 75.0	75.1 to 80.0	80.1 and over
< 6	250	250	250	250	250	250	250	250	250	250	250
6–9	310	310	310	310	310	295	295	280	280	280	275
10–12	285	285	285	285	285	280	280	280	280	280	275
13–15	190	190	190	215	215	245	245	245	245	245	245
16–18	150	150	150	150	150	165	165	165	165	165	165
> 18	120	120	120	120	120	135	135	135	135	135	135

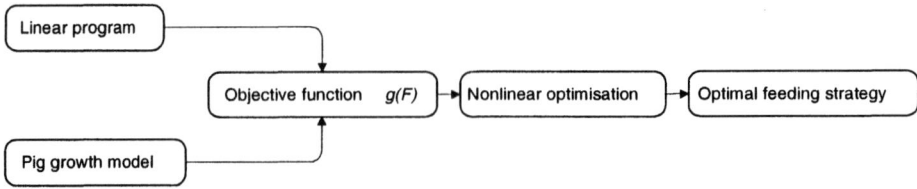

Figure 1.2. The components involved in maximisation of gross margin per pig place per year.

terministic approach is provided by the Nelder-Mead algorithm (Nelder and Mead, 1965). They typically yield, for the particular problem used, a gross margin of around NZ$280 per pig place per year. Pure random search, however, is not successful. Figure 1.3 shows the progress of a genetic algorithm on the problem, compared with that of Pure Random Search. In 30 seconds a genetic algorithm can find a feeding strategy equal to that reached by pure random search in over 100 hours! Pure random search, in that time, reaches a solution of approximately NZ$250 per pig place per year.

Table 1.3 displays the final results of a successful run. It lists the values of d, r and p in the best solution; the cost in cents per kilogram of each diet is also given. Slaughter date x was selected as 63 days, so the parameters for the tenth week were not used and are not shown. The linear program then provides the ingredients for each diet in the feeding strategy, together with the optimal proportion of each ingredient by weight. For example, a kilogram of the first diet in Table 1.3 contains 510g barley, 280g soybean and 100g wheat by-product, with the remainder made up of meat and bone meal, soya oil, salt and synthetic amino acids. (The unusual diet administered in week 7 allows the pig to maintain near maximum protein deposition with virtually zero lipid deposition, permitting continued weight gain but restricting fat content in the pig. Not all solutions found by the optimisation techniques contain this kind of aberration.)

4.   The nature of the objective function

A question which always vexes the global optimiser is whether the optimum has been reached. In order to investigate this question in this context, further information about the nature of the objective function is obtained in two ways:

1 By examining the value of the objective function along randomly taken domain cross-sections through the best known solution. This provides evidence that the optimum has been reached and suggests

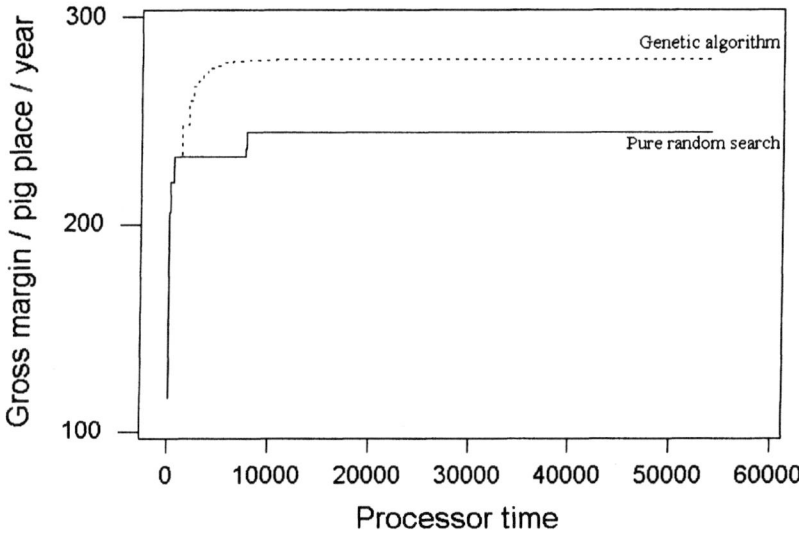

Figure 1.3. Comparison of the time in seconds required by a genetic algorithm and pure random search to attain certain dollar values of gross margin per pig place per year.

Table 1.3. Weekly values of d, r and p and cost per kilogram of diet for a solution found using a genetic algorithm.

Week	d	r	p	Cost (c/kg)
1	14.51	0.791	0.961	41.24
2	14.57	0.627	1.000	35.93
3	14.40	0.554	1.000	33.10
4	13.14	0.568	0.840	30.61
5	14.34	0.493	0.873	31.00
6	14.18	0.478	0.850	30.20
7	13.58	0.758	0.510	37.55
8	13.14	0.550	0.670	30.07
9	14.14	0.528	0.680	31.66

that the objective function takes the form of a single peaked, but very craggy, high-dimensional volcano.

2 By comparing the performance of pure random search with the
  expected performance if the form was broadly that of this "craggy
  volcano".

We now report on these two investigations into the nature of the objective function.

Figure 1.4 shows the typical shape of the objective function along a
random cross-section through the putative argmax. Overall, the function
appears to be unimodal. Jumps in the function are caused by the step
function nature of the price schedule and the fact that the pig must grow
a whole number of days.

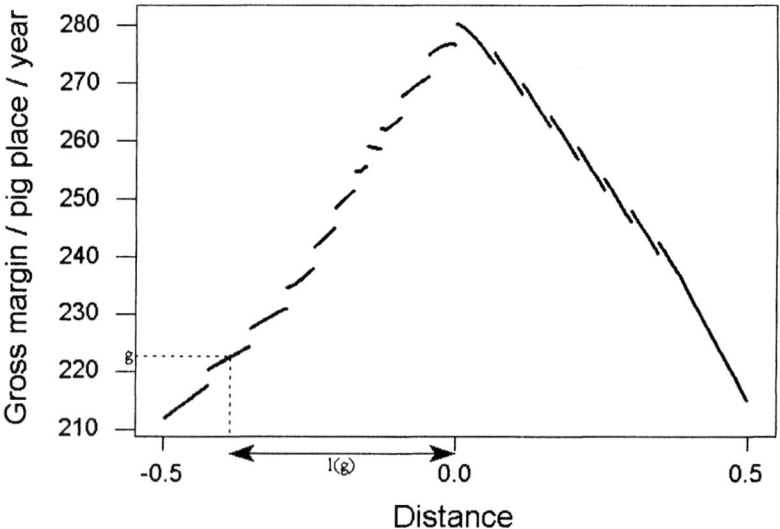

Figure 1.4. The approximately volcano shape of a section through the objective
function at the best known solution.

Through examination of many such cross-sections an average relationship is set up between g, the gross margin per pig place per year, and
$l(g)$, the distance between the corresponding place in the domain and the
best known solution (as illustrated in Figure 1.4). The expected number of iterations until convergence to a particular level for pure random
search will be the reciprocal of the probability that a given iteration falls

in the associated level set. This probability is the relative volume of the level set. Assuming roughly hyper-spherical level sets for the objective function, the probability is

$$\frac{\pi^{15} l(g)^{30}}{15!V}$$

where V is the volume of the hypercuboid domain.

Figure 1.5 compares this theoretical expected number of pure random search iterations to reach a given gross margin per pig place per year with a step function recording progress of a particular run. Evidently

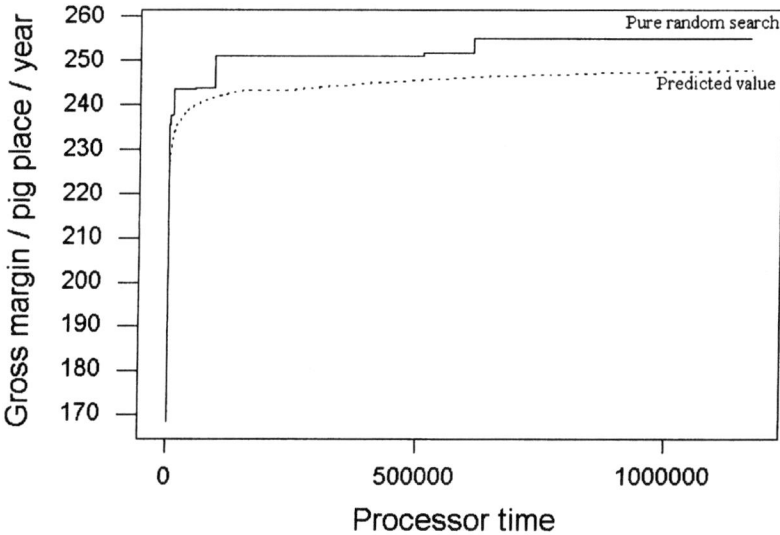

Figure 1.5. A comparison of the expected number of iterations to reach a value of the objective function (dotted line), under the assumption of hyperspherical level sets, with the number required in a particular run of pure random search (solid line).

the forms of the two curves are similar. The actual run is higher than the curve based on the estimated relationship between g and $l(g)$ with the assumption of hyperspherical level sets. This may well indicate that smaller peaks exist. There are also some cross-sections where the objective function does not decrease very rapidly. Overall there is evidence that the level sets have slightly greater relative volume than predicted.

It's time to ask an important question: Is this a theoretical solution with no practical value? In practice, feeding regimens cannot be administered precisely. Allowing for an achievable tolerance in feeding of 0.05 in each parameter of each diet, however, it is evident that the optimal solution is still only one part of the space in 10^{40}. An intelligent search is needed in order to find solutions approaching this optimum, as the comparison with pure random search has shown. Even with only three distinct diets (as is common in practice today), the optimal solution is one part in 10^{12}. A useful, practical solution region remains a needle in a high dimensional haystack. As the world moves to continuous feeding, the dimension of the problem will increase and the need for intelligent optimisation methods will become even more pressing.

5.   Summary

The development of pig growth models, together with the availability of nonlinear optimisation tools on high speed computers, has made it possible to extend the traditional use of optimisation in pig nutrition. Linear programming to determine least cost diets is now a component of a larger optimisation routine, where, for example, the challenge of diet formulation to maximise gross return per pig place per year can be answered. Optimisation methods show considerable promise for increasing the efficiency of the worldwide industry of pig feeding.

Acknowledgments

Professor Paul Moughan is thanked for his encouragement and pertinent questions during the development of this material at Massey University. The contributions of Lindsay Alexander in providing programming advice and of Christopher Clark in developing relevant software are also much appreciated.

References

DeLange, C.F.M. (1995). *Framework for a simplified model to demonstrate principles of nutrient partitioning for growth in the pig*, Modelling Growth in the Pig, EAAP Publication 78 (Moughan, P.J., Verstegen, M.W.A. and Visser-Reyneveld, M.I. (eds)), 71-85.

Holland, J.H. (1975). *Adaptation in natural and artificial systems*. Ann Arbor, MI: The University of Michigan Press.

Metropolis, N., A.W. Rosenbluth, M.N. Rosenbluth, A.H. Teller, and E. Teller. (1953). Equations of state calculations by fast computing machines, *The Journal of Chemical Physics* **21** 1087-1092.

Nelder, J.A., and R. Mead. (1965). A simplex method for function minimisation, *The Computer Journal* **7** 308–313.

Chapter 3

OPTIMIZED DESIGN
OF DYNAMIC NETWORKS
WITH HEURISTIC ALGORITHMS

Valentin Samko

Stephen Hurley

Stuart Allen

Roger Whitaker
Centre for Intelligent Network Design
Cardiff University
United Kingdom
steve@cs.cf.ac.uk

Abstract   This research investigates the design of reliable wireless communication networks in which all communication is carried over the wireless
medium, every node exhibits mobility and is able to establish only a
limited number of point-to-point links with other nodes in its neighbourhood. The design process aims to select a feasible subset of the
potential links in such a way that both network reliability and network
performance are maximised. We describe a graph-theoretic model for
the type of network under consideration, define some graph-theoretic
measures of network reliability in terms of this model, present methods for the optimisation of these reliability measures, and indicate their
performance under simulation.

1.   Introduction

Recently there has been growing interest in the design of backbone
networks for connecting a number of separate, small-scale, conventional

ad-hoc networks into a single large network [9, 14, 19, 20, 23]. Such networks can be implemented by locating one or more long-range communications devices, which we call nodes, within each of the smaller networks. Each backbone node contains one or more high gain (narrow beam) steerable directional antennae [1, 7, 12]. A multihop wireless backbone network can then be constructed by establishing point-to-point links between these nodes [2–1, 4, 15]. Among the advantages of using directional antennas are a higher signal-to-noise ratio, which results in longer transmission ranges and/or lower power requirements, and reduced radio interference, which leads to improved utilisation of the wireless medium and increased network capacity. In hostile environments, further benefits include lower probabilities of detection and greater robustness to jamming.

So far, most research into ad-hoc networks has focused on routing protocols that are robust to changes in network topology resulting from mobility, although methods of improving the energy efficiency of mobile ad hoc networks have also been investigated.

The networks we study exhibit mobility in the sense that they can move from place to place. At any particular instant, each node is able to establish a limited number of point-to-point links to other nodes in its neighbourhood. The main function of our network is to facilitate communication between its constituent nodes. The primary goal of our research is to develop algorithms for the design and maintenance of feasible network topologies so that network performance is optimised as far as possible.

One aspect of our research is to identify effective (and efficient) graph-theoretic measures of network reliability. Since the primary function of our network is to facilitate communication, the effectiveness of any such measure must ultimately be assessed against network performance. In fact, we have implemented several of such measures, and designed algorithms to simulate various types of mobility and traffic that are likely to occur in dynamic networks. These will be used to investigate the correlation between abstract measures of network reliability and various empirical measures of network performance. Any reliability measures that are found to be highly correlated with good network performance will subsequently be exploited in order to develop effective network design algorithms.

It is worth pointing out that topological considerations do not play a major role in traditional multihop wireless networks, where each node employs an omni-directional antenna to establish broadcast links with other nodes in its neighbourhood. The topology of these networks is usually assumed to be determined autonomously, and most research in this area [6, 8, 16, 21] has thus focused on developing routing protocols that are responsive to changes in network topology, occurring for example as a result of node mobility, node failure and link failure. Despite the fact that network topology is critical in determining network performance [5], the problem of designing and maintaining desirable network topologies that allow for effective routing has received little attention. In fact, the only method of achieving topology control in this type of network is to adjust the transmission power at the network nodes – this is addressed in [18, 17].

2. Problem setup

A network management scheme for the type of network under consideration must also include algorithms for maintaining the *topology* of the network – such algorithms are often called *network design* or *topology control* algorithms. At any particular instant, a network design algorithm must determine which links (among those available) should be constructed by the network in order that the performance of the network is optimised as far as possible. The basic role of a network design algorithm is to provide a network topology that allows the routing algorithms to operate effectively. In this context, desirable topological characteristics (for both static and dynamic networks) include *connectivity* so that a path exists between every pair of nodes, *small diameter* so that short paths exist between each pair of nodes, and the existence of *multiple disjoint paths* between each pair of nodes so that routes continue to exist in the presence of link failure.

We aim to construct a model of dynamic communications networks so that many of the known types of dynamic network (e.g ad hoc networks, bluetooth networks) can be represented by the model. Another aim of our research is to develop a comprehensive simulation environment for evaluating the performance of network design algorithms under a wide range of conditions. In particular, we consider different types of

- *Scenario*: including the number of nodes, terrain characteristics, node distribution and hardware specification (e.g. transmission range).

- *Mobility*: including the node mobility model (e.g. random walk, random waypoint, group mobility models), maximum speed and the rate of acceleration.

- *Traffic*: including the mode of communication (e.g. unicast, multicast, broadcast), packet arrival rate and the distribution of inter-arrival times.

We remark that because this evaluation scheme involves a number of random quantities (e.g. the initial node distribution, the mobility of the nodes and the network traffic), we must ensure that simulation is performed sufficiently many times in order that we may draw valid statistical conclusions from the performance data.

Environment

The environment is defined by the rectangle region in which all the nodes are located. Nodes are assumed not to leave this rectangle. All the mobility scenario generators ensure that nodes bounce off the edges of this region, i.e. if a node reaches the left or right boundary, then its horizontal speed is negated, and the vertical speed is negated if the node reaches top or bottom boundary.

Units of measurement

The software system (DynaNet) which implements this research is not tied to any specific unit measures like seconds, metres, etc. Instead, relative units of measurement are used. All the distance units are equal and all the time units are equal. Therefore, one could treat time units as seconds, or as hours, the result will be still valid (in considered time units).

3.   Characteristics of Dynamic Networks

Although each node is able establish a link with any other node located within its transmission range, hardware constraints impose a limit on the number of links that can actually be supported by a node at any particular instant – we call this the *maximum degree* constraint of the node. A *feasible* topology or design for the network is then defined to be any subset of the potential links for which the maximum degree constraint at each node is satisfied.

Network Characteristics

Given a set of network parameters, a network generator produces a sequence of *network instances*

$$\{N_i : 1 \leq i \leq n_N\} \tag{1.1}$$

A network instance is defined to be a network whose nodes have each been assigned a position vector.

Node distribution.   Network instances are generated in two ways, both of which require that a bounded region R is specified.

1 For the *uniform distribution*, the position of each node is selected uniformly at random from R.

2 For the *clustered distribution*,

    (a) The position x_C of the centre of each cluster $C \subset N$ is selected uniformly at random from R.

    (b) The position x_i of each node $n_i \in C$ is selected uniformly at random from the disc of radius r_C centred at x_C (where r_C is the radius of cluster C).

Node velocity.   Since a node trajectory generated according to one of the random waypoint mobility models (see below) depends on the initial position of the node (i.e. the position at time t_{start}), velocities are not assigned by the network generator. Instead, the mobility generator is required to specify the velocity of each network node at time t_{start}.

Mobility

The trajectory of a node over a time interval $[t_{\text{start}}, t_{\text{end}}]$ will be represented by a sequence of *mobility events*, defined to be a change in the velocity of the node, along with the times at which these events occur. The node is assumed to move at constant velocity between successive events. Thus the trajectory of a node $n_i \in N$ can be represented by the set

$$\{(t, v(t)) : t = t_0, t_1, \ldots, t_n\} \qquad (1.2)$$

where $t_0 = t_{\text{start}}$, $t_n = t_{\text{end}}$ and $v(t_q)$ is the velocity of n_i over the interval $[t_q, t_{q+1})]$ (see Table 1.1).

Parameter	Description
time	The time at which the mobility event occurs
node	The node at which the mobility event occurs
xvel	The horizontal component of the new velocity
yvel	The vertical component of the new velocity

Table 1.1.   Mobility parameters

The advantage of this representation over the traditional method of recording the position of the node at various times in the interval $[t_{\text{start}}, t_{\text{end}}]$ is that it allows more flexibility regarding the choice of time interval between successive updates of the network state during simulation.

Random walk.   In the random walk model, the trajectory of a node over the time interval $[t_{\text{start}}, t_{\text{end}}]$ is generated by first constructing an increasing sequence

$$t_{\text{start}} = t_0 < t_1 < \ldots < t_i < t_n = t_{\text{end}} \qquad (1.3)$$

according to a Poisson process of intensity λ. If we define the *inter-impulse* times by

$$\Delta t_i = t_i - t_{i-1} \quad \text{for} \quad 1 \leq i \leq n \qquad (1.4)$$

then each Δt_i will be identically distributed according to the probability distribution

$$F(x) = P(\Delta t \leq x) = 1 - e^{\lambda x} \qquad (1.5)$$

In the context of network mobility, the parameter λ is called the *impulse rate* since its value determines the average rate at which the velocity of a node changes.

Having generated this sequence, we then generate a velocity for each interval $[t_q, t_{q+1}]$ by selecting the speed uniformly at random from the interval $[0, s_{\text{max}}]$ and a direction uniformly at random from the interval $[0, 2\pi]$, where s_{max} is the maximum speed of the node. In particular, the node is assigned an initial velocity (which it maintains for the duration of the interval $[t_0, t_1)$).

Random waypoint.   The random waypoint mobility model was first described in [6] and later refined in [3]. A node randomly selects a point in a bounded region R and moves towards that point at a speed selected uniformly at random from some bounded range. When the point is reached, the node pauses for some pre-defined time, then the procedure is repeated.

In contrast to the random walk model, trajectories generated according to the random waypoint model

1 require that a bounded region R is specified, and

2 depend on the node positions at time t_{start}.

Consequently, a mobility instance M_{ij} generated according to the random waypoint model depends on the environment E and also on the associated network instance N_i.

In [24], it is noted that selecting the speed from the range $[0, s_{\text{max}}]$ for some maximum value s_{max}, results in a steady decrease in the average speed of the nodes. The reason for this is that if a node is assigned a speed that is close to zero, it will have this small speed for a long time (until it reaches its destination). One solution to this problem, is to define some minimum speed $s_{min} > 0$ and select the speed the uniformly at random from the range $[s_{min}, s_{max}]$.

The trajectory of node $n_i \in N$ over the time interval $[t_{start}, t_{end}]$ is generated according to the procedure outlined in Figure 1.1, where $x_i(t_{start})$ and $y_i(t_{start})$ are respectively the horizontal and vertical coordinates of n_i at time t_{start}, and Δp is the pause time.

---

1. Set $t = t_{start}$, $x = x_i(t_{start})$, $y = y_i(t_{start})$
WHILE $(t < t_{end})$
    2.1 Select a point (x', y') uniformly at random within the region R.
    2.2 Select a speed s uniformly at random from the range $[s_{min}, s_{max}]$.
    2.3 Compute the distance $d = ((x' - x)^2 + (y' - y)^2)^{1/2}$ between (x, y) and (x', y').
    2.4 Record the mobility event $(t, s(x' - x)/d, s(y' - y)/d)$.
    2.5 Compute the time $\Delta t = d/s$ to reach (x', y') from (x, y) at speed s.
    2.6 Set $t = t + \Delta t$, $x = x'$, $y = y'$
    2.7 Record the mobility event $(t, 0, 0)$.
    2.8 Set $t = t + \Delta p$
END WHILE

Figure 1.1. The random waypoint mobility model

Group mobility. In the case where the nodes are partitioned into clusters, we also consider mobility models in which all nodes belonging to a particular cluster remain close together.

Let $C \subset N$ denote a cluster of nodes, let r_C denote its radius and let $x_C(t)$ denote the position of its centre at time t. If the nodes are initially distributed according to a clustered distribution, the initial position of the cluster centres $x_C = x_C(t_{start})$ are distributed uniformly within the region R, and the position of each node $x_i \in C$ (at time t_{start}) is uniformly distributed within a disc of radius r_C centred at x_C.

We define analogues of the random walk and random waypoint mobility models, called *group random walk* and *group random waypoint* respectively, by first generating trajectories for each cluster centre using the (simple) random walk and random waypoint models described above. Consider the mobility events that describe the trajectory of the centre of cluster C,

$$\{(t, v(t)) : t = t_0, t_1, \ldots, t_n\} \tag{1.6}$$

To generate the trajectory of node $n_i \in C$, for each mobility event $(t_i, v(t_i))$ with $i > 0$, we

1. choose a point (x', y') uniformly at random within the disc of radius r_C centred at $x_C(t_i)$, where $x_C(t_i)$ is the position of the cluster centre at time t_i.

2. compute the required velocity so that node n_i reaches the point (x', y') at time t_i (this depends on the position of n_i at time t_{i-1}).

However, we must ensure that the required velocity does not exceed the maximum velocity of node n_i. To this end, let $S_{\max}(C)$ denote the maximum speed of the cluster and let

$$s_{\max}(C) = \min\{s_{\max}(n_i) : n_i \in C\} \tag{1.7}$$

where $s_{\max}(n_i)$ is the maximum speed of node $n_i \in N$. Then we require a lower bound on the length of the time interval Δt between successive changes in the velocity of the cluster centre, given by

$$\Delta t \geq 2r_C / S_{\max}(C) \tag{1.8}$$

where r_C is the radius of cluster C. Having selected some Δt according to this constraint, we then define

$$\hat{S}_{\max}(C) = \min\{S_{\max}(C), s_{\max}(C) - 2r_C/\Delta t\} \tag{1.9}$$

and choose the speed of the cluster centre for the interval $[t, t + \Delta t]$ uniformly from the range $[0, \hat{S}_c^{\max}]$. Note that for this to work we need that $s_{\max}(C) > S_{\max}(C)$ for every cluster C.

Traffic

The traffic associated with a given node over some time interval $[t_{\text{start}}, t_{\text{end}}]$ will be represented by a sequence of *traffic events*, along with the times at which these events occur. As illustrated in Table 1.2, a traffic event is defined by

1. the time at which the event occurs

2. a set of destination nodes

3. a message size

Parameter	Description
time	The time at which the traffic event occurs
source	The node at which the traffic event occurs
target	The node(s) to which the message is to be delivered
messageSize	The size of the message to be delivered

Table 1.2. Traffic parameters

The traffic associated with a node over the time interval $[t_{\text{start}}, t_{\text{end}}]$ is generated by first generating a sequence

$$t_{\text{start}} = t_0 < t_1 < \ldots < t_i < t_n = t_{\text{end}} \qquad (1.10)$$

according to a Poisson process of intensity λ. If we define the *inter-arrival* times by

$$\Delta t_i = t_i - t_{i-1} \quad \text{for} \quad 1 \leq i \leq n \qquad (1.11)$$

then each Δt_i will be identically distributed according to the probability distribution

$$F(x) = P(\Delta t \leq x) = 1 - e^{\lambda x} \qquad (1.12)$$

In the context of network traffic, the parameter λ is called the *arrival rate* since its value determines the average rate at which traffic events occur at a node. Having generated this sequence for node $n_i \in N$, we then generate a traffic event for each time t_i by

1 selecting a set of destination nodes.

- For *unicast* traffic events, the set of destination nodes contains exactly one node, selected uniformly at random from among the other nodes in the network.

- For *broadcast* traffic events, the set of destination nodes contains all other nodes in the network.

2 selecting a message size according to a *burst distribution*. This will be either a normal distribution or a uniform distribution. The mean of the burst distribution determines the expected size of the message.

4.   Initial data for network design: *network hints*

If we know the future network traffic or mobility in advance, we may use this information designing the network. We have implemented two design algorithms which exploit such data. One of them assumes knowledge of the centre of each node's trajectory, and another requires the knowledge of the maximum distance between this centre and other trajectory positions, for each node. These algorithms appear to produce better results, than basic greedy algorithms we have described in section 10.0. This leads to a conclusion, that the quality of the network design may significantly depend on whether we have any information about future network mobility and traffic.

In the design algorithms presented later use will be made of so-called *hints*. These simply refer to values or characteristics that could provide some knowledge or insight to the optimisation process.

Mobility hints

Any node ξ can have a number of parameters associated with it:

- Maximum range, the node can transmit signal over $\xi_{maxrange}$.

- Initial node position ξ_{pos}.

- Initial node velocity ξ_{vel}.

- Centre of the area where the node is allowed to move $\xi_{areacentre}$.

- Maximum distance between the above mentioned centre and positions in the node trajectory $\xi_{arearadius}$.

- Minimum node speed $\xi_{minspeed}$.

- Maximum node speed $\xi_{maxspeed}$.

- Average turn degrees per time unit (0 - 360) $\xi_{avgtrndegr}$.
  For example, if this value is 360, then the node can turn up to 360 degrees every time unit. If this value is zero, the node never turns. If this value is 3, then on average the node turns 3 degrees each time unit, i.e., for instance, the node can go straight for 20 time units, and then turn 60 degrees.

Traffic hints

Traffic hints could include the following:

- Average number of bytes sent from node per time unit ξ_{avgtrf}.

- Average number of nodes it sends messages to during one time unit $\xi_{avgdest}$.

Environment hints

We denote the environment by Ω. Some characteristics of the environment could include:

- Minimum distance between any two nodes $\Omega_{mindist}$.

- Average distance between all the pairs of nodes $\Omega_{avgdist}$.

5.   Mobility, network states

The network simulator reads all instances from input, where each instance consists of the initial network state, mobility model and traffic model. It also reads all the corresponding network design from the design input file. Then it executes the mobility loop for each instance. During each iteration of the mobility loop it does the following:

- reconnect failed links (depending on configuration file parameter);

- update node positions and velocities (according to the mobility model);

- disconnect failed links;

- update routing tables;

- record network state in the output file;

- flush node buffers (depending on configuration file parameter) i.e. discard any message at a node and link

- run traffic loop.

Link failure

The link fails when the distance between two nodes surpasses the maximum transmitting range of these nodes.

There are two modes on how we threat failed links.

- Once some link fails, it is never considered again, and is never reconnected even if the nodes get close enough for the link to be established.

- If the nodes constituting a failing link move back close enough for the link to be established, the link is reestablished and is valid again.

6.   Routing

Perhaps the most important aspect of network management is *routing*. Routing algorithms must determine paths (routes) for packets to travel across the network so that the packets are delivered quickly and reliably. To minimise the delay experienced by a packet, the path is generally chosen to be one of the shortest paths between the associated source and destination nodes. A simple routing algorithm maintains a *routing table* at each node. For any given destination, the table records the link (among the set of links supported by the node) along which a packet must be sent in order to reach that destination.

One attractive feature of implementing hierarchical network architectures is that the size of the routing tables maintained at each node can be reduced. Each (conventional) node in a cluster maintains a routing table only for nodes within its cluster, while all inter-cluster traffic is routed through the backbone nodes. In this context, the backbone nodes are often called *gateway* nodes [10, 11].

A routing algorithm aims to find a "good" path from a source u_s to a destination u_d. For each link there is an associated cost of sending a packet down the link. In particular, the cost could represent a measure of the congestion on the link (e.g the buffer queue length) and hence the expected waiting time for packets wishing to use it. A *least cost path* between u_s and u_d is defined to be a path for which the sum of the costs over each link is minimum. If all link costs are equal, this is also called the *shortest path*.

A *dynamic* routing algorithm is one that changes routing paths in response to changes in traffic patterns and network topology.

Link state (LS) routing algorithms use variants of Dijkstra's algorithm to compute the least cost path from a node to all other nodes (provided these exist), and has a worst case running time of $O(n^2)$. LS algorithms are centralised in the sense that they need to know the state (cost) of each link. In a distributed setting, each node must therefore broadcast the cost of each of its adjacent links to *all* other nodes in the network (resulting in high control traffic overheads).

We should note that the routing algorithms we use may produce different routes if we run it several times for the same network. This happens if there are several routes from one node to another with the same minimum weight. In such a case, the specific route may be chosen pseudo-randomly, depending on the used memory allocation schema, etc.

7.   Traffic processing

In each traffic loop we inject packets (corresponding to the traffic model), process these packets (a packet can only move from one node to the next and only if there is available link capacity between the source and target nodes).

Each node and link have packet buffers, where packets are stored temporary, until transmitted further. The buffers are limited by specified size, and packets are dropped if the destination packet buffer is full.

8.   Performance measures

The performance of any network management algorithm must be measured by the performance of a network under its control. For packet switched networks, examples of network performance measures include *network throughput* and *network delay*, defined to be the average number of packets delivered successfully per unit time and the average time taken for a packet to reach its destination respectively.

The notion of network reliability corresponds to how well the network performs *over a given time interval* under dynamic conditions. A particular network design may perform well (e.g. has high throughput) at time $t = 0$, but is unable to maintain this performance level throughout the time interval $[0, T]$. Another design having inferior performance at $t = 0$ might be considered to be 'more reliable' if its average performance over $[0, T]$ is better than that of the original design.

A simple indicator of network reliability is provided by the *mean link duration* (taken over all links) over a given time interval. Other measures which can be calculated include the following:

- **packets_injected** - Number of packets injected into the network (generated according to the traffic model) during the given time interval.

- **packets_dropped** - Number of packets dropped during the given time interval. Packets are dropped, in the following cases:

  - when the simulator attempts to inject a packet into some node input queue, and the node queue is full;

  - when simulator attempts to push the packet to the link buffer, but the buffer is full.

This measure is equivalent to the number of packets which do not reach their destinations.

- **packet_loss_ratio** - Ratio of **packets_dropped** to the total number of generated packets (this number is reset to zero after each mobility step if the node buffers are emptied).

- **packets_arrived** - Number of packets which arrived to the destination node during the given time interval.

- **delay** - Ratio of **packets_arrived** to the total time between generation and arrival of all these arrived packets i.e.

$$\sum_{arrived\_packets} packet_{arrival\_time} - packet_{generation\_time}$$

- **throughput** - Ratio of **packets_arrived** to the length of the time interval.

- **packets_in_transit** - Number of packets in the network, which have neither arrived, nor dropped. These are the packets in node and link buffers waiting to be sent to the next node according to the routing table.

9.   Graph-theoretic reliability measures

The task of finding good designs based on global information is a difficult one, since the number of candidate designs increases exponentially with the number of nodes. This makes an exhaustive search of the design space computationally infeasible. Thus we adopt meta-heuristic search techniques in an attempt to obtain 'near-optimal' network designs.

Given that each step of a meta-heuristic search requires a number of 'neighbouring solutions' to be evaluated, in order that our search is effective we need a method for *quickly* evaluating candidate designs, a requirement that excludes the use of the simulation scheme outlined above.

This need for rapid evaluation motivates the first objective of our research, which is to identify rapidly computable *graph-theoretic* measures of reliability for dynamic networks. For example, we can define a graph to represent the network whose edges are weighted according to the robustness of the corresponding link. A possible definition of link robustness is given by the length of the link relative to the transmission range of the node that supports the link. Intuition suggests that the minimum edge cut in the graph somehow corresponds to the 'weakest point' of the network with regard to node mobility.

The first step along this line of enquiry has been to propose a number of such graph-theoretic measures, evaluate the performance of a set of randomly generated network designs under simulation, and investigate whether there exists any correlation between the best network designs (as determined by simulation) and the proposed measures. Having identified a set of promising graph-theoretic measures in this way, we refine the study by actively seeking 'good' network designs corresponding to each measure, again using meta-heuristic search. The performance of these networks under simulation should then allow us to decide which among the set of measures provides the best indicator of network reliability under various dynamic conditions.

Generic measures

- **average_node_degree** - Ratio of twice number of edges to the number of vertices in the graph.

- **minimum_node_degree** - Minimum node degree. This is always zero for disconnected graphs.

- **number_of_components** - One divided by the number of disconnected components in the graph. This value is always 1 for connected graphs, and it is less then 1 otherwise.

- **number_of_nodes_in_largest_component** - Number of nodes in the largest graph component. This is equivalent to the number of vertices for connected graphs.

- **clustering_coefficient** - the probability that a path of length 2 is part of a triangle (a 3-cycle). That is, if one has edges ab and bc, the probability that a and c are joined by an edge. This is calculated by counting the triangles in the graph and also the number of paths of length 2. Then the coefficient is the ratio of 3 times the number of triangles to the number of paths of length 2.

Measures for connected graphs

The following measures return zero for disconnected graphs.

- **diameter** - One over the maximum length of the shortest path between two vertices, $\max_{u,v} d(u, v)$ where $d(u, v)$ is the length of the shortest path between u and v.

- **diameter_weighted_by_slack** - Same as **diameter**, but $d(u, v)$ is the minimum weight path. Edge weight is defined as the ratio of distance between nodes u and v to their range.

- **average_separation** - Same as **diameter**, but instead of the maximum length of the shortest path we calculate the average length of the shortest path $avg_{u,v}d(u,v)$.

- **average_separation_weighted_by_slack** - Same as **average_separation**, but $d(u,v)$ is the minimum weight path. Edge weight is defined as the ratio of distance between nodes u and v to their range.

- **edge_connectivity** - The minimum number of edges one has to disconnect to make the graph disconnected. This is equivalent to the minimum number of edge disjoint paths among all the pairs of vertices, $\min_{u,v} p(u,v)$ where $p(u,v)$ is the number of edge disjoint paths between u and v.

- **minimum_edge_cut_weighted_by_slack** - The minimum sum of edge weights, such that disconnecting these edges we make the graph disconnected. Edge weight is defined as the ratio of distance between nodes u and v to their range.

- **vertex_connectivity** - The minimum number of vertex disjoint paths among all the pairs of vertices, $\min_{u,v} r(u,v)$ where $r(u,v)$ is the number of vertex disjoint paths between u and v.

10.   Network design algorithms

The primary purpose of a communications network is to facilitate communication between its constituent nodes, and the role of a network management algorithm is to enable the network to perform this function as effectively as possible. We investigate algorithms for the design and maintenance of feasible network topologies so that network performance is optimised as far as possible.

Feasible links

We consider a link as feasible if and only if the minimum transmitting range of nodes constituting this link is greater than the distance between these nodes.

All the network design algorithms we provide produce designs which only contain links, which are feasible at initial network state (time 0). Optionally, we could use links which are not feasible at time 0, but there is a high probability that they will be feasible at later steps. Experiments have indicated that if we assume the former (only using links feasible at time 0), then all the measures are very high at time 0, but generally small elsewhere. If we assume the later (using any reasonable links), the measure values are not so good at time 0, but are much better elsewhere.

Minimum spanning tree with maximum node degree constraint

This problem is known to be NP-complete and so we provide a heuristic method which produces spanning tree with maximum node degree constraint, but this is not always the one with minimum weight. We have implemented one design algorithm **MinSpanningTreeDesign** which only builds this minimum spanning tree with maximum node degree constraint.

Heuristic approach.

1. Build minimum spanning tree using the Kruskal algorithm.

2. Iterate through nodes with degree higher than the specified maximum.

3. For each such node, sort links by weight (starting from the greatest weight) and iterate through these links.

4. For each such link, try disconnecting the link (thus separating the tree into two clusters) and finding a feasible link (with the smallest possible weight) connecting these clusters so that degrees of source and target nodes of this link are not higher than the specified maximum degree.

5. If such feasible link is not found, connect the link back, and proceed. Otherwise, if such a link is found, use it instead of the disconnected one.

· Greedy network design

All the greedy network design algorithms follow the same schema and only differ in the link weight assignment procedure.

1. Assign weights to all the feasible links.

2. Build minimum weight spanning tree with maximum node degree constraint.

3. Sort feasible links by weight, with minimum weight in the beginning.

4. Iterate through feasible links and add those which do not break the maximum node degree constraint.

Link weights.   We propose several greedy design algorithms, defined by their weight assignment procedures. Let l be a link connecting nodes x and $\tilde{x}$, $range(x)$, $range(\tilde{x})$ be the maximum transmitting ranges of these nodes and $vel(x)$, $vel(\tilde{x})$ be their velocity vectors. Then we define $s(l)$ by

$$s(l) = \min(range(x), range(\tilde{x})) - ||x - \tilde{x}|| \qquad (1.13)$$

and $ss(l)$ as the positive root of the equation

$$||vel(x) - vel(\tilde{x})||^2 t^2 + 2 * (x - \tilde{x}) \cdot (vel(x) - vel(\tilde{x}))t + ||x - \tilde{x}||^2 \quad (1.14)$$

$$= \min(range(x), range(\tilde{x}))^2 \qquad (1.15)$$

in respect to t. Here $ss(l)$ corresponds to the time that will elapse before the link fails under the assumption that the velocity of each node will not change over this time period.

Thus we have the following design algorithms:

- **SlackGDA** - links weighted according to $s(l)$;

- **SuperSlackGDA** - links weighted according to $ss(l)$;

- **AntiSlackGDA** - link weighted by $-s(l)$;

- **RandomGDA** - link weighted randomly, this is used for tests and comparisons;

- **TestDesign** - calculates weighted (by time spent at these positions) average positions over the time of each node and then uses distances between these positions, instead of initial positions.

- **TestDesign_MaxRad** - modification of the **TestDesign** which also exploits more information about node trajectories (maximum distance between all the trajectory points and the average position).

Heuristic network design

This method accepts a set of reliability measures on input, which are used in this method to compare the effectiveness of the designed network. The main idea behind this network design approach is to first define some network design algorithm dependent on a set of input parameters (e.g. weights assigned to different network characteristics) and then to run a multidimensional minimisation routine in order to determine which set of parameters corresponds to good values of specified reliability measures of the designed network.

The base network design algorithm depending on input parameters first calculates weights of all the feasible links and builds a minimum spanning tree with a maximum node degree constraint. Then it repeatedly recalculates all the weights and adds feasible links with the minimum weight to the network design, while there are any feasible links (i.e. links which do not invalidate the maximum node degree constraint).

Weight construction. Let us define a set of weights $w_\alpha \in W$ dependent on a number (N) of discrete parameters $\alpha = \{\alpha_i\}_{i=0}^{N} \in A$, where $\alpha_i \in D_i \subset R$. So, we would have a N-dimensional discrete space of weights. Then we build a set of functions $H_k(W_\alpha)$, $k = 1, ...H$ corresponding to H different reliability measures.

Having this, the next step would be to build heuristic minimisation algorithms to find a set of $\{\tilde{\alpha}_i\}_{i=0}^{N}$ such that $H_k(w_{\tilde{\alpha}})$ are small.

One could minimise

$$H(w_\alpha) = \sum_{k=0}^{H} H_k(w_\alpha)$$

or

$$H(w_\alpha) = \max_{k=0}^{H} H_k(w_\alpha)$$

or some other combination.

Forming generic weights. Let us assume, we have a set of parameters associated with each node. It would be natural to split them into several groups:

- set P_1 of parameters such that a link is generally better when the difference between values of these parameters of source and target nodes is small;

- set P_2 of parameters such that a link is generally better when the difference between values of these parameters of source and target nodes is great;

- set P_3 which contains parameters such that the quality of the link only depends on the minimum value of this parameter of source and target nodes (for example maximum node range).

- set P_4 which includes all the remaining parameters.

Now, we construct a generic form of weight $w_c(\xi, \tilde{\xi}) =$

$$\sum_{p_i \in P_1} c_{1,i} |p_i - \tilde{p}_i| + \sum_{p_i \in P_2} c_{2,i} \frac{1}{|p_i - \tilde{p}_i|} + \sum_{p_i \in P_3} c_{3,i} \min(p_i, \tilde{p}_i) + \sum_{p_i \in P_4} c_{4,i} (p_i + \tilde{p}_i)$$

where p_i and $\tilde{p}_i$ are parameters associated with nodes ξ and $\tilde{\xi}$ respectively.

Adapting generic weights to our parameters.   Next, we note that we have the following parameters we can consider while forming weights for each feasible link:

- static parameters - $\xi_{pos} \in P_1$, $\xi_{maxrange} \in P_3$, $\xi_{areacentre} \in P_1$, $\xi_{arearadius} \in P_4$;

- velocity parameters - $\xi_{vel} \in P_4$, $\xi_{minspeed} \in P_4$, $\xi_{maxspeed} \in P_4$;

- mobility parameters - $\xi_{avgtrndegr} \in P_4$;

- traffic parameters - $\xi_{avgtrf} \in P_2$, $\xi_{avgdest} \in P_2$;

So, the final formula would be

$$w_c(\xi, \tilde{\xi}) = \quad c_1|\xi_{pos} - \tilde{\xi}_{pos}| + c_2|\xi_{areacentre} - \tilde{\xi}_{areacentre}|$$

$$+ \quad \frac{c_3}{|\xi_{avgtrf} - \tilde{\xi}_{avgtrf}|} + \frac{c_4}{|\xi_{avgdest} - \tilde{\xi}_{avgdest}|}$$

$$+ \quad c_5 \min(\xi_{maxrange}, \tilde{\xi}_{maxrange})$$

$$+ \quad c_6(\xi_{vel} + \tilde{\xi}_{vel})$$
$$+ \quad c_7(\xi_{arearadius} + \tilde{\xi}_{arearadius}) + c_8(\xi_{minspeed} + \tilde{\xi}_{minspeed})$$
$$+ \quad c_9(\xi_{maxspeed} + \tilde{\xi}_{maxspeed}) + c_{10}(\xi_{avgtrndegr} + \tilde{\xi}_{avgtrndegr})$$

Dynamic weight construction.   We use the following approach to calculate the link weights. Note that we include the length of the minimum path between source and target nodes of the link (if there is such a path) and combine this value with $w_c(\xi, \tilde{\xi})$.

1 Calculate $t = w_c(\xi, \tilde{\xi})$ for each feasible link.

2 Iterate through feasible links (only feasible links, ones which are not present in the graph).

3 If source or target node of a feasible link has maximum allowed degree, continue iterating.

4 Calculate "flow" between source and target nodes ("flow" depends on the number of disjoint paths between source and target nodes, and on the weights of the links in these paths).

5 Combine and save "flow" and weight of the feasible link.

6 When the iterations are completed, find the feasible link (preferably, one which connects disjoint paths) with the smallest weight and establish it in the graph.

Multidimensional minimisation. Here we present our version of a modified simulated annealing algorithm [13], [22].

- 1. start with random α_0, set $i = 0$, calculate $q_0 = H(\alpha)$, and set $\beta = \pi$;

- 2. pick random unit vector v_i in space A such that the angle between v_i and v_{i-1} is not greater than β;

- 3. find optimal $\tilde{a} > 0$ by minimising a one dimensional function $F(a) = H(\alpha_i + av_i)$;

- 4. calculate $q_{i+1} = H(\alpha + \tilde{a}v_i)$;

- 5. if $q_{i+1} \geq q_i$, increase β by $\frac{\pi}{8}$ and go to 2;

- 6. decrease β by $\frac{\pi}{8} sign((q_i - q_{i+1}) - (q_{i-1} - q_i))$;

- 7. set $\beta = 2\pi$ if $\beta > 2\pi$

- 8. increase i by one and set $\alpha_i = \alpha_{i-1} + av_i$;

- 9. go to 2.

11. Experiments and Comparisons

We focus our experiments on the heuristic network design algorithms (section 10.0), since they are more flexible than the greedy methods due to being able to design networks with respect to the given network reliability measures. In Figures 1.2 to 1.9 we present graphs which demonstrate the effectiveness of proposed heuristic approach. Each figure presents the values of the reliability measures presented in section 9.0 plotted against simulation time (results are averaged over all mobility models). Each figure corresponds to the heuristic approach (section 10.0) optimising for one of the reliability measures. The abbreviations used indicate the appropriate reliability measure i.e.

- AND - average node degree

- AS - average separation

- ASWBS - average separation weighted by slack

- D - diameter

- DWBS - diameter weighted by slack

- EC - edge connectivity

- MECWBS - minimum edge cut weighted by slack

- VC - vertex connectivity

- MULTI - a combination of all of the above reliability measures

The figures demonstrate that the heuristic network design algorithms indeed generate network designs that most optimise the specified network reliability measures. One can see that in each graph the maximum measure value is obtained by the design algorithm optimised for this measure. This is to be expected, however the "MULTI" approach of using all the reliability measures produces networks which perform reasonably well over all types of reliability measure.

We have identified a number of issues that need to be considered in the design and maintenance of point-to-point multihop wireless networks. Although only distributed topology control algorithms based on local information are likely to be useful in practice, we have shown that tailored global optimisation algorithms also have a useful role. In obtaining network designs based on network-level information, we have also indicated the importance of identifying rapidly computable, graph-theoretic indicators of topological stability for dynamic networks under simulation. Future work will need to consider how network performance (in terms of traffic) behaves in relation to the different reliability measures.

Acknowledgments

The authors acknowledge the United Kingdom's EPSRC for funding this work.

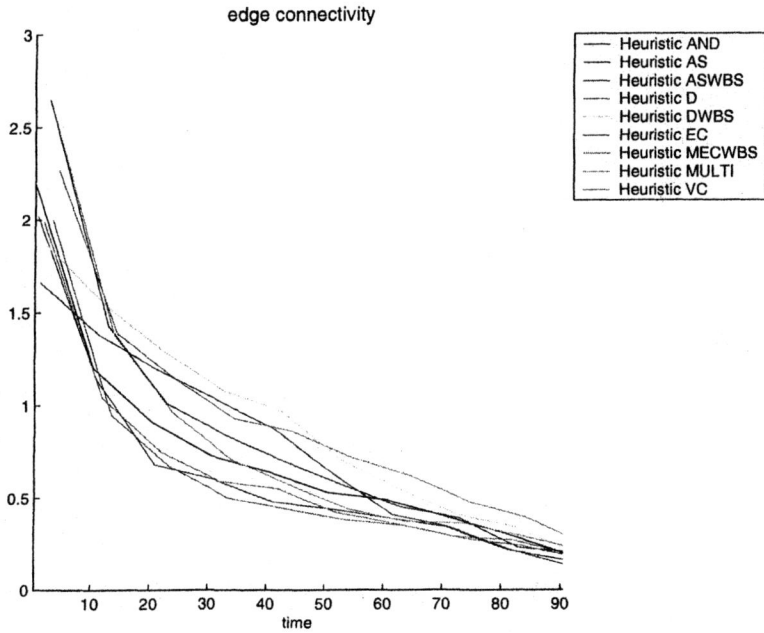

Figure 1.2.   Edge connectivity

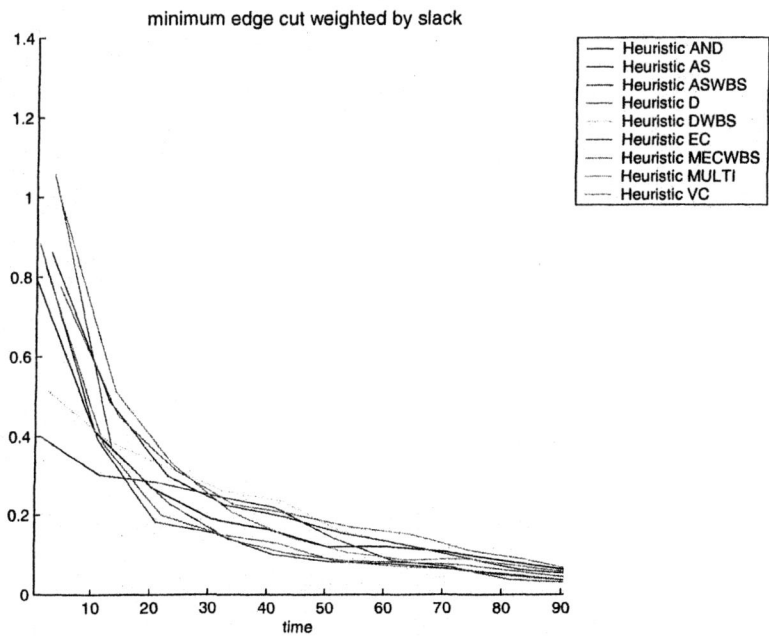

Figure 1.3.   Minimum edge cut weighted by slack

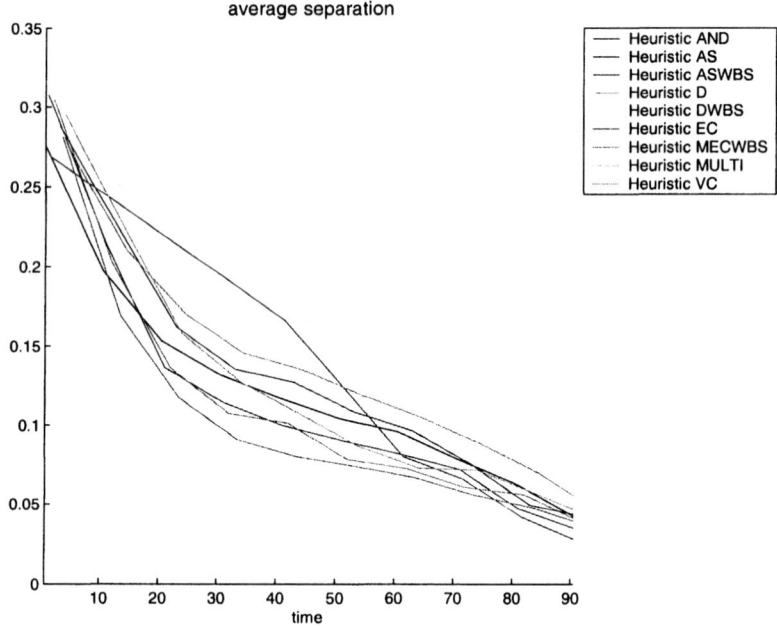

Figure 1.4.   Average separation

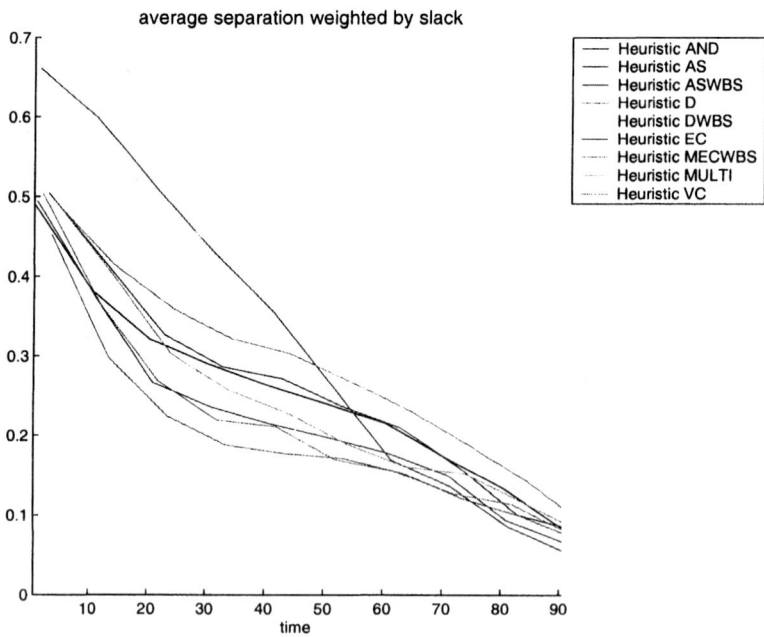

Figure 1.5.   Average separation weighted by slack

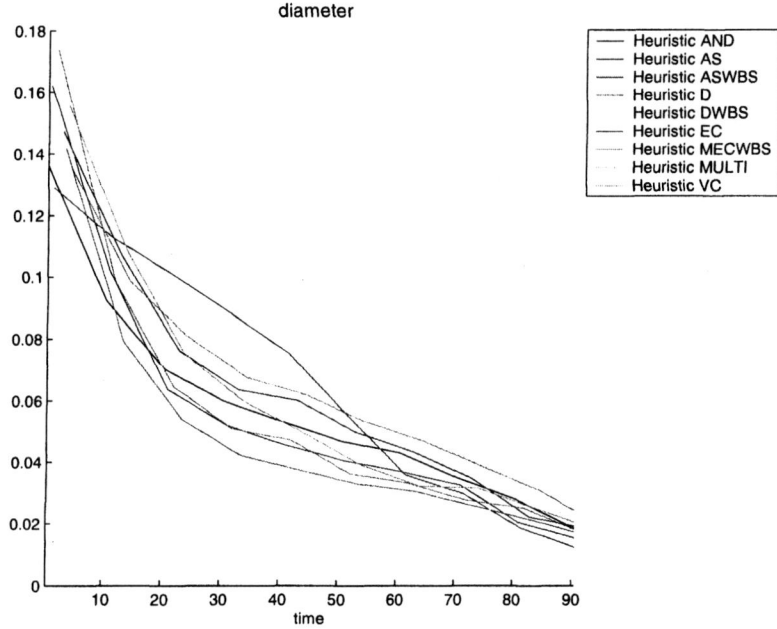

Figure 1.6.   Diameter

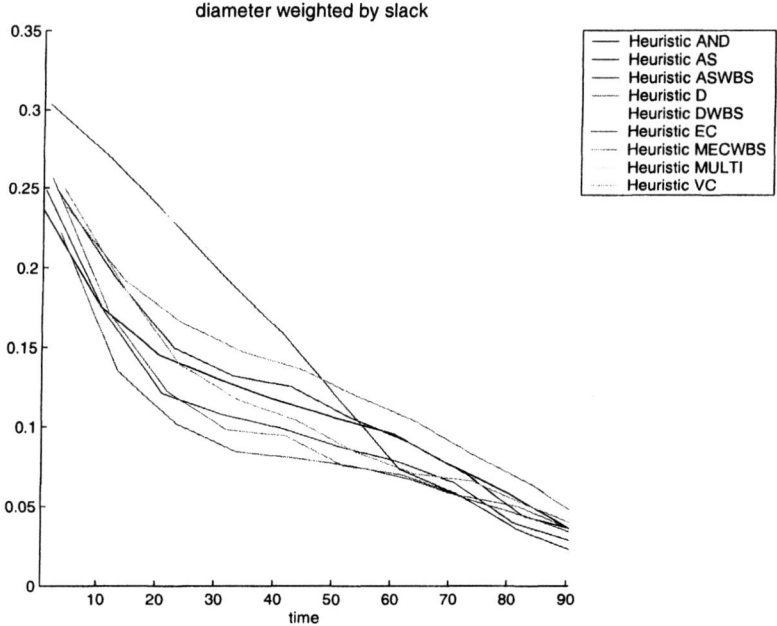

Figure 1.7.   Diameter weighted by slack

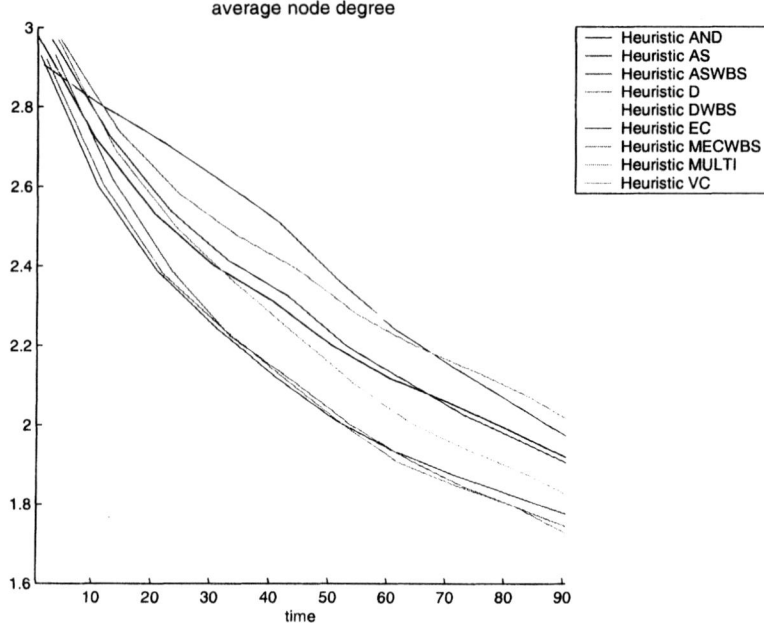

Figure 1.8.   Average node degree

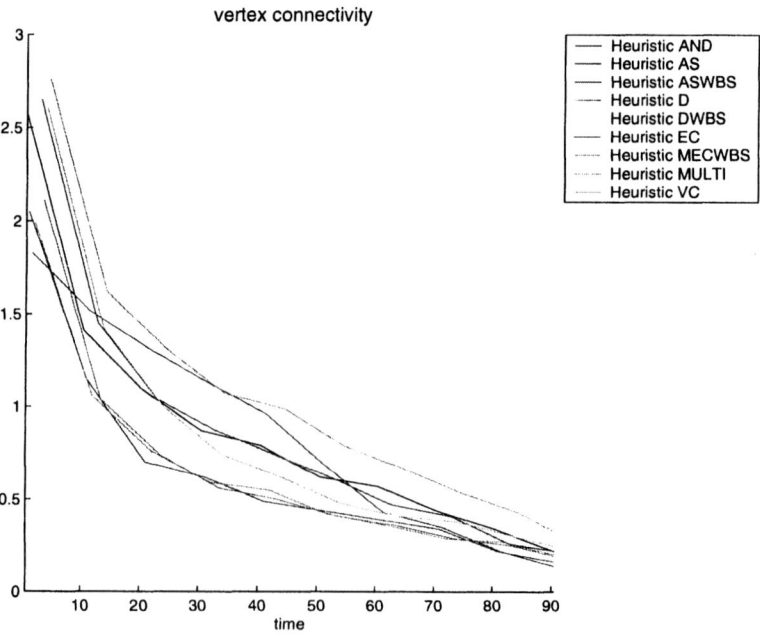

Figure 1.9.   Vertex connectivity

[1] S. M. Allen, D. Evans, S. Hurley, and R. M. Whitaker. Optimisation
   in the design of mobile ad-hoc networks with steerable directional
   antennae. In *Proc 1st Int Conf Ad-hoc Networks And Wireless*, pages
   151Ũ–159, 2002.

[2] S. Bandyopadhyay, K. Hasuike, S. Horisawa, and S. Tawara. An
   adaptive mac protocol for wireless ad hoc community network (wac-
   net) using electronically steerable passive array radiator antenna. In
   Proceedings of the IEEE Global Telecommunications Conference),
   volume 5, pages 2896–2900, 2001.

[3] J. Broch, D. A. Maltz, D. B. Johnson, Y.-C. Hu, and J. Jetcheva.
   A performance comparison of multi-hop wireless ad hoc network
   routing protocols. *Mobile Computing and Networking (MobiCom)*,
   pages 85–97, 1998.

[4] R. R. Choudhury, X. Yang, N. H. Vaidya, and R. Ramanathan.
   Using directional antennas for medium access control in ad hoc networks. In *Proceedings of the eighth annual international conference
   on Mobile computing and networking (MOBICOM '02)*, pages 59–
   70, 2002.

[5] P. Gupta and P. R. Kumar. The capacity of wireless networks. *IEEE
   Transactions on Information Theory*, 46(2):388–404, 2000.

[6] D. B. Johnson and D. A. Maltz. Dynamic source routing in ad hoc
   wireless networks. In T. Imielinski and H. Korth, editors, *Mobile
   Computing*, chapter 5, pages 153–181. Kluwer Publishing Company,
   1996.

[7] Y-B. Ko, V Shankarkumar, and N. H. Vaidya. Medium access control protocols using directional antennas in ad hoc networks. In
   *Proceedings of the Nineteenth Annual Joint Conference of the IEEE
   Computer and Communications Societies*, volume 1, pages 13–21,
   2000.

[8] Y-B. Ko and N. H. Vaidya. Location aided routing (LAR) in mobile
   ad hoc networks. In *Proceedings of ACM/IEEE MOBICOM*, 1998.

[9] U. C. Kozat, G. Kondylis, B. Ryu, and M. K. Marina. Virtual
   dynamic backbones for mobile ad hoc networks. In *Proc. IEEE Int.
   Conf. on Comm.*, pages 250–255, 2002.

[10] C. R. Lin and M. Gerla. Adaptive clustering for mobile wireless
   networks. *IEEE J. Sel. Areas in Comm.*, 15(7):1265–1275, 1997.

[11] H-C. Lin and Y-H. Chu. A clustering technique for large multihop
   mobile wireless networks. In *Proceedings of IEEE VTC*, pages 1545–
   1549, 2000.

[12] A. Nasipuri, J. Mandava, H. Manchala, and R. E. Hiromoto. On-demand routing using directional antennas in mobile ad hoc networks. In *Proceedings of the Ninth IEEE International Conference on Computer Communication and Networks*, pages 535–541, 2000.

[13] Otten and van Ginneken. Review of the annealing algorithm. *ACM Computing Reviews*, 31(6):296–298, 1990.

[14] G. Pei and M. Gerla. Mobility management in hierarchical multihop mobile wireless networks. In *Proceedings of IEEE ICCCN*, pages 324–329, 1999.

[15] G. Pei, D. B. Lofquist, R. Lin, and J. Erickson. Network topology management for mobile ad hoc networks with directional links. In *Proc. IEEE Military Communications Conference (MILCOM '02)*, volume 2, pages 1476–1480, 2002.

[16] C. E. Perkins and E. M. Royer. Ad-hoc on-demand distance vector routing. In *Proceedings of 2nd IEEE Workshop on Mobile Computing Systems and Applications*, 1999.

[17] R. Ramanathan. Making ad hoc networks density adaptive. In *Proc. IEEE Military Communications Conference (MILCOM)*, volume 2, pages 957–961, 2001.

[18] R. Ramanathan and R. Rosales-Hain. Topology control of multihop wireless networks using transmit power adjustment. In *Proc. Nineteenth Annual Joint Conference of the IEEE Computer and Communications Societies (INFOCOM)*, volume 2, pages 404–413, 2000.

[19] I. Rubin and P. Vincent. Effective backbone architectures for mobile wireless networks. Technical report, UCLA, 1998.

[20] I. Rubin and P. Vincent. Design and analysis of mobile backbone networks. Technical report, UCLA, 2001.

[21] C-K. Toh. Associativity-based routing for ad-hoc mobile networks. *Wireless Personal Communications*, 4(2):1–36, 1997.

[22] W.T. Vettering, S.A. Teukolsky, and W.H. Press. *Numerical Recipes*. Cambridge University Press, 1993.

[23] K. Xu, X. Hong, and M. Gerla. An ad hoc network with mobile backbones. In *Proceedings of IEEE ICC*, pages 324–329, 2001.

[24] J. Yoon, M. Liu, and B. D. Noble. Random waypoint considered harmful. In *Proceedings of INFOCOM '03*, volume 2, pages 1312–1321, 2003.

Chapter 4

A NEW SMOOTHING-BASED GLOBAL OPTIMIZATION ALGORITHM FOR PROTEIN CONFORMATION PROBLEMS

Aqil M. Azmi

Department of Computer Science, College of Computer and Information Systems, King Saud University, Riyadh 11543, Saudi Arabia.

Richard H. Byrd, Elizabeth Eskow, Robert B. Schnabel

Computer Science Department, University of Colorado , Boulder, Colorado 80309, USA.
*

Abstract   To help solve difficult global optimization problems such as those arising in molecular chemistry, smoothing the objective function has been used with some efficacy. In this paper we propose a new approach to smoothing. First, we propose a simple algebraic way to smooth the Lennard-Jones and the electrostatic energy functions. These two terms are the main contributors to the energy function in many molecular models. The smoothing scheme is much cheaper than the classic spatial averaging smoothing technique. In computational tests on the proteins polyalanine with up to 58 amino acids and metenkephalin, smoothing is very successful in finding the lowest energy structures. The largest case of polyalanine is particularly significant because the lowest energy structures that are found include ones that exhibit interesting tertiary as opposed to just secondary structure.

Keywords: Global Optimization, Molecular Chemistry, Polyalanine, Smoothing Techniques, Protein Folding

*Research Supported by Air Force Office of Scientific Research Grant F49620-97-1-0164, Army Research Office Contracts DAAG55-98-1-0176 and DAAD19-02-1-0407, and National Science Foundation Grants CDA-9502956 and CHE-0205170.

1.   Introduction

The topic of this paper is the development of new smoothing methods for large-scale global optimization problems which frequently arise in molecular chemistry applications. The prediction of molecular conformation by energy minimization gives rise to very difficult global optimization problems. To aid in solving these large problems, both the chemistry and the optimization communities have relied on different techniques, smoothing being one of these techniques. Our overall goal is to develop new smoothing techniques that are both inexpensive and effective in the molecular context, and to integrate them into sophisticated global optimization algorithms. This paper is an extension of an earlier work that dealt with smoothing for molecular cluster configuration problems [24].

The protein folding problem is defined in [2] as the problem of finding the native state of a protein in its normal physiological milieu. In other words, we want to determine the three dimensional structure of a protein, called its "tertiary structure," just from the sequence of amino acids that it is composed of (its "primary structure"). Under the assumption that in the native state the potential energy of a protein is globally minimized, the protein folding problem can be regarded as equivalent to solving the problem

$$\min_{x \in R^{3n}} E(x) \tag{1.1}$$

where $E(x)$ is the value of the potential energy function (1.6) for a configuration of an n atom protein described by the $3n$ dimensional vector x.

To handle this problem, an optimization algorithm has to solve problems with many variables, since even the smallest proteins have a large number of free variables. Apart from that, the potential energy function is known to have many local minimizers. For proteins, it is speculated that the potential energy function has at least 3^n local minima, n being the number of free variables [15] [28]. Thus, the protein folding problem belongs to the class of NP-hard problems.

A large scale global optimization algorithm that does not utilize the solution structure of the cluster has been developed by some of the authors in the past few years [8]. It has been successfully used to solve Lennard-Jones molecular cluster problems with up to 76 atoms as well as more complex water cluster problems of up to 21 molecules. However, in practice, it is expected that it will generally be too expensive – if not impossible – to solve problems of such size, such as polyalanine, by using the same global optimization algorithm directly. In fact [26] was unsuccessful in solving for polyalanine $N > 40$ using the global optimization algorithm of [8]. This realization motivates approaches that seek to improve the effectiveness of global optimization algorithms via transformations of the objective function.

One transformation method is to use a parameterized set of smoothed objective functions. The smoothed functions are intended to retain coarse structure of the original objective function, but have fewer local minima. By selecting different smoothing parameters, objective functions with different degrees of smoothness can be derived. However, it is quite possible that as we vary smoothing parameters, the trajectory from the global minimizer of the smoothed problem will not lead to the global minimizer of the original problem. Indeed, the later situation which we term *order flips* appears to be a common and fundamental problem in smoothing, and one that we must deal with. The algorithm described in this paper handles this problem by applying a global optimization algorithm to the smoothed function, and following the trajectories of several of the best local minimizers.

In the rest of this paper we will describe in Section 2 the nature of the problem and the structure of the potential energy function. In Section 3 we will describe our efficient approach for smoothing the energy function. How the smoothed function is used as part of a global optimization strategy is explained in Section 4. Section 5 will present some computational results of this approach for two different proteins.

2. Protein Structure and Potential Energy

A naturally occurring protein is a bonded chain of different amino acids. All amino acids (except proline) have the same underlying structure. A central carbon atom (C_α), to which are attached: a hydrogen atom (H), an amino group (NH_2), a carboxyl group $(COOH)$ and a residue R. Residues are what distinguishes one amino acid from another. Figure 1.1 is the primary structure of a protein, the repeating chain $-NC_\alpha C'-$ is known as the backbone. Overall there are twenty possible different residues, thus we have that many amino acids.

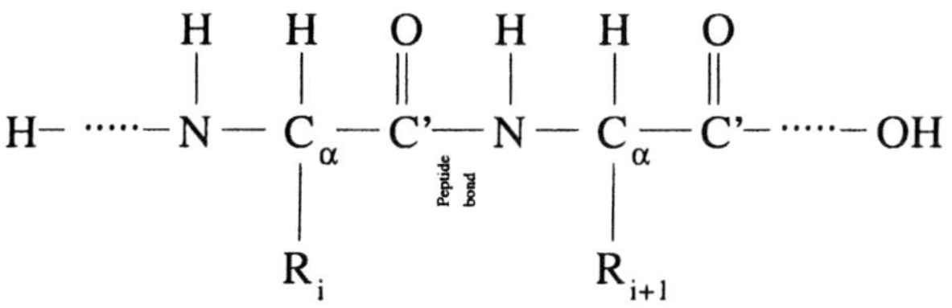

Figure 1.1. The primary structure of a protein. The peptide bond links two amino acids.

2.1   Protein Geometry

There are two different but equivalent coordinate systems used to describe the conformation of a protein: internal and the external (or Cartesian) coordinates. The position of any group of atoms in space can be equally specified in either one of these representations. In external coordinates, each atom is represented by its x, y and z coordinates. In the internal coordinates system, which is more closely related to the structure of a protein, we use the *bond length*, *bond angle* and *dihedral angle* notions to specify the coordinates of the atom. The bond length is simply defined as the Euclidean distance between two consecutive atoms and bond angle is the angle between three consecutive atoms, assuming the atoms are located in a single plane. In the sequence of four atoms, the dihedral angle is the angle between the plane defined by the first three atoms in sequence and the last three atoms in the sequence. This is known as the *proper dihedral angle*. Ramachandran [20] has showed that there isn't a lot of variation in the values of the bond angles and the bond lengths. In fact, for many proteins, we have only three free variables per amino acid, corresponding to the three dihedral angles in an amino acid. Furthermore the *peptide bond* (Figure 1.1) is very rigid, and may be kept fixed. This leaves us with just two free variables per amino acid. So for proteins, use of internal coordinates is computationally more efficent than the external coordinate system. It is easier to keep bond lengths and bond angles fixed in internal coordinates than in external coordinates.

This paper considers the performance of new smoothing methods to solve the molecular conformation of two different proteins: an artificial protein known as polyalanine [9] [21], and metenkephalin [17]. Polyalanine is a molecular structure consisting of N alanine (ALA) amino acids, which amounts to $10N - 8$ atoms. Polyalanine is a good molecular conformation test problem for two reasons. First, the problem is a difficult global optimization problem due to its sheer size, polyalanine 58 consists of 572 atoms. Secondly, some of the larger polyalanine structures are known to posses more than a single optimal structure [13]. The minimum energy state of polyalanine 58 is known both as straight and bent with very close minimum energy values (Figure 1.2). On the other hand, metenkephalin is a much smaller naturally occuring protein with just 75 atoms. But unlike polyalanine, metenkephalin consists of four different amino acids. The objective function for proteins contains two terms common to many molecules, namely the Lennard-Jones potential and the electrostatic potential energy function. Thus for this reason, the

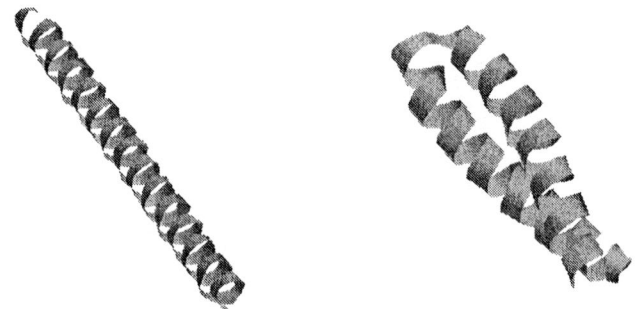

Figure 1.2. The two minimum energy states of polyalanine 58, straight and bent with corresponding energy values (based on CHARMM function) of -1559.494 and -1567.240.

techniques developed in this paper can be extended to cover smoothing in a wide range of chemical problems.

2.2 Modeling the Potential Energy

As with molecular cluster problems, it is believed that in the native state of a protein, the potential energy function of the protein is in its global energy minimum [1]. Therefore, we need a potential energy function to model the energy of a protein. Several potential energy functions have been developed to model proteins. Three of the most widely used are ECEPP [16] [18], AMBER [27] and CHARMM [4].

For this research we used the CHARMM and later AMBER potential energy functions. Due to differences in constants and other fine details, each reports a different energy for the same configuration. The CHARMM potential energy function is given by

$$E_{\text{CHARMM}} = E_b + E_\theta + E_\varphi + E_\omega + E_{\text{vdW}} + E_{\text{ES}} \qquad (1.2)$$

in which the first four terms are the separable internal coordinate terms, and the last two are the pairwise nonbonded interaction terms. The formula for each of these terms, as well as the relevant definitions, are given below.

- E_b is the bond potential, which equals $\sum k_b(r - r_0)^2$, where k_b is a bond force constant, dependent on the type of the atoms involved in the bond. The actual bond length is r, while r_0 is the equilibrium bond length, i.e the ideal bond length for the type of atoms involved in the bond.

- E_θ is the bond angle potential, which equals $\sum k_\theta(\theta - \theta_0)^2$, such that k_θ is an angle force constant which depends on the type of the atoms that constitutes the angle. θ and θ_0 are the actual bond angle and the equilibrium bond angle respectively.

- E_φ is the proper dihedral angle potential. It is given by $\sum |k_\varphi - k_\varphi \cos(n\varphi)|$, where k_φ is a dihedral angle force constant, dependent on the type of the atoms that constitutes the proper dihedral angle, n is a multiplication factor that can have the values $2, 3, 4$ or 6, and φ is the actual proper dihedral angle.

- E_ω is the improper dihedral angle potential, which equals $\sum k_\omega(w - w_0)^2$, where k_ω is an improper dihedral angle force constant that depends on the types of the atoms that constitute the improper dihedral angle (i.e. dihedral angle involving 3 backbone atoms and one residue atom). ω and ω_0 are the actual and the equilibrium improper dihedral angle, respectively.

- E_vdw is the van der Waals potential, which is a repulsive-attractive force that is very repulsive at very short distances, most attractive at an intermediate distance, and a very weak attractive force at longer distances. We represent these pairwise interactions between atoms i and j using the Lennard-Jones potential of the form

$$e_{ij}\left[\left(\frac{\sigma_{ij}}{d_{ij}}\right)^{12} - 2\left(\frac{\sigma_{ij}}{d_{ij}}\right)^6\right] \tag{1.3}$$

where d_{ij} is the Euclidean distance between atoms i and j, $\sigma_{ij} = \sigma_i + \sigma_j$ where σ is the van der Waals radius, and $e_{ij} = \sqrt{e_i e_j}$ where e_i is the potential well depth.

In this formulation, the Lennard-Jones pairwise equilibrium distance (the distance of greatest attraction) is scaled to σ_{ij}, and its minimum energy is scaled to $-e_{ij}$.

- E_ES is the (Coulomb-Fekete) electrostatic potential. Two atoms with charges of same sign repel, and attract if the charges have opposite signs. The pairwise electrostatic energy for atoms i and j is given by

$$\frac{q_{ij}}{4\pi\epsilon_0 d_{ij}} \tag{1.4}$$

where ϵ_0 is the vacuum permittivity, and $q_{ij} = q_i q_j$ where q is the charge. Usually, C is used to denote the constant term $1/4\pi\epsilon_0$.

E_b is summed over all pairs of bonded atoms. E_θ is summed over all bond angles. E_φ and E_ω are summed over all proper dihedral angles, and all improper dihedral angles respectively. Finally, E_{vdw} and E_{ES} are summed over all pairs of nonbonded atoms. The first four terms force the local structures (bond length, bond angle, proper dihedral angle and improper dihedral angle) into their ideal values. The last two terms account for the long range attractive and repulsive interaction forces.

Our main objective is to find the minimum energy structure of a protein using the CHARMM potential energy function. If we define the position of the protein structure by

$$x = (x_1, x_2, \cdots, x_n) \tag{1.5}$$

where x_i is a three dimensional vector denoting the coordinates of the i-th atom, then the overall potential energy function is

$$
\begin{aligned}
E(x) &= E_b + E_\theta + E_\varphi + E_\omega + E_{\mathrm{vdw}} + E_{\mathrm{ES}} \\
&= E_\bullet + \sum_{i \neq j}^{n} e_{ij} \left[\left(\frac{\sigma_{ij}}{d_{ij}} \right)^{12} - 2 \left(\frac{\sigma_{ij}}{d_{ij}} \right)^{6} \right] + C \frac{q_{ij}}{d_{ij}}
\end{aligned} \tag{1.6}
$$

where $d_{ij} = \|x_i - x_j\|_2$ is the distance between atoms i and j. The term $E_\bullet$ denotes the sum of bonded interactions, $E_b + E_\theta + E_\varphi + E_\omega$.

In equation (1.6) there are two points worth noting. First, the σ_{ij}, e_{ij} and q_{ij} occur in tuples (Table 2.1 of [3]). These tuples take completely different values depending on whether there is an attractive or a repulsive force between atoms i and j. Also, different proteins will have different entries. Secondly, the total contribution of $E_b + E_\theta + E_\varphi + E_\omega$ does not exceed 5-6% of the total potential.

In Figure 1.3 we show the total pairwise interaction $E_{\mathrm{vdw}} + E_{\mathrm{ES}}$ for two different pairs with an attractive electrostatic potential. Each of the two plots corresponds to one σ_{ij}, e_{ij} and q_{ij} tuple. As we have quite a few σ_{ij}, e_{ij} and q_{ij} tuples, the pairwise equilibrium distance would vary accordingly.

For the attractive electrostatic, near distance $d = 0$, the Lennard-Jones potential and the electrostatic forces are in opposing directions. But as we can see from Figure 1.3, the Lennard-Jones is the dominant term due to its higher growth rate. For $d \gg 0$ the situation is reversed; the electrostatic potential drops more slowly than the Lennard-Jones potential and thus is the dominant force.

The interaction of Lennard-Jones with a repulsive electrostatic is simple. The sum of the two terms is repulsive for all r so that there is no finite stable distance between two such atoms.

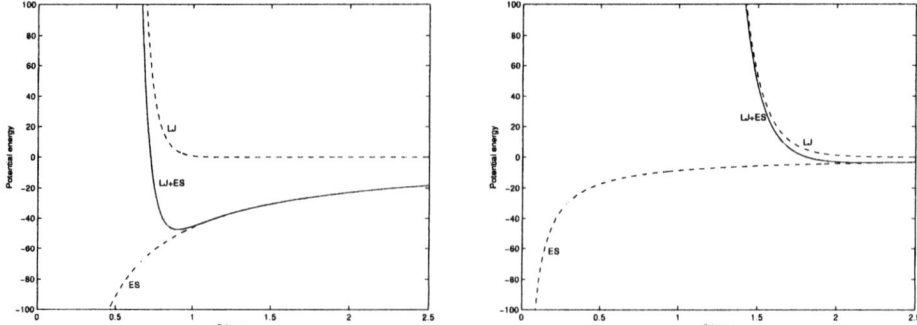

Figure 1.3.   Pairwise interactions in polyalanine 58. Potential of Lennard-Jones (LJ), attractive electrostatic (ES), and the overall potential energy (LJ+ES). Two different senarios corresponding to different values of energy constants; see text for details.

3.   A Smoothing Technique and its Behavior

The basic idea of smoothing a function is to modify it by reducing abrupt function value changes and fine grain fluctuations, while retaining the large scale structure of the original function. As a result, nearby minimizers should merge after sufficient smoothing is applied to remove the barriers between them. Therefore, smoothing reduces the total number of minima in the problem.

The Lennard-Jones and electrostatic potential energy functions for a pair of atoms have a pole at distance zero, and thus very large derivative values for distances near zero. The poles and large gradient values create huge barriers that separate similarly structured minimizers in the overall potential energy function (1.6). That is a fundamental reason why this and similar problems that include Lennard-Jones potential and/or electrostatic potential, have so many minima, and are so difficult to solve. A technique that smoothes Lennard-Jones should be able to remove these barriers in some effective way.

Most smoothing techniques generate a family of smoothing functions that is parameterized by one or more smoothing parameters. Such a family can be represented as

$$\tilde{E}_s : \mathcal{D} \to I\!\!R, \quad \ni \mathcal{D} \subset I\!\!R^n \text{ and } s \in I\!\!R^d \qquad (1.7)$$

where D is some closed region and d is the number of smoothing parameters. By varying the smoothing parameters, one can create a series of functions that gradually smoothes the original function. In our case, a family of smooth problems can be constructed

$$\min_{x \in \mathcal{D}} \tilde{E}_s(x) \tag{1.8}$$

where s is some smoothing parameter set and $\tilde{E}_s$ is a smoothed potential function (1.6). The number of minima should be reduced gradually as the objective function becomes smoother.

One general smoothing technique is spatial averaging, which has been widely studied [10] [11] [12] [14] [23] [29]. The fundamental idea of this technique is that the smoothed function value at each point is given by a weighted average of the energy function in a neighborhood of the point using a distribution function centered at this point. The Gaussian distribution function is commonly used to provide the weighting. In this case, the smoothing transformation is

$$\tilde{f}_{\gamma,\lambda}(x) = \int H(f(\acute{x}), \gamma) e^{-\|x - \acute{x}\|^2/\lambda^2} d\acute{x} \tag{1.9}$$

where λ and γ are the smoothing parameters. The parameter λ determines the scale of the Gaussian distribution, while the parameter γ is used with the function H to transform the original function $f(x)$ into a function with no poles. The transformation $H(f, \gamma)$ is necessary to make the function integrable, and also further dampens the function. This transformation takes on different forms, in the case of [14] it consists of approximating $f(x)$ by a sum of Gaussian functions, while in the work of [29] the transformation consists of truncating $f(x)$ to some fixed maximum value.

In this section, we will discuss a new family of smoothing functions. As opposed to spatial averaging techniques, this family of smoothing functions does not involve integration, but makes an explicit algebraic modification to the more wildly varying terms of the (1.6) potential energy function. and removes poles algebraically. The advantage of this algebraic approach compared to spatial averaging is simplicity and low evaluation cost.

3.1 A New Smoothing Scheme

To algebraically smooth the potential energy function we focus on the terms in (1.6) containing poles, and replace the term σ/d with an expression that is finite at zero. We also vary the Lennard-Jones exponents. As a result (1.6) is changed to the smoothed function

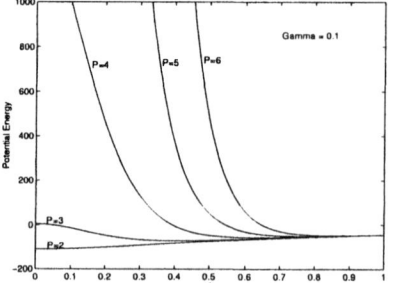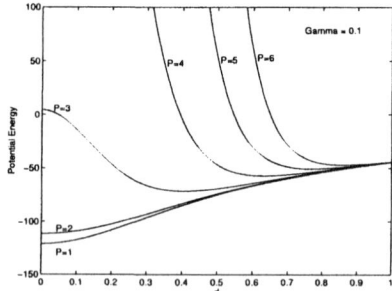

Figure 1.4.   Pairwise smoothed potential energy curves between two atoms of opposite charges on different setting of P at $\gamma = 0.1$. Two different vertical scales are shown.

$$\tilde{E}_{<\gamma,P>}(x) = E_{\bullet} + \sum_{i \neq j}^{n} e_{ij} \left[\left(\frac{1+\gamma}{r_{ij}^2 + \gamma} \right)^P - 2 \left(\frac{1+\gamma}{r_{ij}^2 + \gamma} \right)^{\frac{P}{2}} \right]$$

$$+ C \frac{q_{ij}}{\sigma_{ij}} \sqrt{\frac{1+\gamma}{r_{ij}^2 + \gamma}} \quad (1.10)$$

where $E_{\bullet}$ is the bonded interaction, γ and P are two smoothing parameters, and the scaled distance $r_{ij} = d_{ij}/\sigma_{ij} = \|x_i - x_j\|/\sigma_{ij}$. The smoothing parameter $\gamma > 0$ crops the poles at $r = 0$; the larger γ is, the lower the smoothed potential is at distance zero. The other smoothing parameter, P, is used to widen (stretch) the minimum's region of attraction. Note that (1.10) reverts to equation (1.6) if we pick $\gamma = 0, P = 6$, in other words, we turn off the smoothing. The smoothing of the Lennard-Jones term is similar to the smoothed Lennard-Jones potential proposed in [7], but the algebraic form is different to make it more compatible with the electrostatic term here. We also tried a variant of (1.10) where the smoothing was applied to the Lennard-Jones term only, and which is discussed in the Computational Results section.

Examples of (1.10) with $\gamma = 0.1$ are shown in Figure 1.4. The curves for $P = 3, 4, 5$ and 6 show smoother behavior than the unsmoothed function. For $P = 2$, however, there is a difficulty. The pairwise smoothed potential curve for smoothing $\gamma = 0.1, P = 2$ has a local minimum at $d = 0$; thus the pairwise local minimum of this term is for two atoms to coincide. This difficulty is due to oversmoothing of the Lennard-Jones function, and we can avoid such cases by insisting that pairwise smoothed curve must have a minimum for $d > 0$, which can be assured by having a maximum at $d = 0$.

We develop conditions on the smoothing parameters that must be satisfied in order to have a maximum at zero. Consider two atoms that are distance d apart. Let

$$y = (r^2 + \gamma)/(1 + \gamma), \tag{1.11}$$

where $r = d/\sigma$. Then we can re-write a single pairwise interaction of the smooth potential in terms of y

$$\tilde{E}(r) = e\left[1/y^P - 2/y^{P/2}\right] + C\frac{q}{\sigma}\sqrt{\frac{1}{y}} \tag{1.12}$$

It is easily shown that at $d = 0$, $\partial\tilde{E}/\partial r = 2r\sigma/(1+\gamma)\partial\tilde{E}/\partial y|_{r=0} = 0$. Hence we always have a critical point at zero. The second derivative at zero is

$$\frac{\partial^2 \tilde{E}}{\partial r^2}\bigg|_{r=0} = \frac{2}{1+\gamma}\left[\frac{\partial\tilde{E}}{\partial y} + \frac{2r^2}{1+\gamma}\frac{\partial^2\tilde{E}}{\partial y^2}\right]_{r=0} \tag{1.13}$$

$$= \frac{2}{1+\gamma}\frac{\partial\tilde{E}}{\partial y}\bigg|_{r=0} \tag{1.14}$$

$$= \frac{2}{1+\gamma}\left[\frac{eP}{y_0^{P/2+1}}(1 - y_0^{-P/2}) - \frac{1}{2}C\frac{q}{\sigma}y_0^{-3/2}\right] \tag{1.15}$$

$$= \frac{-2}{1+\gamma}\left[eP\beta y_0^{-P-1} + \frac{1}{2}C\frac{q}{\sigma}y_0^{-3/2}\right] \tag{1.16}$$

where $y_0 = y\big|_{r=0} = (1+\gamma)^{-1}$ and $\beta = -y_0^{P/2}(1 - y_0^{-P/2}) = 1 - (1+\gamma)^{P/2}$. The possibility of a local minimum at zero is ruled out if this quantity is negative, or equivalently if

$$\frac{1}{\gamma} > \left[\frac{-Cq}{2\sigma eP\beta}\right]^{1/(P-0.5)} - 1. \tag{1.17}$$

Since we are considering smoothing parameter values in the range $0 \le \gamma < 1$ only, then $\beta = 1 - (1+\gamma)^{P/2} \ge 1 - 2^{-P/2}$, and (1.17) holds if

$$\frac{1}{\gamma} > \left(\frac{1}{2(1 - 2^{-P/2})P}\right)^{1/(P-0.5)} \times \left(\frac{-Cq}{\sigma e}\right)^{1/(P-0.5)} - 1 \tag{1.18}$$

We want (1.18) to hold for q and σ corresponding to all atom pairs (i, j) for the protein under consideration. This is the case if (1.18) is satisfied for the pair (i, j) for which $q_{ij}/(\sigma_{ij}e_{ij})$ is largest, since $q < 0$. The values corresponding to the largest ratio, which is 366.703 for polyalanine 58, are: $C = 332.167$ (a constant), $q = -0.1375, \sigma = 1.248$ and $e = 0.0998$ (Table 2.1 of [3]). Substituting these values into (1.18) we get

$$\frac{1}{\gamma} > \left(\frac{183.3515}{(1 - 2^{-P/2})P}\right)^{1/(P-0.5)} - 1. \qquad (1.19)$$

The bound (1.19) is specific to polyalanine 58 and may vary for other proteins. It is worth noting that we only consider attractive electrostatic potentials in deriving the condition (1.18). This should be clear, since when the electrostatic force is repulsive the smoothed Lennard-Jones potential cannot allow a minimum at zero. In fact $\max\{-Cq/\sigma e\}$ provides a guarantee that this is the lowest interception point (with the potential energy axis) we can have, and any other pair of attractive atoms will have an interception point that is greater or equal to this. Equation (1.19) is a sufficient condition for a maximum at distance zero.

Figure 1.5 plots the maximum γ for a given P based on (1.19). On the other hand, in (1.18) if we take $\min\{-Cq/\sigma e\}$ then we have a necessary condition for any pair to have a maximum at zero distance.

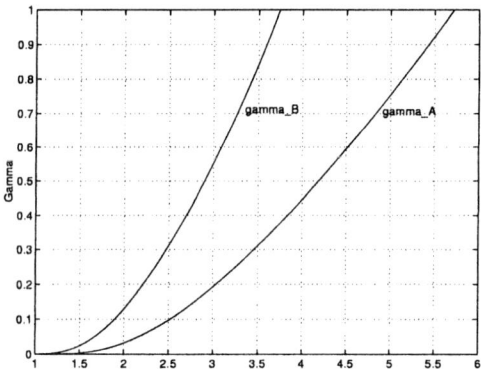

Figure 1.5. Maximum γ for a given P based on equation (1.19) for pairwise local optimization to converge. γ_A and γ_B are the maximum value of γ for sufficient and necessary conditions, respectively.

For example, in Figure 1.4 it is clear that $\gamma = 0.1, P = 2$ is not a good choice (since its has a minimum point at zero). In that case, according to (1.19) we must have $\gamma < 0.032$ in order to change it to a maximum at distance zero.

For problems with many atoms this type of smoothing also has the effect of simplifying the objective function and reducing the number of local minimizers. This is demonstrated in [24] and [7], and in Section 5.

The smoothing is also simple and inexpensive to perform. However, the problem is still not easy, because the global minimum of the smoothed function may not correspond to a local minimizer of the original objective function that is far from global. Indeed, as the experiments in [24] demonstrate, varying the smoothing parameters can cause the order of function values of local minimizers to change extensively. There is no reason to expect this effect to be different for our smoothing technique than for others. Therefore because of these order flips, we still require an effective global optimization algorithm to use with the smoothed energy function.

4. Using Smoothing in a Global Optimization Algorithm

In this section, we discuss a new global optimization algorithm that incorporates the smoothing function described in Section 3. It is based on the global optimization approach proposed in [8], which has been successfully tested on other molecular conformation problems, including water clusters [5] and proteins [7]. The algorithm described here is a modification of that of [8] using smoothing.

The essential idea of our approach is to use the global optimization algorithm of [8] on a smoothed version of the potential energy function to find several local minimizers, and then find related minimizers of the original function by local optimization. Although it is possible to work with several different levels of smoothing, we have found it effective to work with simply a single smoothed function, and the original function. The global optimization algorithm in [8] consists of two phases. Phase I (sample generation phase) uses random sampling and local minimization to build an initial conformation of the protein. In phase II we improve upon the initial conformation. This phase is where most of the computational effort lies. All the local minimizations were performed using the BFGS method in the UNCMIN package [22].

As mentioned earlier, we have two different schemes to represent a protein, the internal and the Cartesian coordinate systems. In this algorithm, we use internal coordinates which simplifies the task of fixing the values that are natural to the proteins, such as bond lengths and bond angles. Also, we reduce the dimensionality of the problem by about $\frac{1}{15}$ (in case of polyalanine 58) of the original problem[1], without limiting the folded states a protein can attain.

Algorithm 4.1 – Framework of the Large-Scale Global Optimization Algorithm for Protein Conformation

1 **Phase I (Sample Generation)**

  (a) **Protein sample point buildup:** Build up k sample configurations from one end of the protein to the other by sequentially generating each dihedral angle in the protein. Randomly sample the current dihedral angle a fixed number of times and select the dihedral angle that gives the lowest energy function value (in smooth domain) for the partial protein generated so far.

  (b) **Start point selection:** Select $\ell < k$ sample points from step 1a to be start points for local minimizations.

  (c) **Full-Dimensional local minimizations (smooth):** Perform a local minimization from each start point selected in step 1b.

2 **Desmooth Minimizers:** Do a full-dimensional local minimization for each generated smooth minimizer in step 1c. Collect m (usually 10) of the best of these non-smooth minimizers to initialize a list $\mathcal{L}$ for improvement in Phase II.

3 **Phase II (Improvement of Local Minimizers):** For some number of iterations, do:

  (a) **Select a minimizer:** From list $\mathcal{L}$, select one conformation, and a small subset of ~ 5 dihedral angles from that local minimizer to be optimized.

  (b) **Global optimization (smooth) on a small subset of variables:** Apply a fairly exhaustive small-scale global optimization algorithm to the smoothed energy of the selected configuration using the selected small subset of the dihedral angles as variables, and keeping the remaining angles temporarily fixed.

  (c) **Full-Dimensional local minimizations (smooth):** Apply a local minimization procedure, with all dihedral angles as variables, to the lowest $\hat{m}$ (~ 25) configurations that resulted from the global optimization of the step 3b.

4 **Desmooth Minimizers:** Do a full-dimensional local minimization for each generated smooth minimizer in step 1c. Collect m (usually 10) of the best of these non-smooth minimizers to initialize a list $\mathcal{L}$ for improvement in Phase II.

5 **Phase II (Improvement of Local Minimizers):** For some number of iterations, do:

  (a) **Select a minimizer:** From list $\mathcal{L}$, select one conformation, and a small subset of ~ 5 dihedral angles from that local minimizer to be optimized.

  (b) **Global optimization (smooth) on a small subset of variables:** Apply a fairly exhaustive small-scale global optimization algorithm to the smoothed energy of the selected configuration using the selected small subset of the dihedral angles as variables, and keeping the remaining angles temporarily fixed.

(c) **Full-Dimensional local minimizations (smooth):** Apply a local minimization procedure, with all dihedral angles as variables, to the lowest $\hat{m}$ (~ 25) configurations that resulted from the global optimization of the step 3b.

(d) **Merge the new local minimizers:** Merge the new lowest configurations into the existing list of local minimizers $\mathcal{L}$.

(e) **Retire the selected minimizer:** Remove the selected conformation (the one we picked up at step 3a) from $\mathcal{L}$.

6 **Desmooth Minimizers:** Do a full-dimensional local minimization of the unsmoothed energy for each generated smooth minimizer in step 3d.

The first phase of the algorithm starts by generating several initial sample conformations, using a smoothed potential energy function with a reasonable value of the smoothing parameters. The values used here were determined by trial and error, but we have had some success in applying the same values to different proteins. The conformation of the protein is built up one dihedral angle at a time. Each dihedral angle is sampled a number of times, and for every sample the corresponding atoms are added to the polymer and the partial (smoothed) energy is evaluated. The dihedral angle resulting in the best partial energy is chosen. Sampling is then continued with the next dihedral angle. From these sample points, start points for local minimizations are selected and a local minimization is performed from each selected sample point using the smoothed energy. In selecting the start points, we pick the ℓ sample points that are lowest in potential energy. All the smooth minimizers generated by phase I undergo another local minimization, this time to remove the smoothness (desmoothing). The m (~ 10) lowest unsmooth local minimizers found are passed on to phase II.

In the second phase, we use the m initial conformations (out of phase I) to initialize a working list, and try to improve them. Note, in this phase as in phase I, all function evaluations and local minimizations including the full-dimensional are done in the smooth domain. A heuristic (described shortly) is used to select a configuration for improvement from the list. Next, we select a number of dihedral angles. The selection of these dihedral angles is based on another heuristic. Then, rather than just sampling on the selected dihedral angles as in the sample point improvement procedure in phase I, a complete global optimization algorithm is applied to find the best new positions for the selected dihedral angles within the selected configuration, with the remainder of the configuration temporarily fixed. The global optimization method used is the stochastic method of [6] [25], which is a very effective global optimization method for problems with small numbers of variables. When doing the global optimization on the selected dihedral angles, the rest of the dihedrals are kept fixed. So generally, these best new positions found for the selected dihedral angles by the small global optimization algorithm lead to configurations in the basin of attraction of new local minimizers for the entire problem. Therefore, a full-dimensional local minimization algorithm is then applied to "polish" the $\hat{m}$ (~ 25) best of these new configurations. The reason for choosing the $\hat{m}$ best rather than just the best is because we have found that sometimes the best polished solution does not come from the best unpolished solution. These new full-dimensional local minimizers are then merged into the working list and the entire process is repeated a fixed number of times. In our experience, this phase is able to identify significantly improved local minimizers and leads to the success of the method. Finally, all the smooth minimizers generated by this phase are passed on for a full-dimensional local minimization in the non-smooth domain.

The heuristic used to determine which configuration to select at each iteration of the second phase is the following. We consider an initial configuration and any configurations generated from it to be related, such that the latter is a "descendent" of the former. For some fixed number of iterations, the work in this phase is balanced over each of the m sets of configurations consisting of an initial minimizers and all of that minimizer's descendants. In this "balancing phase", the same number of minimizers is selected from each set, and within a set we select the minimizer with lowest potential energy that has not been selected before. The remaining iterations of the local minimizer improvement phase constitute the "non-balancing phase" in which we select the best (lowest potential energy) configuration that has not been selected before, regardless of where it is descended from. We have found that the combination of the breadth of search of the configuration space that the balancing phase provides with the depth of search that the non-balancing phase allows is useful to the success of our method.

In our experiments we tested several different criteria for selecting the dihedral angles to be varied in the small-scale global optimizations of Phase II. These criteria are described in [26]. The method described below was proven to be the most successful for our target, and is the one we ended up using in all of the runs. The main effect of varying a given dihedral angle is on the interaction energies between atoms to the right and to the left of the given angle. Therefore, in this heuristic we compute for each backbone dihedral the total left-right interaction energy and normalize this energy value by the product of the number of atoms to the left times the number to the right. Some specified number (generally five) of dihedral angles with the highest normalized interaction energies are then selected.

The complete framework for the global optimization algorithm for protein conformation is outlined in Algorithm 4.1. For further information on the algorithm and details on the criteria to select dihedral angles, see [26].

5.   Computational Results

In this section we report on some of the experiments we conducted as a first step in assessing the effectiveness of the new smoothing algorithm presented in this paper. These experiments give a preliminary indication of the effectiveness of integrating smoothing techniques within a powerful global optimization algorithm. They also provide some insight on the choice of the smoothing parameters γ and P. We will study the effect of smoothing in phase I and in phase II of the algorithm. The smoothing has been tested on two different proteins: polyalanine and metenkephalin. We report the effect of smoothing in terms of individual phases for polyalanine. Later we repeat the same for metenkephalin.

5.1   Smoothing polyalanine in phase I

Polyalanine-N is a molecular structure that consists of $N - 2$ alanine amino acids, which translates into $10N - 8$ atoms. The phase I algorithm was applied to polyalanine of sizes 3, 5, 10, 20, 30, 40 and 58 amino acids. We will discuss the smoothing in terms of polyalanine of sizes larger than 20 amino acids. This is because our global algorithm without smoothing [26] was able to find the optimal configuration for polyalanine of sizes up to 20 in phase I (albeit not so easily) without employing any of the smoothing techniques.

Our procedure was to generate 1200 sample points and then save the 60 configurations with best smoothed energy. Since the ordering of smooth energy values differs from that of unsmooth energy values, we do a full scale local minimization on the unsmoothed function with these smoothed minimizers as starting points. We then pick the best m configurations based on the unsmooth values (usually 10) as input to phase II algorithm.

The results for phase I are in Table 1.1. For each of the tabulated γ, P combinations, the results are based on five runs (300 smooth minimizations total). The (γ, P) combinations shown were chosen based on prior experience with smoothing Lennard-Jones clusters [7]. In the last column, we consider two unsmooth minimizers to be the same if the difference is less than 10^{-5}. This last column indicates that num-

| N | smooth parameters | | minimizer value (unsmooth) | | | # diff. |
	P	γ	min (occurrence)	max	avg	mins
30	no smoothing		-758.046 (3)	-619.615	-686.12	297
	5.0	0.00	-760.498* (96)	-659.159	-737.88	137
		0.05	-760.498* (119)	-672.960	-740.10	101
		0.10	-760.498* (165)	-680.830	-747.30	53
40	no smoothing		-1040.723 (1)	-841.812	-925.31	300
	5.0	0.00	-1045.412* (24)	-885.104	-986.28	235
		0.05	-1045.412* (47)	-889.416	-997.19	172
		0.10	-1045.412* (74)	-848.061	-1002.20	132
58	no smoothing		-1437.084 (1)	-1199.966	-1321.48	300
	5.25	0.10	-1559.494* (9)	-1284.743	-1414.35	278
		0.20	-1559.494* (9)	-1267.475	-1416.20	250
	5.0	0.05	-1559.494* (9)	-1312.634	-1433.73	272
		0.10	-1559.494* (14)	-1262.362	-1435.35	238
		0.15	-1559.494* (20)	-1214.783	-1435.93	223
		0.20	-1559.494* (14)	-1235.224	-1429.89	230
		0.25	-1559.494* (11)	-1170.262	-1412.83	255
		0.30	-1559.494* (8)	-1232.790	-1401.75	266
		0.35	-1559.494* (6)	-1053.287	-1359.97	288
	4.75	0.05	-1559.494* (26)	-1232.026	-1455.04	235
		0.10	-1559.494* (30)	-1288.639	-1455.48	206
		0.20	-1559.494* (37)	-1257.514	-1447.97	225
	4.50	0.05	-1559.494* (57)	-1258.894	-1476.74	181
		0.10	-1559.494* (60)	-1282.753	-1484.21	164
		0.15	-1559.494* (61)	-1173.177	-1467.65	185

Table 1.1.   Summary of results of phase I for polyalanine-N (Energy is in kcal/mole). The results are based on five runs (overall 300 smoothed minimizers out of phase I undergoing local minimization in non-smooth surface). The unsmooth values marked with '*' are those of the best known straight helical minimizer.

ber of minimizers found, given 300 starting points, tends to decrease as we increase smoothing (although not monotonically in γ. This fact would seem to indicate that smoothing does indeed tend to decrease the number of local minimizers, as was clearly shown for Lennard-Jones clusters in [7]. It is noteworthy that for these values of N, phase I without smoothing is unable to find the best known minimizer, while it is found several times using smoothing. For polyalanine size 58, the best minimizer found, with unsmooth value of -1559.494 kcal/mole is the best known minimizer with a *straight* α-helix configuration (Figure 1.2). but, phase I smoothing failed to discover the lowest energy state of polyalanine 58, the bent structure. Interestingly, in one case a smoothing scheme with a pole at distance zero ($\gamma = 0$), but with $P < 6$ fared better than not smoothing.

We also tried a variation of the smoothing scheme (1.10) where we smoothed the Lennard-Jones term but kept the electrostatic term in the unsmoothed form (i.e Cq_{ij}/d_{ij}). A drawback of this smoothing scheme is that attractive pairs have a negative pole at zero due to the electrostatic term. This can yield local minimizers where two atoms coincide. In phase I, this smoothing scheme found the best straight helix more often than the one just presented, and it had an even more pronounced tendency to reduce the number the number of distinct local minimizers found. However, this scheme did poorly in phase II. See [3] for further details.

5.2   Smoothing polyalanine in phase II

This phase was only applied to polyalanine 58, since phase I was successful in finding the low energy states of polyalanines up to 40 and one of the low energy states of polyalanine 58 (the straight α-helix configuration). So our goal in this phase is to find the other low energy state of polyalanine 58 (the bent α-helix configuration, Figure 1.2).

All runs in phase II had the same input. For input to this phase, we picked the best 10 configurations (in terms of unsmooth energies) from the phase I run with $E_{<0.25,5>}$. This set included the -1559.494 (best straight α-helix) configuration. The phase II parameters were: 4 balancing and 20 non-balancing iterations. Each phase II small-scale minimization generated 50 configurations, from which we selected the lowest 25 configurations for full-dimensional local minimization. We used the smoothing parameter values that seemed effective in Phase I, as well as some neighboring values.

Table 1.2 tabulates the results of phase II on polyalanine 58. To get a better overall picture, we did five runs for each smoothing parameter combination. These are labeled run 1, run 2, ... in Table 1.2 and ordered with best performance first. For polyalanine 58, the global minimizer is -1567.240. There are two points worth noting. First, for most of the smoothing parameters, we were able to reach the global at least 20% of the time (a single run out of five). In some cases we have even a better chance, for example $\gamma = 0.10, P = 4.75$ (a success ratio of 80%). In some cases we have multiple occurrences of the global minimizer in a single

| parameters | | best found minimizer (unsmooth) | | | | |
P	γ	run 1	run 2	run 3	run 4	run 5
no smooth		-1559.753	◊	◊	◊	◊
5.5	0.20	-1567.240*	-1567.240*	-1561.748	◊	◊
	0.26	-1567.240* (2)	-1567.240* (2)	-1567.240*	◊	◊
5.0	0.00	-1561.944	-1560.066	-1560.066	◊	◊
	0.05	-1567.240*	-1562.291	-1561.645	◊	◊
	0.10	-1567.240*	-1561.976	-1560.593	-1559.753	◊
	0.15	-1567.240* (2)	-1567.240*	-1567.240*	-1561.944	◊
	0.16	-1567.240* (3)	-1567.240* (2)	-1567.240*	-1563.720 (2)	◊
	0.20	-1561.536	-1559.678	◊	◊	◊
	0.25	◊	◊	◊	◊	◊
4.75	0.10	-1567.240* (2)	-1567.240*	-1567.240*	-1567.240*	◊
	0.15	-1564.817	-1561.748	-1561.438	-1559.678	◊
	0.20	-1561.748	◊	◊	◊	◊
4.50	0.05	-1561.438	-1560.593	-1559.986	-1559.678 (2)	◊
	0.10	-1567.240* (3)	-1567.240* (3)	-1567.240* (2)	-1567.240*	-1561.976
4.25	0.10	◊	◊	◊	◊	◊

Table 1.2. Summary of results of phase II on polyalanine 58. The same starting configuration was used for all runs. Runs shown are arranged on best, second best, ... The ◊ means we could not improve over the minimum (best) input potential energy. Number inside the parenthesis indicate multiple occurance of that particular minimizer. Values marked '*' means these are the best known global minimizer.

run, e.g runs 1-3 in $\gamma = 0.10, P = 4.5$. Second, if we oversmooth then the results are inferior to no smoothing, e.g $\gamma = 0.15, P = 4.5$.

To further understand smoothing, we did extensive testing using the single smoothing parameter choice, $\gamma = 0.15, P = 5$. Again, all runs had the same input configurations to start with and the same limit on balancing/non-balancing iterations. In 30 runs out of a total of 38 runs we saw improvement in the minimizer value at end of phase II. In fact 23 runs out of these 30 runs were successful in finding the global minimizer. We kept track of the iteration number when a specific minimizer is first found. In one instance, the -1567.240 was found after only 10 iterations (4 balancing and 6 non-balancing iterations), on average it was generated on the 15-16th iteration.

It is interesting to trace the chains of minimizers produced by the algorithm. Since each minimizer produced by phase II is a 'child' of some minimizer found earlier, the local minimizers form a set of trees

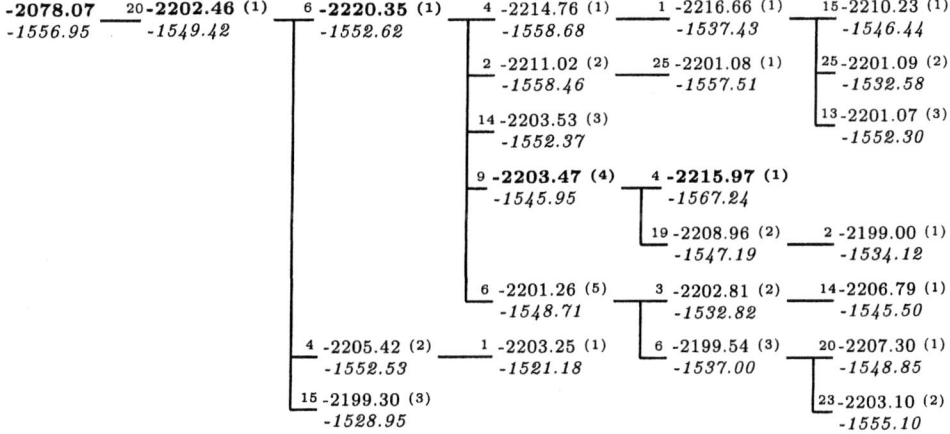

Figure 1.6. A trimmed tree of the full trace of one of the minimizers throughout the life of phase II ($\gamma = 0.15, P = 5$) which yielded the global (smoothing minimizers on path to the global, -1567.240, are boldfaced). The full tree has 178 nodes (minimizers) and is 6 levels deep. To reduce page cluttering, we only show the non-leaf nodes in the full tree. The value at top is the value of smooth minimizer, the number beneath (italicized) is that after desmoothing the minimizer. The labels at left and right are the ranking of the smooth minimizer within the set of minimizers produced by the small-dimensional optimization before and after the polishing (step 3c in Algorithm 4.1), respectively.

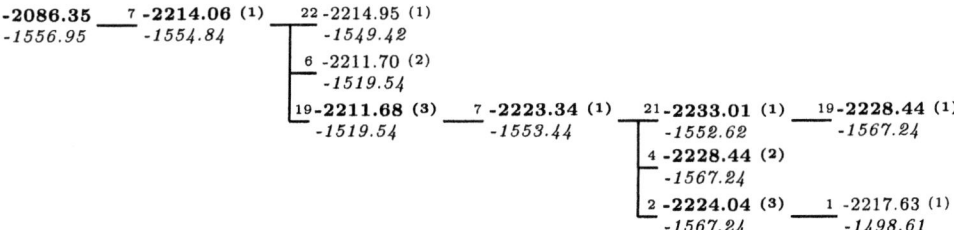

Figure 1.7. Another trimmed tree of phase II ($\gamma = 0.16, P = 5$). This run found the global thrice (minimizers on the path to the global are boldfaced). The full tree has 149 nodes and is 7 levels deep.

rooted at the minimizers input to phase II. Although such a tree is quite large, we can display the most important part of it by showing only those minimizers which generated good minimizers. Recall that at step 3c of Algorithm 4.1), we perform 25 full dimensional minimizations, and thus find up to 25 different minimizers, which can be considered children of the minimizer used to begin that step. A new minimizer is added to the list if it is among the top 200 found so far. We show in

Figure 1.6 a trace of a single starting conformation -1556.95 (a straight α-helix). It shows those minimizers in the final list which produced children good enough to be added to the list. The full tree has 178 nodes (minimizers) and is six levels deep. We also show, in figure 1.7 a trace from a run that resulted in finding the global energy minimum configuration three times, interestingly from two different parents and from different smoothed minimizers. These trees show the non-monotone nature of the algorithm, and the occurence of order flips in smoothing.

As an estimate for the cost of running the algorithm, we report the number of function evaluations of the parallel version of Algorithm 4.1 for polyalanine 58. The parallel version of phase I is slightly different than its sequential counterpart, and so the results will vary. The same is true in phase II which is implemented as a three layered structure as opposed to two in phase I. For further details, see [3] and [26]. For phase I, the average number of function evaluations is 2349 per sample. In phase II each iteration requires 21617 function evaluations in addition to the gradient evaluations. In terms of time, phase II takes about 3–4 times as long as phase I to finish[2]. The final desmoothing phase required about 492 function evaluations per smoothed minimizer.

5.3   Smoothing metenkephalin in phase I

We also tried Algorithm 4.1 on metenkephalin, a small protein with 75 atoms and five amino acids (Figure 1.8). Though it is a smaller protein than polyalanine, it is heterogeneous (different amino-acids), so it is a good test for our smoothing technique.

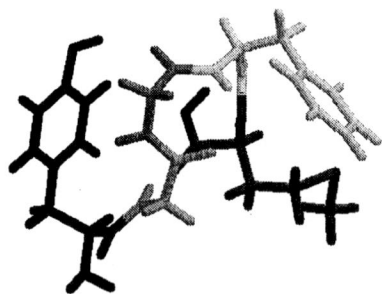

Figure 1.8.   The tertiary structure of metenkephalin. It has a total of 75 atoms.

The input parameters to smoothing were as follow: generate 600 sample points, then output the best 80 smoothed configurations. At the end of this phase, we desmooth the smoothed minimizers by doing full scale

| smooth params | | overall | avg of best | worst | average |
P	γ	best min	minima	minimum	minimum
no smoothing		-13.05571	-10.67496	14.17403	0.96144
5.50	0.00	-11.59565	-10.49555	21.73974	1.95112
	0.05	-16.90684	-13.22026	52.96502	3.14996
	0.10	-12.63449	-9.63245	34.17817	3.17763
5.25	0.00	-15.82020	-11.76161	23.03342	2.62739
	0.05	-16.90684	-13.60065	49.39080	3.94882
	0.10	-13.92418	-12.51320	47.20828	4.74853
5.00	0.00	-11.09422	-8.65001	36.35550	4.14277
	0.05	-11.02481	-8.79110	36.98082	4.72494
	0.10	-17.44337	-14.33585	58.78065	4.72565
	0.15	-17.09287	-13.42397	40.46203	4.97514
	0.20	-11.05957	-9.18396	48.71783	5.67317
	0.25	-13.15117	-11.82889	62.58349	6.25258
4.75	0.00	-15.80765	-10.99920	37.74938	4.09213
	0.05	-12.87557	-10.99422	46.34991	4.58853
	0.10	-14.95210	-11.61584	44.97417	5.10956
	0.15	-15.60696	-12.95194	48.13361	6.67790
	0.20	-15.38735	-11.24722	50.50901	6.64184
	0.25	-12.27669	-9.59790	44.89006	6.78987
4.50	0.00	-13.97437	-11.60363	47.81042	4.64018
	0.05	-14.89564	-9.42210	38.91343	4.50206
	0.10	-17.09297	-13.10101	48.14509	5.38676
	0.15	-15.80756	-12.05443	51.10003	6.44918
	0.20	-13.22904	-11.18331	52.99727	7.58794
	0.25	-16.90807	-11.17747	66.32735	8.58450

Table 1.3. Summary of results of phase I on metenkephalin. The results are based on five runs. The column overall best min corresponds to the overall best minimum among the five runs, while average of best minima is the average of the best in each of the five runs.

local minimization. Afterward, we pick few of the best (usually 10), unsmoothed minimizers. These will be used as input to the phase II algorithm.

The results for phase I are shown in Table 1.3. The results for each γ, P combination are based on five runs, that is 400 smoothed minimizers. Unlike phase I in polyalanine (Table 1.1), which was successful in finding the global minimizer for polyalanines up to 40, here it never did find the global minimizer. Considering the fact that metenkephalin has fewer atoms than polyalanine 10, this confirms that this protein's heterogeneity of amino-acids leads to a more challenging problem. From Table 1.3 we observe two different patterns. In most of the instances, smoothing was more likely to find a better best minimum than no smoothing. This is clear from the second column in the table. The other pattern is that with smoothing we are more likely to generate bad minima. This is evident from columns 3 and 4 in the Table 1.3. In other words, with smoothing we generated a wider range of minima. The best ones are better than those generated without smoothing, and at the same time the worst ones are worse than those generated with no smoothing.

5.4   Smoothing metenkephalin in phase II

To be consistent throughout, every run in phase II used the same initial set of minimizers. The input for this phase was 10 unsmoothed minimizers from phase I with smoothing parameters $\gamma = 0.1, P = 5$. The best input configuration had energy -17.09286. The run time parameters were: 4 balancing and 20 non-balancing iterations. This is the same as we used for phase II in polyalanine. Phase II kept a sorted list of the best 250 smoothed minimizers. When a good minimizer is found, it is added to the list, and a minimizer from the end of the list is discarded.

The results for phase II are tabulated in Table 1.4. We did six runs for each smooth parameter combination, with the best run labeled run 1, etc. Since even the algorithm without smoothing was twice successful in reaching the global minimum (-18.8475), we include more information to judge between different smoothings. Besides listing the best unsmoothed minimizer in each run, we list the iteration in which it was first generated, and the number of times it was generated throughout phase II. These two values are enclosed in brackets in front of the best unsmoothed minimizer (Table 1.4).

Overall it is clear that most of the smoothing runs did a better job than the no smoothing run. For example: all six runs of $\gamma = 0.04, P = 4.5$ were successful in reaching the global more than once, but these runs usually found the global minimizer later than the no-smoothing run (when the no-smoothing run actually found it). However, there is clearly significant random variation. Some smoothing runs were inferior to the no smoothing runs, e.g $\gamma = 0.10, P = 5.5$ and $\gamma = 0.05, P = 5$, where none of the runs reached the global. However, using the slightly different values, $\gamma = 0.04, P = 5$, resulted in reaching the global in five out of six runs.

6. Conclusions and Future Research

The protein conformation problem belongs to the class of NP-hard problems. By taking advantage of the internal structure of the protein, the algorithm of [8] performs well. However, this algorithm, without smoothing, failed to find the minimum energy state of polyalanine larger than 20. In the first part of the paper we introduced a new family of smoothing functions to use in conjunction with this algorithm. In our experiments, this algorithm was able to find the apparent low energy state(s) of all sizes of polyalanines we tried. For polyalanine 58 we found two different low energy states, a straight and a bent α-helical structure. The algorithm also performed well on a smaller, but more complex protein, metenkephalin.

There are some points worth future consideration. The algorithm we present uses smoothing to simplify the energy function, but minimizes the harmful effect of order flips by using more computational effort to track more local minimizers. An important issue is how this extra effort will scale when we tackle larger proteins. It is also desirable to achieve a better understanding of the mathematical behavior of the smoothing function as smoothing increases, especially in comparison with spatial averaging techniques. The smoothing parameter values that worked best were somewhat different for the two proteins we studied; how to estimate the best values in advance is another important topic for future work.

smooth params		best unsmoothed minimizer found (iteration generated/# of times found)					
P	γ	run 1	run 2	run 3	run 4	run 5	run 6
no smoothing		-18.8475* (6/1)	-18.8475* (12/1)	-18.8468 (8/1)	-18.8468 (8/1)	-18.8468 (8/1)	-18.8468 (8/1)
5.50	0.01	-18.6468 (9/3)	-18.4477 (19/3)	-18.4477 (19/3)	-18.4477 (19/2)	-18.4477 (19/1)	-18.4477 (19/1)
	0.05	-18.6461 (1/7)	-18.6461 (1/6)	-18.6461 (1/6)	-18.6460 (13/3)	-18.6460 (13/3)	-18.6460 (13/2)
	0.10	-18.6469 (13/2)	-18.6469 (13/1)	-18.6469 (13/1)	-18.6461 (1/4)	-18.6460 (13/3)	-18.6460 (13/2)
	0.15	-18.8461 (1/4)	-18.8461 (1/4)	-18.6468 (9/3)	-18.6468 (9/1)	-18.6441 (14/3)	-18.6441 (14/2)
	0.16	-18.8475* (7/2)	-18.8475* (7/2)	-18.8475* (7/2)	-18.8475* (7/2)	-18.8475* (7/2)	-18.8475* (7/1)
	0.17	-18.8475* (7/1)	-18.8469 (7/5)	-18.8469 (7/3)	-18.8469 (7/1)	-17.5954 (9/2)	-17.5954 (9/2)
	0.18	-18.8475* (7/2)	-18.8475* (7/2)	-18.8475* (7/2)	-18.8475* (7/1)	-18.8469 (7/1)	-18.8469 (7/1)
	0.19	-18.8469 (13/2)	-18.8469 (7/3)	-18.8461 (1/3)	-18.8461 (1/3)	-18.8461 (1/2)	-18.8461 (1/1)
5.00	0.01	-18.8469 (7/1)	-18.8469 (7/4)	-18.8469 (7/3)	-18.8469 (7/5)	-17.2385 (1/3)	-17.2385 (1/2)
	0.02	-18.8475* (7/1)	-18.8469 (7/3)	-18.8469 (7/3)	-18.8469 (7/1)	-18.6468 (9/4)	-18.6460 (13/1)
	0.03	-18.8475* (7/1)	-18.8475* (7/1)	-18.8475* (7/1)	-18.8475* (7/1)	-18.8469 (7/2)	-18.8469 (7/1)
	0.04	-18.8475* (7/1)	-18.8475* (7/2)	-18.8475* (17/2)	-18.8475* (17/2)	-18.8461 (1/6)	-18.8461 (1/1)
	0.05	-18.8475* (17/2)	-18.8475* (17/2)	-18.8469 (13/1)	-18.8475* (17/2)	-18.6468 (13/1)	-18.6455 (17/2)
	0.08	-18.8469 (13/1)	-18.8469 (13/1)	-18.8475* (17/2)	-18.8469 (13/1)	-18.6455 (17/2)	-18.8469 (13/1)
	0.09	-18.8475* (17/2)	-18.8475* (17/2)	-18.8475* (24/2)	-18.8475* (13/1)	-18.8475* (17/1)	-18.8475* (17/1)
	0.10	-18.8475* (17/2)	-18.8475* (17/2)	-18.6475 (16/7)	-18.6475 (16/7)	-17.7719 (20/1)	-18.6468 (13/1)
	0.11	-18.8475* (13/1)	-18.8469 (16/8)	-18.8469 (13/3)	-18.8468 (13/1)	-18.6468 (13/1)	-18.6468 (13/1)
	0.12	-18.8475* (17/3)	-18.8475* (17/3)	-18.8469 (13/1)	-18.8475* (13/1)	-18.6475 (16/6)	-17.4455 (13/1)
	0.15	-18.8475* (17/3)	-18.8475* (17/2)	-18.8475* (17/1)	-18.6468 (13/1)	-18.6468 (13/1)	-18.8461 (1/4)
4.50	0.01	-18.8475* (17/2)	-18.8475* (17/2)	-18.8475* (17/1)	-18.8469 (13/1)	-18.8469 (13/1)	-18.8461 (1/6)
	0.02	-18.8475* (17/2)	-18.8475* (17/2)	-18.8475* (17/2)	-18.8475* (17/2)	-18.8461 (1/5)	-18.8461 (1/5)
	0.03	-18.8475* (17/2)	-18.8475* (17/2)	-18.8475* (17/2)	-18.8475* (17/2)	-18.6449 (14/1)	-18.6449 (14/1)
	0.04	-18.8475* (17/3)	-18.8475* (17/3)	-18.8475* (17/3)	-18.8475* (17/2)	-18.8475* (17/2)	-18.8475* (17/2)
	0.05	-18.8475* (17/3)	-18.8475* (17/3)	-18.8475* (17/3)	-18.8475* (17/2)	-18.8475* (17/1)	-18.8475* (17/1)
	0.10	-18.8475* (17/3)	-18.8475* (17/3)	-18.8475* (17/2)	-18.8475* (17/2)	-18.8475* (17/1)	-18.8475* (17/1)
	0.15	-18.8475* (8/2)	-18.8475* (24/1)	-18.6460 (13/2)	-18.4463 (14/4)	-17.7712 (5/1)	-17.7712 (5/1)

Table 1.4. Summary of results of phase II on metenkephalin. The same input was used for all runs. The "*" refers to the global minimizer. Runs are sorted by best, second best, ...

References

[1] ABAGYAN, R.A., *Towards Protein Folding by Global Energy Minimization. FEBS Letters*, 325(1,2):pp. 17-22, June 1993.

[2] ANFINSEN, C.B., *Principles that Govern the Folding of Protein Chains. Science*, **181**, pp. 223-230, July 1973.

[3] AZMI, A.M., *Use of Smoothing Methods with Stochastic Perturbation for Global Optimization (a study in the context of molecular chemistry). Ph.D thesis, University of Colorado*, December 1998.

[4] BROOKS, B.R., BRUCCOLERI, R.E., OLAFSON, B.D., STATES, D.J., SWAMINATHAN, S., AND KARPLUS, M., *A Program for Macromolecular Energy, Minimization, and Dynamics Calculations. J. Comp. Chem.*, **4**(2), pp. 187-217, 1983.

[5] BYRD, R.H., DERBY, T., ESKOW, E., OLDENKAMP, K., AND SCHNABEL, R.B. *A New Stochastic/Perturbation Method for Large-Scale Global Optimization and its Application to Water Cluster Problems*, in Large-Scale Optimization: State of the Art, W. Hager, D. Hearn, and P. Pardalos, eds., Kluwer Academic Publishers, Dordrecht, The Netherlands, pp. 71–84, 1994.

[6] BYRD, R.H., DERT, C.L., RINNOOY KAN, A.H.G., AND SCHNABEL, R.B *Concurrent stochastic methods for global optimization, Mathematical Programming*, **46**, 1-29, 1990.

[7] BYRD, R.H., ESKOW, E., VAN DER HOEK, A.,SCHNABEL, R.B, SHAO, C-S., AND ZOU, Z. *Global optimization methods for protein folding problems*, Proceedings of the DIMACS Workshop on Global Minimization of Nonconvex Energy Functions: Molecular Conformation and Protein Folding, P. Pardalos, D. Shalloway, and G. Xue, eds., American Mathematical Society, **23**, pp. 29-39, 1996.

[8] BYRD, R.H., ESKOW, E., AND SCHNABEL, R.B., *A New Large-Scale Global Optimization Method and its Application to Lennard-*

Jones Problems. Technical Report CU-CS-630-92, Dept. of Computer Science, University of Colorado, revised, 1995.

[9] CHAN, H.S., AND DILL, K.A., *The Protein Folding Problem. Physics Today*, pp. 24-32, February 1993.

[10] COLEMAN, T., SHALLOWAY, D. AND WU, Z. *Isotropic Effective Energy Simulated Annealing Searches for Low Energy Molecular Cluster States*, Technical Report CTC-92-TR113, Center for Theory and Simulation in Science and Engineering, Cornell University, 1992.

[11] COLEMAN, T., SHALLOWAY, D. AND WU, Z. *A Parallel Build-up Algorithm for Global Energy Minimizations of Molecular Clusters Using Effective Energy Simulated Annealing*, Technical Report CTC-93-TR130, Center for Theory and Simulation in Science and Engineering, Cornell University, 1993.

[12] COLEMAN, T., AND WU, Z. *Parallel Continuation-Based Global Optimization for Molecular Conformation and Protein Folding*, Technical Report CTC-94-TR175, Center for Theory and Simulation in Science and Engineering, Cornell University, 1994.

[13] HEAD-GORDON, T., AND STILLINGER, F.H. *Enthalpy of Knotted Polypeptides, J. Phys. Chem.*, **96**, 7792-7796, 1992.

[14] KOSTROWICKI, J., PIELA, L., CHERAYIL, B.J., AND SCHERAGA, A. *Performance of the Diffusion Equation Method in Searches for Optimum Structures of Clusters of Lennard-Jones Atoms, J. Phys. Chem.*, **95**, 4113-4119, 1991.

[15] LI, Z., AND SCHERAGA, H.A., *Monte Carlo Minimization Approach to the Multiple-Minima Problem in Protein Folding. Proceedings of National Academic Sceince USA*, **84**, pp. 6611-6615, October 1987.

[16] MOMANY, F.A., MCGUIRE, R.F., AND SCHERAGA, H.A., *Energy Parameters in Polypeptides. VII. Geometric Parameters, Partial Atomic Charges, Nonbounded Interactions, Hydrogen Bond Interactions, and Intrinsic Torsional Potentials for the Naturally Occuring Amino Acids. J. Chem. Phys.*, **79**(22), pp. 2361-2381, 1975.

[17] MONTCALM, T., CUI, W., ZHAO, H., GUARNIERI, F., AND WILSON S.R., *Simulated Annealing of Met-enkephalin: Low Energy*

States and their Relevance to Membrane-bound, Solution and Solid-sate Conformation. J. Molecular Structure (Theochem), **308**, pp. 37-51, 1994.

[18] NÉMETHY, G., POTTLE, M.S., AND AND SCHERAGA, H.A., *Energy Parameters in Polypeptides. 9. Updating of Geometrical Parameters, Nonbounded Interactions, and Hydrogen Bond Interactions for the Naturally Occuring Amino Acids. J. Chem. Phys.*, **87**, pp. 1883-1887, 1983.

[19] ORESIC, M., AND SHALLOWAY, D., *Hierarchical Characterization of Energy Landscapes using Gaussian Packet States. J. Chem. Phys.*, **101**, pp. 9844-9857, 1994.

[20] RAMACHANDRAN, G.N., ET AL, *Biochim Biophysics*, **74**, pp. 359:298-302, 1974.

[21] RICHARDS, F.M., *The Protein Folding Problem. Scientific American*, pp. 54-63, January 1991.

[22] SCHNABEL, R.B., KOONTZ, J.E., AND WEISS, B.E., *A Modular System of Algorithms of Unconstrained Minimization., ACM Trans on Math Software*, **11**, pp. 419-440, 1985.

[23] SHALLOWAY, D. *Packet annealing: a deterministic method for global minimization, with application to molecular conformation. Preprint, Section of Biochemistry, Molecular and Cell Biology, Global Optimization*, C. Floudas and P. Pardalos, eds., Princeton University Press, 1992.

[24] SHAO, C.-S., BYRD, R.H., ESKOW, E., AND SCHNABEL, R.B., *Global Optimization for Molecular Clusters Using a New Smoothing Approach, in Large-Scale Optimization with Applications, Part III: Molecular Structure and Optimization*, L. Biegler, T. Coleman, A. Conn and F. Santosa, eds., Springer, pp. 163-199, 1997.

[25] SMITH, S.L., ESKOW, E., AND SCHNABEL, R.B. *Adaptive, asynchronous stochastic global optimization algorithms for sequential and parallel computation, In Proceedings of the Workshop on Large-Scale Numerical Optimization*, T.F. Coleman and Y. Li, eds. SIAM, Philadelphia, 207-227, 1989.

[26] VAN DER HOEK, A., *Parallel Global Optimization of Proteins. Master's thesis, University of Colorado*, March 1996.

[27] WEINER, S.J., KOLLMAN, P.A., NGUYEN, D.T., AND CASE D.A., *An All Atom Force Field for Simulations of Proteins and Nucleic Acids. J. Comp. Chem.*, **7**(2), pp. 230-252, 1986.

[28] WILSON, S.R., CUI, W., MOSKOWITZ, J.W., AND SCHMIDT, K.E., *Conformational Analysis of Flexible Molecules: Location of the Global Minimum Energy Conformation by the Simulated Annealing Method. Tetrahedron Letters*, **29**(35), pp. 4373-4376, 1988.

[29] WU, Z. *The Effective Energy Transformation Scheme as a Special Continuation Approach to Global Optimization with Application to Molecular Conformation*, Technical Report CTC-93-TR143, Center for Theory and Simulation in Science and Engineering, Cornell University, 1993.

Chapter 5

PHYSICAL PERSPECTIVES ON THE GLOBAL OPTIMIZATION OF ATOMIC CLUSTERS

Jonathan P. K. Doye

University Chemical Laboratory, Lensfield Road, Cambridge CB2 1EW, United Kingdom
jpkd1@cam.ac.uk

Abstract   In this chapter the physical aspects of the global optimization of the geometry of atomic clusters are elucidated. In particular, I examine the structural principles that determine the nature of the lowest-energy structure, the physical reasons why some clusters are especially difficult to optimize and how the basin-hopping transformation of the potential energy surface enables these difficult clusters to be optimized.

Keywords: atomic clusters, basin-hopping, multiple funnels, energy landscapes

1.   INTRODUCTION

Global optimization (GO) is essentially a mathematical task. Namely, for the class of GO problems I will be particularly considering here, it is to find the absolute minimum (or maximum) of a cost function, $f(\mathbf{x})$, where $\mathbf{x}$ belongs to a subset D of Euclidean n-space, $\mathcal{R}^n$ [1], i.e.

$$\text{find } \mathbf{x}^* \text{ such that } f(\mathbf{x}^*) \le f(\mathbf{x}) \qquad \forall \mathbf{x} \in D \subset \mathcal{R}^n. \qquad (1.1)$$

Although the applications of global optimization span a wide range of fields— from the economics of business in the travelling salesman problem to biophysics in the lowest-energy structure of a protein—this does not take away from the essentially mathematical nature of the optimization problem.

So why do I wish to discuss the *physical* aspects of global optimization? To begin with we should realize that even for GO problems that do not correspond to a physical system, physical properties can be associated with the system by thinking of the cost function as a potential energy function, $E(\mathbf{x})$. This allows the thermodynamics of the system to be defined. When the system is at equilibrium at a temperature T each point $\mathbf{x}$ in configuration space will be sampled with a probability proportional to its Boltzmann weight, $\exp(-E(\mathbf{x})/kT)$, where k is the Boltzmann constant. Furthermore, for systems with continuous coordinates the forces, $F(\mathbf{x})$, associated with each coordinate can be obtained from the gradient of the cost function, i.e. $F(\mathbf{x}) = -\nabla E(\mathbf{x})$. Once masses are associated with each coordinate, the dynamics are then defined through Newton's equations of motion. If one wishes, the system's dynamics can then be simulated by integrating these equations of motion, as in the molecular dynamics method [2]. Even when the coordinates can only take discrete values, Monte Carlo (MC) simulations can still provide a pseudo-dynamics with the number of steps taking the role of time.

Of course, this connection to physics is most transparent, and most natural, when the system being optimized is a physical system, which has a real (and potentially observable) thermodynamics and dynamics. Furthermore, in those cases where the cost function does truly correspond to the potential energy of the system, there is another physical dimension to the problem—how is the structure of the global minimum determined by the physical interactions between the atoms and molecules that make up $E(\mathbf{x})$?

Given that we have established that physical properties can be associated with any system being optimized, what relevance does this physics have to the task of global optimization? Firstly, many GO algorithm have drawn their inspiration from physics. Most famously, simulated annealing is analogous to the slow cooling of a melt to allow the formation of a near perfect crystal, the idea being that if equilibrium is maintained in the simulation as the system is cooled, then at zero temperature it must end up in the global minimum [3]. There are many other physically-motivated GO approaches. The extension of statistical thermodynamics to systems with non-extensive thermodynamics through the use of Tsallis statistics [4, 5] has led to a generalized simulated annealing [6, 7] which is no longer tied to the Boltzmann distribution and is often more efficient than standard simulated annealing. Genetic algorithms imitate the biophysical evolution of the genome [8]. And I could go on.

However, this is not the link between physics and global optimization that is my focus here. Rather, I wish to show how the ease or difficulty of global optimization is often intimately linked to the physics of the system. The insights obtained from understanding the physical basis for the success or failure of an algorithm not only provide an understanding of the limitations of the method and a basis for assessing the likelihood of success in future applications, but also aid the development of new algorithms by identifying the main physical challenges that need to be overcome to enable greater efficiency and suggesting the type of physical behaviour that would need to be incorporated into an improved algorithm.

I will attempt to achieve this aim by concentrating on one class of problems, namely the global minimization of the potential energy of an atomic cluster.

Furthermore, I will mainly concentrate on model systems where the cost function is computationally cheap to evaluate, enabling the physical properties of these systems to be comprehensively examined and understood. As outlined by Hartke elsewhere in this book [9], this class of problems is of great general interest to the chemical physics community, because the identification of a cluster's structure is often a prerequisite for understanding its other physical and chemical properties.

In this chapter I start at the 'end', first showing the structures of the putative global minima[1] for a number of cluster systems in order that the reader can understand some of the physical principles that determine the structure and how these relate to the interatomic interactions. Furthermore, the structure provides a basis for understanding a cluster's thermodynamic and dynamic properties, especially when, as in some of our examples, the competition between different structural types plays an important role.

I then consider some of the GO algorithms that are most successful for these systems focussing on those that use the basin-hopping transformation of $E(\mathbf{x})$ [10] and on how the performance of these algorithms depend on the system and the cluster size. I then look at the physical properties of some of the clusters, relating these back to the ease or difficulty of global optimization. I firstly examine the topography of the multi-dimensional surface defined by $E(\mathbf{x})$ (the so-called potential energy surface (PES) or energy landscape), then the thermodynamics and dynamics. Finally, I show why basin-hopping is able to locate the global minimum in those clusters where the PES has a multiple-funnel topography, and make some suggestions as to how further gains in efficiency might be secured.

2.   CLUSTER STRUCTURE

In this section I mainly concentrate on the structures of model clusters, where the interactions have simple idealized forms that are isotropic, thus favouring compact geometries. The models have been chosen so that they span a wide range of structural behaviour that is likely to be relevant to rare gas, metal and molecular clusters bound by dispersion forces, but not to clusters with directional covalent bonding or molecular clusters with directional intermolecular forces, as with the hydrogen bonding in water clusters.

Most of the clusters we consider have only pair interactions, i.e.

$$E(\mathbf{x}) = \sum_{i<j} V(r_{ij}), \tag{1.2}$$

where V is the pair potential and r_{ij} is the distance between atoms i and j. In this case we can partition the energy into three terms [16]:

$$E = -n_{nn}\epsilon + E_{strain} + E_{nnn} \tag{1.3}$$

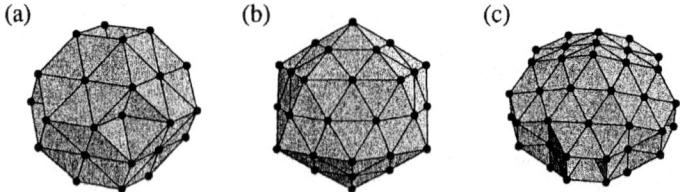

Figure 1.1   Three examples of the structures clusters can adopt: (a) a 38-atom truncated octahedron, (b) a 55-atom Mackay icosahedron, and (c) a 75-atom Marks' decahedron. These clusters have the optimal shape for the three main types of regular packing seen in clusters: close-packed, icosahedral and decahedral, respectively. The Mackay icosahedron is a common structure and is observed for rare gas [11] and many metal [12] clusters. The truncated octahedron has been recently observed for nickel [13] and gold [14] clusters, and the Marks decahedron for gold clusters [15].

where ϵ is the pair well depth, n_{nn} is the number of nearest neighbours,

$$E_{strain} = \sum_{i<j, r_{ij}<r_0} (V(r_{ij}) - V(r_{eq})), \qquad E_{nnn} = \sum_{i<j, r_{ij}\geq r_0} V(r_{ij}), \quad (1.4)$$

and r_{eq} is the equilibrium pair distance. Two atoms are defined as nearest neighbours if $r_{ij} < r_0$. r_0 should lie between the first and second coordination shells and a typical value would be $1.35\, r_{eq}$, although the exact value is somewhat arbitrary. The first term in Equation 1.3 is the ideal pair energy if all n_{nn} nearest-neighbour pairs lie exactly at the equilibrium pair distance, the strain energy is the energetic penalty for the deviation of nearest-neighbour distances from the equilibrium pair distance and E_{nnn} is the contribution to the energy from non-nearest neighbours.

E_{nnn} is usually smaller than the other two terms and is relatively independent of the detailed structure. Therefore, the global minimum usually represents the best balance between maximizing n_{nn} and minimizing E_{strain}. For an atom in the interior of a cluster this is usually achieved through the atom having a coordination number of twelve. This can be achieved as in close-packing, but another possibility is an icosahedral coordination shell. n_{nn} is further increased through the cluster having a compact spherical shape, and through the surface mainly consisting of faces with a high co-ordination number. For example, an atom on a face-centred-cubic (fcc) $\{111\}$ face has nine nearest neighbours, whereas an atom on a $\{100\}$ face has eight nearest neighbours.[2]

The three main types of cluster structure found for systems with isotropic interactions, namely icosahedral, decahedral and close-packed[3] structures, are depicted in Figure 1.1. These examples have the optimal shape for each structural type, and all have been identified experimentally.

One of the unusual properties of clusters is that they can exhibit non-crystallographic symmetries, because there is no requirement for translational periodicity. Decahedral clusters have a single five-fold axis and are based on

(a)   (b)

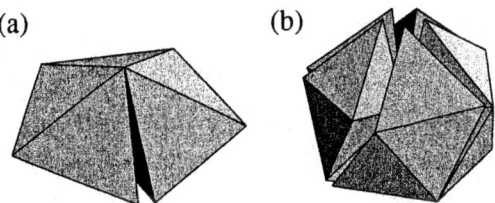

Figure 1.2   Examples of the strain involved in packing tetrahedra. (a) Five regular tetrahedra around a common edge produce a gap of 7.36°. (b) Twenty regular tetrahedra about a common vertex produce gaps equivalent to a solid angle of 1.54 steradians.

a pentagonal bipyramid that can be thought of as five strained fcc tetrahedra sharing a common edge. The symmetry axis corresponds to this common edge. The Marks decahedron [17], which represents the optimal shape for this structural type, can be formed from a pentagonal bipyramid by exposing {100} faces at the equatorial edges then introducing reentrant {111} faces. Mackay icosahedra [18] have six five-fold axes of symmetry and can be thought of as twenty strained fcc tetrahedra sharing a common edge. The fcc cluster represented in Figure 1.1a is simply a fragment of the bulk fcc lattice.

Icosahedral structures generally have the largest n_{nn} because of their spherical shape and {111} faces, and close-packed clusters the smallest n_{nn} because of their higher proportion of {100} faces. By contrast, close-packed clusters can be unstrained, whereas, as Figure 1.2 illustrates, decahedra and icosahedra are increasingly strained. The strain energy is proportional to the volume of the cluster, but differences in n_{nn} are due to surface effects. Therefore, icosahedra are likely to be found at small sizes, but at sufficiently large size, the cluster must take on the bulk structure. At intermediate sizes decahedra can be most stable. The sizes at which the crossovers between structural types occur is system dependent.

The structures illustrated in Figure 1.1 involve exactly the right number of atoms to form a cluster of the optimal shape. For example, complete Mackay icosahedra can be formed at $N = 13, 55, 147, 309, \ldots$ At intermediate sizes clusters have an incomplete surface layer. Marks decahedra with square {100} faces occur at $N = 75, 192, 389, \ldots$. Other complete Marks decahedra that are less spherical can be found in between these sizes, e.g. at $N=101$ and 146. Fcc truncated octahedra with regular hexagonal {111} faces can be found at $N = 38, 201, 586 \ldots$, and other less spherical truncated octahedra can be found, for example, at $N=79$, 116, 140 [19]. Furthermore, because the energy of a twin plane is often small, close-packed structures with other forms can also be particularly stable. Four examples are given in Figure 1.3 for $N < 100$. The 26-atom structure has a hexagonal close-packed (hcp) structure; the 50-atom structure consists of two fragments of the 38-atom structure joined at a twin plane; the 59-atom structure consists of a 31-atom fcc

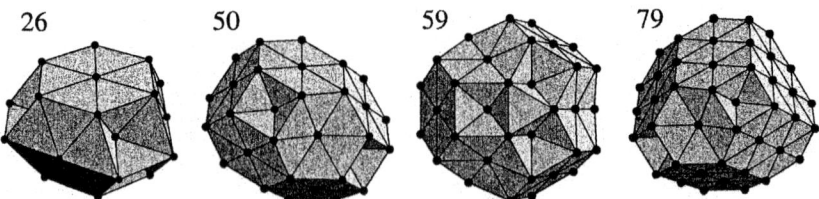

Figure 1.3 Four examples of stable close-packed structures for $N < 80$. The sizes are as labelled. The 59-atoms structure has T_d point group symmetry, and the rest D_{3h}.

truncated tetrahedron with each face covered by a 7-atom hexagonal overlayer that occupies the hcp surface sites; and the 79-atom structure, which is similar to the 50-atom structure, is formed by the introduction of a twin plane into the 79-atom truncated octahedron [20].

Recently, a new structural type called a Leary tetrahedron [21] has been discovered. An example with 98 atoms is illustrated in Figure 1.4. At the centre of this structure is an fcc tetrahedron. To each of the faces of this tetrahedron, further fcc tetrahedra (minus an apical atom) are added, to form a stellated tetrahedron. Finally, the edges of the original tetrahedron are covered by 7-atom hexagonal overlayers. The coordination along the edges of the central tetrahedron is the same as along the symmetry axis of the decahedron and so the strain energy of this structure is intermediate between icosahedra and decahedra. It is not yet clear how general this class of structures is. The 98-atom example is the global minimum for a model potential [21] and mass spectroscopic studies of clusters of C_{60} molecules suggest that $(C_{60})_{98}$ has this structure [22]. However, it may be that the stability of this structural class is restricted to $N=98$, because this size results in a particularly spherical shape, and that equivalent structures at larger sizes (e.g. $N=159$, 195) are never competitive.

2.1   LENNARD-JONES CLUSTERS

In this section I focus on clusters bound by the Lennard-Jones (LJ) potential [23]:

$$E = 4\epsilon \sum_{i<j} \left[\left(\frac{\sigma}{r_{ij}} \right)^{12} - \left(\frac{\sigma}{r_{ij}} \right)^{6} \right], \qquad (1.5)$$

where ϵ is the pair well depth and $2^{1/6}\sigma$ is the equilibrium pair separation. The potential is illustrated in Figure 1.5a, and provides a reasonable description of the interatomic interactions of rare gases, such as argon. LJ clusters have become probably the most common test system for GO algorithms for configurational problems. The number of papers with applications to this model system is now very large, but unfortunately many are distinctly unimpressive, only reporting results for small sizes or failing for relatively simple cases. I

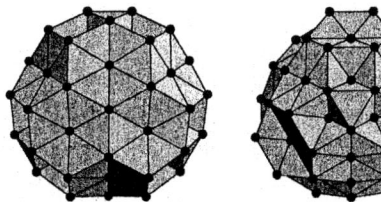

Figure 1.4   Front and back views of the 98-atom Leary tetrahedron.

do not attempt to review this literature, but instead refer the interested reader elsewhere [24].

At small sizes the LJ potential is able to accommodate the strain associated with icosahedral packing relatively easily [25]. Indeed, only for $N > 1600$ are the majority of global minimum expected to be decahedral and the crossover to fcc clusters has been estimated to occur at $N \approx 10^5$ [20]. This preference for icosahedral packing is also evident from Figure 1.6 where I compare the energies of icosahedral, decahedral and close-packed clusters. Sizes where complete Mackay icosahedra are possible (N=13, 55) stand out as particularly stable. The icosahedra are least stable when the overlayer is roughly half-filled. Therefore, when especially stable non-icosahedral clusters coincide with these sizes there is a possibility that the global minimum will be non-icosahedral. There are eight such cases for $N \leq 147$. At N=38 the global minimum is the fcc truncated octahedron [16, 26, 27]; at N=75–77 [16] and 102–104 [19] the global minima are Marks decahedra; and at N=98 the global minimum is a Leary tetrahedron [21]. At these sizes the lines for the decahedral or close-packed structures in Figure 1.6 dip just below the line for the icosahedra. For $148 \leq N \leq 309$ there are a further eight non-icosahedral global minima [9, 28], all of which are decahedral and which divide into two sets that are based on the complete Marks decahedra possible at N=192 and 238. For $310 < N < 561$ the global minima are all icosahedral [29], but for $562 < N < 1000$ there are 41 decahedral globlal minima [30].

2.2   MORSE CLUSTERS

In this section I focus on clusters bound by the the Morse potential [31]:

$$V_M = \epsilon \sum_{i<j} e^{\rho(1-r_{ij}/r_{eq})} \left(e^{\rho(1-r_{ij}/r_{eq})} - 2 \right), \tag{1.6}$$

where ϵ is the pair well depth and r_{eq} is the equilibrium pair separation. In reduced units there is a single adjustable parameter, ρ, which determines the range of the interparticle forces. Figure 1.5b shows that decreasing ρ increases the range of the attractive part of the potential and softens the repulsive wall, thus widening the potential well. Values of ρ appropriate to a range of materials have been catalogued elsewhere [32]. The LJ potential has the same curvature

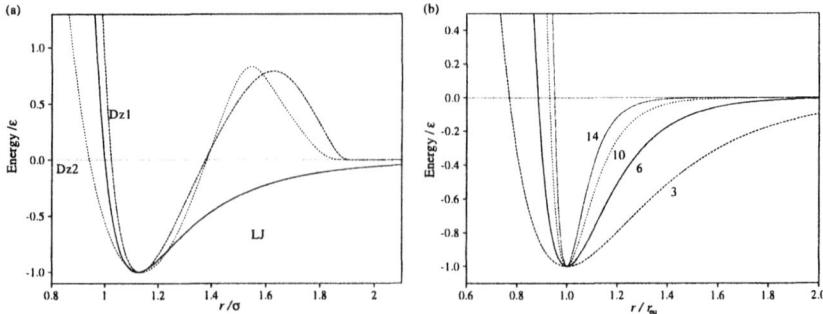

Figure 1.5 (a) A comparison of the Lennard-Jones (LJ) potential with the Dzugutov potential (Dz1) and a modified version of it (Dz2). (b) The Morse potential for several values of the range parameter, ρ.

at the bottom of the well as the Morse potential when $\rho=6$. Girifalco has obtained an intermolecular potential for C_{60} molecules [33] that is isotropic and short-ranged relative to the equilibrium pair separation with an effective value of $\rho=13.62$ [34]. The alkali metals have longer-ranged interactions, for example $\rho=3.15$ has been suggested for sodium [35].

The global minima for this system have been found as a function of ρ for all sizes up to $N=80$ [16, 36, 37]. Equation (1.3) enables us to understand the effect of ρ on cluster structure. As ρ increases and the potential well narrows, the energetic penalty for distances deviating from the equilibrium pair separation increases. Thus, E_{strain} increases for strained structures, and so icosahedral and decahedral structures become disfavoured as ρ increases. This is illustrated in Figure 1.7, which shows how the structure of the global minimum depends on N and ρ. The global minimum generally change from icosahedral to decahedral to close-packed as ρ is increased. It can be seen that the value of ρ appropriate for the LJ potential lies roughly in the middle of the icosahedral region of Figure 1.7.

Alternatively, the effect of ρ can be thought of in terms of its effect on the crossover sizes at which a particular structural type becomes dominant. As ρ increases, the less strained structures become dominant at smaller sizes. These effects can also be found in real materials. For example, sodium clusters have been shown to exhibit icosahedral structures up to at least 22 000 atoms [38], whereas the thermodynamically stable structure of clusters of C_{60} molecules have recently been shown to be non-icosahedral for $N > 30$ [22].

As well as these trends, Figure 1.7, of course, also reflects the specifics of the structures that are possible at each size, so the boundaries between structural types are not smooth lines but show a lot of detailed structure. For example, the range of ρ values for which icosahedral structures are most stable is a local maximum at $N=55$ because of the complete Mackay icosahedra possible at this size. At sizes where close-packed structures have a greater or equal number

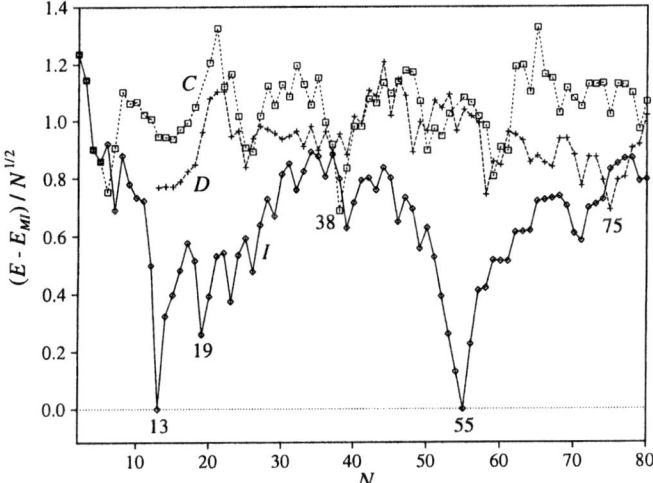

Figure 1.6 Comparison of the energies of icosahedral (I), decahedral (D) and close-packed (C) LJ$_N$ clusters. The energy zero is E_{MI}, a function fitted to the energies of the first four Mackay icosahedra at N=13, 55, 147 and 309:

of nearest neighbours to the best decahedral structure, the global minimum changes directly from icosahedral to close-packed.

A new class of structures appears in the bottom right-hand corner of Figure 1.7. They are polytetrahedral clusters with disclination lines running through them. A polytetrahedral structure can be decomposed into tetrahedra without any interstices. The 13-atom icosahedron is an example, and one of the possible ways of adding atoms to the surface of the icosahedron—the so-called anti-Mackay overlayer—continues the polytetrahedral packing. This overlayer does not lead to the next Mackay icosahedron but instead to the 45-atom rhombic tricontahedron (Figure 1.8a), which can be thought of as an icosahedron of interpenetrating icosahedra. If one imagines adding regular tetrahedra to the form in Figure 1.2b one soon realizes that the 45-atom structure must be extremely strained, and for this reason it is the global minimum only at low ρ, where this strain can be accommodated. For $N > 45$ similar polytetrahedral clusters can be formed but based not on the 13-atom icosahedron but on polyhedra with a higher coordination number. The two examples in Figure 1.8a have a 14- and 16-coordinate central atom. These structures can be described in terms of disclination lines, where the lines pass along those nearest-neighbour contacts that are the common edge for six tetrahedra [39]. These types of structures might be thought to be fairly esoteric, but they form the basis for the crystalline Frank-Kasper phases [40, 41] where atoms of different size create a preference for coordination numbers higher than 12, and so they might be good candidate structures for certain mixed metal clusters. Furthermore, they have recently been found in models of clusters of heavy metal atoms [42] and aluminium

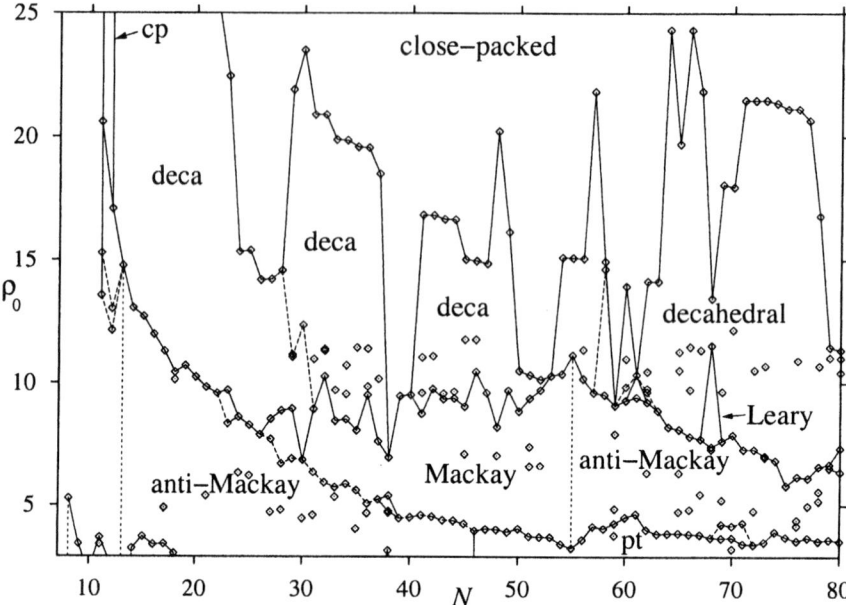

Figure 1.7 Zero temperature 'phase diagram' showing the variation of the lowest-energy structure with N and ρ. The data points are the values of ρ at which the global minimum changes. The lines joining the data points divide the phase diagram into regions where the global minima have similar structures. The solid lines denote the boundaries between the four main structural types—icosahedral, decahedral, close-packed and the polytetrahedral (pt) structures associated with low ρ (L)—and the dashed lines are internal boundaries within a structural type, e.g. between icosahedra with different overlayers (Mackay and anti-Mackay), or between decahedra with different length decahedral axes. There is also a small region where structures based on the Leary tetrahedron are most stable.

clusters [43], and recent experimental diffraction and electron microscopy data for small cobalt clusters can best be modelled by a disclinated polytetrahedral structure that is a fragment of a Frank-Kasper phase [44]. However, at the size corresponding to this experiment ($N \approx 150$) the long-ranged Morse clusters have disordered polytetrahedral global minima.

2.3   DZUGUTOV CLUSTERS

In contrast to the potentials that we have so far examined, the Dzugutov potential [45, 46] has a maximum that penalizes distances near to $\sqrt{2}$ times the equilibrium pair distance (Figure 1.5a), the distance across the diagonal of the octahedra in close-packed structures. This maximum loosely resembles the first of the Friedel oscillations [47] often found in effective metal potentials. The potential was originally designed to suppress crystallization in bulk simulations so that the properties of supercooled liquids and glasses could more easily be

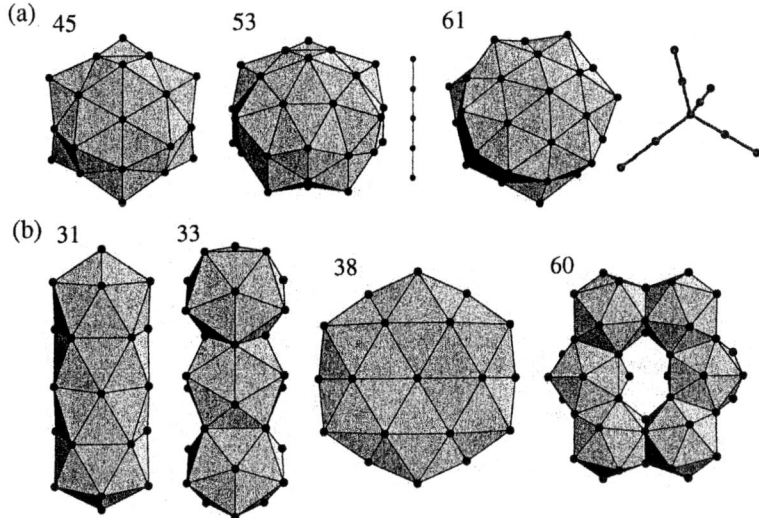

Figure 1.8 Polytetrahedral structures that are the global minima for the (a) long-ranged Morse potential and (b) Dzugutov potential. Sizes as labelled. The point group symmetries of the structures in (a) are I_h, D_{6d} and T_d and in (b) are D_{5h}, D_{3d}, D_{6h} and C_s. The disclination networks associated with the 53- and 61-atom structures are illustrated next to the clusters.

studied in a one-component system [45, 48]. However, under certain conditions it was found that a dodecagonal quasicrystal could be formed on freezing [49].

For clusters the potential will penalize close-packed, decahedral and Mackay icosahedral structures (the latter two because octahedra are found within the fcc tetrahedral units from which the structures are made) and will favour polyte-trahedral clusters. Therefore, one might think that this potential would provide a good model for small cobalt clusters. However, as can be seen from Figure 1.5a the potential is narrower than the LJ potential, and matching the second derivative at the equilibrium pair separation to that of the Morse potential gives an effective value of ρ of 7.52. Therefore, the potential cannot accommodate the strain in compact polytetrahedral clusters. Instead, the global minima are non-compact polytetrahedral structures, such as the needles, disc and torus illustrated in Figure 1.8b [50]. These structures are made up of face-sharing or interpenetrating 13-atom icosahedral units. In terms of Equation (1.3) they represent the best balance between maximizing n_{nn} whilst minimizing both E_{strain} and E_{nnn}, where the latter now corresponds to the total energetic penalty for distances close to $\sqrt{2}\, r_{eq}$.

In order to generate a model that exhibits ordered compact polytetrahedral clusters a modified Dzugutov potential was constructed with an effective value of ρ of 5.16, allowing it to accommodate more strain (Figure 1.5a). Indeed, the global minima do have the desired structural type, and so this system should be useful in generating realistic candidate structures to compare with the cobalt experiments [51].

2.4   COMPARISON WITH EXPERIMENT

One of the remarkable features of these simple model potentials is that the structures they exhibit do provide good candidates for the structures of real clusters. Indeed, frequently the structures first identified for these model systems are subsequently identified in experiments. For example, the special stability of the structures exhibited by the fcc Ni_{38} [13] and Au_{38} [14], the decahedral Au_{75} [15] and the tetrahedral $(C_{60})_{98}$ [22] were first identified through calculations on LJ and Morse clusters [16, 21].

Furthermore, as experiments can rarely identify a cluster's structure directly, but often have to rely on comparison with properties calculated using candidate structures, it is extremely useful to have databases of plausible structures available. This is the philosophy behind internet repositories such at the Cambridge Cluster Database (http://www-wales.ch.cam.ac.uk/CCD.html), which contains the global minima for all the potentials described here, and the Birmingham Cluster website (http://www.tc.bham.ac.uk/bcweb/).

In comparisons between experiment and theory the role of temperature and kinetics should be remembered. The global minimum is only rigorously the equilibrium structure at zero temperature. At higher temperatures other structures may become more stable due to entropic effects [52, 53] as we will see in Section 4.1. Furthermore, it is not always clear whether equilibrium has been achieved under the experimental conditions, especially for clusters formed at low temperature [22, 54, 55].

3.   GLOBAL OPTIMIZATION APPROACHES

The type of GO algorithms in which I am interested are those that find global minima, not those that are also able to *prove* that the best structure found is in fact truly global. Unsurprisingly, the latter is a much more demanding task. For example, for LJ clusters good putative global minima have been found up to $N=309$, but only up to $N=7$ have these structures been proven to be global [56]. Of course, the problem with settling for obtaining putative global minima is that it is difficult to know when to give up looking for a lower-energy solution. For example, to my surprise, at least, a new putative global minimum was recently found for LJ_{98} [21], even though powerful GO algorithms had previously been applied to this cluster [10, 57, 58, 59]. The failure of these previous attempts to locate the global minimum was not because the algorithms are unable to locate the Leary tetrahedron, but simply because the computations had been terminated too soon.

I also wish to concentrate on GO algorithms that are unbiased, i.e. those that do not artificially bias the system towards those structures that physical insight would suggest are low in energy, for example, by seeding the algorithm with fragments of a certain structural type [59] or by searching on a lattice for a specified structural type [25]. Of course, such biased algorithms are usually more efficient. For example, most of the LJ global minima were first found using such methods [25, 28]. However, they cannot cope with 'surprising' structures that fall outside the expected categories, and lack transferability to other systems, because they have sacrificed generality for greater efficiency in the specific problem instances. Furthermore, they require a sufficient prior understanding of the structure. This may be possible for the model potentials we consider here, but it is a much more difficult task with the complex interactions that are often necessary to realistically describe a system.

Virtually all the global optimization algorithms that are most successful at locating the global minima of clusters have a common feature. Namely, they make extensive use of local minimization. All GO algorithms require elements of both local and global search. The algorithm has both to be able to explore all regions of configuration space (overcoming any energy barriers that might hinder this) whilst also sufficiently sampling the low-energy configurations within each region. Performing local minimizations from configurations generated by a global search is one way of combining these two elements.

Simulated annealing provides a perhaps more traditional way of achieving this goal. In simulated annealing, by varying a parameter, the temperature, the nature of the search is changed from global (at high temperature) to local (as $T \rightarrow 0$). However, this approach has a number of weaknesses. There is effectively only one local minimization, so if the configuration does not become confined to the basin of attraction of the global minimum as the temperature is reduced the algorithm will fail, even if the system had passed through that basin of attraction at higher temperature. This condition for success is unnecessarily restrictive and leads to inefficiency.

There is a further element to the most successful algorithms, namely that the energies of the local minima, not of the configurations prior to minimization, are the basis for comparing and selecting structures. This approach was first used in 1987 by Li and Scheraga in the application of their 'Monte Carlo plus minimization' to polypeptides [60]. However, despite this approach being independently adopted a number of times subsequently [61, 62], it was only in 1997 that it was realized that in this approach one is effectively searching a transformed PES, $\tilde{E}(\mathbf{x})$, where the energy associated with each point in configuration space is that of the minimum obtained by a local minimization from that point [10], i.e.

$$\tilde{E}(\mathbf{x}) = \min\{E(\mathbf{x})\}, \qquad (1.7)$$

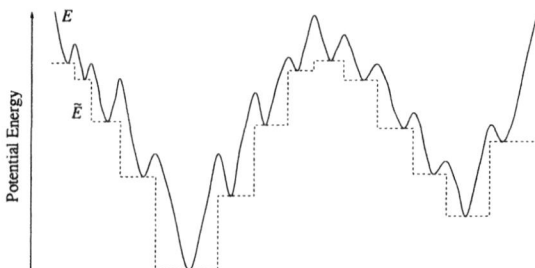

Figure 1.9 A schematic diagram illustrating the effects of the basin-hopping potential energy transformation for a one-dimensional example. The solid line is the potential energy of the original surface and the dashed line is the transformed energy, $\tilde{E}$.

where 'min' signifies that an energy minimization is carried out starting from **x**. Unlike many PES transformations proposed in the name of global optimization, this 'basin-hopping' transformation is guaranteed to preserve the identity of the global minimum. The transformation maps the PES onto a set of interpenetrating staircases with plateaus corresponding to the basins of attraction of each minimum (i.e. the set of configurations which lead to a given minimum after optimization). A schematic view of the staircase topography that results from this transformation is given in Figure 1.9.

The potential advantages of using the basin-hopping transformation become clear when we contrast the inter-minimum dynamics on the original and transformed PESs. In molecular dynamics simulations on the original PES much time is wasted as the system oscillates back and forth within the well surrounding a minimum, waiting for the kinetic energy to become sufficiently localized along the direction of a transition state valley to enable the system to pass into an adjacent minimum. A similar dynamical (albeit as a function of steps rather than time) picture holds for MC when only local moves are used. The biased random walk is confined to the well around a minimum, frequently being reflected back off the walls of this well, until by chance the system happens to wander over a transition valley into a new minimum. However, a completely different picture is appropriate to the dynamics in simulations (using MC or discontinuous molecular dynamics[4] [63]) on the transformed PES. The transformation removes vibrational motion (the Hessian has no positive eigenvalues) and transitions out of a basin are possible anywhere along the boundary of the basin. Therefore, steps in any direction can lead directly to a new minimum. Furthermore, downhill transitions are now barrierless. However, as Figure 1.9 illustrates, significant barriers between low-energy minima can remain if they are separated by high-energy intervening minima.

Consequently, on $\tilde{E}(\mathbf{x})$ the system can hop directly between basins; hence the name of this transformation. Furthermore, much larger MC steps can be taken on $\tilde{E}(\mathbf{x})$; such steps would virtually always be rejected on the original

PES because atoms would become too close and an extremely high energy would result. After the transformation atoms can even pass through each other.

The method of searching $\tilde{E}(\mathbf{x})$ is of secondary importance compared to the use of the transformation itself. Indeed, the performance is fairly similar for the two main methods used, genetic algorithms and the constant temperature MC used in basin-hopping. Here, we mainly concentrate on the basin-hopping approach and refer readers to Hartke's chapter for more detail on the genetic algorithm methodology [9].

In the basin-hopping or Monte Carlo plus minimization method, standard Metropolis MC is used, i.e. moves are generated by randomly perturbing the coordinates, and are always accepted if $\tilde{E}$ decreases and are accepted with a probability $\exp(-\Delta\tilde{E}/kT)$ if $\tilde{E}$ increases. Using constant temperature is sufficient, since there is no great advantage to using an annealing schedule because the aim is not to trap the system in the global minimum, but just to visit it at some point in the simulation. One of the advantages of this method is its simplicity—there are few parameters to adjust. It is usually satisfactory to dynamically adjust the step size to produce a 50% acceptance ratio. An appropriate temperature also needs to be chosen, but fortunately the temperature window for which the method is effective is usually large, and can be quickly found after some experimentation. Furthermore, there is a well-defined thermodynamics associated with the method [64] that makes understanding the physics behind the approach easier, as we shall see in Section 4.3.

Typically, a series of basin-hopping runs of a specified length will be performed starting from a random geometry. This is advantageous over a single longer run because it can provide a loose gauge of success. If all the runs return the same lowest-energy structure one would imagine that the true global minimum had been found. It can also often prove useful to perform runs starting from the best structures at sizes one above and below, with the lowest-energy atom removed or an atom added, respectively.

The local minimization method that we have found to be most efficient for clusters is a limited memory BFGS algorithm [65]. The basin-hopping approach is also found to be more efficient when the configuration is reset to the configuration of the local minimum after each accepted step [66]; this avoids problems with evaporation of atoms from the cluster since the basin-hopping transformation also reduces the barriers to dissociation. In addition to the usual steps, it is advantageous to have occasional angular steps for low-energy surface atoms. These have a similar aim to the directed mutations introduced by Hartke into his genetic algorithm [58]; they both enable the best arrangement of the surface atoms to be found more rapidly. For biopolymers other system-specific step types have been introduced to increase efficiency [67].

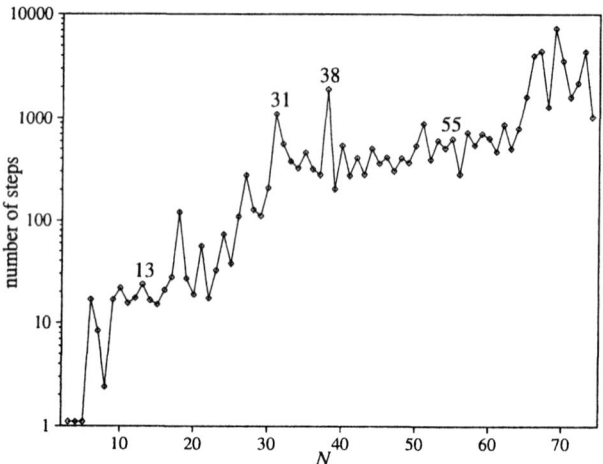

Figure 1.10   Mean number of steps to reach the LJ$_N$ (up to $N=74$) global minimum from a random starting point with the basin-hopping approach. The average is over a hundred runs at $T = 0.8\epsilon k^{-1}$. Reproduced from Ref. [69].

One of the main differences between the basin-hopping and genetic algorithms is that only local moves are used in basin-hopping, whereas genetic algorithms employ co-operative crossover moves in which a new structure is formed from fragments of two 'parent' clusters. However, recently non-local moves have been introduced into a variant of basin-hopping by rotating or reflecting a fragment of the structure [68].

Two examples of the performance of basin-hopping algorithms for LJ clusters as a function of size are given in Figures 1.10 and 1.11. The basin-hopping transformation must lead to a considerable speed-up (in terms of steps) if the method is to be cost-effective, because the transformation is computationally expensive and requires many evaluations of the energy and the forces at each step. Clearly, one would not want to use the many millions of steps and cycles that are typically used in molecular dynamics and MC simulations on the original PES. However, the results in Figure 1.10 show that the number of steps required to find the global minimum is remarkably few, only of the order of hundreds or thousands of steps. This is even more remarkable when the number of minima on the PES is considered. In line with theoretical expectations [70] the number of minima for small LJ clusters increases exponentially with size [71]. Extrapolating this trend provides, for example, an estimate of 10^{21} minima for LJ$_{55}$. Therefore, a Levinthal-type paradox [72][5] can be formulated for locating the global minimum of a cluster: the number of minima of an atomic cluster quickly becomes so large that beyond a fairly small size, if these minima were *randomly* searched, even at an extraordinarily fast rate, it would take an unfeasibly long time to locate the global minimum—so how is

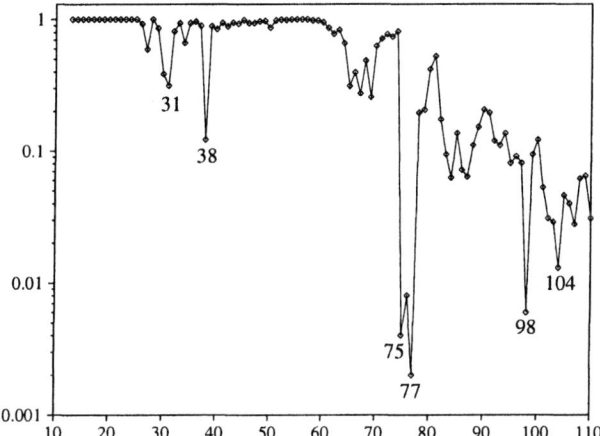

Figure 1.11 Observed probability of hitting the global minimum in a monotonic sequence basin-hopping run starting from a random starting point, averaged over 1000 runs. Reproduced from Ref. [77].

it possible to find the global minimum? Yet the basin-hopping algorithm using optimal parameters finds the LJ_{55} global minimum from a random starting configuration on average within 150 steps [69]. The fallacy in Levinthal's paradox has long been known to be the assumption of random searching [73, 74, 75, 76] however it does emphasize the extremely non-random nature of basin-hopping for LJ_{55}—the runs are extremely biased towards the global minimum.

Figure 1.10 shows some interesting variations in the ease of global optimization with cluster size. The 38-atom global minimum particularly stands out as being difficult to locate, suggesting that competition between different structural types makes global optimization more difficult. Indeed for LJ_{75}, the next largest cluster with a non-icosahedral global minimum, the number of steps required is so large that it was not possible to obtain good enough statistics to be included in Figure 1.10. Subsequent calculations on a super-computer by Leary suggest that the mean first passage time is of the order of 10^7 steps [77]. The other six non-icosahedral global minima for $N < 150$ are of roughly similar difficulty to locate. It is also noteworthy that the global minimum for LJ_{31} is relatively difficult to find. In this case, there is some structural competition between the two types of icosahedral overlayer—it is the first size at which the overlayer that leads to the next Mackay icosahedron is lowest in energy.

An alternative perspective on these effects can be obtained from Figure 1.11, which shows the probability that a 'monotonic sequence' basin-hopping run ends at the global minimum. In this variation of the basin-hopping algorithm [77] only downhill steps are accepted (i.e. $T{=}0$) and the run is stopped after there is no further improvement for a certain number of steps. For those sizes with non-icosahedral global minima there is a much smaller probability of the run ending in the global minimum, and so again global optimization is more difficult. In these cases the majority of runs end at low-energy icosahedral structures. Those examples at larger sizes are again more than an order of magnitude more difficult than LJ$_{38}$, but interestingly the Marks decahedra at N=102–104 are somewhat more easy to locate than those at N=75–77.

These results enable us to comment on the use of LJ clusters as a test system for GO methods. They show that the icosahedral global minima are relatively easy to locate, and in these cases optimization only starts to become more difficult as N approaches 100 (Figure 1.11). As for the non-icosahedral global minima, the number of unbiased GO methods that have found the LJ$_{38}$ global minimum is now quite large [10, 27, 57, 58, 68, 77, 78, 79, 80, 81, 82, 83, 84] but those that can find the LJ$_{75}$ global minimum is still small [10, 58, 68, 77, 83]. Therefore, a good test for a GO method is to attempt to find all the global minima up to N=110. Any GO method 'worth its salt' for clusters should be able find all the icosahedral global minima and the truncated octahedron at N=38 . Success for the other non-icosahedral global minima would indicate that the method has particular promise. However, far too many GO algorithms have only been tested on cluster sizes where global optimization is relatively trivial.

The weakness of LJ clusters as a test system is that they have a relatively uniform structural behaviour. Morse clusters could provide a much more varied test system, as Figure 1.7 illustrates. A suitable test would be to aim to find all the global minima at ρ=3, 6, 10 and 14 up to N=80, as putative global minima have been tabulated for this size and parameter range [16, 36, 37]. One would generally expect the difficulty of global optimization to increase with ρ because the number of minima increases [85, 86] and the energy landscape becomes more rough [86, 87]. The system also provides many examples of structural competition, particularly for the short-ranged potentials where decahedral and close-packed clusters can have similar energies. A number of studies have begun to use Morse clusters as a test system [88, 89].

4.   MULTIPLE-FUNNEL ENERGY LANDSCAPES

The aim of this section is to provide a physical perspective that can help us understand why the global optimization of a system is easy or difficult, for example, to explain the size-dependence of Figures 1.10 and 1.11. As mentioned in Section 3, an equivalent of Levinthal's paradox, which was originally formulated to capture the difficulty of a protein folding to its native state, can be applied to a cluster locating its global minimum. The flaw in that paradox is its assumption that conformations will be sampled randomly, i.e. all configurations are equally likely, because we know that in an equilibrium physical sampling of the conformation space, say in the canonical ensemble, each point will be sampled with a probability proportional to the Boltzmann weight, $\exp(-E(\mathbf{x})/kT)$. Thus, the Boltzmann factor favours low-energy conformations. Therefore, we can begin to see the vital role played by the potential energy surface. This role extends beyond purely thermodynamic considerations to the dynamics: does the topography and connectivity [90] of the PES naturally lead the system towards or away from the global minimum?

The Levinthal assumption of random sampling is equivalent to assuming that the energy landscape has the topography of a perfectly flat putting green with no thermodynamical or dynamical biases towards the global minimum at the bottom of the 'hole'. Similarly, the NP-hard character of the global optimization of atomic clusters [91], which results in part from the exponential increase in the search space with size, considers a general case where no assumptions about the topography of the PES can be made. In the protein folding community, after the fallacy in the Levinthal paradox was recognized, attention focussed on the more important question of how does the topography of the PES differ for those polypeptides that are able to find their native states from those that cannot [75]. Here, I address similar questions for the global optimization of clusters.

One of the topographical features of the energy landscape that the protein folding community has found to be common is, what has been termed, a 'funnel' [75, 92]. By this they mean a region of configuration space that can be described in terms of a set of downhill pathways that converge on a single low-energy structure or a set of closely-related low-energy structures. As its name suggest a protein PES with a single funnel converging on the native state will be a good folder because the topography helps in guiding the protein towards that native state.

If these ideas are to be useful, one needs a way of depicting the physically-relevant aspects of the topography of a complicated $3N$-dimensional energy landscape. One technique that has proven to be helpful in characterizing the PESs of proteins [93, 94, 95, 96] and clusters [96, 97, 98] is the disconnectivity graph. This graph provides a representation of the connectivity of the multi-dimensional energy landscape and by depicting the effective barriers between minima it is especially useful in the interpretation of dynamics.

To construct a disconnectivity graph, at a series of energy levels the minima on the PES are divided into sets which are connected by paths that never exceed that energy level. In the graph each set is represented by a node at the appropriate energy and lines connect a node to the sets at higher and lower energy which contain the minima corresponding to the original node. A line always ends at the energy of the minimum it represents.

Disconnectivity graphs can be understood by analogy to the effects of the water level in a geographical landscape. The number of nodes in a graph at a given energy is equivalent to the number of distinct seas and lakes for a

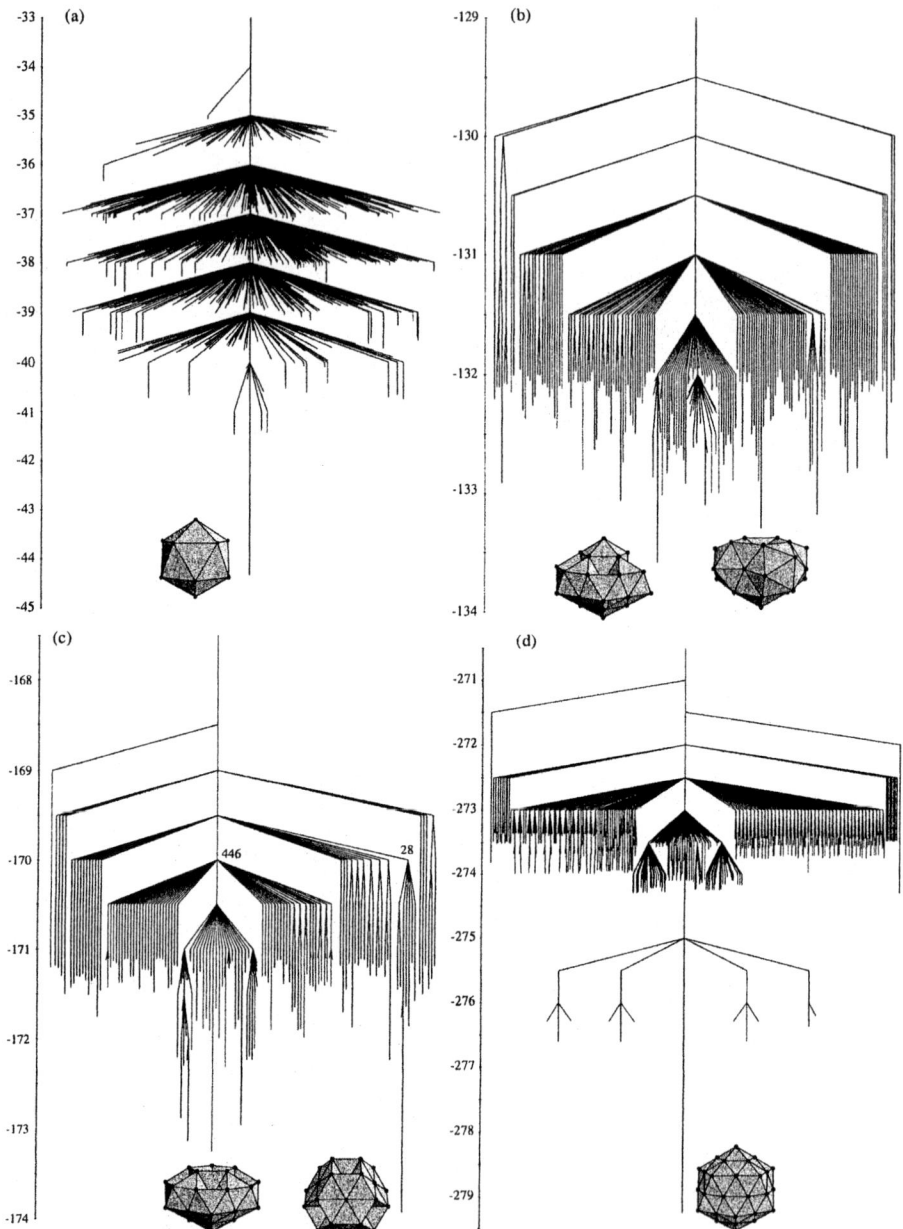

Figure 1.12 Disconnectivity graphs for (a) LJ_{13}, (b) LJ_{31}, (c) LJ_{38}, (d) LJ_{55}, (e) LJ_{75} and (f) LJ_{102}. In (a) all the minima are represented. In the other parts only the branches leading to the (b) 200, (c) 150, (d) 900, (e) 250 and (f) 200 lowest-energy minima are shown. The numbers adjacent to the nodes indicate the number of minima the nodes represent. Pictures of the global minimum, and sometimes the second lowest-energy minimum, are adjacent to the corresponding branch. The units of the energy axes are ϵ.

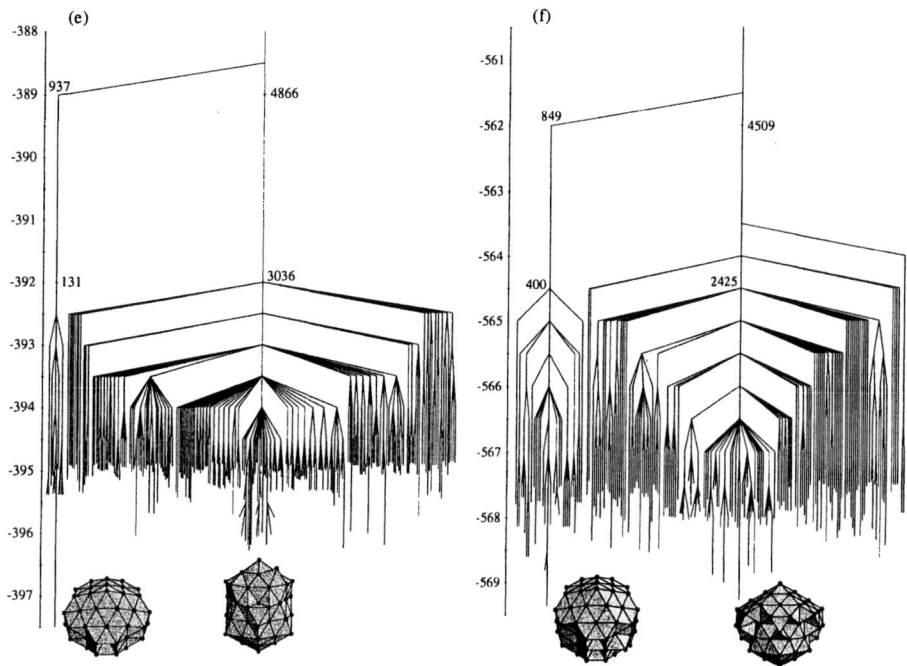

Figure 1.12   cont.

given water level. Inspection of actual disconnectivity graphs, e.g. Figure 1.12, also helps to clarify these ideas. At sufficiently high energy, all minima are mutually accessible and so there is only one node, but as the energy decreases sets of minima become disconnected from each other and the graph splits, until at sufficiently low energy, again there is only one node left, that of the global minimum. The pattern of the graph can reveal particularly interesting information about the PES topography. For a PES with a single funnel there is a single dominant stem with the other minima branching directly off it as the energy is decreased. By contrast for 'multiple-funnel' PESs the graph is expected to split at high energy into two or more major stems.

In Figure 1.12 disconnectivity graphs for a selection of LJ clusters are presented, in particular some of those clusters that Figures 1.10 and 1.11 indicated are more difficult to optimize. In the graph for LJ_{13} all the minima in our near-exhaustive sample are represented. The graph shows the form for an ideal single-funnel PES, because the icosahedral global minimum is particularly low in energy, and dominates the energy landscape. The PES has a remarkable connectivity: 911 distinct transition states are connected to the global minimum and all minima are within three rearrangements of the global minimum [90]. The disconnectivity graph for LJ_{55}, another 'magic number' LJ cluster, also has a single-funnel. Unlike for LJ_{13}, we are unable to represent

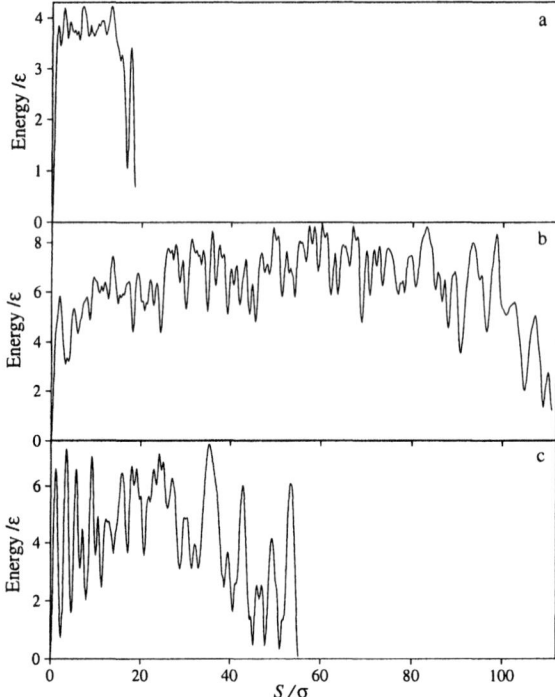

Figure 1.13   The lowest-energy path from the global minimum to the second lowest-energy minimum for (a) LJ_{38}, (b) LJ_{75} and (c) LJ_{102}. In each case the zero of energy corresponds to the energy of the global minimum.

all the minima that we found on the graph, so instead we concentrate on the lower-energy minima in our sample. Indeed, the two bands of minima in the graph represent Mackay icosahedra with one or two defects. The graph only reveals the bottom of a funnel which extends up into the liquid-like minima [98].

In contrast to these two clusters, the bottom of the LJ_{31} PES is much flatter, and there are significant barriers between the low-energy minima, in particular between the two lowest-energy minima, which are icosahedral structures, but with different types of surface overlayer. These effects of structural competition are found in more extreme form in the graphs of those clusters with non-icosahedral global minima. The graphs of LJ_{38}, LJ_{75} and LJ_{102} split at high energy into stems associated with icosahedral and fcc or decahedral structures, and so these energy landscapes have two major funnels. This splitting is most dramatic for LJ_{75} and LJ_{102}, where the barrier between the two funnels is much greater than between any of the other sets of minima. The barriers between the two lowest-energy minima are 4.22ϵ and 3.54ϵ for LJ_{38}, 8.69ϵ and 7.48ϵ for LJ_{75} and 7.44ϵ and 7.36ϵ for LJ_{102}. The corresponding lowest-barrier paths are

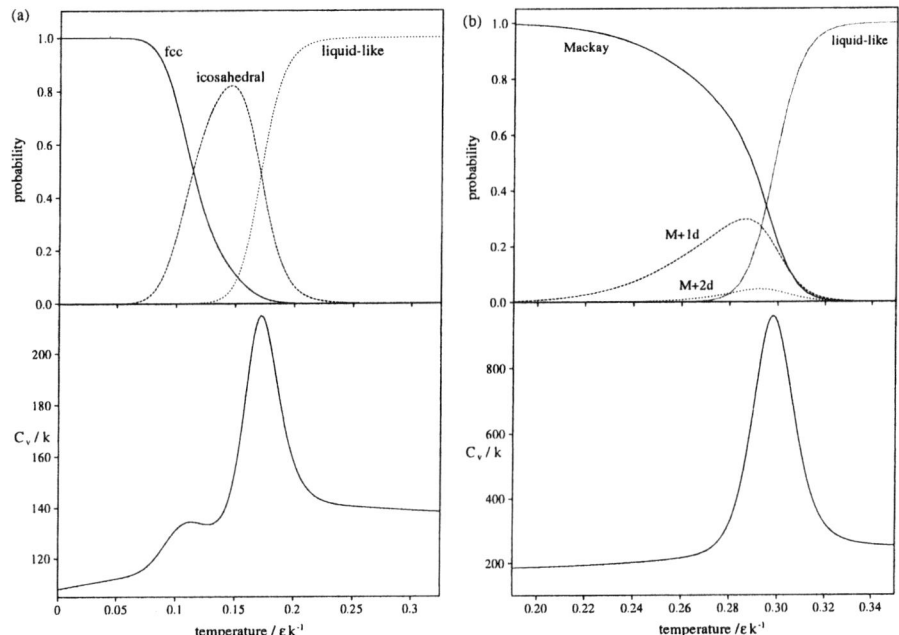

Figure 1.14   Equilibrium thermodynamic properties of (a) LJ$_{38}$ and (b) LJ$_{55}$ in the canonical ensemble. Both the probability of the cluster being in the labelled regions of configuration space and the heat capacity, C_v are depicted. The label M+nd stands for a Mackay icosahedron with n surface defects.

represented in Figure 1.13. These pathways pass over many transition states—13, 65 and 30 for LJ$_{38}$, LJ$_{75}$ and LJ$_{102}$, respectively. There are many other pathways connecting the funnels, but they are either longer or involve higher effective barriers. For LJ$_{38}$ the pathway passes through disordered liquid-like minima. However, for the two larger clusters all the minima along the pathways are ordered and the main structural changes are achieved by rearrangements that involve cooperative twists around the five-fold axis of the decahedron— the conservation of this axis throughout the structural transformation has also been observed in simulations of the decahedral to icosahedral transition in gold clusters [15]. For LJ$_{75}$ the Marks decahedron is oblate whilst the low-energy icosahedral minima are prolate, so the pathway involves a greater amount of reorganization of the surface layer either side of these cooperative transitions than for LJ$_{102}$. For the latter cluster the decahedral and icosahedral structures have fairly similar shapes; hence the shorter pathway for LJ$_{102}$ (Figure 1.13), even though it is larger.

As expected from the general dominance of icosahedral structures in this size range, there are many more low-energy icosahedral minima than low-energy decahedral or fcc minima for these three clusters. There are many low-energy arrangements of the incomplete surface layer of the icosahedral structures, whereas the decahedral or fcc structures have especially compact structures and so any alteration in structure leads to a significant increase in energy. The number of minima in our samples that lie within the respective funnels is marked on Figure 1.12 and is indicative of the greater width of the icosahedral funnels.

From characterizing the energy landscapes of these clusters, the physical origins of some of the differences in difficulty for GO algorithms should be becoming apparent. The single funnels of LJ_{13} and LJ_{55} make the global minimum particularly accessible, and the system is strongly directed towards the global minimum on relaxation down the PES. For LJ_{31}, once the system has reached a low-energy structure, the flatness and the barriers at the bottom of the PES mean that further optimization is relatively slow compared to LJ_{13} and LJ_{55}. For the clusters with non-icosahedral global minima, the icosahedral funnel is much more accessible because of its greater width. Furthermore, after entering the icosahedral funnel, subsequent escape into the fcc or decahedral funnel is likely to be very slow because of the large barriers that need to be overcome. The icosahedral funnel acts as a kinetic trap hindering global optimization. These effects will come out even more clearly in the next two sections as we look at the thermodynamics and dynamics associated with these clusters. From the disconnectivity graphs one would expect trapping to be a significantly greater hindrance to global optimization for LJ_{75} and LJ_{102} than for LJ_{38}, and the longer interfunnel pathway for LJ_{75} provides a possible explanation of why LJ_{102} seems to be somewhat less difficult to optimize than LJ_{75}.

4.1   THERMODYNAMICS

The typical thermodynamic properties of a cluster are illustrated in Figure 1.14b for LJ_{55}. The heat capacity peak is associated with a melting transition that is the finite-size analogue of a first-order phase transition [99] and up to melting the structure is based on the global minimum, perhaps with some surface defects. The thermodynamics of the clusters with non-icosahedral structures are significantly different. Now, as well as the melting transition there is a further transition associated with a transition from the global minimum to the icosahedral structures that gives rise to a second lower-temperature feature in the heat capacity (Figure 1.14a and 1.15). For LJ_{38} the transition occurs fairly close to melting,[6] however as the size increases the transition temperature generally decreases [53].

These solid-solid transitions have a number of implications for global optimization. Firstly, on cooling from the melt it is thermodynamically more favourable for the cluster to enter the icosahedral funnel than that associated with the global minimum. Secondly, the transitions, particularly those for the

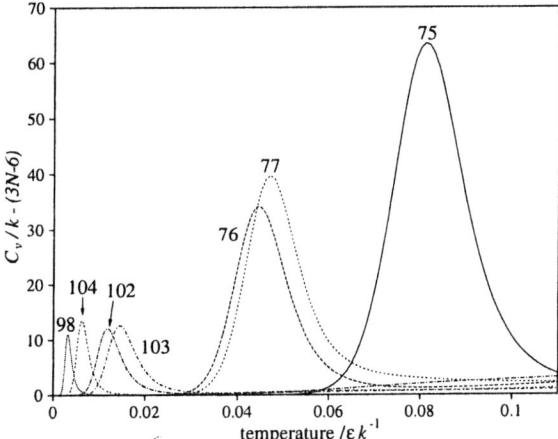

Figure 1.15 Canonical heat capacity peaks associated with the structural transitions from the global minimum to icosahedral structures for the LJ clusters with non-icosahedral global minima. The sizes are as labelled.

larger clusters, lie below the 'glass transition' temperature where the cluster is effectively trapped in the current local minimum. This presents a nightmare scenario for simulated annealing. On cooling the cluster would first enter the icosahedral funnel, where it would then become trapped, even when the global minimum becomes thermodynamically more stable, because of the large free energy barriers (relative to kT) for escape from this funnel [100].

In the protein folding literature, good folders have been shown to have a large value of the ratio of folding temperature to the glass transition temperature, T_f/T_g, because this ensures the kinetic accessibility of the native state of the protein at temperatures where it is thermodynamically most favoured [75]. By contrast, these clusters have effective T_f/T_g values less than one and are archetypal 'bad folders'.

4.2 DYNAMICS

It is impractical to examine the interfunnel dynamics by standard molecular dynamics simulations because of the extremely long time scales involved. However, it is possible to calculate the rate of interfunnel passage by applying a master equation approach[7] to the large samples of minima and transition states used to construct the disconnectivity graphs [87, 96]. The rate constants for LJ_{38} have been computed and show that the interfunnel dynamics obeys an Arrhenius law (i.e. the rate is proportional to $\exp(-E_a/kT)$, where E_a is the activation energy) well with the activation energies corresponding to the barriers associated with the lowest-energy pathway between the two funnels [87, 96]. For LJ_{38} this gives a value of $43\,\mathrm{s}^{-1}$ for the interfunnel rate constant at the centre of the fcc to icosahedral transition (using parameters appropriate

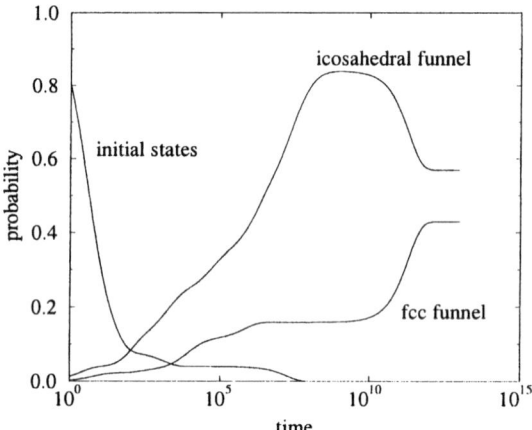

Figure 1.16   Relaxation of LJ_{38} from high-energy minima showing the fast and slow contributions to the final probability of the fcc funnel. The time is in units of $(m\sigma^2/\epsilon)^{1/2}$.

for Ar). As exected, this is beyond the time scales accessible to molecular dynamics simulations. The equivalent rate constants for the larger clusters are much slower because of the larger activation energies and the lower transition temperatures.

Figure 1.16 illustrates for LJ_{38} the dynamics of relaxation from high-energy states. The initial relaxation is relatively rapid and the majority of the population enters the icosahedral funnel (the approximations in this calculation actually lead to an overestimation of the probability of initially entering the fcc funnel). These processes are separated by a couple of decades in time from the subsequent equilibration between the two funnels.

The combined effects of the thermodynamics and dynamics can be illustrated by some simulated annealing results. For LJ_{55} the probability of reaching the global minimum in annealing simulations of 10^6 and 10^7 MC cycles is 29% and 94%. The equivalent values for LJ_{38} are 0% and 2%, and for LJ_{75} the annealing simulations were never able to locate the global minimum.

4.3   OPTIMIZATION SOLUTIONS

The previous sections illustrate the difficulty of finding the global minimum of the LJ clusters with non-icosahedral global minima if the natural thermodynamics and dynamics of the system are followed. However, optimization approaches do not have to be restricted to this behaviour. For example, as we mentioned in Section 3, the basin-hopping transformation accelerates the dynamics, allowing hops directly between basins. However, the transformation only reduces the interfunnel energy barriers by 0.68ϵ for LJ_{38}, 0.86ϵ for LJ_{75} and 0.89ϵ for LJ_{102}. Therefore, multiple-funnels are still potentially problematic.

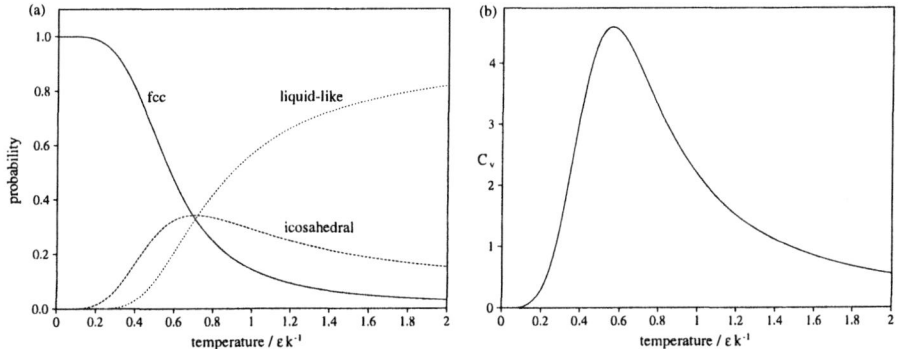

Figure 1.17 Equilibrium thermodynamic properties of LJ$_{38}$ on the transformed PES in the canonical ensemble. (a) The probability of the cluster being in the fcc, icosahedral and liquid-like regions of configuration space. (b) The configurational heat capacity, C_v.

Figure 1.17 shows that the thermodynamic properties of LJ$_{38}$ are dramatically changed by the transformation. The transitions have been significantly broadened. There is now only a single heat capacity peak and a broad temperature range where all states are populated. In particular, the global minimum now has a significant probability of occupation at temperatures where the free energy barriers between the funnels can be surmounted. This effect is illustrated by the basin-hopping simulations in Figure 1.18. At low temperature the system is localized in one of the funnels with transitions between the two funnels only occurring rarely. However, at higher temperatures transitions occur much more frequently [64]. As we noted earlier the rate of interfunnel passage is proportional to $\exp(-E_a/kT)$. Although the transformation only reduces E_a by a small amount, the temperature for which the occupation probability for the global minimum still has a significant value is now over ten times larger. Hence, the increased interfunnel rates.

The transformation has a second kinetic effect: it increases the width of the funnel of the global minimum. On the original LJ$_{38}$ PES only 2% of the long annealing runs entered the icosahedral funnel, whereas 13% of Leary's downhill basin-hopping runs ended at the global minimum (Figure 1.11). On relaxation down the energy landscape the system is much more likely to enter the fcc funnel on the transformed PES, thus making global optimization easier.

These changes to the thermodynamics and dynamics also reduce the difficulty of global optimization for the larger non-icosahedral clusters, making it possible, if still very difficult, to reach the global minimum. The results of Leary indicate that for these clusters the increased accessibility of the funnel of the global minimum is the more important effect of the transformation. He found that it is more efficient to restart a run when it gets stuck in a funnel (and hope that it enters the funnel of the global minimum next time) than to wait for the cluster to escape from that funnel [77].

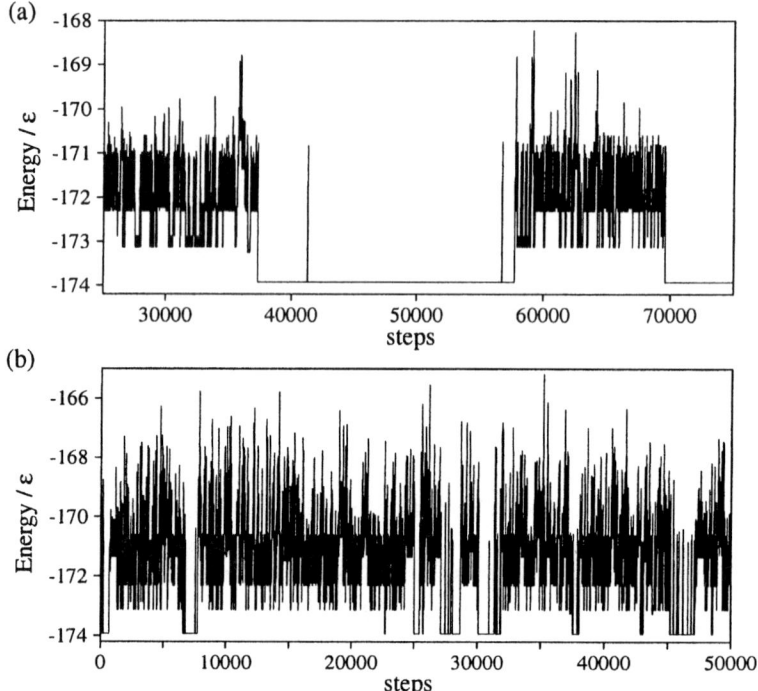

Figure 1.18 $\tilde{E}$ as a function of the number of steps in basin-hopping runs for LJ$_{38}$ at (a) $T=0.4\epsilon k^{-1}$ and (b) $T=0.9\epsilon k^{-1}$.

We can understand why the PES transformation so dramatically changes the thermodynamics of the system, by examining p_i, the occupation probability of a minimum i. For the untransformed PES within the harmonic approximation $p_i \propto \exp(-\beta E_i)/\overline{\nu_i}^{3N-6}$, where E_i is the potential energy of minimum i and $\overline{\nu_i}$ is the geometric mean vibrational frequency. For the transformed PES $p_i \propto A_i \exp(-\beta E_i)$, where A_i is the hyperarea of the basin of attraction of minimum i. The differences between these expressions, the vibrational frequency and hyperarea terms, have opposite effects on the thermodynamics. The higher-energy minima are generally less rigid and so the vibrational term entropically stabilizes the icosahedra and, even more so, the liquid-like state, pushing the transitions down to lower temperature and sharpening them. By contrast the hyperarea of the minima decreases with increasing potential energy, thus stabilizing the lower-energy states and broadening the thermodynamics.

5.   CONCLUSIONS

In this chapter I hope to have shown how physical insight can play an important role in understanding the behaviour of global optimization algorithms and hope that these insights can help provide a firmer physical (rather than just empirical or intuitive) foundation for the design of new improved algorithms.

I also hope that it will encourage some to think about the physical aspects of other types of GO problems.

In particular I have highlighted some examples where a multiple-funnel energy landscape strongly hinders global optimization. Such challenging cases are likely to be a common feature for any cluster system where there is not a single strongly dominant structural type. More generally, multiple funnels probably represent the most difficult problem for this class of GO problems, in which the lowest-energy configuration of a system is sought. For example, for this reason, the main criterion in the design of polypeptides that fold well is the avoidance of multiple funnels by optimizing the energy gap between the native state and competing low-energy structures [101].

I have also shown why the basin-hopping transformation of the energy landscape makes global optimization for these multiple-funnel cases easier. The method's success results from a broadening of the thermodynamics, so that the occupation probability of the global minimum is significant at temperatures where the interfunnel free energy barriers can be surmounted. This idea should act as a design principle in the development of any new GO algorithms that hope to overcome the challenge of multiple funnels.

Although basin-hopping is successful for the LJ multiple-funnel examples, further improvements in efficiency are required before one can hope to succeed for similar cases at larger size or for potential energy functions that are significantly more computationally intensive to evaluate. A number of avenues by which this may be achieved suggest themselves. Firstly, the basin-hopping algorithm searches the transformed potential energy surface using simple MC. However, there are a whole raft of methods that have been developed in order to speed up the rate of rare events in simulations on the untransformed PES, which could potentially be applied to the transformed PES. These include parallel tempering [102], jump-walking [103], and the use of non-Boltzmann ensembles, such as Tsallis statistics [7].

Secondly, the basin-hopping transformation can be combined with other PES transformations, of which there have been many suggestions [104], to get a double transformation (and hopefully a greater simplification) of the PES [105]. The potential problem with this type of approach is that as well as smoothing the PES, most transformations also change the relative stabilities of different structures. This can work in one's favour if the transformation stabilizes the global minimum. For example, a recent transformation proposed by Locatelli and Schoen, which favours compact spherical clusters, stabilizes the non-icosahedral LJ global minima [83]; for the 38-atom cluster the PES, when sufficiently deformed, has a single-funnel topography with the truncated octahedron at its bottom [105]. But just as often a transformation will destabilize the global minimum—this is why the performance of many PES-transformation GO methods is erratic, perhaps solving some 'hard' instances while failing for 'easier' examples. However, the proposed approach could be run alongside standard basin-hopping runs, and so would perhaps succeed in instances where the standard algorithm struggles.

Thirdly, Hartke recently proposed a modification to the genetic algorithm approach, in which a diversity of structures is maintained in the population [58]. This leads to significant increases in efficiency for the LJ clusters with multiple-funnels, because it prevents the whole population being confined (and trapped) within the icosahedral funnel. Such an approach could also potentially increase the efficiency of other algorithms. For example, a diversity of structures could be maintained between a set of parallel basin-hopping runs.

Acknowledgments

The author is a Royal Society University Research Fellow. I would like to acknowledge the role played by David Wales and Mark Miller in this research program and to thank Bob Leary for providing the data for Figure 1.11.

Notes

1. As I cannot prove the optimality of the lowest-energy minima that I find, I should refer to the lowest known structures only as *putative* global minima, but for convenience I usually drop this adjective.

2. The fcc $\{111\}$ planes corresponds to the close-packed planes, where the atoms in the planes form a grid of equilateral triangles. The atoms in the $\{100\}$ planes form a square grid.

3. I use the term *close-packed* to refer to any structure where all the interior atoms of the cluster have a face-centred-cubic or hexagonal close-packed coordination shell. This definition allows for any stacking sequence of close-packed planes, but does not admit any configuration of twin planes that must involve strain.

4. The forces are zero except when the system reaches the edge of a basin of attraction at which point the cluster receives an impulse. Such force fields can be handled in discontinuous molecular dynamics [63]), a method that has often been used in dynamical simulations of systems of hard bodies.

5. A copy of Levinthal's difficult-to-locate citation classic (Ref. [72]) can now be found on the web at http://www-wales.ch.cam.ac.uk/~mark/levinthal/levinthal.html.

6. The thermodynamic properties illustrated in Figures 1.14 and 1.15 have been calculated using a method where the thermodynamic properties of the individual minima are summed [96, 106]. However, recent simulations using parallel tempering indicate that for LJ_{38} the two transitions are slightly closer than indicated by Figure 1.14 and so the fcc to icosahedral transition results only in a shoulder in the heat capacity curve [84].

7. In the master equation approach the occupation probabilities of all the minima can be followed as a function of time, given a set of rate constants between adjacent minima. These rate constants can be approximated using standard rate theories [107].

References

[1] R. Horst and P. M. Pardalos, *Handbook of Global Optimization* (Kluwer Academic, Dordrect, 1995).

[2] D. Frenkel and B. Smit, *Understanding Molecular Simulation* (Academic Press, San Diego, 1996).

[3] S. Kirkpatrick, C. D. Gelatt and M. P. Vecchi, *Optimization by simulated annealing*, Science **220**, 671 (1983).

[4] C. Tsallis, *Possible generalization of Boltzmann-Gibbs statistics*, J. Stat. Phys. **52**, 479 (1998).

[5] C. Tsallis, *Nonextensive statistics: Theoreitical, experimental and computational evidences and connections*, Braz. J. Phys. **29**, 1 (1999).

[6] C. Tsallis and D. A. Stariolo, *Generalized simulated annealing*, Physica A **233**, 395 (1996).

[7] I. Andricoaei and J. E. Straub, *Generalized simulated annealing algorithms using Tsallis statistics: Application to conformational optimization of a tetrapeptide*, Phys. Rev. E **53**, R3055 (1996).

[8] D. E. Goldberg, *Genetic Algorithms in Search, Optimization, and Machine Learning* (Addison-Wesley, Reading, 1989).

[9] B. Hartke, *Efficient global geometry optimization of atomic and molecular clusters*, in *Global Optimization—Selected Case Studies*, edited by J. D. Pinter (Kluwer Academic, Dordrecht, 2005).

[10] D. J. Wales and J. P. K. Doye, *Global optimization by basin-hopping and the lowest energy structures of Lennard-Jones clusters containing up to 110 atoms*, J. Phys. Chem. A **101**, 5111 (1997).

[11] I. A. Harris, R. S. Kidwell and J. A. Northby, *Structure of charged argon clusters formed in a free jet expansion*, Phys. Rev. Lett. **53**, 2390 (1984).

[12] T. P. Martin, *Shells of atoms*, Phys. Rep. **273**, 199 (1996).

[13] E. K. Parks, G. C. Niemann, K. P. Kerns and S. J. Riley, *Reactions of Ni_{38} with N_2, H_2 and CO: Cluster structure and adsorbate binding sites*, J. Chem. Phys. **107**, 1861 (1997).

[14] M. M. Alvarez, J. T. Khoury, T. G. Schaaff, M. Shafigullin, I. Vezmar and R. L. Whetten, *Critical sizes in the growth of Au clusters*, Chem. Phys. Lett. **266**, 91 (1997).

[15] C. L. Cleveland, W. D. Luedtke and U. Landman, *Melting of gold clusters: Icosahedral precursors*, Phys. Rev. Lett. **81**, 2036 (1998).

[16] J. P. K. Doye, D. J. Wales and R. S. Berry, *The effect of the range of the potential on the structures of clusters*, J. Chem. Phys. **103**, 4234 (1995).

[17] L. D. Marks, *Surface-structure and energetics of multiply twinned particles*, Philos. Mag. A **49**, 81 (1984).

[18] A. L. Mackay, *A dense non-crystallographic packing of equal spheres*, Acta Crystallogr. **15**, 916 (1962).

[19] J. P. K. Doye and D. J. Wales, *Magic numbers and growth sequences of small face-centred-cubic and decahedral clusters*, Chem. Phys. Lett. **247**, 339 (1995).

[20] B. Raoult, J. Farges, M.-F. de Feraudy and G. Torchet, *Comparison between icosahedral, decahedral and crystalline Lennard-Jones models containing 500 to 6000 atoms*, Philos. Mag. B **60**, 881 (1989).

[21] R. H. Leary and J. P. K. Doye, *Tetrahedral global minimum for the 98-atom Lennard-Jones cluster*, Phys. Rev. E **60**, R6320 (1999).

[22] W. Branz, N. Malinowski, H. Schaber and T. P. Martin, *Thermally induced structural transitions in $(C_{60})_n$ clusters*, Chem. Phys. Lett. **328**, 245 (2000).

[23] J. E. Jones and A. E. Ingham, *On the calculation of certain crystal potential constants, and on the cubic crystal of least potential energy*, Proc. R. Soc. London, Ser. A **107**, 636 (1925).

[24] L. T. Wille, *Lennard-Jones clusters and the multiple-minima problem*, in *Annual Reviews of Computational Physics VII*, edited by D. Stauffer (World Scientific, Singapore, 2000).

[25] J. A. Northby, *Structure and bonding of Lennard-Jones clusters:* $13 \leq N \leq 147$, J. Chem. Phys. **87**, 6166 (1987).

[26] S. Gomez and D. Romero, *Two global methods for molecular geometry optimization*, in *Proceedings of the First European Congress of Mathematics*, volume III, pp. 503–509 (Birkhauser, Basel, 1994).

[27] J. Pillardy and L. Piela, *Molecular-dynamics on deformed potential-energy hypersurfaces*, J. Phys. Chem. **99**, 11805 (1995).

[28] D. Romero, C. Barrón and S. Gómez, *The optimal geometry of Lennard-Jones clusters: 148-309*, Comp. Phys. Comm. **123**, 87 (1999).

[29] Y. Xiang, H. Jiang, W. Cai and X. Shao, *An efficient method based on lattice construction and the genetic algorithm for optimization of large Lennard-Jones clusters*, J. Phys. Chem. A **108**, 3586 (2004).

[30] Y. Xiang, L. Cheng, W. Cai and X. Shao, *Structural distribution of Lennard-Jones clusters containing 562 to 1000 atoms*, J. Phys. Chem. A, in press (2004).

[31] P. M. Morse, *Diatomic molecules according to the wave mechanics. II. Vibrational levels*, Phys. Rev. **34**, 57 (1929).

[32] D. J. Wales, L. J. Munro and J. P. K. Doye, *What can calculations employing empirical potentials teach us about bare transition metal clusters?*, J. Chem. Soc., Dalton Trans. p. 611 (1996).

[33] L. A. Girifalco, *Molecular-properties of C_{60} in the gas and solid-phases*, J. Phys. Chem. **96**, 858 (1992).

[34] D. J. Wales and J. Uppenbrink, *Rearrangements in model face-centred-cubic solids*, Phys. Rev. B **50**, 12342 (1994).

[35] L. A. Girifalco and V. G. Weizer, *Application of the Morse potential function to cubic metals*, Phys. Rev. **114**, 687 (1959).

[36] J. P. K. Doye and D. J. Wales, *Structural consequences of the range of the interatomic potential: A menagerie of clusters*, J. Chem. Soc., Faraday Trans. **93**, 4233 (1997).

[37] J. P. K. Doye, R. H. Leary, M. Locatelli and F. Scoen, *The global optimization of Morse clusters by potential energy transformations*, IN-FORMS J. Comput. **16**, in press (2004).

[38] T. P. Martin, T. Bergmann, H. Göhlich and T. Lange, *Observation of electronic shells and shells of atoms in large Na clusters*, Chem. Phys. Lett. **172**, 209 (1990).

[39] D. R. Nelson and F. Spaepen, *Polytetrahedral order in condensed matter*, Solid State Phys. **42**, 1 (1989).

[40] F. C. Frank and J. S. Kasper, *Complex alloy structures regarded as sphere packings. I. Definitions and basic principles.*, Acta Crystallogr. **11**, 184 (1958).

[41] F. C. Frank and J. S. Kasper, *Complex alloy structures regarded as sphere packings. II. Analysis and classification of representative structures*, Acta Crystallogr. **12**, 483 (1959).

[42] L. C. Cune and M. Apostol, *Ground-state energy and geometric magic numbers for homo-atomic metallic clusters*, Phys. Lett. A **273**, 117 (2000).

[43] J. P. K. Doye, *A model metal potential exhibiting polytetrahedral clusters*, J. Chem. Phys. **119**, 1136 (2003).

[44] F. Dassenoy, M.-J. Casanove, P. Lecante, M. Verelst, E. Snoeck, A. Mosset, T. Ould Ely, C. Amiens and B. Chaudret, *Experimental evidence of structural evolution in ultrafine cobalt particles stabilized in different polymers—From a polytetrahedral arrangement to the hexagonal structure*, J. Chem. Phys. **112**, 8137 (2000).

[45] M. Dzugutov and U. Dahlborg, *Molecular-dynamics study of the coherent density correlation-function in a supercooled simple one-component liquid*, J. Non-Cryst. Solids **131-133**, 62 (1991).

[46] M. Dzugutov, *Monatomic model of icosahedrally ordered metallic glass formers*, J. Non-Cryst. Solids **156-158**, 173 (1993).

[47] D. G. Pettifor, *Bonding and Structure of Molecules and Solids* (Clarendon Press, Oxford, 1995).

[48] M. Dzugutov, *Glass-formation in a simple monatomic liquid with icosahedral inherent local order*, Phys. Rev. A **46**, R2984 (1992).

[49] M. Dzugutov, *Formation of a dodecagonal quasicrystalline phase in a simple monatomic liquid*, Phys. Rev. Lett. **70**, 2924 (1993).

[50] J. P. K. Doye, D. J. Wales and S. I. Simdyankin, *Global optimization and the energy landscapes of Dzugutov clusters*, Faraday Discuss. **118**, 159 (2001).

[51] J. P. K. Doye and D. J. Wales, *Polytetrahedral clusters*, Phys. Rev. Lett. **86**, 5719 (2001).

[52] J. P. K. Doye and D. J. Wales, *Thermodynamics of global optimization*, Phys. Rev. Lett. **80**, 1357 (1998).

[53] J. P. K. Doye and F. Calvo, *Entropic effects on the size dependence of cluster structure*, Phys. Rev. Lett. **86**, 3570 (2001).

[54] F. Baletto, C. Mottet and R. Ferrando, *Reentrant morphology transition in the growth of free silver clusters*, Phys. Rev. Lett. **84**, 5544 (2000).

[55] F. Baletto, J. P. K. Doye and R. Ferrando, *Evidence of kinetic trapping in clusters of C_{60} molecules*, Phys. Rev. Lett. **88**, 075503 (2002).

[56] C. D. Maranas and C. A. Floudas, *A global optimization approach for Lennard-Jones microclusters*, J. Chem. Phys. **97**, 7667 (1992).

[57] D. M. Deaven, N. Tit, J. R. Morris and K. M. Ho, *Structural optimization of Lennard-Jones clusters by a genetic algorithm*, Chem. Phys. Lett. **256**, 195 (1996).

[58] B. Hartke, *Global cluster geometry optimization by a phenotype algorithm with niches: Location of elusive minima, and low-order scaling with cluster size*, J. Comp. Chem. **20**, 1752 (1999).

[59] M. D. Wolf and U. Landman, *Genetic algorithms for structural cluster optimization*, J. Phys. Chem. A **102**, 6129 (1998).

[60] Z. Li and H. A. Scheraga, *Monte-Carlo-minimization approach to the multiple-minima problem in protein folding*, Proc. Natl. Acad. Sci. USA **84**, 6611 (1987).

[61] G. L. Xue, *Molecular conformation on the CM-5 by parallel two-level simulated annealing*, J. Global Optim. **4**, 187 (1994).

[62] D. M. Deaven and K. M. Ho, *Molecular-geometry optimization with a genetic algorithm*, Phys. Rev. Lett. **75**, 288 (1995).

[63] B. J. Alder and T. E. Wainwright, *Studies in molecular dynamics. I. General methods*, J. Chem. Phys. **31**, 459 (1959).

[64] J. P. K. Doye, D. J. Wales and M. A. Miller, *Thermodynamics and the global optimization of Lennard-Jones clusters*, J. Chem. Phys. **109**, 8143 (1998).

[65] D. Liu and J. Nocedal, *On the limited memory BFGS method for large scale optimization*, Mathematical Programming B **45**, 503 (1989).

[66] R. P. White and H. R. Mayne, *An investigation of two approaches to basin hopping minimization for atomic and molecular clusters*, Chem. Phys. Lett. **289**, 463 (1998).

[67] P. Derreumaux, *Ab initio polypeptide structure prediction*, Theor. Chem. Acc. **104**, 1 (2000).

[68] I. Rata, A. A. Shvartsburg, M. Horoi, T. Frauenheim, K. W. M. Siu and K. A. Jackson, *Single-parent evolution algorithm and the optimization of Si clusters*, Phys. Rev. Lett. **85**, 546 (2000).

[69] D. J. Wales and H. A. Scheraga, *Global optimization of clusters, crystals and biomolecules*, Science **285**, 1368 (1999).

[70] F. H. Stillinger, *Exponential multiplicity of inherent structures*, Phys. Rev. E **59**, 48 (1999).

[71] C. J. Tsai and K. D. Jordan, *Use of an eigenmode method to locate the stationary points on the potential energy surfaces of selected argon and water clusters*, J. Phys. Chem. **97**, 11227 (1993).

[72] C. Levinthal, *How to fold graciously*, in *Mössbauer Spectroscopy in Biological Systems, Proceedings of a Meeting Held at Allerton House, Monticello, Illinois*, edited by J. T. P. DeBrunner and E. Munck, pp. 22–24 (University of Illinois Press, Illinois, 1969).

[73] R. Zwanzig, A. Szabo and B. Bagchi, *Levinthal's paradox*, Proc. Natl. Acad. Sci. USA **89**, 20 (1992).

[74] R. Zwanzig, *Simple model of protein folding kinetics*, Proc. Natl. Acad. Sci. USA **92**, 9801 (1995).

[75] J. D. Bryngelson, J. N. Onuchic, N. D. Socci and P. G. Wolynes, *Funnels, pathways, and the energy landscape of protein folding: A synthesis*, Proteins **21**, 167 (1995).

[76] J. P. K. Doye and D. J. Wales, *On potential energy surfaces and relaxation to the global minimum*, J. Chem. Phys. **105**, 8428 (1996).

[77] R. H. Leary, *Global optimization on funneling landscapes*, J. Global Optim. **18**, 367 (2000).

[78] J. A. Niesse and H. R. Mayne, *Global geometry optimization of atomic clusters using a modified genetic algorithm in space-fixed coordinates*, J. Chem. Phys. **105**, 4700 (1996).

[79] K. Michaelian, *A symbiotic algorithm for finding the lowest energy isomers of large clusters and molecules*, Chem. Phys. Lett. **293**, 202 (1998).

[80] R. V. Pappu, R. K. Hart and J. W. Ponder, *Analysis and application of potential energy smoothing and search methods for global optimization*, J. Phys. Chem. B **102**, 9725 (1998).

[81] J. Pillardy, A. Liwo and H. A. Scheraga, *An efficient deformation-based global optimization method (self-consistent basin-to-deformed basin mapping). Application to Lennard-Jones atomic clusters*, J. Phys. Chem. A **103**, 9370 (1999).

[82] D. B. Faken, A. F. Voter, D. L. Freeman and J. D. Doll, *Dimensional strategies and the minimization problem: Barrier avoiding algorithm*, J. Phys. Chem. A **103**, 9521 (1999).

[83] M. Locatelli and F. Schoen, *Fast global optimization of difficult Lennard-Jones clusters*, Comput. Optim. and Appl. **21**, 55 (2001).

[84] J. P. Neirotti, F. Calvo, D. L. Freeman and J. D. Doll, *Phase changes in 38 atom Lennard-Jones clusters.I: A parallel tempering study in the canonical ensemble*, J. Chem. Phys. **112**, 10340 (2000).

[85] J. P. K. Doye and D. J. Wales, *The effect of the range of the potential on the structure and stability of simple liquids: from clusters to bulk, from sodium to C_{60}*, J. Phys. B **29**, 4859 (1996).

[86] M. A. Miller, J. P. K. Doye and D. J. Wales, *Structural relaxation in Morse clusters: Energy landscapes*, J. Chem. Phys. **110**, 328 (1999).

[87] M. A. Miller, J. P. K. Doye and D. J. Wales, *Structural relaxation in atomic clusters: Master equation dynamics*, Phys. Rev. E **60**, 3701 (1999).

[88] C. Roberts, R. L. Johnston and N. T. Wilson, *A genetic algorithm for the structural optimization of Morse clusters*, Theor. Chem. Acc. **104**, 123 (2000).

[89] H. Xu and B. J. Berne, *Multicanonical jump-walking annealing: An efficient method for geometric optimization*, J. Chem. Phys. **112**, 2701 (2000).

[90] J. P. K. Doye, *The network topology of a potential energy landscape: A static scale-free network*, Phys. Rev. Lett. **88**, 238701 (2002).

[91] L. T. Wille and J. Vennik, *Computational-complexity of the ground-state determination of atomic clusters*, J. Phys. A **18**, L419 (1985).

[92] P. E. Leopold, M. Montal and J. N. Onuchic, *Protein folding funnels: A kinetic approach to the sequence structure relationship*, Proc. Natl. Acad. Sci. USA **89**, 8271 (1992).

[93] O. M. Becker and M. Karplus, *The topology of multidimensional potential energy surfaces: Theory and application to peptide structure and kinetics*, J. Chem. Phys. **106**, 1495 (1997).

[94] Y. Levy and O. M. Becker, *Effect of conformational constraints on the topography of complex potential energy surfaces*, Phys. Rev. Lett. **81**, 1126 (1998).

[95] M. A. Miller and D. J. Wales, *Energy landscape of a model protein*, J. Chem. Phys. **111**, 6610 (1999).

[96] D. J. Wales, J. P. K. Doye, M. A. Miller, P. N. Mortenson and T. R. Walsh, *Energy landscapes of clusters, biomolecules and solids*, Adv. Chem. Phys. **115**, 1 (2000).

[97] D. J. Wales, M. A. Miller and T. R. Walsh, *Archetypal energy landscapes*, Nature **394**, 758 (1998).

[98] J. P. K. Doye, M. A. Miller and D. J. Wales, *Evolution of the potential energy surface with size for Lennard-Jones clusters*, J. Chem. Phys. **111**, 8417 (1999).

[99] P. Labastie and R. L. Whetten, *Statistical thermodynamics of the cluster solid-liquid transition*, Phys. Rev. Lett. **65**, 1567 (1990).

[100] J. P. K. Doye, M. A. Miller and D. J. Wales, *The double-funnel energy landscape of the 38-atom Lennard-Jones cluster*, J. Chem. Phys. **110**, 6896 (1999).

[101] R. Goldstein, Z. Luthey-Schulten and P. G. Wolynes, *Optimal protein-folding codes from spin-glass theory*, Proc. Natl. Acad. Sci. USA **89**, 4918 (1992).

[102] E. Marinari and G. Parisi, *Simulated tempering: A new Monte-Carlo scheme*, Europhys. Lett. **19**, 451 (1992).

[103] D. D. Frantz, D. L. Freeman and J. D. Doll, *Reducing quasi-ergodic behaviour in Monte Carlo simulations by J-walking: Applications to atomic clusters*, J. Chem. Phys. **93**, 2769 (1990).

[104] S. Schelstrate, W. Schepens and H. Verschelde, *Energy minimization by smoothing techniques: a survey*, in *Molecular Dynamics: From Classical to Quantum Mechanics*, edited by P. B. Balbuena and J. M. Seminario, pp. 129–185 (Elsevier, Amsterdam, 1999).

[105] J. P. K. Doye, *The effect of compression on the global optimization of atomic clusters*, Phys. Rev. E **62**, 8753 (2000).

[106] J. P. K. Doye and D. J. Wales, *Calculation of thermodynamic properties of small Lennard-Jones clusters incorporating anharmonicity*, J. Chem. Phys. **102**, 9659 (1995).

[107] W. Forst, *Unimolecular Reactions* (Cambridge University Press, Cambridge, 2003).

Chapter 6

EFFICIENT GLOBAL GEOMETRY OPTIMIZATION OF ATOMIC AND MOLECULAR CLUSTERS

Bernd Hartke

Institut für Physikalische Chemie
Christian-Albrechts-Universität
Olshausenstraße 40
24098 Kiel
GERMANY

hartke@phc.uni-kiel.de

Abstract   After a short survey of the general chemical context of the global cluster geometry optimization problem, several of its aspects are discussed, touching upon computational complexity, links to ab-initio calculations and experiment, benchmark systems, and some of the solution methods applied so far. Our current method, a variant of the standard Genetic Algorithms, is presented, discussing several aspects crucial for its efficiency. Applications of this method to both benchmark and real-life examples of atomic and molecular clusters are shown, including Lennard-Jones clusters and pure and mixed water clusters.

Keywords: global cluster geometry optimization, random search, genetic algorithms

1.   Introduction

1.1   Computational Complexity in Theoretical Chemistry

The complexity of many-body systems is one of the main challenges in modern science. Due to the interaction between the constituent particles, the computational effort has to increase exponentially with system size in complete and exact simulations. Even if one is not interested in a dynamical description of processes but settles for a search for the optimal configuration (of lowest energy) of such a system, arriving at a global optimization task, the same basic problem prevails: The search space typically increases exponentially with system size, and an exact treatment of systems with a non-trivial number of particles seems to be beyond reach. This problem is ubiquitous also in theoretical chemistry and molecular physics: The many-electron problem of quantum chemistry scales exponentially in a fully correlated, exact treatment (full configuration interaction). Drastic approximations had to be introduced to make this problem tractable at all for any system of chemical interest: For example, the famous Hartree-Fock self-consistent-field method ignores electron correlation completely (keeping only Fermi correlation due to electron spin) [1]. Only recently it was recognized that a local treatment of electron correlation [2, 3] could bring down the impracticable N^4 to N^7 scaling of usual approximate correlation treatments to an ideal linear scaling. As their name implies, all these electronic structure methods solve the many-electron problem as a static problem, and one might expect the presence of multiple minima – however, although this is sometimes acknowledged, it does not appear to be a cause for concern in usual computational practice in this field.

In the usual Born-Oppenheimer approximation, after solving the electronic structure problem (and hence having the forces between atoms and molecules at their disposal), theoretical chemists tackle the remaining half of their field: the structure and dynamics of atomic and molecular arrangements, including simulations of actual chemical reactions. A proper, fully quantum mechanical approach to the latter is hit by the curse of dimensionality very soon: Only systems involving three to four atoms can be treated exactly even with the best methods available today [4–6]. If one largely ignores the quantum aspect of the problem and focuses on structural questions, several still unsolved many-body problems feature prominently in theoretical chemical physics, constituting true challenges for global optimization methods:

In biochemistry, we have the problem of fitting small molecules to larger molecular receptors (molecular docking) and the question of how proteins (and similar polymeric macromolecules) fold after their natural synthesis (or after denaturation in vitro) into their biologically active, native three-dimensional form. The native form is apparently given solely by the sequence of amino acids (the primary structure). Nevertheless, for a given primary structure, a complete search through configuration space for the native form can be ruled out, as simulation procedure (even with the usual restriction of keeping bond distances and bond angles fixed and varying only dihedral angles, absence of water solvation, etc.) and as hypothesis for the actual natural process: In both cases this would simply take much too long (Levinthal's paradoxon) [7–9].

Any bound arrangement of several atoms and/or molecules may be called a "cluster". Very many of these arrangements are well known to organic and inorganic chemists, as documented in numerous textbooks. They are usually called "molecules", and chemists have developed well working sets of rules and practical intuitions for their structure, energetics, and reactions, and this is what those textbooks are all about. In recent decades, however, several experimental techniques made it possible to synthesize and study other arrangements that fall outside the usual realm of traditional textbook chemistry; the discovery of carbon clusters of the "fullerene" type is only one popular example [10,11]. For these clusters, there are usually no rules or intuitions to predict their structure; all we know is that the structure most likely to be found experimentally will be that of lowest potential energy (with some reservations, see below). Since the overall potential energy depends in a complicated and non-convex fashion on the coordinates of the particles in the cluster, prediction of cluster structure again constitutes a complicated global optimization problem.

All of these problems are related: In all of them, one tries to optimally arrange a given set of constituents, which have an internal structure that is flexible to a certain degree (but which is often modeled as rigid), in three-dimensional space according to a given rule (the potential function) that tells us how to evaluate the cost function (the potential energy of the whole system) for each arrangement of the constituents. Therefore, it is tempting to postulate the existence of an optimal solution strategy for all of these problems. However, the author believes that efficient solution methods will have to be tailored to specific features of each problem, since there are also major structural differences between them: For example, in the case of protein folding, the constituents are always forced to be linked to each other in a given sequence (the primary structure); this is not the case in the cluster geometry optimization problem.

Similarly, it seems that problems of this kind are intimately related to three-dimensional tiling. In fact, some results of abstract 3D tiling are currently making their way into chemistry [12]. However, these results typically concern only periodic tilings with rigid tiles, while here we have flexible tiles and no periodicity (and even surface effects instead). This is a decisive difference, as exemplified also by the status of the (again, related) problem of optimal hard-sphere packing, which is solved for infinite packings (in 3D, but also in many other dimensions) but unsolved for finite ones [13].

1.2 Cluster Geometry Optimization

Experimental and theoretical cluster research is a rapidly growing field (for reviews see Refs. [14–20]), fueled by the realization that clusters play central roles in such diverse areas as chemical vapor deposition [21], aerosol chemistry and earth climate research [22–24] and nanotechnology [25–28]. However, prior to applications like these, some basic issues have to be addressed: Of primary interest is the role of clusters as intermediaries between single atoms or molecules and the infinite solid. It turned out very soon that clusters are typically not simply small pieces of solid, neither in their structure nor in their physical or chemical properties. Instead, with the number n of cluster constituents growing from $n = 1$ to "$n = \infty$", one usually observes several structural transitions [29] and property changes. The onset of the solid state structure and properties occurs at different values of n for different systems, and even at different values of n for the same system if one looks at different properties. Also, particular values of n, so-called "magic numbers" (their values again depending on the system under study) often show enhanced stability or a particularly low reactivity, which is usually explained as a shell closure, following some particular structural principle. None of these basic issues has been properly resolved or understood as yet.

Again, questions for cluster properties and reactivities cannot be addressed before solving the problem of cluster structure. As in the case of many other global optimization problems, it has been shown that finding the structure of an atomic cluster with globally minimal energy is an NP-hard task [30–32]. These proofs do not rely upon particular forms of the potential energy function, but argue independently of it. In fact, even related problems without the presence of a potential energy function have been shown to be NP-hard, for example finding the three-dimensional arrangement of points given their pairwise distances [33]. Thus it may seem that there is no hope for solving the global geometry optimization problem of clusters for any values of n relevant to physical or chemical applications. However, complexity proofs are typically worst-case scenarios [32], and one may arrive at solutions for NP-complete problems in polynomial time if one looks only at the non-nasty cases [34] (which may not be discernible from the difficult ones a priori) or if one imposes additional conditions [35]. It is well known that the potential energy function has a strong influence on number and nature of the local minima in the cluster geometry optimization problem (see below). One may hope that physically relevant potential energy functions might be a simplifying additional condition. However, mixed experience with potential transformation methods indicates that this is presumably not the case in practice.

If one is willing to sacrifice exactness or the guarantee for finding the global minimum, better size scaling and shorter computer times are possible: Finding the optimal tour for the famous traveling salesman for about 7000 cities took many hours on 60 workstations in parallel, but an approximate tour only 30% longer than that can be found in only 2 seconds on a single personal computer. In fact, is has been shown that approximate tours can be found in $O(n \log(n))$ time [36,37]. In the case of cluster geometry optimization, no such general results are available yet, but obviously the configuration space to be covered may be cut down drastically by applying external prior knowledge (using physical or chemical arguments) or information gathered during the present program run.

Ideally, in local and global geometry optimizations of clustes, the forces between the atoms and molecules should be calculated ab-initio, i.e. by using the electronic structure methods mentioned above to solve the time-independent Schrödinger equation for the electrons. However, this is orders of magnitude more expensive than using empirical model potential functions for this purpose.

Unfortunately, number and nature of the local minimum cluster structures depend sensitively on details of the potential. This was pointed out already almost 20 years ago by Hoare and McInnes [38] and was emphasized again recently by Doye and Wales [39]. Therefore, global geometry optimization of clusters on model potentials runs the risk of producing physically meaningless results, while global optimization on the ab-initio level is simply too expensive for systems of practically interesting sizes. In the general global optimization community, to the knowledge of the author, there are no systematic attempts to design global optimization methods for the case of excessively expensive cost functions – presumably because the task is considered impossible from the outset; however, exactly this problem has to be solved here, as described.

Several groups have decided to ignore the presence of this problem and to apply brute-force simulated annealing directly on density functional or ab-initio potentials [40–43], but in most cases some of the intended goals cannot be reached: Either the level of the ab-initio treatment used is not sufficiently accurate, or the reliability in locating global minima remains remains in doubt, or the systems considered are simply too small to make the use of global optimization methods really necessary. In this situation, the author has recently presented a way to circumvent this problem in a non-rigorous fashion [44]: The global cluster geometry optimization is still done on a model potential, now serving as a guiding function. To ensure a close proximity between the model potential and the true ab-initio surface, the model potential is adapted to a growing set of ab-initio single-point calculations, at cluster geometries corresponding to important model potential minima. In a final step, the best resulting geometries are locally optimized on the ab-initio potential. This procedure has been shown to yield the true ab-initio global minima with high probability in actual applications to silicon clusters [45] and water clusters [46]. However, in spite of the practical success of this method, the requirements on the model potential to work successfully as guiding function in such an approach still remain to be investigated.

In comparison to experimental data, further problems arise: The classical-mechanical global minimum of a potential energy surface does not necessarily correspond to the cluster structure most likely to be found experimentally. A famous example for this is the water hexamer: In Saykally's experiments [47], the cage form was found, but most ab-initio calculations agree upon the prism as classical-mechanical global minimum [46,48] and put the cage in second place. Depending on the accuracy level of theory, quantum mechanical zero-point energy corrections alone [49] or in combination with entropy and finite-temperature corrections [48] apparently reverse the theoretical ordering and bring it into accord with experiment. Of course, also the experimental cluster preparation conditions can influence the resulting cluster structure: Under different conditions, Nauta and Miller [50] recently found the 6-ring form of the water hexamer. Therefore, the ultimate theoretical modelling of experiment has to include all these effects, up to a simulation of the preparation process. On a sufficiently accurate level, this is clearly beyond the capabilities of current supercomputers.

It can be argued that the latter difficulties abate somewhat through the observation that classical-mechanical global minimum structures surprisingly often do correspond to experimentally observed structures, for example in silicon clusters [45, 51, 52]. Also, since zero-point energy and finite-temperature corrections are easily applied a posteriori, and since a simulation of cluster preparation as well as a proper global optimization directly on the ab-initio potential are beyond current capabilities anyway, the location of classical-mechanical minima on model potential energy surfaces is a valid research goal at present, offering a staggering degree of complexity in itself. Therefore, the remainder of this chapter will focus solely on this problem.

Model potentials for every conceivable chemical system have been developed en masse during the last decades. Therefore, a review is again not possible here; instead the reader is referred to excellent literature on this topic (for example Ref. [53] and references therein). Several simpler potentials of the early days have survived as benchmark systems, most notably the Lennard-Jones potential. It was originally intended to model systems of rare gas atoms; today we know that it is not even fit enough for this purpose: It does not have the correct form [53] and erroneously neglects many-body effects (in fact, it does not even lead to the correct face-centered cubic (fcc) structure in the infinite solid, but to hexagonal close-packed (hcp) [54]). Recent research has made much better potentials available [55]. Nevertheless, in the present context it is very valuable, since global minimum energies and structures for Lennard-Jones clusters are known with a fair amount of certitude up to $n = 150$, and good proposals have been made up to $n = 309$; also, all these data are freely available on the internet [56, 57]. Similarly, a recent large-scale calculation of Wales and Hodges [58] on water clusters employing the simple TIP4P potential has established another benchmark system for which the global minima are fairly certain up to $n = 21$. In this chapter, calculations for both systems will be shown.

Wille has recently presented an extensive review [32] of global optimization methods applied to the standard benchmark system of Lennard-Jones clusters, and there are also several other reviews with different emphasis and scope (for example Ref. [59]). Therefore, instead of giving another methods review, the author presents a brief, personally biased list of several approaches to the problem of global cluster geometry optimization:

Already the early studies of Hoare et al. [60] pointed out that the number of local minima increases exponentially with size in Lennard-Jones clusters, after a brute-force enumeration of all local minima up to $n = 13$. It was quickly recognized that Lennard-Jones clusters follow an icosahedral growth pattern proposed by Mackay [61]. This was systematically exploited in the landmark paper by Northby [62]: By literally growing larger clusters from smaller ones by adding additional atoms in optimal places on the icosahedral grid and subsequently relaxing the whole clusters he arrived at an impressive list of proposed global minima up to $n = 150$. Only a few of these proposals could be improved upon later by other authors, and disturbingly in most of these cases other structural patterns were involved: decahedral forms at $n = 75 - 77$ and $n = 102 - 104$, and at $n = 38$ even the fcc paradigm that was expected to come into play only at the transition to the infinite solid structure. Doye, Miller and Wales [63] recently managed to locate the lowest-energy member of each of these three classes throughout the range $n = 5 - 80$, revealing that there is always a rather close competition between decahedral and fcc types which is hidden over wide size ranges by a clear dominance of the icosahedral pattern – however, in between shell closures the icosahedral forms tend to loose this dominance, so far that at the few values of n mentioned above one of the other forms wins by a narrow margin.

The presence of qualitatively different packings can be rationalized by combining packing arguments with the softness of the Lennard-Jones potential: In fcc forms, all interparticle distances are the same (this is a true hard-sphere packing) and can attain their minimum value as dictated by the pair potential. In icosahedral clusters, however, the number of nearest neighbors is larger, at the expense of some elongated and some compressed interparticle distances (the decahedrals are in-between the other two forms in these respects). This can be illustrated by five tetrahedra with a common edge resulting in an almost but not quite closed pentagonal bipyramid; similarly, twenty tetrahedra can be assembled roughly to an icosahedron, leaving some wider gaps. In both cases, the gaps can be closed by deviations from the ideal pair distance. Obviously, as long as the pair potential function does not penalize these deviations too much, the icosahedral form will win over fcc (cf. Wales and Doye's study of Morse clusters [39], demonstrating this very nicely).

Of course, Northby's growth strategy based upon the icosahedral pattern alone could not find the decahedral or fcc forms. Therefore, Romero et al. [57] used grids for all three forms in their extended search up to 309 Lennard-Jones particles. Ironically, not even a year later, a new tetrahedral form was discovered as global minimum for 98 Lennard-Jones particles [64], increasing the number of basic spatial grids to be considered in a grid-based search from three to four.

All this simply illustrates our acute lack of actual understanding in the field of global cluster geometry optimization. Although we are able to rationalize some structure–potential relationships a posteriori (as mentioned above), we are unable to predict a complete list of general structural motifs that could or should be present in global cluster minimum structures, for a given interparticle potential — not even for the Lennard-Jones pair potential, which is widely considered as a simplistic toy problem in the physical chemistry community. More realistic potentials include not just terms depending solely on pair distances, but also terms depending on many-particle geometry, up to 4-body or even 5-body terms [53,65,66]. This may complicate the problem even further, and only few authors have attempted to include many-body terms in systematic global cluster geometry studies and to understand their effects [67]. However, the present author is of the opinion that once the Lennard-Jones problem is properly understood in the above sense, many-body terms and similar complications will turn out to be a minor challenge only (in fact, they may even simplify the problem in giving it more structure).

Therefore, approaches that restrict configuration space by a priori assumptions on structures to be expected do give us the speedups we need to tackle larger clusters, but today's a priori information simply is not accurate and complete enough. On the other hand, more exact and complete solution methods have been tried on the global cluster geometry optimization problem, but either they have met mixed success or they scale so badly with cluster size that they cannot be applied to clusters in the experimentally interesting size range. For example, deformation methods (homotopy methods, aiming at transforming the original problem into a related simpler one and then tracking the solution of the simple problem through the back-transformation to the original problem) appear to work pretty well for several sizes of the Lennard-Jones cluster benchmark system [68], but surprisingly fail for certain small and seemingly simple sizes [69,70]. As another example, a DC transformation approach [71] could be applied as such only up to the size $n = 7$ (which is still well within the range studied already by Hoare et al. [60]); an extension up to $n = 24$ was only possible by combining it with a growth strategy, i.e. claiming that the structure of the cluster with n atoms can be obtained from that with $n-1$ atoms by simply adding an atom — this would have failed at $n = 38$ where this hypothesis is not true. Further attempts in this general direction are documented in Wille's review [32].

Given the mixed success of competing methods, it is not too surprising that simple methods based on random search, possibly enhanced by suitable search heuristics, fare comparatively well. One of the most successful of these is "basin–hopping", initially introduced by Li and Scheraga [72] for protein folding. There, local optimizations alternate with purely random search steps; the latter are typically "Monte Carlo" steps, i.e. downhill moves are always accepted while uphill moves are accepted with a probability exponentially depending on the energy difference (Metropolis criterion [73]). Although this acceptance criterion still contains a "temperature" as control parameter, this is kept constant at a suitable value and not changed as in simulated annealing [74]. Some applications of this method are given in Ref. [59]; in the present context, the application by Wales and Doye [75] is important: They managed to find all the global minima known at that time for Lennard-Jones clusters up to $n = 110$, without external a priori knowledge and within reasonable computer time. Published computer times for applications of this method to the Lennard-Jones benchmark system [59,76] are impressive for small clusters, but indicate a size scaling no better than n^5.

Until very recently, the only other approach capable of finding all Lennard-Jones minima in this range with reasonable efficiency is based on the idea of "Genetic Algorithms" (GAs). GAs are search heuristics mimicking some aspects of natural evolution, for textbooks and reviews see Refs. [77–81]. In contrast to basin-hopping, they typically include not only random moves ("mutations") but also recombination of parts of previous solution candidates to form new solution candidates ("crossover"). Thus, conceptually, they contain the implicit assumption that there is at least some degree of separability in the problem, such that there is a good chance that parts taken from two (or several) good solution candidates produce an even better solution candidate upon combination, i.e. that these parts constitute so-called "building blocks" for better solutions and ultimately also for the global optimum. If this is the case, a GA gathers information about the problem during its solution and is able to combine things it has "learned" in disjoint regions of configuration space. Therefore, it can be expected to have an edge over pure random walk strategies like basin-hopping, which can explore only the immediate neighborhood of a given solution candidate, in the case of small steps, or blindly jump into the void, in the case of large steps. However, if there is no separability of this kind, the crossover operation effectively degenerates to a complicated way of doing mutations. The degree of separability hinges upon the way the problem is represented in the GA or (which amounts to the same) upon the way the crossover and mutation operators are implemented. Unfortunately, in practice, in many applications this issue is apparently neglected.

After GAs had been used for the global optimization of dihedral angles in small biomolecules [79,82,83], they were employed first by the present author [84] and then also by other groups [85] for the global optimization of all degrees of freedom in atomic and molecular clusters. In these early papers, an actually "genetic" representation of the optimization problem was used by concatenating the cartesian coordinates of all particles in the cluster, and by operating with standard forms of crossover and mutation on these coordinate "strings". As already pointed out in Ref. [84], this representation does not achieve the best possible separability. Nevertheless, adding in local search steps, we were able to push the size scaling of such a GA down to $n^{4.5}$, for the benchmark system of Lennard-Jones clusters [86], but we had to add in a seed growth method to get beyond $n = 20$ within reasonable computer time. Deaven and Ho [87] managed to improve this standard representation by eliminating the representation issue altogether, i.e. they applied crossover and mutation operators directly on the clusters in physical space (since there are no genetic strings anymore, the present author prefers to call this variant "phenotype algorithm"). This enabled them to treat Lennard-Jones clusters up to $n = 100$ [88], but they arrived at wrong solutions for $n = 75, 76, 77$. Other groups [89] introduced minor variations to their basic recipe, but still could not solve these hard cases without seeds, i.e. without a priori external knowledge. By employing the concept of niches, the present author [90] demonstrated that a phenotype algorithm is able to find all currently accepted global minima of Lennard-Jones clusters up to at least $n = 150$ without any use of external prior knowledge. At the same time, the size scaling of the method could be pushed below n^3, further improving the access to larger clusters. Recently, the same methodology was extended to molecular clusters and applied to the benchmark system of TIP4P water clusters [91].

In section 2, the basic algorithm of this phenotype method will be explained. In section 3, new applications to pure and mixed atomic and molecular clusters will be described.

2.   Algorithm

The phenotype algorithm used here has already been described in some detail in Refs. [90,91]; for completeness, its main features are sketched here. A fixed number of cluster geometries, the "population", is evolved in discrete steps, the "generations". Population size depends on problem size; although there have been some attempts within the GA community to find a relation between these two, this is still a matter of trial-and-error in practice, in particular since population size also strongly depends on details of the GA method used. Deaven and Ho [87,88] used very small populations down to only four members. Here several niches have to be accomodated within the population (see below), each of about this size; therefore, we typically use 10–30 individuals per generation.

Generation zero is chosen at random and then locally optimized. Alternatively, one can introduce desired seed structures at this point; we never do this. Nevertheless, some faint degree of external prior information does enter here almost inevitably: We draw random numbers for the particle coordinates of a cluster within some preset range, thus favoring compact structures. We also reject particles being placed too close to each other, but this merely serves to avoid numerical problems with the at short distances typically steeply repulsive branches of the potential during the subsequent local optimization.

In propagating one generation m to $m + 1$, we first generate an intermediate pool of geometries: Each possible pair of "parents" from generation m is formed (even allowing self-pairing, and disregarding all niches), and two "children" are generated from each pair. In each of these steps, a crossover is performed by cutting each parent cluster in two parts along a plane. As described in Ref. [90], it is advantageous to randomly choose either an optimized orientation of this cutting plane or a random orientation. Additionally, as described in Ref. [91], in difficult cases it is advisable not to cut the clusters in exact halves always, but also to allow non-symmetric cuts, with a Gaussian distribution about the exact halves. With a possibilty of 15%, each resulting child is mutated by moving a randomly determined number of particles away from their original positions by random vectors. Irrespective of mutations, each child is locally optimized and then moved into the intermediate pool, to which we also add all unchanged parent geometries.

From this pool, the next generation $m + 1$ is selected. In order to maintain diversity, a combination of energy and geometry criteria are used: The pool is ordered according to the energy. Starting from the lowest-energy member, clusters are drawn from this pool and inspected. A cluster is selected into the next generation if it is different in geometry from the other clusters selected before, according to one or several niching criteria; it thus opens a niche of its own. If another cluster with similar geometry is drawn from the pool later, it will be accepted also, provided the number of clusters in this niche is below a given limit and their energy differences are larger than a prescribed value (to avoid filling niches with almost identical geometries). Additionally, there is always a niche for mutants that has to be filled, in order to guarantee a minimum amount of exploration. This process is continued until the fixed number of population members has been reached for the next generation.

Thus, our niches exist only at the moment of selection from the pool into the next generation, and solely for the purpose of maintaining diversity (i.e. avoiding the common problem of premature convergence of the whole population to identical or closely related copies of a single solution which is probably not (yet) the global optimum). Also, they are fully automatic and dynamic in that the only parameters given are the maximum number of members allowed per niche and a "width" for each niche, measuring the geometrical variance allowed within a niche; the total number of niches is not fixed, nor are they required to be the same from one generation to the next. As described in Refs. [90, 91], niching can increase the performance of the algorithm by an order of magnitude or more in notoriously difficult cases.

Of course, the actual niche criteria do contain some external insight into crucial features of the specific clusters under study: For Lennard-Jones clusters [90], it is essential to differentiate the basic geometry types described in section 1.2. For TIP4P water clusters [91], it turned out to be useful to discern between clusters containing varying amounts of bond angles close to 90 degrees. In contrast to optimization approaches using fixed, prescribed grids in physical space, as the icosahedral growth sequence in Northby's study [62] or the icosahedral, the decahedral, and the fcc lattice in the work of Romero et al. [57], these niching criteria are much "softer" and less specific; typically, one of our niching criteria can be fulfilled by qualitatively different geometries. Also, we can derive these criteria without actual knowledge of any global minimum structures; it suffices to inspect the results of preliminary runs using no niches and to strive for maximum diversity in introducing niche criteria. Obviously, however, this discovery and testing of niche criteria is hard to implement in algorithmic form; we still do it "by hand".

After the members of the next generation have been selected as described above, the optimization history over the past generations is examined. If a population member has failed to change its energy recently, a number of post-processing attempts are launched, in this order: For molecular clusters, the positions of the molecules are fixed and their orientations (Euler angles) are globally optimized by running a copy of the whole algorithm described so far, but this time operating only on the orientation coordinates. Alternatively, for small molecular clusters, it is possible to do a full enumeration of chemically reasonable orientations, as described in Ref. [91]. As another measure, a "directed mutation" step is tried several times: A small number of particles (typically between one and four) that contribute least to the overall cluster energy is removed and successively reintroduced into the best vacancy sites in the cluster. This operation is even more successful if the cluster is expanded by 10% before reintroduction of the removed particles.

As in many other global optimization algorithms of this type, there is no reliable stopping criterion. Common experience indicates that the energy minimization progress has an $\exp(-x)$ form: A rapid decrease in the beginning gradually turns into an almost levelling-off — however, this general behavior is difficult to exploit in practice, since the presence of the random mutation operator makes sudden, small improvements still possible even after long generations of no change. Other indicators are of a similarly limited value: For example, there is usually a general trend in the energy increments obtained by successively adding single particles to a cluster, but there are also deviations from this trend, indicating "magic numbers" and particularly stable or unstable cluster sizes. Therefore, one typically uses some ill-defined combination of these and similar criteria for stopping.

3. Results

3.1 Lennard-Jones Clusters

We have extended our studies on the benchmark system of Lennard-Jones clusters beyond the range $2 \leq n \leq 150$ reported in Ref. [90]. A suitable database for comparison is available by the work of Romero et al. [57], who employed a GA-based placement search on the given icosahedral, decahedral, and fcc lattices. Dropping any claims for completeness, we performed an explorative study by limiting the number of generations to 100, and by stopping the algorithm after 20 generations without any improvement in the best energy. Nevertheless, up to $n = 190$, we managed to reproduce about 50% of the energies given by Romero et al.; the largest cluster for which we have found agreement so far is $n = 250$. Computer times for finding these values are still tolerable (at most a couple of days on a single-processor PC) and well within the approximate n^3 size scaling found in Ref. [90]. This is already remarkable, since our algorithm does not use any a priori knowledge (apart from the niching, which is not crucial in this size range), while the algorithm of Romero et al. already starts from the known lattices.

More interestingly, we managed to improve upon the global minima given originally by Romero et al. for $n = 185, 186, 187$. They originally proposed decahedral structures for these three cases, with the energies given in Table 1.1; their structures for $n = 185, 186$ constituted a new geometry subtype for Lennard-Jones global minima, namely that of a decahedral core with an outer layer in fcc positions. Our unbiased search managed to find lower-lying minima of the more usual icosahedral type, thus eliminating this new structural subtype. Our proposal for $n = 185$ was recently again improved by Leary.

Table 1.1. Improving Lennard-Jones minima; energies are given in units of the pair potential well depth

size n	Romero et al.	this work	refinements
185	-1125.299820	-1125.304876	-1125.4938 [a]
186	-1132.503199	-1132.669966	
187	-1139.240017	-1139.455696	

a) reported by R. H. Leary, Jan. 21, 2000.

It is also instructive to look at the cases where our algorithm failed to find the structure and energy of Romero et al. (or a better one) within the prescribed 100 generations: Our directed mutation, as described in section 2, is very efficient in removing badly placed atoms from the outer layer into holes in the outer core; it is vital to introduce such an operator since the standard operators of crossover and mutation take a long time to achieve this. However, our directed mutation is actually overdoing it: Contrary to the original expectations by Northby [62], it was discovered that there are global minimum structures with incomplete cores [92], constituting deviations from any conceivable growth sequence. Up to $n = 150$, there are only a few isolated examples for this phenomenon, but in larger clusters there are more cases. For example, for $n = 169, 170, 171, 172$ Romero et al. proposed structures with 4,3,2,1 holes in the core. Of these, our algorithm correctly found $n = 169$ and $n = 171$ within 100 generations, but constructed complete cores for the other two cases, leading to higher energies. Presumably, a less effective version of this operator or a counterbalancing operator that removes atoms from the core and places them into the outer layer could remedy the situation.

For $n = 163$ and $n = 164$, our algorithm finished at a peculiar structural type higher in energy than the geometries proposed by Romero et al.. The same structural type also occured as low-lying local minimum in our optimization runs at many other values of n. At the time of this study, we had not yet heard about the new tetrahedral structural type discovered by Leary and Doye [64] for $n = 98$. Comparing our structures with theirs revealed that they follow exactly the same pattern. Thus, unintentionally and before their discovery, we have shown that our algorithm is able to find also this new, fourth basic structure of Lennard-Jones clusters.

Since Lennard-Jones clusters are considered a mere toy problem in chemistry, we have only used it as a benchmark for our algorithm and did not try to push this application to the limit. Indeed, by now, the Lennard-Jones cluster results reported above for basin-hopping and our phenotype algorithm have been extended to still larger cluster sizes by other authors and other approaches [93,94], but this is beyond the scope of the present text.

3.2   Water Clusters

3.2.1   Pure Water Clusters.
In an extension of our work reported in Ref. [46], we have successfully located the global minimum and several low-lying local minima on the ab-initio LMP2 potential energy surface of the water heptamer. Geometries and energies are available on the internet [95]. The heptamer global minimum can be obtained in a rather obvious fashion from the (classical mechanical) trigonal prism geometry of the hexamer global minimum, by capping one of the triangle edges. This geometry agrees qualitatively with the one obtained by Buck, Buch et al. [96] on a good empirical potential; these authors also calculated an approximate OH-stretch vibrational spectrum which is in very good agreement to their experimental spectrum. As opposed to the infamous case of the water hexamer (alluded to in the introduction section 1.2), in the heptamer there are no severe difficulties with cluster isomers of qualitatively different geometry but only marginally different energy.

The hard part of our water heptamer calculations is on the ab-initio side, not in the global geometry optimization part. In contrast to a GA study by Niesse and Mayne [97], we find that the global geometry optimization of $(H_2O)_n$ on model potentials for sizes up to and including $n = 7$ can still be handled easily with a brute-force application of a standard GA using a genetic string representation; this starts to run into difficulties at about $n = 8$. In order to be able to tackle larger molecular clusters, we have therefore extended our phenotype approach to molecular clusters, as described here in section 2. We have recently [91] applied this method to water clusters on the TIP4P potential

[98]. The virtue of this potential is not its accuracy, but rather its simplicity in combination with qualitatively correct results for liquids and small clusters, its still widespread use in molecular dynamics simulations of liquids, solutions, and water-solvated biomolecules, and the availability of a large-scale global cluster geometry optimization study by Wales and Hodges [58] on this potential, employing the basin-hopping method.

The performance of our phenotype algorithm on this system [91] is satisfactory: We could confirm all the minima proposed by Wales and Hodges, and the computational expense appears to be approximately equal to theirs. In both studies, however, a breakdown of performance was experienced at $n = 21$, apparently due to a massive proliferation of qualitatively different structures within a very small energy band [91], indicating the need for further method development. In fact, this benchmark is obviously much more difficult than the Lennard-Jones benchmark, in several respects: The exponentially increasing positional configuration space is extended by the also exponentially increasing orientational configuration space [91]. Furthermore, a closer inspection of the global minima and the most important low-lying local minima indicates that there is no general, simple growth sequence in the range $n = 10-21$: In no case, the global minimum for $n + 1$ can be obtained by a simple addition of another water molecule to a suitable position in the cluster of n molecules. Instead, such simple addition geometries survive at best as low-lying local minima. The only discernible generalization is a preference for structures consisting of fused cubes and pentagonal prisms (although $n = 17, 19, 21$ violate this rule, they can be classified as distorted realizations of this principle). In how far this observation could serve as a basis for an improved global optimization algorithm for this particular application case remains an issue of future research.

In fact, the pure water cluster problem also turns out to suffer from the high sensitivity to small errors in the potential energy model mentioned in the introduction. In a recent extension of our studies to the highly accurate but also 20 times more expensive TTM2-F water potential [99], we found [100] that while TIP4P yields global minimum structures that are qualitatively surprisingly reliable for small cluster sizes up to about 12 water molecules, it then starts to deviate slightly for larger sizes, arriving at strong differences to the TTM2-F model at $n = 17$ and $n = 21$. There, surprisingly, TIP4P still predicts all water molecules to reside on the cluster surface. TTM2-F is more in line with chemical intuition: It starts to incorporate a first water molecule into the cluster interior at these cluster sizes. Obviously, from here it is still a long way to go until the first formation of an ice core inside a water cluster, but even this question is now directly being investigated [101].

3.2.2 Water Hetero Clusters. Our ability to perform global cluster geometry optimizations for clearly non-trivial values of n allows first, tentative chemical applications, one of which is briefly sketched here:

While the structure of liquid water and of small water clusters in the gas phase is an intriguing problem in itself, even more interesting from a chemist's perspective is the structure of water in the presence of other molecules and ions. Recent experimental developments [102, 103] have made it possible to transport water-solvated species into the gas phase and study them in isolation. A favorite structural proposal typically coming up in these studies is that of a "clathrate" structure, which ideally consists of a dodecahedral arrangement of 20 water molecules around the hetero molecule or ion. This model is usually ascribed to Castleman [104], but actually goes back to still earlier sources, see references in Refs. [105, 106]. Solid clathrates involving such cages enclosing many different guest species, including CO_2 and methane, are almost ubiquitous, from deep-sea sediments on Earth [107] to the pole caps on Mars [108]. The actual occurence of such dodecahedral arrangements in solvation is subject to intense debate, in particular for hydrophobic hetero molecules [109–111]. Unfortunately, tests of this model are usually done using large-scale molecular dynamics simulations from which only averaged data like radial distribution functions or statistics about distributions of 5-membered rings can be extracted in a meaningful fashion. There are also some calculations on semiempirical and ab-initio levels [112], but these are done on locally optimized geometries starting from intuitive guesses, an approach that is dangerously error-prone even for rather small clusters [44]. Thus, global geometry optimization seems to be completely lacking in this area, although it may give important complementary insights.

In an attempt to remedy this situation, we are currently studying simple model systems for isolated, water-solvated species, focussing on alkaline and alkaline earth ions and methane in water clusters, using the simple TIP4P/OPLS potentials [98] often used in molecular dynamics simulations of these systems [113,114]. Mg^{2+}, which is not in the standard parameter set of these potentials, is modelled by a charge of $+2$ on the Mg-atom and a repulsive term of the form $a \cdot \exp(-d \cdot r)$ between the Mg-atom and the O-atom of the water molecules, with $a = 73822.82E_h$ and $d = 3.86a_0$, as a function of the distance r between these centers. In addition, we include an artificial repulsion term of the form $\exp(-r)/r^7$ between the hetero ion or molecule and all O-atoms and H-atoms. This term does not influence the physically meaningful interactions modelled by the other terms at normal inter-site distances r, but prevents "cold fusion" of the hetero ion or molecule and a water molecule at unphysically short distances. This is usually not necessary in molecular dynamics, where this extremely short distance regime cannot be reached, but it can be accessed in our algorithm via recombination of cluster halves in the crossover operation. For $Mg^{2+}(H_2O)_{20}$, our best structure found so far is depicted in Fig. 1.1. Clearly, it does not even bear a faint resemblance to a dodecahedral clathrate. A clathrate-like structure also does not show up as important local minimum in our optimizations.

Figure 1.1. Proposal for the global minimum of $Mg^{2+}(H_2O)_{20}$

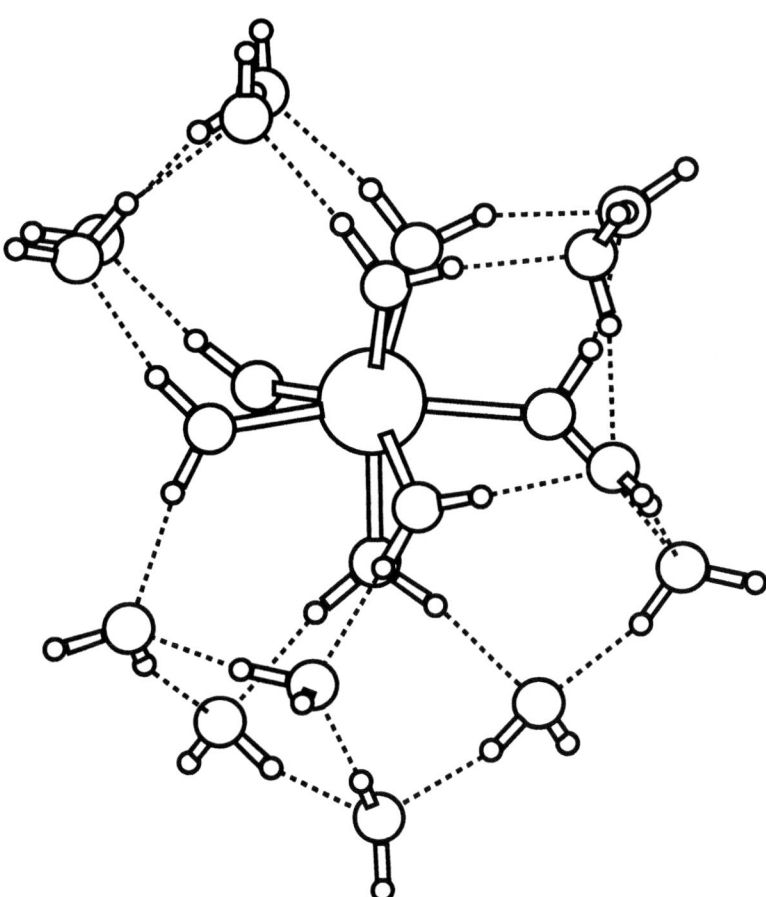

This is in line with our findings for the alkaline cations Na^+, K^+ and Cs^+. The sodium cation simply is too small to fill the dodecahedral clathrate hull. Also, again due to its small size, its orienting effect on the immediately surrounding water molecules is particularly strong. All this results in totally different microhydration cluster structure characteristics, with the dodecahedron playing no important role whatsoever. Actually, these findings may already have some support by experiment [115]. For the potassium and cesium cations, microhydration cluster structures resembling the dodecahedral ideal actually are globally optimal at a size of 20 water molecules in a mathematical sense, but different structures containing also 4- and 6-rings are degenerate in energy within the error bars of the potential.

For the case of the simple model methane in TIP4P water, it is interesting to calculate the energy of "hand-generated" ideal dodecahedral water shells around the methane, after local optimization, for varying van-der-Waals radius of methane. It turns out that the binding energy for this model clathrate is maximal at the usual literature value of the methane radius, while at this point the water molecules in the clathrate hull have almost the same distances they also have in complete absence of the methane. The "methane" becomes too large to fit into the dodecahedral hull only at about double its literature radius. From this, one is tempted to expect that a methane clathrate could be a favorable structure for this system, since it does not seem to have a disturbing influence on the water structure (interestingly, the literature parameters were not adjusted to this situation but rather to molecular dynamics simulations of liquids). However, actual global geometry optimization with our phenotype algorithm also used in the other examples results in the predominant formation of "surface states", exhibiting not a methane surrounded by water molecules but a segregation of the two species, with the methane sitting on the surface of a compact water cluster. Our best species of this kind obtained so far is shown in Fig. 1.2. It has an energy of -859.204247 kJ/mol; this has to be compared to -841.290266 kJ/mol for the best "hand-constructed" dodecahedral clathrate shown in Fig. 1.3 and also to -872.988754 kJ/mol for the Wales/Hodges global minimum of pure $(H_2O)_{20}$.

Therefore, neither the ideal clathrate nor the surface states are stable with respect to dissociation into a pure water cluster with an infinitely separated methane (actually, this "phase separation" is in accord with the chemist's intuition of methane as a hydrophobic species). Using a new niche criterion based on the distance between the center of mass of the water molecules and the methane coordinates, we have tried to enforce the presence of clathrate-like states in the optimization, and indeed they are duly found by this method — but on the energy scale they are simply not competitive to several types of surface states.

We have also checked methane-water clusters of different sizes, $n = 19$ and $n = 24$, but again the same picture prevails: By enforcing the water molecules and the model methane to be at least in neighboring regions of space, and by using the niche criterion mentioned above, our method finds both surface states and clathrate structures, but several variants of surface states are lower

Figure 1.2. Proposal for the global minimum of $CH_4(H_2O)_{20}$. The position of the methane is indicated by the single large circle at the bottom.

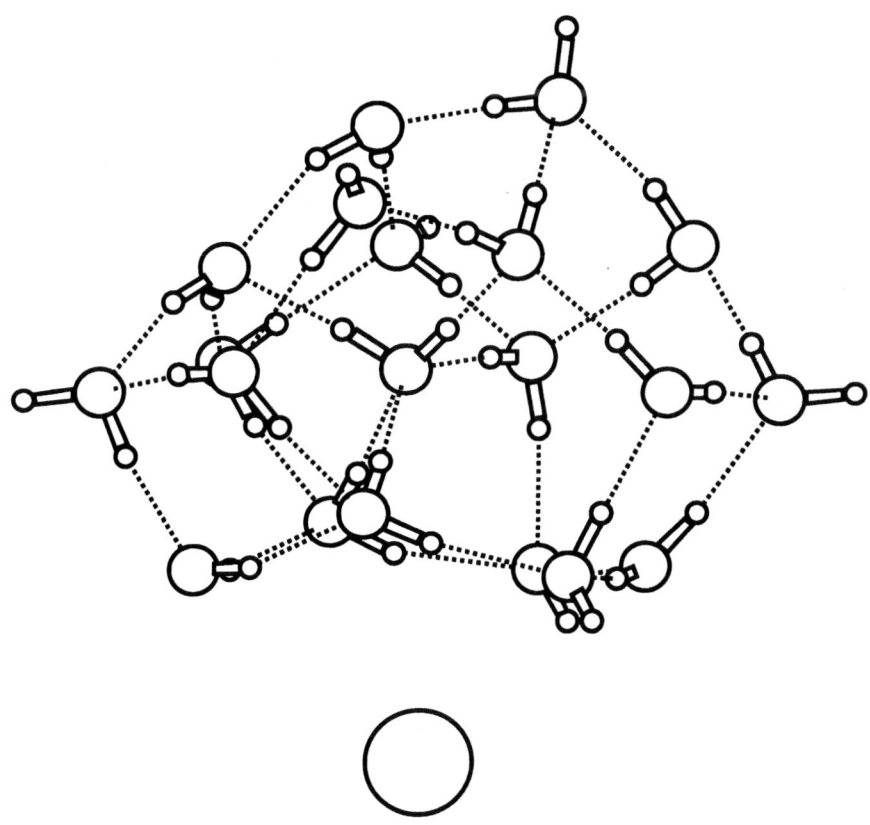

in energy than the best clathrates, and all these states are higher in energy than the corresponding pure water clusters without the model methane.

Of course, water clathrates with methane and other hydrophobic guest molecules are an experimental fact — however, there they are typically stable only under high pressures and in the bulk. Neither pressure nor bulk packing effects are taken into account in our modelling (besides other approximations, as the absence of polarizability in the TIP4P model, or as the simplistic picture of methane in the "united atom approximation", i.e. as a single structureless, isotropic Lennard-Jones site, as indicated in the Figures). Nevertheless, our model studies serve as a warning that seemingly obvious hypotheses (possibly

Figure 1.3. The best dodecahedral clathrate structure of $CH_4(H_2O)_{20}$. The position of the methane is indicated by the single large circle in the center.

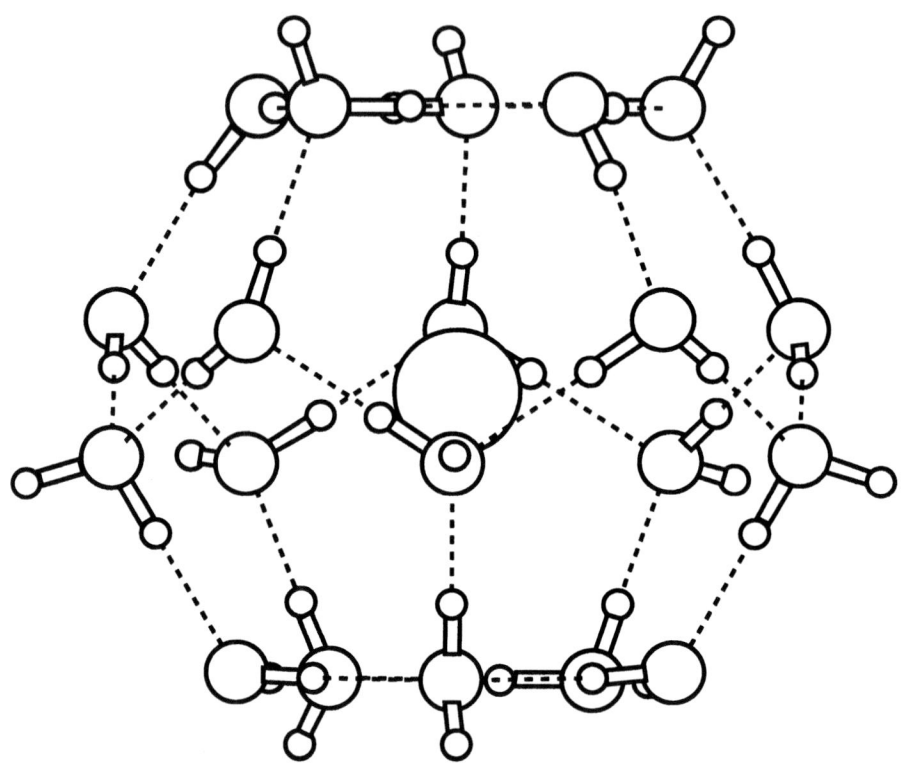

derived from analogies to the bulk) simply may not work for structures on the nanoscale regime.

4.   Conclusions

In the application examples shown here and in our previous publications, it has been demonstrated that the task of global geometry optimization of atomic and molecular clusters can be approached successfully with our phenotype strategy. In fact, this method is capable of tackling cluster sizes of experimental interest, and therefore we are now moving from "simple" benchmark systems to more sophisticated application examples.

As all GA-based methods, our current strategy does not pretend to be complete, exhaustive and exact. On the one hand, this is its main strength, since the efficiency of the method mainly stems from its strong exploitation of search heuristics (both the "automatic" search heuristics implicit in the standard GA concept and the additional guidance imposed by the niches), disregarding their tendency to narrow down on limited regions of search space. On the other hand, this is also its main weakness, leading to the absence of a reliable stopping criterion and of any type of measure or estimate of "distance" remaining to the global minimum or of the amount of search space not yet covered. In practice, this is an annoying drawback, only partially compensated by the impressively low size scaling demonstrated for our method.

Therefore, the author does not believe for a moment that GAs (or any derivative of this concept, as the one presented here) are the ultimate answer to the problem of global cluster geometry optimization. Progress is likely to come from more sophisticated combinations of methods: Efficient heuristics guide the search to the presumably most interesting regions of configuration space which are then covered completely by more exact methods (probably of the branch-and-bound type, as in Ref. [117]). Also, the method presented here is still almost a general global optimization strategy; the only elements specific for clusters are some implementation details of the crossover and mutation operators and the various niching criteria. Hence, in our quest to design global optimization methods applicable to still larger clusters in still shorter computer times, it will be advisable to make the algorithm still more problem specific. To this end, a thorough study of potential energy surfaces in typical clusters is likely to reveal common characteristics. On the one hand, this can be used to design better and more general guiding potentials for our method of combining global optimization with highly expensive ab-initio calculations [44–46]. On the other hand, more generally, this information can then actually be used in the global optimization process — at present, either no assumptions are made about the potential energy function at all, or the particles are distributed on a few fixed lattices in physical space (which is equivalent to the assumption that the potential energy function would have guided the search to these solutions anyway); clearly, some intermediate strategy between these two extremes is more efficient. Also, as indicated in the introduction, it may turn out to be fruitful to join the outcome of such an improved understanding of potential energy functions with what pure mathematicians have already found out about three-dimensional tiling.

References

[1] I. N. Levine: "Quantum Chemistry", 4th Edition, Prentice Hall, Englewood Cliffs, New Jersey, 1991.

[2] S. Sæbø and P. Pulay, Ann. Rev. Phys. Chem. 44 (1993) 213.

[3] M. Schütz, G. Hetzer and H.-J. Werner, J. Chem. Phys. 111 (1999) 5691.

[4] R. D. Levine and R. B. Bernstein: "Molecular Reaction Dynamics and Chemical Reactivity", Oxford University Press, New York, 1987.

[5] J. F. Castillo, B. Hartke, H.-J. Werner, F. J. Aoiz, L. Bañares and B. Martínez-Haya, J. Chem. Phys. 109 (1998) 7224.

[6] D. H. Zhang and J. Z. H. Zhang, J. Chem. Phys. 101 (1994) 1146.

[7] A. Neumaier, SIAM Rev. 39 (1997) 407.

[8] C. M. Dobson, A. Šali and M. Karplus, Angew. Chem. 110 (1998) 908.

[9] H. A. Scheraga, J. Lee, J. Pillardy, Y.-J. Ye, A. Liwo and D. Ripoll, J. Global Opt. 15 (1999) 235.

[10] H. W. Kroto, J. R. Heath, S. C. O'Brien, R. F. Curl and R. E. Smalley, Nature 318 (1985) 162.

[11] H. W. Kroto, A. W. Allaf and S. P. Balm, Chem. Rev. 91 (1991) 1213.

[12] U. Müller, Angew. Chem. 112 (2000) 513.

[13] J. H. Conway and N. J. A. Sloane: "Sphere packings, lattices, and groups", Springer, Berlin, 1993.

[14] M. Moskovits, Ann. Rev. Phys. Chem. 42 (1991) 465.

[15] G. Schmid, Chem. Rev. 92 (1992) 1709.

[16] D. M. P. Mingos, Contemp. Phys. 35 (1994) 181.

[17] R. S. Berry, J. Phys. Chem. 28 (1994) 6910.

[18] D. Braga and F. Grepioni, Acc. Chem. Res. 27 (1994) 51.

[19] D. L. Freeman and J. D. Doll, Annu. Rev. Phys. Chem. 47 (1996) 43.

[20] T. P. Martin, Phys. Rep. 273 (1996) 199.

[21] M. T. Swihart and S. L. Girshick, J. Phys. Chem. B 103 (1999) 64.

[22] F. Q. Yu, R. P. Turco and B. Karcher, J. Geophys. Res. – Atmos. 104 (1999) 4079.

[23] C. F. Clement and I. J. Ford, Atmos. Environ. 33 (1999) 489.

[24] A. R. Bandy and J. C. Ianni, J. Phys. Chem. A 102 (1998) 6533.

[25] R. W. Siegel, Nanostruct. Matls. 3 (1993) 1.

[26] S. Datta, D. B. Janes, R. P. Andres, C. P. Kubiak and R. G. Reifenberger, Semicond. Sci. Technol. 13 (1998) 1347.

[27] R. Tenne, M. Homyonfer and Y. Feldman, Chem. Mat. 10 (1998) 3225.

[28] S. Subramoney, Adv. Mater. 10 (1998) 1157.

[29] B. Hartke, Angew. Chem. Int. Ed. 41 (2002) 1468.

[30] L. T. Wille and J. Vennik, J. Phys. A 18 (1985) L419.

[31] G. W. Greenwood, Z. Phys. Chem. 211 (1999) 105.

[32] L. T. Wille, in: "Annual Reviews of Computational Physics VII", D. Stauffer (Ed.), World Scientific, Singapore (2000): p. 25.

[33] B. A. Hendrickson, SIAM J. Opt. 5 (1995) 835.

[34] D. S. Johnson, J. Algorithms 5 (1984) 284.

[35] V. G. Deineko, R. Rudolf and G. J. Wöginger, SIAM Discrete Math. 11 (1998) 81.

[36] J. L. Bentley, Orsa J. Comput. 4 (1992) 387.

[37] G. Das, S. Kapoor and M. Smid, Algorithmica 19 (1997) 447.

[38] M. R. Hoare and J. A. McInnes, Adv. Phys. 32 (1983) 791.

[39] J. P. K. Doye and D. J. Wales, J. Chem. Soc. Faraday Trans. 93 (1997) 4233.

[40] D. Hohl, R. O. Jones, R. Car and M. Parrinello, J. Chem. Phys. 89 (1988) 6823.

[41] U. Röthlisberger and W. Andreoni, J. Chem. Phys. 94 (1991) 8129.

[42] B. Hartke and E. A. Carter, J. Chem. Phys. 97 (1992) 6569.

[43] B. Hartke and E. A. Carter, Chem. Phys. Lett. 216 (1993) 324.

[44] B. Hartke, Chem. Phys. Lett. 258 (1996) 144.

[45] B. Hartke, Theor. Chem. Acc. 99 (1998) 241.

[46] B. Hartke, M. Schütz and H.-J. Werner, Chem. Phys 239 (1998) 561.

[47] K. Liu, M. G. Brown and R. J. Saykally, J. Phys. Chem. A 101 (1996) 501.

[48] J. Kim and K. S. Kim, J. Chem. Phys. 109 (1998) 5886.

[49] K. Liu, M. G. Brown, C. Carter, R. J. Saykally, J. K. Gregory and D. C. Clary, Nature (London) 381 (1996) 501; J. K. Gregory and D. C. Clary, J. Phys. Chem. 100 (1996) 18014.

[50] K. Nauta and R. E. Miller, Science 287 (2000) 293.

[51] E. C. Honea, A. Ogura, C. A. Murray, K. Raghavachari, W. O. Sprenger, M. F. Jarrold and W. L. Brown, Nature 366 (1993) 42.

[52] K.-M. Ho, A. A. Shvartsburg, B. Pan, Z.-Y. Lu, C.-Z. Wang, J. G. Wacker, J. L. Fye and M. F. Jarrold, Nature 392 (1998) 582.

[53] A. J. Stone: "The theory of intermolecular forces", Clarendon Press, Oxford, 1996.

[54] B. W. van de Waal, Phys. Rev. Lett. 76 (1996) 1083.

[55] C. Nyeland and J. P. Toennies, Chem. Phys. 188 (1994) 205.

[56] The Cambridge Cluster Database, D. J. Wales, J. P. K. Doye, A. Dullweber and F. Y. Naumkin, http://brian.ch.cam.ac.uk/CCD.html

[57] D. Romero, C. Barrón and S. Gómez, Comput. Phys. Comm. 123 (1999) 87;
http://www.vetl.uh.edu/~cbarron/LJ_cluster/LJpottable.html

[58] D. J. Wales and M. P. Hodges, Chem. Phys. Lett. 286 (1998) 65.

[59] D. J. Wales and H. A. Scheraga, Science 285 (1999) 1368.

[60] M. R. Hoare and P. Pal, Adv. Phys. 20 (1971) 161; M. R. Hoare, Adv. Chem. Phys. 40 (1979) 49.

[61] A. L. Mackay, Acta Cryst. 15 (1962) 916.

[62] J. A. Northby, J. Chem. Phys. 87 (1987) 6166.

[63] J. P. K. Doye, M. A. Miller and D. J. Wales, J. Chem. Phys. 111 (1999) 8417.

[64] R. H. Leary and J. P. K. Doye, Phys. Rev. E 60 (1999) R6320.

[65] B. C. Bolding and H. C. Andersen, Phys. Rev. B 41 (1990) 10568.

[66] I. L. Garzon, G. Kaplan, R. Santamaria and O. Novaro, J. Chem. Phys. 109 (1998) 2176.

[67] D. J. Wales, J. Chem. Soc. Faraday Trans. 86 (1990) 3505.

[68] J. Pillardy and L. Piela, J. Phys. Chem. 99 (1995) 11805.

[69] J. Kostrowicki, L. Piela, B. J. Cherayil and H. A. Scheraga, J. Phys. Chem. 95 (1991) 4113.

[70] H. A. Scheraga, Int. J. Quant. Chem. 42 (1992) 1529.

[71] C. D. Maranas and C. A. Floudas, J. Chem. Phys. 97 (1992) 7667.

[72] Z. Li and H. A. Scheraga, Proc. Natl. Acad. Sci. USA 84 (1987) 6611.

[73] N. Metropolis, A. W. Rosenbluth, M. N. Rosenbluth, A. H. Teller and E. Teller, J. Chem. Phys. 21 (1953) 1087.

[74] S. Kirkpatrick, C. D. Gelatt, Jr., and M. P. Vecchi, Science 220 (1983) 671; P. J. M. van Laarhoven and E. H. L. Aarts: "Simulated Annealing: Theory and Applications", Reidel, Dordrecht (1987).

[75] D. J. Wales and J. P. K. Doye, J. Phys. Chem. A 101 (1997) 5111.

[76] D. J. Wales, private communication.

[77] J. H. Holland: "'Adaption in Natural and Artificial Systems'", University of Michigan Press, Ann Arbor (1975).

[78] D. E. Goldberg: "'Genetic Algorithms in Search, Optimization, and Machine Learning'", Addison-Wesley, Reading (1989).

[79] R. S. Judson, Rev. Comput. Chem. 10 (1997) 1.

[80] M. Mitchell: "'An Introduction to Genetic Algorithms'", MIT Press, Boston (1996).

[81] "The Hitchhiker's Guide to Evolutionary Computation", J. Heitkötter and D. Beasley (Eds.),
http://alife.santafe.edu/~joke/encore/www/

[82] R. S. Judson, M. E. Colvin, J. C. Meza, A. Huffer and D. Gutierrez, Int. J. Quant. Chem. 44 (1992) 277.

[83] D. B. McGarrah and R. S. Judson, J. Comp. Chem. 14 (1993) 1385.

[84] B. Hartke, J. Phys. Chem. 97 (1993) 9973.

[85] Y. Xiao and D. E. Williams, Chem. Phys. Lett. 215 (1993)

[86] S. K. Gregurick, M. H. Alexander and B. Hartke, J. Chem. Phys. 104 (1996) 2684.

[87] D. M. Deaven and K. M. Ho, Phys. Rev. Lett. 75 (1995) 288.

[88] D. M. Deaven, N. Tit, J. R. Morris and K. M. Ho, Chem. Phys. Lett. 256 (1996) 195.

[89] M. D. Wolf and U. Landman, J. Phys. Chem. A 102 (1998) 6129.

[90] B. Hartke, J. Comput. Chem. 20 (1999) 1752.

[91] B. Hartke, Z. Phys. Chem. 214 (2000) 1251..

[92] C. Barrón, S. Gómez and D. Romero, Appl. Math. Lett. 10 (1997) 25.

[93] J. Lee, I.-H. Lee and J. Lee, Phys. Rev. Lett. 91 (2003) 080201.

[94] Y. Xiang, H. Jiang, W. Cai and X. Shao, J. Phys. Chem. A 108 (2004) 3586.

[95] ftp://ravel.phc.uni-kiel.de/pub/lmp2_water/heptamer/

[96] J. Brudermann, M. Melzer, U. Buck, J. Sadley, J. K. Kazimirski and V. Buch, J. Chem. Phys. 110 (1999) 10649.

[97] J. A. Niesse and H. R. Mayne, J. Comput. Chem. 18 (1997) 1233.

[98] W. L. Jorgensen, J. Chem. Phys. 77 (1982) 4156; W. L. Jorgensen, J. Chandrasekhar, J. D. Madura, R. W. Impey and M. K. Klein, J. Chem. Phys. 79 (1983) 926; W. L. Jorgensen, J. D. Madura and C. J. Swenson, J. Am. Chem. Soc. 106 (1984) 6638.

[99] C. J. Burnham and S. S. Xantheas, J. Chem. Phys. 116 (2002) 5115.

[100] B. Hartke, Phys. Chem. Chem. Phys. 5 (2003) 275.

[101] J. K. Kazimirski and V. Buch, J. Phys. Chem. A 107 (2003) 9762.

[102] S.-W. Lee, P. Freivogel, T. Schindler and J. L. Beauchamp, J. Am. Chem. Soc. 120 (1198) 11758.

[103] F. Sobott, A. Wattenberg, H.-D. Barth and B. Brutschy, Int. J. Mass Spectrom. 185/186/187 (1999) 271.

[104] P. M. Holland and A. W. Castleman, Jr., J. Chem. Phys. 72 (1980) 5984.

[105] J. L. Kassner and D. E. Hagen, J. Chem. Phys. 64 (1976) 1860.

[106] J. Q. Searcy and J. B. Fenn, J. Chem. Phys. 64 (1976) 1861.

[107] T. Appenzeller, Science 252 (1991) 1790.

[108] A. Dobrovolskis and A. P. Ingersoll, Icarus 26 (1975) 353; D. A. Brain and B. M. Jakosky, J. Geophys. Res. – Planets 103 (1998) 22689.

[109] T. Head-Gordon, Proc. Natl. Acad. Sci. USA 92 (1995) 8308.

[110] L. A. Lipscomb, F. X. Zhou and L. D. Williams, Biopolymers 38 (1996) 177.

[111] W. Blokzijl and J. B. F. N. Engberts, Angew. Chem. 105 (1993) 1610.

[112] A. Khan, Chem. Phys. Lett. 217 (1994) 443; J. Phys. Chem. 99 (1995) 12450; J. Chem. Phys. 106 (1997) 5537; Chem. Phys. Lett. 319 (2000) 440.

[113] O. K. Førrisdahl, B. Kvamme and A. D. J. Haymet, Mol. Phys. 89 (1996) 819.

[114] P.-L. Chau, T. R. Forester and W. Smith, Mol. Phys. 89 (1996) 1033.

[115] B. Hartke, A. Charvat, M. Reich and B. Abel, J. Chem. Phys. 116 (2002) 3588.

[116] F. Schulz and B. Hartke, Chem. Phys. Chem. 3 (2002) 98.

[117] J. D. Pintér: "Global Optimization in Action – Continuous and Lipschitz Optimization", "Nonconvex Optimization and its Applications", Vol. 6, Kluwer Academic Publishers, Dordrecht (1996).

Chapter 7

COMPUTATIONAL ANALYSIS OF HUMAN DNA SEQUENCES: AN APPLICATION OF ARTIFICIAL NEURAL NETWORKS.

Artemis Hatzigeorgiou[1,2,3] and Molly Megraw[1,2]
[1] *Center for Bioinformatics,* [2] *Department of Genetics, School of Medicine, and* [3] *Department of Computer and Information Science, School of Engineering, University of Pennsylvania, Philadelphia, PA 19104, USA*

Abstract: In this chapter we give an introduction to the area of bioinformatics handling the nucleotide sequence analysis problem. We give a brief introduction to the nature of DNA and RNA and use one of many topics - the Translation Initiation Start (TIS) problem - to explain a computational prediction of motifs on biological sequences. Correct identification of the Translation Initiation Start (TIS) in cDNA sequences is an important issue for genome annotation. Here we describe a computational method for TIS identification based in a combination of statistics and Artificial Neural Networks (ANNs). This method makes use of two modules, one sensitive to the conserved motif and the other sensitive to the coding/non-coding potential around the start codon. Finally by applying a method inspired by molecular biology, the simplified method of the ribosome scanning model improves the prediction significantly.

Keywords: Neural networks; sequence analysis; TIS prediction; translation initiation start; coding potential.

1.   INTRODUCTION

The genome sequence of an organism is an information resource unlike any other that biologists have previously had access to. But the value of a genome is only as good as its annotation, which bridges the gap from the sequence to the biology of the organism. More than 300 genomes have been sequenced in the last decade, most of them from prokaryotes. Now that most of the human genome has been sequenced, several new eukaryotic genome projects have been started for a variety of organisms, including animals, plants, fungi, and numerous pathogenic protozoa. In the ideal case, genome analysis should be provided automatically after a significant amount of sequence assembly for a new genome is available. This is not presently possible, however. Although a large number of gene prediction programs exist, all include organism-specific parameters that must be determined from training examples. The usual process is to find experimentally some genes, and then to use these data to design a new gene prediction algorithm, or retrain an existing one.

This chapter describes for us an example of an Artificial Neural Network (ANN) application to the mathematical structure analysis of complimentary DNA (cDNA) - a subproblem of gene prediction. First we give some biological background on DNA and then we describe how this information is presented to the ANN's.

2.   DNA AND GENES

2.1   DNA

In humans, as in other higher organisms, a DNA molecule consists of two strands that wrap around each other to resemble a twisted ladder whose sides, made of sugar and phosphate molecules, are connected by rungs of nitrogen-containing chemicals called bases. Each strand is a linear arrangement of repeating similar units called nucleotides, which are each composed of one sugar, one phosphate, and a nitrogenous base (Fig. 1.1). Four different bases are present in DNA: Adenine (A), Thymine (T), Cytosine (C), and Guanine (G).

The two DNA strands are held together by weak bonds between the bases on each strand, forming base pairs (bp). Strict base- pairing rules are adhered to:

- adenine will pair only with thymine (an A- T pair) and

- cytosine with guanine (a C- G pair).

Genome size is usually stated as the total number of base pairs; the human genome contains roughly 3 billion bp. The particular order of the bases arranged along the sugar- phosphate backbone is called the DNA sequence. In other words a DNA sequence can be described as a very long word over a four letter alphabet $\mathcal{A} = \{A, C, G, T\}$, a letter for every base.

2.2   GENES

Each DNA molecule contains many genes, the basic physical and functional units of heredity. The number of human genes is still an unclear issue. It is estimated to be between 24,000 and 50,000.

Human genes vary widely in length, often extending over thousands of bases, but less than 5% of the genome is known to include the protein coding sequences of genes called *exons*. Exons are interrupted by many intron sequences without coding function.

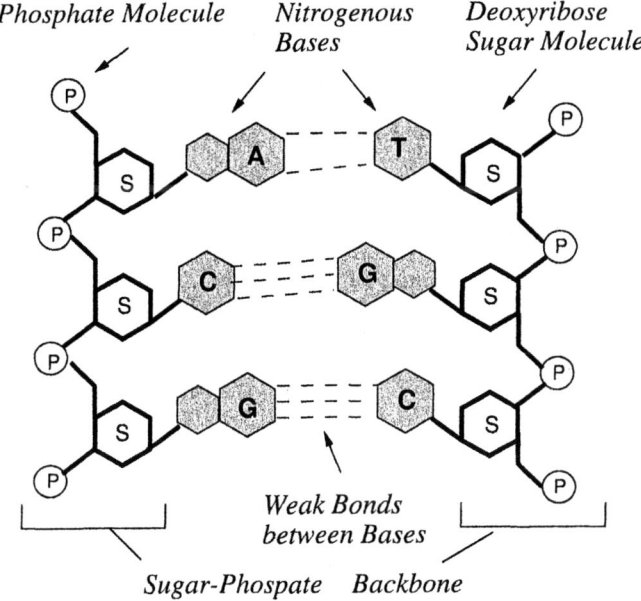

Figure 1.1.   The four nitrogenous bases of DNA are arranged along the sugar- phosphate back-bone in a particular order (the DNA sequence), encoding all genetic instructions for an organism. Adenine (A) pairs with Thymine (T), while Cytosine (C) pairs with Guanine (G). The two DNA strands are held together by weak bonds between the bases.

For the information within a gene to be expressed, a complementary RNA strand is produced (a process called *transcription*) from the DNA. This strand is a molecule, called messenger ribonucleic acid (mRNA). similar to a single strand of DNA. It is also made up of four nucleotides: Adenine (A), Cytosine (C), Guanine (G) and Uracil(U), which replaces Thymine (T). Before the mRNA serves as a template for protein synthesis the introns are removed through the splicing machinery. Then the mRNA arrives at the ribosome, a protein- synthesizing machinery, which reads the instructions from the mRNA to build a protein (Casey, 1992).

2.3   THE GENETIC CODE AND PROTEINS

Proteins are large, complex molecules made up of long chains of subunits called *amino acids*. There are twenty different kinds of amino acids. Each amino acid is encoded through triplets of nucloetides called codons (genetic code).

Out of the 4 nucleotides, $64(4^3)$ different codons can be built. 61 of them encode 20 amino acids. Three codons (TAA, TAG and TGA) cause protein transcription to cease. These are known as stop codons.

Since every triplet encodes an amino acid or a stop codon, many pairs of codons, that differ only in the third position base, code for the same amino acid. A translation example of nucleotides (codons) to amino acids is:

$$\underbrace{ATG}_{Meth} \quad \underbrace{TAC}_{Tyr} \quad \underbrace{TGC}_{Cys} \quad \underbrace{GGC}_{Gly} \quad C...$$

A shift of one letter in reading the same nucleic acid sequence results in a very different amino acid sequence:

$$A \quad \underbrace{TGT}_{Cys} \quad \underbrace{ACT}_{Thr} \quad \underbrace{GCG}_{Ala} \quad \underbrace{GCC}_{Pro} \quad ...$$

The phase of codon reading is called the reading frame. There are three different frames in one direction and three more reading frames on the complementary strand.

2.4   COMPLEMENTARY DNA (CDNA)

A gene region of the DNA itself, after the copying mechanism (transcription), contains parts with coding information (Exons) and noncoding information (Introns). The introns are getting through the splicing mechanism. The remaining parts certain only exons and are called messenger RNA (mRNA). The only Exons which contain non-coding regions are the first and the last exon.

In the laboratory, the mRNA molecule can be isolated and used as a template to synthesize a complementary DNA (cDNA) strand, which can then be used analysed by conventional molecular biology methods.

A cDNA starts with an untranslated region (UTR) followed by a coding region and ends again with another UTR. The start of the cDNA is called the 5'end, and the end is called 3'end of the cDNA.

The coding part starts with a specific codon (AUG), which is called the start codon. The coding region stops with one of the stop codons. A start codon is a nucleotide triplet (ATG) that translates the amino acid Methionin. The same triplet (ATG) can also occur in the middle of a coding frame. In this case it just encodes Methonin and does not act as a start codon. A start codon leads the coding frame, that is always part of an open reading frame (ORF). An ORF is a frame that translates amino acids without the interruption of stop codons. In other words, an open reading frame is a frame between two stop codons - read always in steps of three nucleotides. The first nucleotide of the codon defines the translation initiation start (TIS).

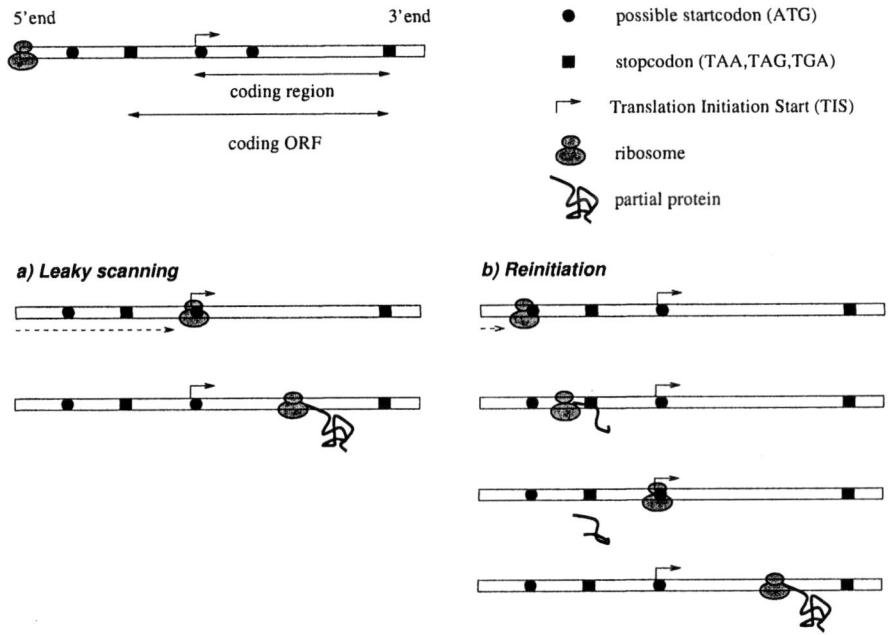

Figure 1.2. Description of the ribosome scanning model on the cDNA. See also text.

2.5   TRANSLATION INITIATION START (TIS)

According to the ribosome scanning model (Kozak, 1996), the ribosome first attaches to the 5'end of the mRNA and then *scans* the sequence until the first ATG start codon, which is in optimal nucleotide context. Translation is then started. Although this is true for most mRNA studied, there are some notable exceptions (reviewed in (Kozak, 1996) & (Pain, 1990)). Mentioned below are such cases for eukaryotic non-viral sequences (Fig. 1.2):

- Leaky scanning, when the first ATG codon has less optimal nucleotide context and therefore can be bypassed by the ribosome, which then initiates translation from a start codon in more optimal nucleotide context further downstream.

- *Reinitiation*, when translation starts from any ATG codon in optimal nucleotide context in the 5' UTR and then ends at the first stop codon, normally after a short distance. Scanning then continues until the authentic ATG codon is reached.

In these cases the TIS can be located on the second ATG third ATG, or further downstream. The following part of this chapter is dedicated to the computational recognition of the TIS.

Alignment of experimentally verified translation initiation starts leads to the consensus motif *GCCACCatgG*, where a *G* residue following the ATG codon, and a purine, preferably *A*, three nucleotides upstream, are the two highly conserved positions that exert the strongest effect (Kozak, 1984).

A computational identification of a TIS that we present here recognize such conserved motif before the start ATG and analyze the coding potential around the TIS. It is expected that the region before the TIS has a low coding potential and the region following it has a high coding potential. There are different methods to measure the coding potential of a sequence region. The simplest example is the Open Reading Frame (ORF) i.e., the absence of a stop codon in frame. One common method is to count the codons of the region. It is expected that the frequency of codons that encode a protein is different from the frequency of the codons coming from a noncoding region.

The following sections of this chapter present a method for the prediction of TIS based on statistics and ANN's. The algorithm consists of two modules: one sensitive to the conserved motif before the TIS, and one sensitive to the coding/non-coding potential around the TIS. For the final result the two modules are integrated in an algorithm that simulates, in a simplified form, the biological process for the recognition of the start ATG, the ribosome scanning model.

3. DATA SET

A main problem today with collecting data from genomic databases is the reliability of the data. During a literature search it was found that only 1/5th of the annotated TIS's had experimental verification. Finally it was possible, with the help of human expertise, to retrieve a total of 475 corresponding human cDNAs, completely sequenced, annotated and experimentally verified.

For developing and testing the algorithm these genes are grouped into two categories: one gene pool with 75% of the genes is used for the extraction of the training data, and a second pool with 25% of the genes is used for the extraction of the test data.

For the training of the ANN to the *local* information around the TIS regions, 12 nucleotides are extracted from the genes. Every mRNA provides only 1 positive data example for the TIS (Fig. 1.3). Consequently, there is only a relatively small number of data to be used for the training of this ANN. A total of 325 positive and 325 negative regions are used for training and 155 positive and 155 negative regions for testing. A possible similarity at the nucleotide level between the training and the test genes has no effect on the testing because only small regions of sequence (mainly extracted from non-coding regions) are used.

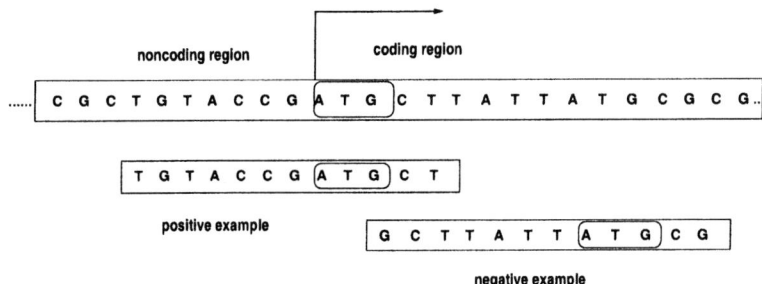

Figure 1.3. Construction of the positive and negative pattern using 12 bases long windows with an ATG starting at the 8th position.

In contrast, for the training of the second ANN for coding potential, it is possible to extract more positive data from every gene. The length of the window used in this case is 54 nucleotides. Here, possible homology between training and test data could influence the result. For this reason, such homologies between the training and test genes are eliminated through pairwise alignment with the full Smith-Waterman algorithm (Waterman, 1995).

Only the genes from the training pool with less than 70% homology to the genes of the test pool are used for extracting the training data for coding potential - a total of 282 genes. Out of these genes, 700 positive and 700 negative sequence windows are extracted from the training pool and 500 windows (half-positive and half-negative) are extracted for testing the performance of the ANN.

For the parametrisation of the integrated algorithm, the whole sequence of genes in the training pool are used. The final evaluation of the algorithm is performed on all the genes from the test pool. This procedure ensures that the test data are never used for training or adjusting any parameters through the entire development of the algorithm.

4.   METHODS
4.1   CONSENSUS - ANN

For the consensus-ANN a window of 12 nucleotides is used. This sequence includes the positions from -7 to $+5$, where $+1$ is the position of the first nucleotide of the coding region (see Fig.1.3). The input is presented to the network through the universal encoding system, where each nucleotide is transformed into a binary 4-digit string (Fig. 1.4).

The different ANN architectures tested during the training are:

■ feed forward nets without hidden units (perceptron),

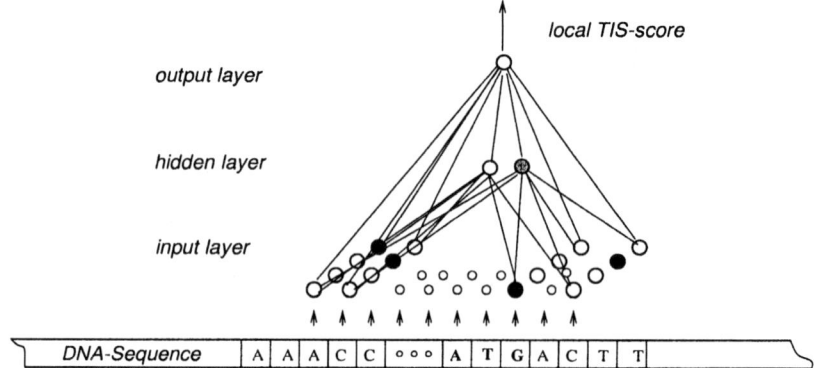

Figure 1.4.   The architecture of the module for the recognition of the consensus motif around
TIS.

- feed forward nets with hidden units and

- feed forward nets with hidden units and short cut connections (direct connections from the input to the output units).

The last mentioned architecture gives the best performance with an accuracy of $76, 4\%$, where the accuracy is taken to be the average of the predictions on the positive and the negative data. In this case, the training is performed with cascade correlation (Fahlman and Lebiere, 1990). In cascade correlation the training starts with a perceptron, which is an ANN with weights only between the input and the output units. After some iterations, part of the weights get frozen (do not change anymore) and a hidden unit is added. In the remaining iteration the new weights are trained to learn examples which were not successfully learned in the first steps of the training.

4.2   CODING - ANN

For the recognition of coding regions, sliding windows of 54 nucleotides are used. Previous investigation has shown that preprocessing the data through a coding measure can significantly improve the performance of the ANN (Hatzigeorgiou et al., 1999). There are different methods for such coding measures. The best results are obtained by applying the codon usage statistic to the sequence window. This leads to a transformation of the sequence window to a vector of 64 units. Every unit gives the frequency (normalized) of the corresponding codon appearing in the window (Fig. 1.5) .

The counting starts with the first nucleotide of the window, counting all non overlapping codons. If the window starts with the first nucleotide of a codon, the ANN has a high score (close to 1), otherwise the score is low (close to 0).

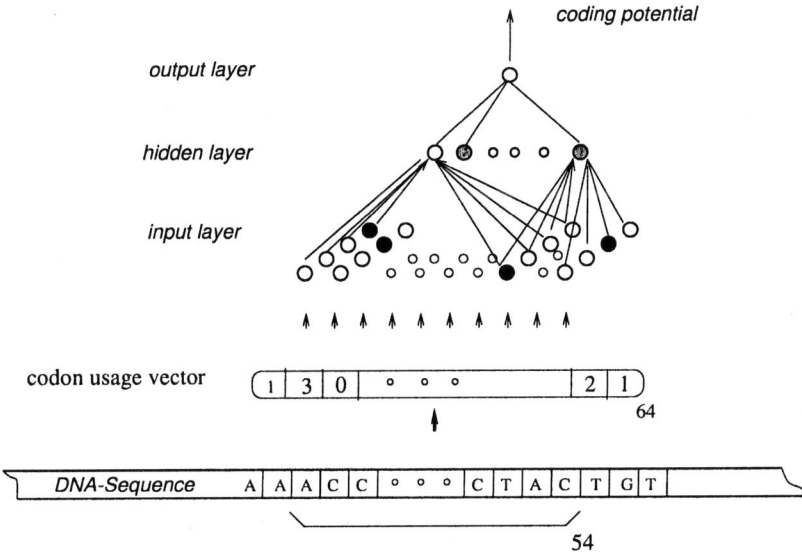

Figure 1.5. The architecture of the module for the recognition of the coding region around TIS. A sequence window of 54 nucleotides length is transformed into a 64 lnog vector, which then is used as an input to the ANN.

The training is done with the algorithm Resilient Backpropagation (RPROP, an improved version of the *classical* Backpropagation) (Riedmiller and Braun, 1993), applied to a feedforward ANN.

There are several variants of such procedures. This work uses the "classical" Backpropagation algorithm (Rumelhart and McClelland, 1986), and its improved version , the Resilient Backpropagation (RPROP) (Riedmiller and Braun, 1993). The main differences of RPROP in comparison with Backpropagation are that:

- the change of the weights depends only on the sign of the potential derivation of the error and not on its size,

- the weight update phase incorporates the current gradient and the gradient of the previous step and

- every weight has its own learning parameter for the changing of the weight value.

The experiments show that RPROP gives better results than Backpropagation. The number of hidden units (2) is examined experimentally (Hatzigeorgiou and Reczko, 1999).

4.3   TRAINING PROCEDURE

A critical point of the training of an ANN is to define the right moment to stop the training. A long training can not only decrease the global error of the training set, but also lead to overtraining of the ANN - it learns the characteristic of every example and not its global nature.

In order to avoid this only 2/3 of the examples of the training set are used for training - the remaining 1/3 is used for evaluating the performance of the ANN after every iteration. The training of the ANN stops once the performance of the *evaluation* group starts to decrease. Notice that the extraction of the evaluation examples is made from genes belonging to the training pool and not to the test pool.

The prediction accuracy of the consensus ANN on the examples of the test set is 76.4% and the prediction of the coding ANN on the examples of the test set is 82.5%. The resulting accuracy is taken to be an average of the correct prediction rate on the positive (true TIS) and negative examples.

All the training of the ANNs is performed by the Stuttgarter Neural Network Simulator (SNNS), publicly available from the University of Stuttgart, Germany (Zell et al., 1993).

4.4   INTEGRATED METHOD

The final algorithm is designed for full sequenced mRNA sequences.

It looks for the longest ORF and gives evidence of the coding potential of this ORF. If this coding score is very low, the user can analyze another ORF within the sequence. This ORF is scanned from the beginning of the sequence for ATG's in frame and their score is investigated.

The score is obtained by combining the output of the two ANNs. Fig. 1.6 gives the prediction of the two modules along the first part of a cDNA sequence. Among many potential ranking strategies, a simple multiplication of the two scores is chosen. The first suitable start codon with a score bigger than 0.2 is examined as the correct start codon. The method gives, for every gene, only one prediction. Out of genes of the test pool 94% of the TIS are correctly predicted and subsequently 6% of the prediction are false positives.

5.   CONCLUDING REMARKS

The problem of gene identification is one of the main tasks of bioinformatics. There has been a great deal of progress in gene identification methods in the last few years. The older coding region identification methods have given way to methods that can suggest the overall structure of genes.

Mapping sequences from cDNA remains the most direct way to characterize the coding parts and provide reliable information for the structural annotation of genes in genomic sequences. This task can not be achieved without proper

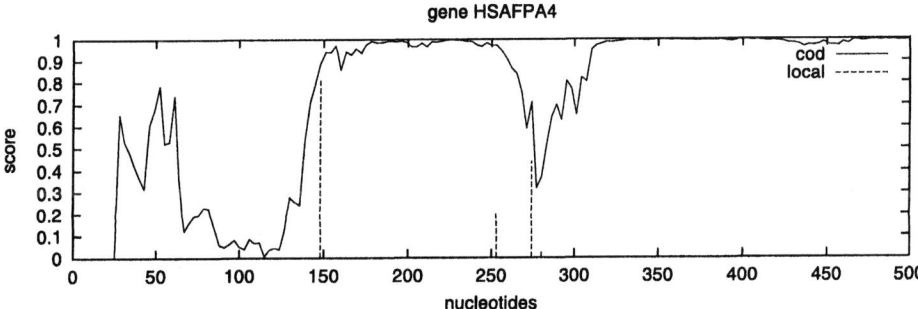

Figure 1.6. The prediction of the two ANNs on a cDNA sequence (part of the gene). The cod-line gives the score of the coding ANN for the coding frame. The local-line gives the position and the score of the consensus ANN for all ATGs in coding frame. The correct TIS is in position 148.

annotation of cDNA itself. A successful method for annotation of cDNA sequences has been demonstrated in this chapter.

After the large sequencing projects (such as the Human Genome Project) are completed the much larger part of the analysis of these sequences begins and computer simulated prediction systems play a major role. Algorithms using sophisticated ANN, fuzzy logic, integrated methods and hybrid systems are in great demand.

Moreover, as we enter the post-genomic era, it is becoming clear that interesting aspects of biology will go far beyond assembling and finding genes, and even beyond predicting the function of endless DNA sequences. Bioinformatics and data generation continue to develop hand in hand to enable us to understand the complexities of cells. It will be exciting to watch the cooperation between bioinformatics and biology in the coming years.

References

Casey, D. (1992). Primer on molecular genetics. Technical report, Human Genome Managment Information System, Oak Ridge National Laboratory.

Fahlman, S. E. and Lebiere, C. (1990). The cascade-correlation learning architecture. In Touretzky, D., editor, *Advances in Neural Information Processing systems II*, pages 524–532, Los Altos, California. Morgan Kaufmann.

Hatzigeorgiou, A., Papanikolaou, H., and Reczko, M. (1999). Finding the reading frame in protein coding regions on dna sequences: a combination of statistical and neural network methods. In Mohammadian, M., editor, *Computational Intelligence: Neural Networks & Advanced Control Strategies*.

Hatzigeorgiou, A. and Reczko, M. (1999). Feature recognition on expressed sequence tags in human dna. In *Proc. of the Intern. Joint Conf. on Neural Networks*. INNS Press.

Kozak, M. (1984). Compilation and analysis of sequences upstream from the translational start site in eukaryotic mrnas. *Nucl. Acids Res.*, 12:857–872.

Kozak, M. (1996). Interpreting cdna sequences: some insights from studies on translation. *Mamalian genome*, 7:563–574.

Pain, V. (1990). Initiaton of proteins synthesis in eukaryotic cells. *Eur. J. Biochem.*, 236:747–771.

Riedmiller, M. and Braun, H. (1993). A direct adaptive method for faster back-propagation learning: The RPROP algorithm. In Ruspini, H., editor, *Proceedings of the IEEE International Conference on Neural Networks (ICNN 93)*, pages 586–591. IEEE, San Francisco.

Rumelhart, D. and McClelland, J. (1986). *Parallel Distributed Processing: Explorations in the Microstructure of Cognition; Vol. 1: Foundations; Vol. 2: Psychological and Biological Models*. MIT Press, Cambridge, Mass.

Waterman, M. (1995). *Introduction to Computational Biology: Sequences, Maps and Genomes*. Chapman Hall.

Zell, A., Mache, N., Hübner, R., Mamier, G., Vogt, M., Herrmann, K. U., Schmalzl, M., Sommer, T., Hatzigeorgiou, A., Döring, S., Posselt, D., Reczko, M., and Riedmiller, M. (1993). SNNS user manual, version 3.0. Technical report, Universität Stuttgart, Fakultät Informatik.

Chapter 8

DETERMINATION OF A LASER CAVITY FIELD SOLUTION USING GLOBAL OPTIMIZATION

Glenn Isenor[1], János D. Pintér[2], and Michael Cada[1]

1 Department of Electrical and Computer Engineering, Dalhousie University
1360 Barrington Street, PO Box 1000, Halifax, NS, Canada B3J 2X4
Glenn.Isenor@dal.ca   and   Michael.Cada@dal.ca

2 Pintér Consulting Services, Inc.
129 Glenforest Drive, Halifax, NS, Canada B3M 1J2
jdpinter@hfx.eastlink.ca   http://www.pinterconsulting.com

Abstract: This chapter explores the steady state characteristics of a distributed feedback (DFB) laser operating in a region above its threshold (turn-on) injection current. Starting with coupled wave theory and using the transfer matrix method, the laser operations problem is formulated in terms of an objective function whose value corresponds to the relative "flatness" of the laser's internal (cavity) field solution. Global optimization techniques implemented in the LGO solver suite are then used to find an optimally flattened steady state field solution for the DFB laser that also satisfies implicitly defined boundary conditions.

Key words: Distributed feedback laser; computational modeling; transfer matrix method; constrained global optimization; LGO solver suite; numerical results.

1.   A BRIEF INTRODUCTION TO LASERS

The history of the laser dates back to the 1958 work of Schawlow and Townes. They proposed a method to synchronize the radiation resulting from a large number of excited atoms by stimulating the exited atoms to emit radiation within a special type of resonant cavity. This was followed in 1960 by the first solid-state ruby laser and helium neon (He-Ne) gas laser. Concurrent research resulted in the 1962 discovery of lasing behavior in semiconductor material: since that time lasers have found increasingly widespread usage. A short – and by no means comprehensive – list of applications would include communication systems, radars, military weapons guidance systems, range finding, medicine, holography, manufacturing, and home entertainment systems.

We present below a brief introduction to distributed feedback lasers, and their theory of operation. For additional information, the reader may wish to consult the excellent books of Agrawal (1992), Ghafouri-Shiraz and Lo (1996), Green (1993), Kawaguchi (1994), Keiser (1993), Tamir (1988), Van Etten and Van der Plaats (1991), and Yariv (1991).

The structure and size of lasers could vary greatly. Semiconductor lasers can be as small as the head of a pin, or can be of room size. Lasing media can consist of various gasses, insulating crystals, liquids, and semiconductor materials. Optical communication systems require a small compact monochromatic (single frequency) laser source, and semiconductor lasers are the choice for this application. However, unless carefully designed, these lasers are not strictly monochromatic.

The achievement of an efficient monochromatic source from a semiconductor laser has been an ongoing research goal. Much of this effort involves numerical modeling of the laser behavior, including modeling the internal optical field shape and intensity. A partial list of references would include the work of Ghafouri-Shiraz and Lo (1996), Kogelnik (1969), Kogelnik and Shank (1972), Makino (1991), Makino and Glinski (1988), Sargent, Swantner, and Thomas (1980), Wang (1974), and Yariv (1973).

The (optical) field solutions must satisfy an increasingly nonlinear set of relations as the laser's optical power increases. The solution space, in turn, is highly nonlinear and multi-extremal. Hence, the required analysis becomes extremely complicated, and the design of optimized laser structures is a challenging task. In our present work, this challenge will be addressed by introducing a numerical approach that integrates the transfer matrix method (TMM) and global optimization (GO) strategies. It will be demonstrated that by applying this approach it is possible to optimize the DFB laser field solution for maximum flatness over varying levels of optical power (injection current).

All types of lasers operate using the same basic principles, which involve the following three processes: photon absorption, spontaneous emission, and stimulated emission. For illustrative purposes consider a simple two state system, with energy levels denoted as E_1 and E_2. The lasing medium's electrons will normally exist in their lowest energy or ground energy state E_1. If an incoming photon has an energy level of $h\nu_{1,2} = E_2 - E_1$, (where h is Planck's constant and ν is frequency), then an electron in the ground state can absorb the photon's energy and be promoted to the excited state E_2. This is an unstable energy state for the electron and in a short time it will spontaneously return to its ground state, while emitting a photon of energy $h\nu_{1,2}$. This process occurs at random and does not rely on an external stimulus. Photons produced in this way are not in phase and assume random polarizations and directions within the lasing medium.

Stimulated emission occurs when an electron in the excited state E_2, is externally induced to make the transition to its ground energy state. This occurs when a photon of energy $h\nu_{1,2}$ interacts with the excited electron. The stimulated photon produced when the electron drops to its ground state is also of energy $h\nu_{1,2}$, but unlike the case of spontaneous emission, this photon is in phase (coherent) with the stimulating photon and has the same polarization.

Normally the level of stimulated emission is negligible unless a condition known as population inversion is achieved. This occurs when the number of electrons occupying the excited state is greater than the number of electrons in the ground state. In this situation stimulated emission may dominate over both absorption and spontaneous emission, and a net optical gain will result. In order to achieve this condition, the laser medium is "pumped" and depending on the type of laser this is done either optically (by an external light source), or electrically (by applying a current or voltage to the lasing medium).

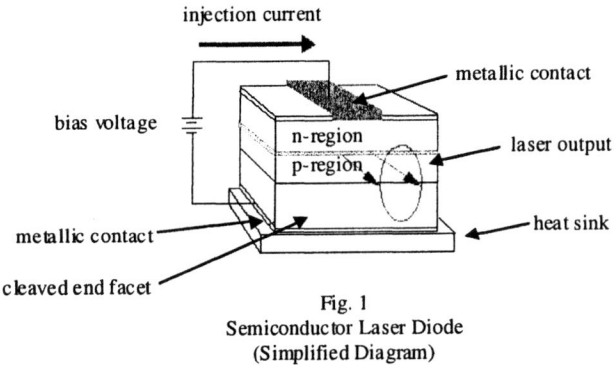

Fig. 1
Semiconductor Laser Diode
(Simplified Diagram)

Fig.1 shows a typical semiconductor laser design. Its approximate dimensions would be in the order of length ≈ 300 μm, width ≈ 200 μm, and thickness ≈100 μm. It consists of a higher refractive index active medium sandwiched between heavily positive (p) and negative (n) doped cladding layers with lower refractive indices. Placing the doped semiconductor materials together in the manner depicted achieves two important results. Firstly, because of the electrical properties associated with the doping (i.e. different band-gap) energies, the electron-hole pairs injected from the electrical current source will be confined to the active region only. This means that when recombination takes place the optical gain is also confined to this region. Secondly, the refractive index difference between the layers acts to confine the field transversely and guide it longitudinally. Therefore in the process of the lasing action, the semiconductor laser medium is pumped by application of an injection current at the device's metallic contacts. This process fills the energy states of the conduction and the valence bands associated with the laser's p-n junction.

Spontaneous and stimulated emissions involve the transition of electrons from the conduction band to the valence band, thus annihilating electron-hole pairs, which is analogous to the E_2 to E_1 energy level transition of the two state system previously introduced. The reader interested in further details may wish to consult one of the suggested references for a detailed discussion of these mechanisms.

Once pumping occurs, all that remains to initiate laser action is to establish a mechanism of optical feedback. In order to set up this situation, the pumped active medium is placed within the confines of an optical resonant cavity. The resonant condition can be achieved in several ways; however, in the case of semiconductor lasers, the earliest methods employed reflective facets at the cavity ends. In very simple terms, as the light bounces back and forth between the reflecting end facets it interacts with the pumped active medium stimulating the emission of photons as the electron-hole pairs recombine. A longitudinal field builds up until internal optical losses are exceeded and the laser begins to exhibit a self–sustained oscillation. This is known as the threshold condition and it is the point that the laser is just "turning on". Less than 100% reflectivity allows a portion of the light to escape through the end facets.

Lasers that employ cavity end reflectivity, whether using reflective facets or external mirrors, are referred to as Fabry–Perot lasers. A semiconductor Fabry-Perot laser has a wide gain spectrum and exhibits multimode longitudinal oscillations. Unlike the simple two-level, single-frequency energy transition model previously discussed, this means that the active medium is capable of emitting radiation across a broad range of frequencies. Because a cavity with facet reflectivity will support oscillations at multiple frequencies, the laser will emit more than one lasing frequency. This makes the Fabry-Perot laser useful, for example, in applications such as a CD player, but it is undesirable for use in optical communication systems.

A single longitudinal mode oscillation resulting in a monochromatic output can be achieved by changing the nature of the cavity feedback. Removing the mechanism of feedback from the cavity ends and distributing it over the entire length of the cavity in the form of a periodic index variation will accomplish this. A semiconductor laser employing this type of feedback mechanism is known as a distributed feedback laser. Please see Fig. 2 for a simplified schematic of a DFB laser.

DFB laser structures are planar and composed of semiconductor materials such as indium gallium arsenide phosphide (InGaAsP) waveguide and active layers as well as n- and p- doped indium phosphide (InP) buffer and substrate layers. Typical dimensions of a DFB laser's active layer are a length of 500 μm, a width of 1.5 μm, and a depth of 0.12 μm.

The grating shown in Fig. 2 is formed using an etching process. These methods are very precise and result in an accurate grating depth, pitch, and period, parameters that determine the feedback and operational characteristics of the laser.

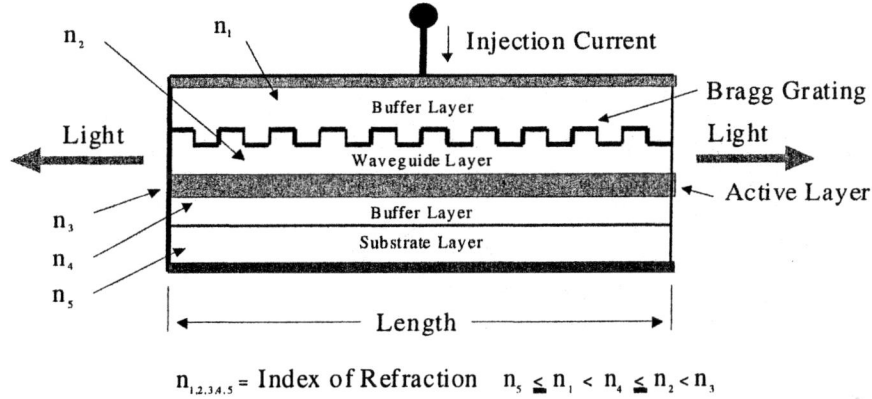

$n_{1,2,3,4,5}$ = Index of Refraction $n_5 \leq n_1 < n_4 \leq n_2 < n_3$

Fig. 2

Index - Coupled DFB Laser

(Simplified Diagram)

The longitudinal periodic index distribution is known as a Bragg grating and the mechanism of feedback is by Bragg diffraction. A wave that is incident on the Bragg grating will result in reflected wavelets at each grating corrugation. If the phase difference between reflecting wavelets is an integer multiple of 2π, then constructive interference will take place. This provides the mechanism for longitudinal mode selectivity. Only those modes will couple constructively that satisfy the condition

(1) $\Lambda = m\left(\lambda_B / 2n_e\right).$

In (1) Λ is the grating period, m is a positive integer, λ_B is the wavelength, and n_e is the effective refractive index of the mode. The example presented in this chapter considers a first order grating, where $m = 1$.

Optical feedback generated in this manner results in two counter-propagating electromagnetic (optical) waves: the right-traveling wave couples energy into the left-traveling wave, at the same time the left-traveling wave couples energy into the right-traveling wave. The strength of this coupling is governed by the coupling coefficient K, a parameter that is determined by the grating pitch, depth, period, and position relative to the active layer.

When the coupling is restricted to a grating that consists of a periodic index perturbation only, this is referred to as index-coupling. Other classes of coupling include mixed or gain-coupling. The mixed-coupled structure has its corrugation layer fabricated on the upper portion of the active layer. Index-coupling is induced through the periodic variation in the refractive index associated with the corrugation layer; however, the active layer thickness is also modulated by the presence of the corrugation. This in turn results in a longitudinal modulation of the amplitude gain, which induces the gain-coupling effect.

Pure gain-coupling is achieved by fabricating a second grating layer on top of the mixed-coupling grating structure. This second grating employs an inverse phase corrugation, which acts to cancel the index-coupling effects of the original grating. These various coupling classes, index, mixed, and gain-coupling, are reflected in K values that are respectively either purely real, complex, or purely imaginary.

Although the forthcoming method could be applied equally well to the other classes of coupling, the index-coupled DFB laser is extensively used, and for the purpose of this chapter consideration will be restricted to the DFB structure, with no fundamental loss of generality.

The DFB laser's ability to produce a single longitudinal mode oscillation resulting in a monochromatic output is of fundamental importance in optical communications applications. Much work has gone into the design of such devices in the effort to achieve narrow spectral line widths, reduced longitudinal spatial hole-burning (LSHB), low threshold (turn-on) current, and efficient power utilization. The introduction of a quarter-wave phase shift at the laser cavity mid-point eliminated a major problem associated with DFB lasers, that of low power mode-degeneracy. This is the appearance of unwanted side modes at an injection current level that is above, but still close to the threshold injection current. This characteristic is due to insufficient gain margin at the low power level. Unfortunately, addition of the quarter-wave phase shift also resulted in a highly concentrated field at the phase shift plane causing LSHB. At this point the intense field of a QWS DFB structure locally depletes the carrier density, and as the laser power increases, this effect ultimately results in mode-degeneracy now occurring at higher operating powers.

DFB laser design, including the reduction of LSHB, has been the target of intensive research. A partial list of references includes the works of Fang, Hsu, Chuang, Tanbun-Ek, and Sergent (1997), Fessant (1997), Morthier and Baets (1991), Morthier, David, Vankwikelberge, and Baets (1990), and Rabinovitch and Fieldman (1989). Recent research by Wang, Cada, and Makino (1998), Wang, Cada, and Sun (1999), and Wang and Cada (2000) has resulted in the development of a novel coupled-power technique, which has also been successfully used to investigate the behavior and the LSHB reduction of a DFB laser.

One method used to reduce LSHB is to introduce a longitudinally distributed coupling coefficient (DCC). The DCC is typically length-normalized and is referred to as the Kappa-L product (KL). By optimally choosing a KL distribution, the field maximum that occurs at the laser cavity midpoint in a QWS structure for a uniform KL is reduced, thereby reducing the LSHB effects. The question then is, given a finite number of (fixed) KL sections, is it possible to select the KL distribution so that the internal field is optimally flattened, or at least approaches a uniform state?

It is also possible to achieve a reduction in the field intensity non-uniformity by moving away from the QWS structure and longitudinally distributing the phase shift (PS) profile. This method may be used alone or in combination with a DCC. Such approaches have been investigated using designs at threshold power, see Ghafouri-Shiraz and Lo (1996), and Yokoyama and Sekino (1998).

The "above threshold" operations approach, discussed in this chapter, uses global optimization in conjunction with the transfer matrix method (TMM) to determine coupling coefficient distributions. This will lead to optimal reductions in the non-uniformity of field intensity, for various levels of above-threshold operation. The advantage of this new method is that it is not limited to a "threshold only" analysis. Rather, it is directly employable in the "above threshold" operating range of the laser, where the relationships modeled are highly non-linear. Key advantages of our approach include the potential to develop, fine tune, and test threshold designs for any desired laser power level. Furthermore, this approach can reveal new and interesting phenomena related to exploring multiple near-optimal design solutions.

2.   COUPLED WAVE THEORY

The pioneering work of Kogelnik and Shank (1972) provided the first theoretical explanation of the operation of a DFB laser structure. Methods based on coupled wave analysis have since been used extensively to model the steady state characteristics of DFB lasers.

Coupled wave theory forms the basis of the TMM as applied to the threshold and above-threshold analysis of DFB lasers: our discussion will be restricted to this approach. Lateral and transverse field confinement imposed by the structural constraints of the DFB laser necessitates that each longitudinal field solution must satisfy the following one-dimensional time independent scalar wave equation

(2) $d^2E(z)/dz^2 + k^2(z)E(z) = 0$.

In (2) z represents the direction of field propagation or longitudinal direction of the laser. The term $E(z)$ is the complex amplitude of a time-harmonic field and is considered to be independent of the width x and thickness or transverse y directions of the laser.

We will consider the general case where the Bragg grating consists of both a periodic refractive index and gain variation. Then, using a first order approximation, the index $n(z)$, and gain $\alpha(z)$ profiles are written as

(3a) $n(z) = n + n_1 \cos(2\beta_o z)$;

(3b) $\alpha(z) = \alpha + \alpha_1 \cos(2\beta_o z)$.

In (3) the parameters $n(z)$ and $\alpha(z)$ are average values, while n_1 and α_1 are the maximum amplitudes of the periodic variations in laser medium refractive index and gain, respectively; and β_o is the Bragg propagation constant. The wave propagation constant, k, for a wave propagating in a complex dielectric is defined as

(4) $$k^2 = k_o^2 n^2(z)\left(1 + j\,\frac{2\alpha(z)}{k_o n(z)}\right).$$

In (4) k_o is the free space propagation constant. Assuming that the perturbations in gain and index are much smaller than their average values, ($n_1 << n$, $\alpha_1 << \alpha$), substituting (3a) and (3b) into (4) results in the following expression for k:

(5) $$k^2 = k_o^2 n^2(z) + j2k_o n(z)\alpha(z) + 4k_o n(z)\left(\frac{\pi n_1}{\lambda} + j\,\frac{\alpha_1}{2}\right)\cos(2\beta_o z).$$

Since the coupling coefficient K is defined as

(6) $$K = \frac{\pi n_1}{\lambda} + j\,\frac{\alpha_1}{2},$$

by replacing $k_o n(z)$ with β, (5) can be re-written in the following form:

(7) $$k^2 = \beta^2 + j2\beta\alpha(z) + 4\beta K \cos(2\beta_o z).$$

Substitution of (7) into (2) yields

(8) $$\frac{d^2 E(z)}{dz^2} + [\beta^2 + j2\beta\alpha(z) + 4\beta K \cos(2\beta_o z)]E(z) = 0.$$

A necessary condition for coupling and propagation is that the Bragg condition must be nearly satisfied. This means that the actual propagation constant β must be sufficiently close to the Bragg propagation constant β_o such that the absolute difference between them is much less than the Bragg propagation constant: $|\beta - \beta_o| << \beta_o$. With this considered, the solution to (8) is the total complex electric field amplitude along the grating and results from a linear superposition of two counter-propagating waves (electric fields).

The expression

(9) $E(z) = R(z)e^{-j\beta_o z} + S(z)e^{j\beta_o z}$

is the trial solution to the scalar wave equation, and is used to construct a general solution by substitution into (8). The field solutions vary slowly in amplitude allowing second order derivatives in $R(z)$ and $S(z)$ to be neglected. Similarly because of $|\beta - \beta_o| \ll \beta_o$ rapidly changing phase terms such as $exp(\pm j3\beta_o z)$ can be ignored. (It is common practice in coupled-mode theory to neglect the higher order terms because there is no coupling between 3rd order traveling waves.) Finally, the following approximation is applied:

(10) $\dfrac{\beta^2 - \beta_o^2}{2\beta_o} \approx \beta - \beta_o = n(\omega - \omega_o)/c = \delta$.

In (10) c is the speed of light in vacuum, ω and ω_o are the longitudinal mode and Bragg frequencies, respectively, and the parameter δ is known as the detuning coefficient. It is a measure of the difference between the Bragg propagation constant and the actual propagation constant of the longitudinal mode. When the terms containing similar exponents are grouped together, the general solution, which consists of the following pair of coupled wave equations, is obtained.

(11a) $-\dfrac{dR(z)}{dz} + (\alpha_o - j\delta)R(z) = jKS(z)$;

(11b) $\dfrac{dS(z)}{dz} + (\alpha_o - j\delta)S(z) = jKR(z)$.

Expressing the trial solution to the coupled wave equations in terms of the complex propagation constant, γ, the following relations are obtained

(12a) $R(z) = R_1 e^{\gamma z} + R_2 e^{-\gamma z}$,

(12b) $S(z) = S_1 e^{\gamma z} + S_2 e^{-\gamma z}$.

In (12) R_1, R_2, S_1, and S_2 are complex coefficients that depend on the field boundary conditions at the left and right hand facets. These boundary conditions in turn depend on facet reflectivity, which is assumed to be zero in this analysis. The dispersion relation determines the complex propagation constant

(13) $\gamma^2 = K^2 + (\alpha - j\delta)^2$.

In the case of zero facet reflectivity, the exact solutions to the coupled wave equations are written as

(14a) $R(z) = \sinh \gamma(z + \dfrac{1}{2}L)$,

(14b) $S(z) = \sinh \gamma(z - \dfrac{1}{2}L)$.

In (14) L is the length of the DFB laser. Eqs. (14a) and (14b) form a set of eigenfunctions with corresponding eigenvalues γ that determine the oscillation modes of the laser, as a function of the length L and the coupling coefficient K.

Substitution of (14a)–(14b) into (11a)–(11b) results in the following threshold equation for zero facet reflectivity (Ghafouri-Shiraz and Lo, 1996):

(15) $j\gamma L = \pm KL \sinh(\gamma L)$.

For a fixed value of the length-normalized coupling coefficient; KL, it is possible to solve (15) for the various oscillation modes of the DFB laser. These modes are expressed in terms of the detuning parameter δ, and the threshold gain α. Typically these parameters are also normalized relative to the length of the DFB laser and are written as δL and αL.

Equations (1) to (15) provide a complete description of the oscillating mode characteristics and resultant field profiles of a DFB laser at threshold, i.e. the point where the laser is just turning on. At threshold, the total quantity of stimulated photons is considered negligible. As the injection current increases, so does the dynamic range of the laser's longitudinal carrier density and the stimulated photon density profiles. This interrelationship is also reflected by an increase in the dynamic range of the laser's internal electric field intensity profile. The non-linear interactions between these profiles must now be considered because of their effects on the laser's oscillating mode characteristics. The next section will illustrate how the transfer matrix method is used to address the above-threshold problem

3.   TRANSFER MATRIX METHOD

The TMM approach presented here uses as its starting point a methodology originally proposed by Ghafouri-Shiraz and Lo (1996).

Beginning at the left-hand facet, and using sectional transfer matrices, the DFB laser's internal electric field is longitudinally propagated through the structure until the right-hand facet is reached. Subsequently, the oscillation mode characteristics for an above-threshold injection current are obtained by matching the internal field's right-hand boundary condition (RBC). This is done iteratively: based on the RBC error in the previous iteration, new potential solution values are selected and evaluated by re-propagating the electric field.

Ghafouri-Shiraz and Lo's heuristic solution strategy utilizes an adaptable numerical grid over the solution space, to select each new set of potential solution values. This method "walks around" the solution space until the RBC error is minimized. Because of the multi-extremal nature of the error function, one has to carefully select a starting point when solving the problem at a given injection current. Solutions are obtained for increasing injection current by using a previous lower injection current result as a starting point. These increases must be sufficiently small to avoid ending up in a local minimum of the error function, which would represent a non-physical field solution.

The methodology presented in this chapter differs from the preceding approach in that it integrates the above-threshold transfer matrix method with robust global optimization (GO) strategies, to search the entire solution space and select the solution that minimizes the RBC error. The GO approach is theoretically insensitive to the starting point. By using this method, it is possible to directly and rapidly solve for injection currents that vary from 1.1 to 5 times the threshold injection current. Ghafouri-Shiraz and Lo (1996) selected this range of injection currents to explore the behavior of a bulk semiconductor lasing around 1550 *nm*. This range was chosen primarily because it covered a sufficient span to permit the material saturation effects to fully mature.

This chapter explores a numerical example by using the combination of the TMM and GO. This approach was fully developed by Isenor (2001), and subsequently discussed by Isenor, Pintér, and Cada (2003). The following exposition draws on both of these works.

The TMM approach requires dividing the DFB laser into a large number of equal-length sections; within each section all physical parameters are considered constant. In the case of the DFB laser's electric field distribution, it allows a simple 2x2 matrix relationship to be developed between the input and output fields for any section. Refer to Fig. 3 for a simplified schematic of an arbitrary (m^{th}) section of a DFB semiconductor laser diode.

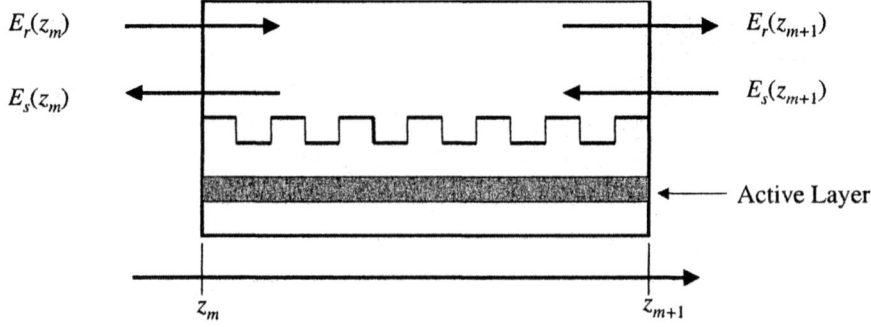

Fig. 3

DFB Section of a Semiconductor Laser Diode

For each laser section it is possible to re-write the trial solutions (12a) and (12b) in such a fashion that the input and output electric fields are related by a matrix equation of the form

(16)
$$
\begin{bmatrix} E_r(z_{m+1}) \\ E_s(z_{m+1}) \end{bmatrix} = \begin{bmatrix} t_{11} & t_{12} \\ t_{21} & t_{22} \end{bmatrix} \cdot \begin{bmatrix} E_r(z_m) \\ E_s(z_m) \end{bmatrix}.
$$

In (16) the matrix elements, $t_{ij,}$ are written as

(17a) $t_{11} = \left(E - \rho^2 E^{-1} \right) \cdot e^{-j\beta_o(z_{m+1}-z_m)} / \left(1 - \rho^2 \right),$

(17b) $t_{12} = -\rho \left(E - E^{-1} \right) \cdot e^{-j\beta_o(z_{m+1}+z_m)} / \left(1 - \rho^2 \right),$

(17c) $t_{21} = \rho \left(E - E^{-1} \right) \cdot e^{j\beta_o(z_{m+1}+z_m)} / \left(1 - \rho^2 \right),$

(17d) $t_{22} = -\left(\rho^2 E - E^{-1} \right) \cdot e^{j\beta_o(z_{m+1}-z_m)} / \left(1 - \rho^2 \right).$

In the relations (17)

(18) $\rho = \dfrac{jK}{(\alpha - j\delta + \gamma)},$

and

(19) $E = e^{\gamma(z_{m+1}-z_m)}, \qquad E^{-1} = e^{-\gamma(z_{m+1}-z_m)}.$

The matrix product of the individual section transfer matrices, written as

(20) $Y\left(z_{M+1} | z_1 \right) = T^M \cdot T^{M-1} \cdots T^2 \cdot T^1$

forms the complete transfer matrix Y, for a DFB laser of M sections. This matrix fully describes the propagation characteristics of the forward and backward traveling waves, for the entire DFB laser.

In the example considered in this chapter, a quarter-wave phase shift (QWS) section is introduced at the midpoint of the laser to ensure single longitudinal mode stability. In a QWS structure, the input and output electric fields are considered continuous, encountering only a shift in phase as they travel through this section. The quarter-wave phase shift section is described by the matrix

$$(21) \quad P = \begin{bmatrix} e^{j\frac{\pi}{2}} & 0 \\ 0 & e^{-j\frac{\pi}{2}} \end{bmatrix}.$$

The phase shift section is easily incorporated into (20) in the following manner

$$(22) \quad Y\!\left(z_{M+1}\big|z_1\right) = T^M \cdot T^{M-1} \cdots T^{\frac{M}{2}+1} \cdot P \cdot T^{\frac{M}{2}-1} \cdots T^2 \cdot T^1.$$

Expression (22) gives the complete transfer matrix representation of a QWS DFB laser. In order to insure symmetry, an even number of equal-length sections is chosen, allowing the phase shift section to be placed at exactly the midpoint of the DFB laser. This consideration is also reflected by (22).

In the numerical example studied here, M is chosen to be 5000, and the phase shift section is placed after section 2500. This section is a plane of quarter-wave phase transition for the longitudinal electric fields, i.e. it has no actual physical length. Using (22) it is now possible to relate the electric fields that appear at the right hand facet of a QWS DFB to the electric fields at the left hand facet by the following matrix equation

$$(23) \quad \begin{bmatrix} E_r(z_{M+1}) \\ E_s(z_{M+1}) \end{bmatrix} = Y\!\left(z_{M+1}\big|z_1\right) \cdot \begin{bmatrix} E_r(z_1) \\ E_s(z_1) \end{bmatrix}.$$

It should be noted that (22) reflects an ideal QWS DFB laser structure with zero facet reflectivity at both facets, as well as no residual grating phase at either facet. Non-zero residual grating phases at the facets as well as facet reflectivity would be accounted for in the transfer matrix formulations, and could easily be incorporated into (22) and (23), if needed. However, for the purposes of the present discussion these considerations are an unnecessary complexity, and may be omitted.

4.   LASER THRESHOLD CONDITION USING THE TMM

A laser at threshold is an optical oscillator. This means that, at threshold, the right-traveling electric field with a value of zero at the left-hand facet would grow in intensity until it reaches the right-hand facet. (The same consideration would apply to the left-traveling electric field). In order for this to occur, the transmission gain of both the left-and right-traveling fields must be theoretically infinite.

Using Eq. 23 the following relationship is obtained for the transmission gain, A_t, of the left-traveling field

$$(24) \quad A_t = \frac{E_s(z_1^-)}{E_s(z_{M+1}^+)} = \frac{1}{y_{22}(z_{M+1}|z_1)},$$

where

$$(25) \quad y_{22}(z_{M+1}|z_1) = 0.$$

When (25) numerically approaches zero, the transmission gain for the left traveling field approaches infinity and the cavity becomes resonant. Solving (25) is analogous to solving (15), and it results in the oscillation mode parameters of the laser at threshold, again expressed in terms of the detuning parameter δ, and the threshold gain α.

5.   ABOVE-THRESHOLD OPTIMIZATION

An analysis of the above-threshold behavior of a DFB laser must incorporate the nonlinear interactions between the longitudinally varying parameters of electric field intensity distribution, carrier concentration, photon density, and refractive index for a given injection current. Because the physical parameters in any given TMM section are assumed to be homogeneous, it is possible to obtain these parameters for that section. By using a sufficiently large number of TMM sections, it is possible to numerically extend the localized results to a continuous distribution.

In the numerical example presented here, we shall study the field flatness optimization of a QWS DFB laser consisting of six equal length sections symmetrically arranged about the laser's midpoint. A schematic of this structure is presented in Fig. 4. Each section has associated length-normalized coupling coefficients, *KL1* to *KL6*; phase shift planes, denoted by *PS1* to *PS5*, separate each section. The structural symmetry permits the following simplifications: *KL1=KL6*, *KL2=KL5*, and *KL3=KL4*. The plane of symmetry is *PS3*, the laser's midpoint, where a 90 degree phase shift is applied. This arrangement reduces the numerical complexity associated with coupling coefficient optimization by three variables. The remaining phase shift planes are indicated for completeness but are considered to be zero for the simple example considered.

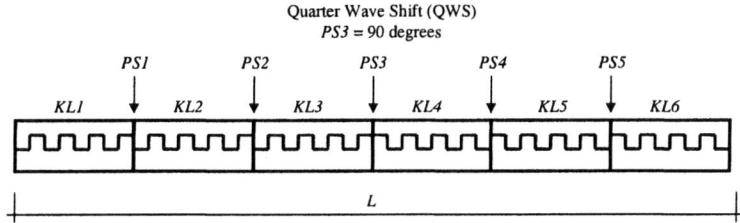

Fig. 4

Schematic of DFB Laser Structure

Therefore, the next step is to apply optimization methodology, to select the coupling coefficient profile *KL* (*KL1 to KL6*) design parameters as well as the lasing wavelength, λ, and the dimensionless coefficient, C_o, such that the field flatness is maximized. Note that C_o, relates the actual total electric field, $\bar{E}$, to the normalized total electric field, E, in the following manner:

$$(26) \quad \bar{E}(z) = C_o E(z) = C_o [E_r(z) + E_s(z)].$$

There are fundamentally two conditions that must be simultaneously considered when solving the above problem. The first condition is the requirement that the field solution must match the right-hand facet boundary condition. The field propagation is chosen to start from the left-hand facet and moves through the laser structure section by section until the right-hand facet is reached. By this choice of the starting point the left-hand facet boundary condition must be automatically satisfied. Assuming that the laser structure is symmetric, and the facets are anti-reflective, at this point the normalized left-traveling field intensity should be zero and the normalized right-traveling field intensity should be one. Under these considerations the right-hand boundary condition is met if (25) is satisfied (Ghafouri-Shiraz and Lo, 1996). One may use this fact directly in the iterative process to find the best solution or opt to consider the actual left-traveling field intensity's requirement to be zero at the right-hand facet. The latter approach was considered in our present investigation. It should be noted that with either approach (25) is still directly or indirectly minimized.

Finding the above-threshold field solution for a fixed KL profile involves the iterative selection of the λ and C_o that results in the best field boundary condition match at the right-hand facet. This in itself is an optimization problem where the objective or merit function to be minimized is the boundary condition error. The second condition to be met is maximizing the actual field flatness. KL values have to be optimized such that the field is maximally flattened while at the same time the boundary condition constraint must be satisfied. The field flatness function F is defined as

$$(27) \quad F = \frac{1}{L} \int_{z_1}^{z_{M+1}} \left(I(z) - I_{avg} \right)^2 dz,$$

and the optimized objective function becomes

$$(28) \quad Objf = F + scal \times RBC\,error^2.$$

In (27) $I(z)$ is the sectional field intensity, I_{avg} is the average field intensity over the laser length, $RBCerror$ is the right-hand boundary condition error, and $scal$ is a suitable scaling (penalty) parameter. This represents one of the simplest possible objective functions for this problem.

The definition of the objective function depends on the problem complexity and the relationships between the solution parameters. If additional considerations are required such as inclusion of a residual phase corrugation and non-zero end facet reflectivity, then it may become necessary to redefine the objective function. However, for the purposes of this discussion, the objective (28) provides a reasonable starting point and yields acceptable results. Based on the outcome of several experimental runs using a selection of scaling parameters, a value of $scal = 1000$ was selected and used in the evaluation of the succeeding numerical examples.

Observe that it is necessary to constrain the solution such that minimization of $RBCerror$ takes precedence over maximizing the field flatness. This consideration is emphasized by specifying $RBCerror=0$, as an explicit independent constraint, which ensures solution validity.

When a laser is operating above–threshold (i.e. it is "turned-on"), the carrier rate equation must be included to properly consider the relationship between the injection current, i, the carrier density, N, the stimulated photon density, S, and the net (material) gain, g, in each laser section. This equation is written as

(29) $\quad \dfrac{i}{qV} = \dfrac{N}{\tau} + BN^2 + CN^3 + \dfrac{v_g gS}{1+\sigma S}.$

The volume, V, is determined from the geometry of the active region. The parameter σ accounts for saturation effects at high photon densities. The remaining equation parameters consist of the bimolecular recombination coefficient B, the Auger recombination coefficient C, the linear recombination lifetime τ, and the group velocity at the Bragg wavelength v_g.

The amplitude gain and detuning in each matrix section depend on the carrier density and are written as

(30) $\quad \alpha = (\Gamma g - \alpha_{loss})/2,$

(31) $\quad \delta = \dfrac{2\pi}{\lambda} n - \dfrac{2\pi n_g}{\lambda \lambda_B}(\lambda - \lambda_B) - \dfrac{\pi}{\Lambda}.$

In (30)-(31) Γ is the optical confinement coefficient, α_{loss} is the internal cavity loss, n is the effective index, n_g is the group refractive index, λ_B is the Bragg wavelength, and λ is the lasing wavelength. The effective index dependence on the carrier density is defined as

(32) $\quad n = n_e + \Gamma \dfrac{\partial n}{\partial N} N,$

where n_e is the effective refractive index at zero injection current, and the term $\partial n/\partial N$ is the differential index. The stimulated photon density in each laser section, S_z, is given by the following expression

(33) $\quad S_z \approx \dfrac{2\varepsilon_o n(z) n_g \lambda}{hc} \cdot C_o^2 \left[\left| E_r(z) \right|^2 + \left| E_s(z) \right|^2 \right],$

where ε_o is the permittivity of free space, h is Planck's constant, and c is the speed of light in vacuum. C_o is the dimensionless normalization coefficient. Both λ and C_o need to be determined in the calculation such that the corresponding field profile matches the boundary conditions at the laser facets. Finally, the following parabolic gain model is used to characterize the active medium's gain

(34) $\quad g = A_0(N - N_o) - A_1[\lambda - (\lambda_o - A_2(N - N_o))]^2.$

At the transparency carrier density N_o, this expression reduces to

(35) $\quad \lambda_o = \lambda_{th} + A_2(N_{th} - N_o).$

The wavelength λ_o is the wavelength at which the material gain is zero, at the transparency carrier density, and is defined as the peak gain wavelength at zero gain transparency. In the above expression, A_0 is the differential gain, and the parameters A_1 and A_2 are associated with the width of the gain spectrum and changes in gain, respectively, that result from shifts in the peak wavelength.

6.   NUMERICAL MODEL EVALUATION

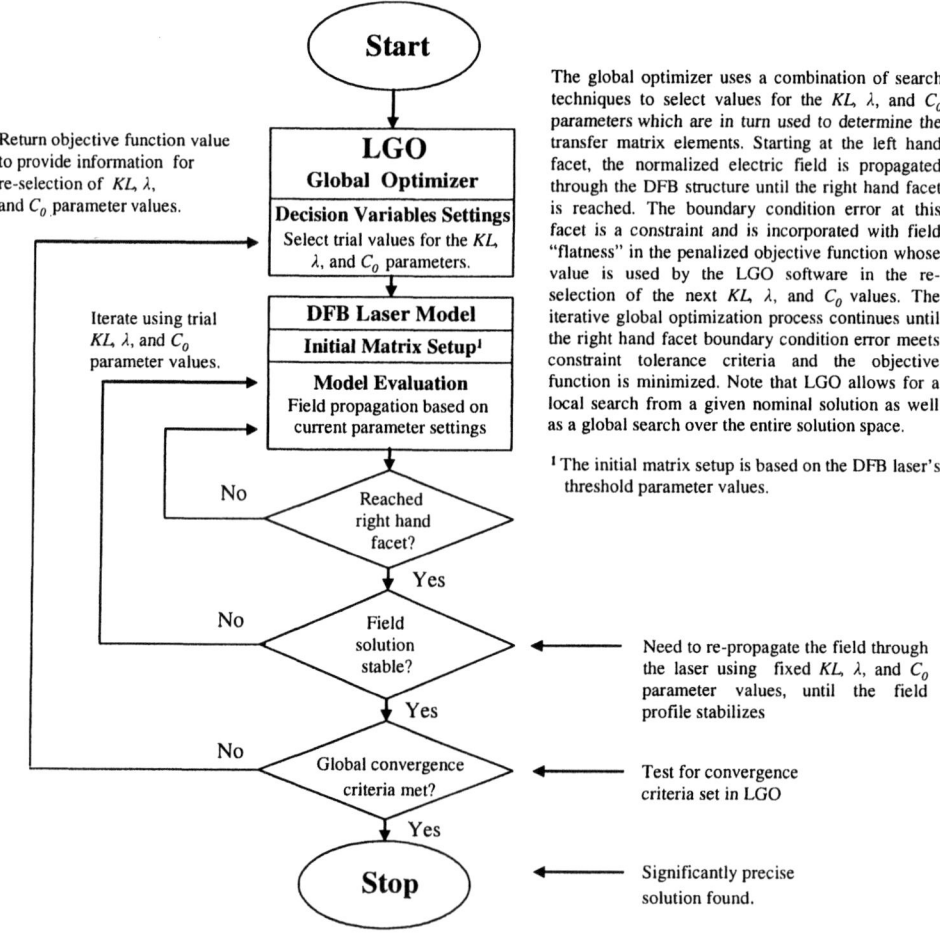

The global optimizer uses a combination of search techniques to select values for the KL, λ, and C_0 parameters which are in turn used to determine the transfer matrix elements. Starting at the left hand facet, the normalized electric field is propagated through the DFB structure until the right hand facet is reached. The boundary condition error at this facet is a constraint and is incorporated with field "flatness" in the penalized objective function whose value is used by the LGO software in the re-selection of the next KL, λ, and C_0 values. The iterative global optimization process continues until the right hand facet boundary condition error meets constraint tolerance criteria and the objective function is minimized. Note that LGO allows for a local search from a given nominal solution as well as a global search over the entire solution space.

[1] The initial matrix setup is based on the DFB laser's threshold parameter values.

Fig. 5

DFB Laser Numerical Modeling and Optimization Procedure

The flow chart shown in Fig. 5 illustrates the use of (the LGO) global optimization software in the solution methodology for obtaining a DFB laser's above-threshold optimized field solution.

The calculation proceeds by first obtaining the laser's threshold condition, α_{th} and δ_{th}, from (25). Next, (26), and (29) to (35) determine all other threshold parameters, N_{th}, n_{th}, λ_o, λ_{th}, and i_{th} , which are used to initialize each sectional transfer matrix. The normalized field is propagated through each laser section, starting at the left-hand facet, until the right-hand facet is reached. The final self-consistent field solution is obtained iteratively such that with each left-to-right pass through a laser section its effective index is updated to reflect the revised sectional carrier density. Because of the complex non-linear interrelationship between the longitudinal field, the longitudinal carrier density, the refractive index, and the photon density profiles, structural iterations must continue until there are no changes in the profiles with each subsequent pass. In so doing, the field solution quickly stabilizes for a given set of KL, λ, and C_o values. Once the profiles have stabilized, the model function values are evaluated and passed along to the optimizer. Based on the search technique employed, the optimizer selects new parameter (KL, λ, and C_o) values.

From the preceding description it is clear that the solution requires the optimization of a possibly highly non-linear "black box" system. The "black box" terminology used indicates the fact that the system model outlined is complicated, and that its evaluation (for each parameter combination studied) requires a computationally intensive procedure. In such cases, there is no guarantee of convexity in the model structure. Therefore the "traditional" repertoire of (local) optimization is insufficient, and a genuine global scope optimization methodology is required.

7.   THE LGO SOLVER SUITE

To find the globally best set of parameters, we use the LGO solver engine. LGO abbreviates a Lipschitz(-Continuous) Global Optimizer that has been designed to handle global optimization models under very general conditions.

Formally, the general continuous global optimization (CGO) model is stated as

(36) $min f(x)$ subject to $x \in D$ $D := \{x: l \le x \le u$ $g_j(x) \le 0$ $j = 1,...,m\}$.

In (36) we apply the following notation and assumptions:
- $x \in R^n$ real n-vector of decision variables,
- $f: R^n \rightarrow R$ continuous (scalar-valued) objective function,

- $D \subset R^n$ non-empty set of feasible solutions, a proper subset of R^n.

The feasible set D is defined by
- l and u finite (component-wise) lower and upper bounds on x,
- $g:R^n \to R^m$ m-vector of continuous constraint functions.

Let us note here that the constraints g_j in (36) could have arbitrary ($\leq$, =, $\geq$) relation signs, and that explicit bounds on the constraint function values can also be imposed. Such – formally more general and/or specialized – models are directly deducible to the form (36). Without going into details not needed here, let us also remark that models with bounded integer variables can also be brought to this form. This, of course, also implies the formal coverage of mixed integer models.

To ensure the numerical solvability of (36), on the basis of a finite sample point sequence from D, the Lipschitz-continuity of the model is often assumed. Recall that a function f is Lipschitz(-continuous) in the set D if the relation

(37) $|f(x_1)-f(x_2)| \leq L\|x_1-x_2\|$

is valid for all vector-pairs x_1, x_2 from D. In the right-hand side of (37), the Euclidean norm is used; the value L is the corresponding (smallest possible) Lipschitz constant. Observe that the value $L=L(D,f)$ is typically unknown, although a suitable value is often postulated or estimated in numerical practice: consult e.g. Pintér (1996). Similar Lipschitz conditions can be postulated with respect to the component constraint functions in g:

(38) $|g_j(x_1) - g_j(x_2)| \leq L_j\|x_1 - x_2\|$ $L_j=L(D,g_j)$ $j=1,...,m$.

The model (36) with a postulated, or proven, Lipschitz structure (37)-(38) is still very general. In fact, it includes most GO problem types that occur in practice. As a consequence, it includes also very difficult problem instances. Obviously, for given model instances, the corresponding "most suitable" solution approach could vary to a considerable extent. On one hand, a "universal" strategy will work for broad model classes, although its efficiency might be lower for certain problem instances. On the other hand, highly tailored algorithms often will not work for models outside of their intended scope.

LGO has been designed to handle, in principle, the entire class of models defined by (36) and (37)-(38), without requiring any further special structure. This design principle and the corresponding choice of component algorithms makes LGO applicable even to "black box" models like the laser design model introduced above. At the same time, models with a given more specific structure (e.g. an indefinite quadratic objective f over convex D) can also be solved by LGO.

The LGO algorithm system is documented in details elsewhere: consult, for instance, Pintér (1996, 2001, 2002, 2005) and further topical references therein. Therefore only a concise review of its features is included here.

The overall solution approach implemented in LGO is based on an integrated suite of theoretically convergent global and efficient local search methods. Currently, the following search algorithms are offered:

- branch-and-bound based global search (BB)
- adaptive global random search (single-start) (GARS)
- adaptive global random search (multi-start) (MS)
- constrained local search (generalized reduced gradient method) (LS).

The global solver modes are mutually exclusive within a given program run. First, the selected global solver is executed and then it is automatically followed by the local solver. The local solver can also be used in a stand-alone mode, started from a user-supplied initial solution.

All three global solvers are gradient-free, since these genuinely need only model function values. Specifically, their operations are based on iteratively calculated values of the exact penalty (merit) function defined as

$$(39) \qquad f(x) + \sum_{j \in E} |g_j(x)|^2 + \sum_{j \in I} \max(g_j(x), 0)^2$$

In (39) the index sets E and I in the summations respectively denote the subsets of equality and inequality constraints g_j, recall (36). The local solver option is also gradient-free, since finite difference based gradient approximations are used. Again, this approach fully supports (also) the optimization of "black box" systems.

In development since some 15 years, LGO has been implemented, tested and supported across a growing number of programming languages and modeling environments. The current list of compiler platforms includes well-tested Fortran (Lahey Fortran 77/90, Lahey-Fujitsu Fortran 95, Digital/Compaq Visual Fortran 95, g77) implementations, with direct connectivity also to C models developed using Borland C/C++, Microsoft Visual C/C++, gcc, and lcc-win32. There is also a range of LGO implementations available for prominent modeling and scientific computing environments. Currently, these include the following:

- LGO solver for the Excel Premium Solver Platform (Frontline Systems and Pintér Consulting Services, 2001)
- LGO solver for GAMS (GAMS Development Corporation and Pintér Consulting Services, 2003)

- MathOptimizer Professional for *Mathematica* (Pintér and Kampas, 2003)
- TOMLAB/LGO for MATLAB (TOMLAB Optimization and Pintér Consulting Services, 2004)
- Global Optimization Toolbox for Maple (Maplesoft and Pintér Consulting Services, 2004)
- LGO solver for MPL (Maximal Software and Pintér Consulting Services, 2005).

LGO has been applied to solve a broad range of optimization models, in a large variety of contexts. The current shipment version can be configured to handle models formulated with thousands of variables and constrains. Note, however, that runtimes may become significant, when solving complex and/or large models. Application areas include engineering design, chemical and process industries, econometrics and finance, medical research, biotechnology, and scientific modeling.

8.   ILLUSTRATIVE RESULTS

The results presented here are based on the analysis of a 500 μm long QWS DFB laser, discussed in further details by Isenor (2001). The injection current is normalized relative to the threshold current i_{th}, and values ranging from $1.1i_{th}$ to $5i_{th}$ are considered.

As a starting point for (local) optimization, all six equal length sections are first initialized at $KL=2$. This value is used to determine the threshold parameters needed to initialize the transfer matrices associated with the laser model. (In our numerical experiments, 5000 such matrices were used.) Because of the symmetry conditions previously discussed, this results in a 5-variable optimization problem formally described as

(40)   *min f(x)*   field flatness function (*F*)

   $g(x) = 0$ right hand boundary condition error (*RBCerror*)

   $xl \leq x \leq xu$ explicit, finite parameter bounds

   $x = \{KL1, KL2, KL3, \lambda, C_o\}$ laser design parameters.

Recall here the preceding discussion regarding the "black box" functions *F* and *RBCerror*, and the related computations as summarized in Fig. 5. The explicit search range considered is based on lower and upper bounds 1.54677 μm to 1.54679 μm for λ, 4.0×10^4 to 1.50×10^6 for C_o. and 0.001 to 3.0 for *KL*.

The laser model was developed and coupled with LGO using the Lahey Fortran (LF95) compiler, from Lahey Computer Systems (2000). The relatively significant runtimes (several hours on an Intel Pentium III 800 MHz processor based personal computer) dictated the restriction of function evaluations. Therefore we set 10,000 or 20,000 model function evaluations, as global search mode termination criterion. The global search phase was then followed by the (generally much faster) local search. According to the authors' extensive numerical experience with LGO, the prescribed numerical search effort is more than sufficient in the global optimization model versions considered here. In our detailed numerical studies solutions were obtainable in less than 5 hours when applying 10,000 global search steps. This lead to feasible solutions with right hand boundary condition errors in the order of 10^{-6}, and the objective (field flatness) function values were significantly improved when compared to earlier results.

In order to first gain some insight into the nonconvex, multi-extremal nature of the objective function, initial studies were performed using the model parameters given in Table 1 (next page), where only two optimization parameters, the oscillating wavelength, λ, and the dimensionless field-scaling coefficient, C_o, were considered. These variables are minimally necessary to achieve a field boundary match at the right hand facet of the laser. In this case $KL1$ to $KL3$ were held fixed at a value of two. Fig. 6 illustrates the unusual multi-extremality of a subspace projection of the objective function.

Next, Fig. 7 illustrates the results achieved at the various normalized injection currents relative to the non-optimized QWS profiles. For clarity, only the non-optimized QWS profiles at the injection current extremes $1.5i_{th}$ and $5i_{th}$ are presented in the graph: the other non-optimized QWS profiles would be drawn in the space between the two QWS graphs shown.

The first fact that is immediately obvious from Fig. 7 is that the longitudinal field intensity profiles are drastically reduced from those of the non-optimized QWS structure, for the entire range of injection currents evaluated. This is very significant, as it clearly indicates that the optimization-based approach works successfully in the above-threshold region of the laser operation and results in significantly flattened profiles. It is also important to note that only minor differences are found between the optimized profiles, regardless of the value of the injection current at which the optimization was carried out. In other words, the optimized normalized field intensity profile of the laser operating at $1.5i_{th}$ is virtually identical to the normalized field intensity profile of the laser operating at any other injection current up to and including $5i_{th}$.

Fig. 8 summarizes the field flatness characteristics for various normalized (i/i_{th}) above-threshold injection current levels.

Table 1   **Model Parameters**

Parameter	Symbol	Value	Unit
Material Parameters			
Spontaneous emission rate	τ^{-1}	2.5×10^8	s^{-1}
Bimolecular recombination coefficient	B	1×10^{-16}	m^3/s
Auger recombination coefficient	C	3×10^{-41}	m^6/s
Differential gain	A_0	2.7×10^{-20}	m^2
Gain curvature	A_1	0.15×10^{20}	m^{-3}
Differential peak wavelength	A_2	2.7×10^{-32}	m^3
Internal cavity loss	α_{loss}	4×10^3	m^{-1}
Refractive index at zero injection	n_e	3.41351524	
Carrier concentration at transparency	N_o	1.5×10^{24}	m^{-3}
Carrier concentration at threshold	N_{th}		m^{-3}
Differential index	$\partial n / \partial N$	-1.8×10^{-26}	m^3
Group velocity at Bragg wavelength	v_g	$3 \times 10^8/3.7$	m/s
Nonlinear gain coefficient	σ	1.5×10^{-23}	m^3
Peak gain wavelength at transparency	λ_o	1.63×10^{-6}	m
Lasing wavelength	λ		m
Lasing wavelength at threshold	λ_{th}		m
Structural Parameters			
Active layer width	d	1.2×10^{-7}	m
Active layer thickness	w	1.5×10^{-6}	m
Coupling coefficient	K	4×10^3	m^{-1}
Laser cavity length	L	500×10^{-6}	m
Optical confinement factor	Γ	0.35	
Grating period	Λ	2.27039×10^{-7}	m
Bragg wavelength	$\lambda_B = 2\Lambda n_o$	1.55×10^{-6}	m
Threshold current	i_{th}		A
Injection current	i		A

The parameters listed in Table 1 have been used extensively by Ghafouri-Shiraz and Lo (1996) in the threshold and the above-threshold analysis of a bulk semiconductor DFB laser, and are considered to be valid for such a device lasing around 1550 *nm*.

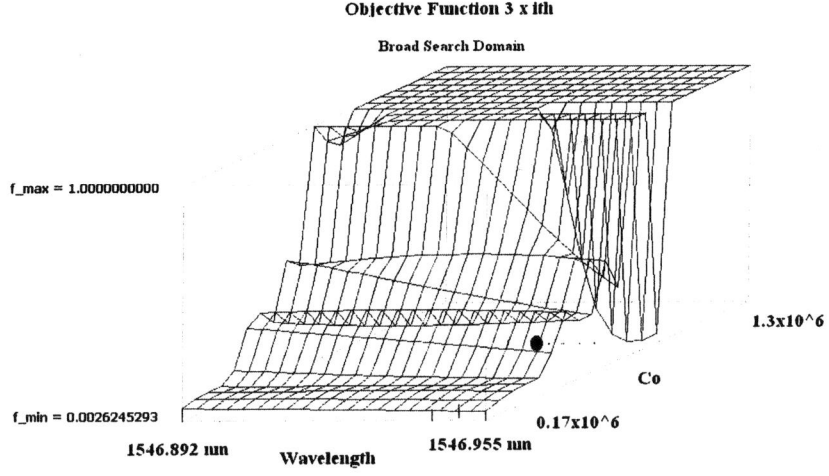

Fig. 6

Objective (Merit) Function versus C_0 and λ for 3 x i_{th}
(All other *KL* parameters held fixed.)

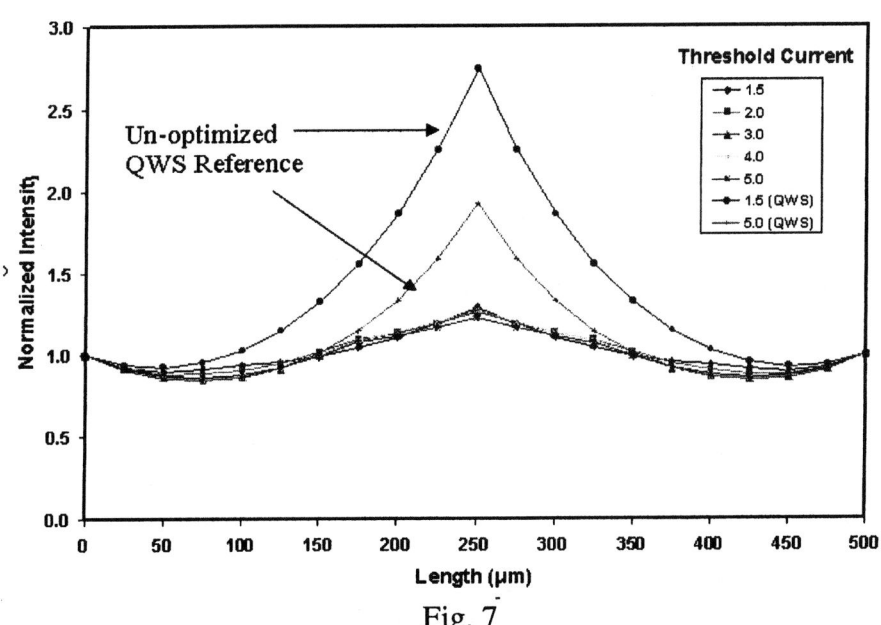

Fig. 7

Coupling Coefficient Optimization vs. Non-Optimized QWS
Structure: Normalized Longitudinal Field Intensities

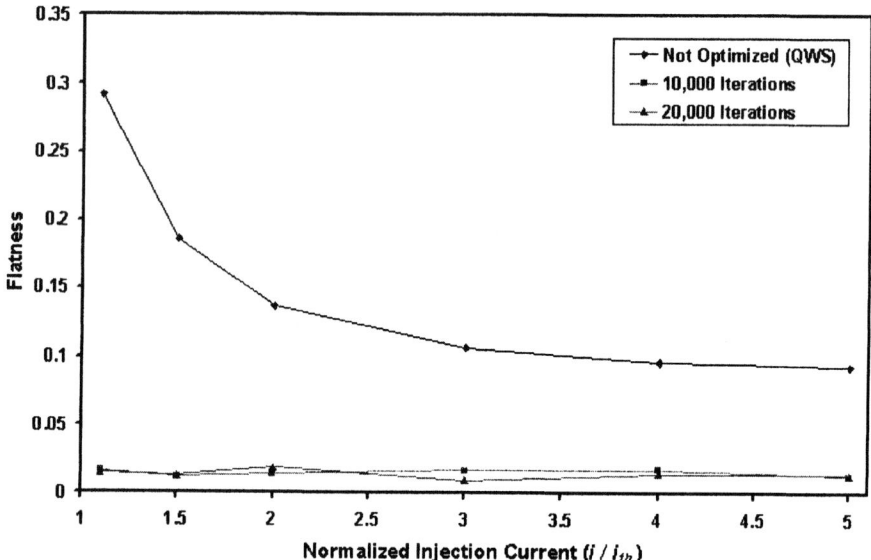

Fig. 8
Field Intensity Flatness Optimization Results

The results presented in Fig. 8 indicate a significant reduction in field flatness over the non-optimized structure, with the greatest improvement at the lower injection currents. This is expected, as the relative normalized field intensity profile of a QWS structure with a constant KL tends to flatten at higher injection currents. There is an average of a 90% improvement in flatness across the range of injection currents evaluated. Little difference is seen between the flatness results obtained by 10,000 and 20,000 global iterations, indicating a stable (numerically) optimal solution.

The coupling coefficient optimized solution values obtained are shown in Tables 2 and 3 (on next page). Examination of these results reveals some interesting differences between the 10,000 and 20,000 iteration based KL values. In some cases, the differences are significant yet the intensity profile solutions as well as the corresponding field intensity flatness shows only minor variations. This behavior is thought to be the result of more than one "near-optimal" solution. This fact may prove very useful to a designer faced by additional constraints (imposed e.g. by an actual laser manufacturing process).

Fig. 9 shows that for runs with 10,000 and 20,000 global scope iterations the average KL over the range of injection currents for the optimized structure remains constant at approximately $KL_{avg} = 1.34$, as opposed to $KL = 2$ corresponding to the initial non-optimized structure. There is little change (0.02) in the average KL between both iteration counts, which is interesting in light of the results presented in Tables 1 and 2. Again, this behavior is thought to be representative of more than one near-optimal solution and may also be indicative of an underlying (so far unknown) relationship between sectional coupling coefficient values and an optimal solution.

Table 2

Optimized KL Parameters (10,000 iterations)

Injection Current	KL1	KL2	KL3	Average KL
$1.1 \times i_{th}$	2.22	0.75	1.18	1.38
$1.5 \times i_{th}$	2.12	1.00	0.84	1.32
$2 \times i_{th}$	2.00	1.36	0.61	1.32
$3 \times i_{th}$	1.85	1.65	0.60	1.37
$4 \times i_{th}$	1.89	1.62	0.64	1.38
$5 \times i_{th}$	1.96	1.47	0.55	1.33
		Overall	Average =	1.35

Table 3

Optimized KL Parameters (20,000 iterations)

Injection Current	KL1	KL2	KL3	Average KL
$1.1 \times i_{th}$	2.11	1.12	0.74	1.32
$1.5 \times i_{th}$	2.21	0.91	0.88	1.33
$2 \times i_{th}$	1.97	1.37	0.85	1.40
$3 \times i_{th}$	2.21	1.30	0.11	1.21
$4 \times i_{th}$	1.93	1.56	0.56	1.35
$5 \times i_{th}$	2.01	1.27	0.85	1.38
		Overall	Average =	1.33

It should be noted that KL_{avg} slightly exceeds 4/3, which is considered as a practical upper limit for overall average coupling; however, at no time does a sectional KL value exceed the considered practical limit of 2.5. See Yokoyama and Sekino (1998). Using this methodology, which incorporates the LGO solver, it would be a simple matter to constrain KL_{avg} to remain below a pre-specified level (such as e.g. 4/3).

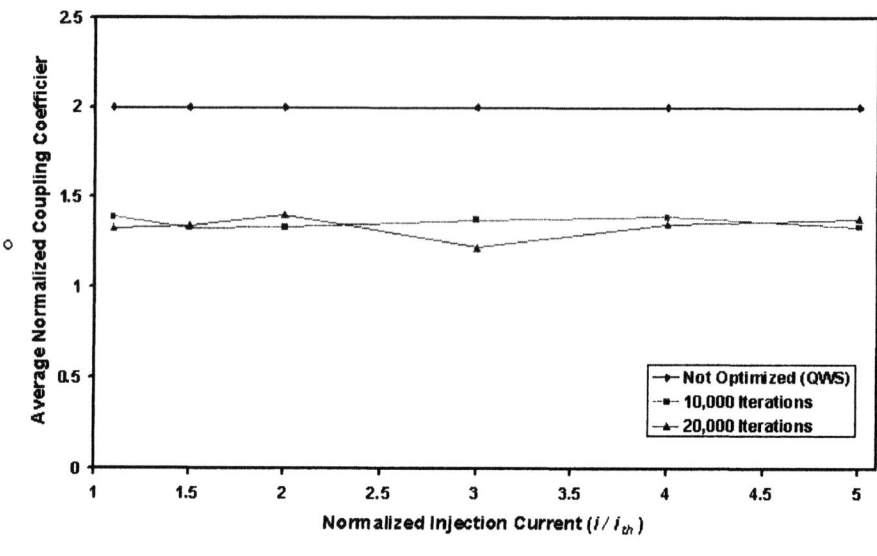

Fig. 9

Optimized Average Normalized Coupling Coefficient *KL*

Fig. 10 (next page) again reveals the highly nonlinear nature of the objective function associated with the QWS DFB laser studied here. This figure shows some of the finer details of the optimal flatness solution sub-space projections for two of the independent design parameters (while the three others are fixed at the best parameterization found). The large "dot" shown indicates the position of the optimal solution estimate in the {*KL1*, *KL3*} subspace.

It appears from Fig. 10 that the objective function is highly sensitive to the value of *KL1*, and less influenced by *KL3*. Recall that the *KL1* (and *KL6*) sections are at the ends of the laser, and include the end facet boundaries. Although the exact mechanism is unclear, one possible explanation for the sensitivity may be related to the fact that the field profiles must accurately satisfy the facet boundary conditions associated with these sections. Note also that other sub-space projections of the solution variables can be seen to exhibit similar type behaviors.

Let us note here that Figs. 6 and 10 were produced by the LGO implementation used in this study. Namely, such figures can be produced upon completion of an optimization run, as a result analysis option. Pintér (2001) provides a detailed description of the LGO integrated development environment that also supports constraint visualization options.

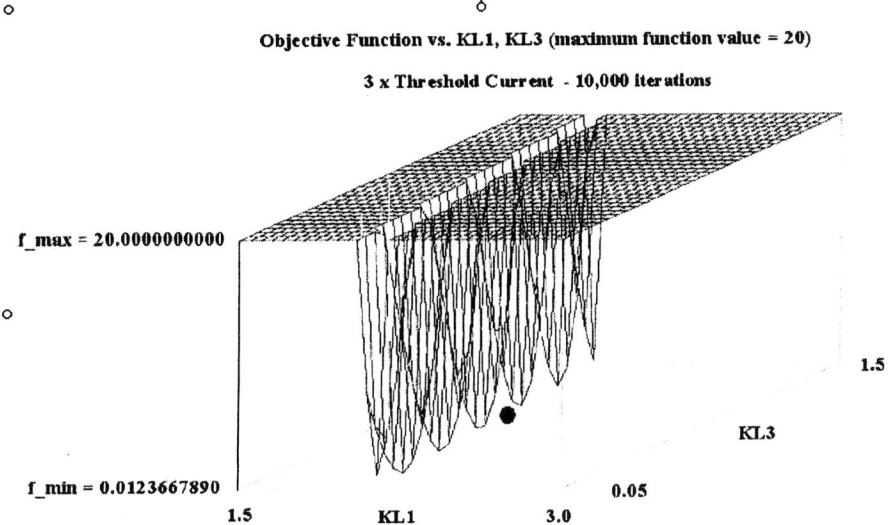

Objective Function vs. KL1, KL3 (maximum function value = 20)

3 x Threshold Current - 10,000 iterations

f_max = 20.0000000000

1.5

KL3

f_min = 0.0123667890

1.5   KL1   3.0   0.05

The image is scaled by minimal and maximal (or cutoff) function values.
The projected location of the solution estimate is denoted by the dot.

Fig. 10
Objective Function versus *KL1* and *KL3*. Note that the graph cut-off function
value is 20: this is set as an LGO visualization option.

9.   SUMMARY AND CONCLUSIONS

In this work we introduced a laser modeling and design methodology
that is based on combining the transfer matrix method with global
optimization. This integrated approach is capable of addressing above-
threshold, nonlinear laser design problems with imposed (physical or
manufacturability) constraints.

Using sectional coupling coefficients as optimization parameters, it is
demonstrated that in all cases considered it is possible to obtain about 90%
reduction in the internal field flatness values over those of the non-optimized
reference QWS DFB laser.

The optimized *KL* parameters demonstrate significant variations with the
different values of injection currents evaluated, as well as between 10,000
and 20,000 iteration based LGO runs. However, the changes in field flatness
always remain negligible. This result is new and not intuitive, and is thought
to support the possibility of multiple near-optimal solutions.

The possible ill-conditioned character of the problem may provide some insight into this behavior. Ill-conditioned behavior points out those situations where one must be very precise, and also where there may be some flexibility in the selection of solution (design) parameters. Changes in what are possibly the "insensitive" solution parameters seem to have a minimal effect on the field flatness value and associated field profile solution.

The visualization presented in Fig. 6 and Fig. 10 reveals some of the complexity of the various sensitivities. It can be seen that along a particular direction in the *KL1 – KL3* solution space, the objective function is relatively insensitive to the values of the solution parameters, while highly sensitive in other directions. Although not included in this chapter, further evidence of this behavior has been demonstrated by the authors in several other objective function subspace visualizations. Additional studies are needed to fully explore the ramifications of this phenomenon, and to confirm whether the data is indeed reflective of the existence of multiple near-optimal solutions.

The need for advanced global optimization is obvious in the entire study. If less sophisticated solution methods are used, such as some form of static or dynamic numerical grid (or some local scope search), then the probability of ending up at a sub-optimal solution is high. This problem becomes even more significant with higher injection currents, necessitating first solving near threshold and then incrementally increasing the injection current and obtaining associated solutions. Using the parameters obtained from the previous iteration as the starting point for the next injection current increment, this process must be continued until the desired injection current is finally reached. In practice these increments are quite small, around 0.2 times i_{th}. Such techniques would become quickly unusable (and unstable) beyond a few degrees of freedom.

Using a genuine global optimization approach, the above considerations are largely unnecessary. A proper global solver will robustly approximate the best possible solution. The presented solutions were obtained by directly solving the optimization problem at the desired injection current. Hence, the need for incremental solutions (by gradually increasing the injection current) was eliminated.

Much work still remains to explore the full capabilities of combining laser design models and global optimization techniques. The potential exists to extend the global optimization approach towards developing a comprehensive advanced methodology for the above-threshold design of DFB lasers. Application of this technique to more sophisticated structures involving combinations of coupling coefficient and phase shift sections, as well as consideration of mode stability, constitute just a few of the related topics for further study and research.

REFERENCES

Agrawal, G. (1992) *Fiber-Optic Communication Systems*. John Wiley & Sons, New York.

Fang, W., A. Hsu, S.L. Chuang, T. Tanbun-Ek, and A.M. Sergent (1997) Measurement and Modeling of Distributed-Feedback Lasers with Spatial Hole Burning, *IEEE Journal of Selected Topics In Quantum Electronics*, Vol. 3, No. 2, pp.547-554.

Fessant, T. (1997) Threshold and above-threshold analysis of corrugation-pitch-modulated DFB lasers with inhomogeneous coupling coefficient, *IEEE Proceedings on Optoelectronics*, Vol. 144, No. 6.

Frontline Systems and Pintér Consulting Services (2001) *Premium Solver Platform – Excel/LGO SolverEngine*. Distributed by Frontline Systems, Inc., Incline Village, NV.

GAMS Development Corporation and Pintér Consulting Services (2003) *GAMS/LGO SolverEngine*. Distributed by the GAMS Development Corporation, Washington, DC.

Ghafouri-Shiraz, H. and B. Lo (1996) *Distributed Feedback Laser Diodes*. John Wiley & Sons, Chichester.

Green, P. Jr. (1993) *Fiber Optic Networks*. Prentice-Hall, New Jersey.

Isenor G. (2001) *A Novel Approach to the Reduction of a Distributed Feedback Laser's Intensity Profile Non-Uniformity Using Global Optimization*. Ph.D. Dissertation, Department of Electrical and Computer Engineering, Dalhousie University, Halifax, N.S.

Isenor, G., J.D. Pintér, M. Cada, (2003) A global optimization approach to laser design. *Optimization and Engineering* 4, 177-196.

Kawaguchi, H. (1994) *Bistabilities and Nonlinearities in Laser Diodes*. Artech House, Boston.

Keiser, G. (1993) *Optical Fiber Communications*. (2nd Edn.) McGraw-Hill, New York.

Kogelnik, H. (1969) Coupled Wave Theory for Thick Hologram Gratings. *The Bell System Technical Journal*, Vol.48, No.9; pp. 2902-2947.

Kogelnik, H., C.V. Shank (1972) Coupled Wave Theory of Distributed Feedback Lasers. *Journal of Applied Physics*, Vol. 43, No. 5; pp. 2327-2335.

Lahey Computer Systems (2000) *Lahey/Fujitsu Fortran 95 User's Guide*. Lahey Computer Systems, Incline Village, NV.

Makino, T. (1991) Transfer-Matrix Formulation of Spontaneous Emission Noise of DFB Semiconductor Lasers. *Journal of Lightwave Technology*, Vol. 9, No. 1; pp. 84-91.

Makino, T. and J. Glinski (1988) Transfer Matrix Analysis of the Amplified Spontaneous Emission of DFB Semiconductor Laser Amplifiers. *IEEE Journal of Quantum Electronics*, Vol. 24, No. 8; pp. 1507-1518.

Maplesoft and Pintér Consulting Services (2004) *Global Optimization Toolbox for Maple*. Distributed by Maplesoft, Waterloo, ON.

Maximal Software and Pintér Consulting Services (2005) *MPL/LGO Solver Engine*. Distributed by Maximal Software, Inc. Arlington, VA.

Morthier, G., and R Baets (1991) Design of Index-Coupled DFB Lasers with Reduced Longitudinal Spatial Hole Burning. *Journal Of Lightwave Technology*, Vol. 9, No. 10, pp. 1305-1313.

Morthier, G., K. David, P. Vankwikelberge, and R. Baets (1990) A New DFB-Laser Diode with Reduced Spatial Hole Burning, *IEEE Photonics Technology Letters*, Vol. 2, No. 6, pp 388-390.

Pintér, J.D. (1996) *Global Optimization in Action*. Kluwer Academic Publishers, Dordrecht.

Pintér, J.D. (2001) *Computational Global Optimization in Nonlinear Systems – An Interactive Tutorial*. Lionheart Publishing, Inc. Atlanta, GA.

Pintér, J.D. (2002) Global Optimization: Software, Test Problems, and Applications. Chapter 15 (pp. 515-569) in: *Handbook of Global Optimization, Volume 2* (Pardalos, P. M. and Romeijn, H. E., Eds.) Kluwer Academic Publishers, Dordrecht.

Pintér, J.D. (2005) *LGO – A Model Development System for Continuous Global Optimization. User's Guide.* (Current revised edition.) Published by Pintér Consulting Services, Inc., Halifax, NS.

Pintér, J.D. and Kampas, F.J. (2003) *MathOptimizer Professional – An Advanced Modeling and Optimization System for Mathematica, Using the LGO Solver Engine. User Guide.* Published by Pintér Consulting Services, Inc., Halifax, NS.

Rabinovich, W.S., and B.J. Feldman (1989) Spatial Hole Burning Effects in Distributed Feedback Lasers. *IEEE Journal Of Quantum Electronics*, Vol. 25, No. 1, pp. 20-30.

Sargent III, M., W. H. Swantner W. H., and J. D. Thomas (1980) Theory of a Distributed Feedback Laser. *IEEE Journal of Quantum Electronics*, Vol. QE-16, No. 4; pp. 465-472.

Tamir, T. (1988) *Guided-Wave Optoelectronics.* Springer-Verlag, Berlin.

TOMLAB Optimization and Pintér Consulting Services (2004) *TOMLAB/LGO Solver Engine.* Distributed by TOMLAB Optimization Inc., San Diego, CA.

Van Etten, W. and J. Van der Plaats (1991) *Fundamentals of Optical Fiber Communications.* Prentice Hall, New York.

Wang, J.-Y., M. Cada, and T. Makino, (1998) Coupled-Power Theory of Nonlinear Distributed–Feedback Lasers, Yielding Reduced Longitudinal Spatial Hole Burning. *Applied Physics Letters*, Vol. 72, No. 25; pp. 3255-3257.

Wang J.-Y., and M. Cada, (2000) Analysis and Optimum Design of Distributed Feedback Laser Using Couple-Power Theory. *IEEE Journal Of Quantum Electronics,* Vol. 36, No. 1, pp. 52-58.

Chapter 9

COMPUTATIONAL EXPERIENCE WITH THE MOLECULAR DISTANCE GEOMETRY PROBLEM

Carlile Lavor

Department of Applied Mathematics (IMECC-UNICAMP), State University of Campinas, CP 6065, 13081-970, Campinas-SP, Brazil.

clavor@ime.unicamp.br

Leo Liberti

DEI, Politecnico di Milano, Piazza L. da Vinci 32, 20133 Milano, Italy.

liberti@elet.polimi.it

Nelson Maculan

COPPE, Universidade Federal do Rio de Janeiro – UFRJ, C.P. 68511, Rio de Janeiro 21945-970, Brazil.

maculan@cos.ufrj.br

Abstract   In this work we consider the molecular distance geometry problem, which can be defined as the determination of the three-dimensional structure of a molecule based on distances between some pairs of atoms. We address the problem as a nonconvex least-squares problem. We apply three global optimization algorithms (spatial Branch-and-Bound, Variable Neighbourhood Search, Multi Level Single Linkage) to two sets of instances, one taken from the literature and the other new.

Keywords: molecular conformation, distance geometry, global optimization, spatial Branch-and-Bound, variable neighbourhood search, multi level single linkage.

1.   Introduction

The Molecular Distance Geometry Problem (MDGP) is the problem of determining the three-dimensional structure of a molecule where a subset of the atomic distances is known. Formally, we need to find vectors $x_1, ..., x_n \in \mathbb{R}^3$, which describe the three-dimensional position of each atom in the molecule, such that:

$$\forall \{i,j\} \in S \quad (\|x_i - x_j\| = d_{ij}),$$

where S is the subset of pairs of atoms $\{i,j\}$ whose distances d_{ij} are known. We address the problem in terms of finding the global minimizer of the function

$$f(x_1, \ldots, x_n) = \sum_{\{i,j\} \in S} \left(\|x_i - x_j\|^2 - d_{ij}^2 \right)^2.$$

It is easy to verify that $x_1, \ldots, x_n \in \mathbb{R}^3$ solve the problem if and only if $f(x_1, \ldots, x_n) = 0$.

The MDGP is an important problem in molecular biology. The objective is to find a molecular conformation satisfying all the constraints imposed by the known distances (i.e., that $\|x_i - x_j\| = d_{ij}$ for all $\{i,j\} \in S$). For some references, see [Crippen and Havel, 1988, Hendrickson, 1995, Moré and Wu, 1997, Moré and Wu, 1999, An, 2003].

The aim of this work is twofold. On the one hand, we present two different methods of generating MDGP instances, and we wish to test which of the methods generates the hardest instances. On the other hand, we want to assess the solution quality and efficiency of three well-known global optimization algorithms applied to the MDGP. The algorithms are: spatial Branch-and-Bound (sBB) [Ryoo and Sahinidis, 1995, Tawarmalani and Sahinidis, 2002, Adjiman et al., 1998, Smith and Pantelides, 1999, Hansen, 1992], Variable Neighbourhood Search (VNS) [Hansen and Mladenović, 2001, Mladenović et al., 2003], and Multi Level Single Linkage (MLSL) [Rinnooy-Kan and Timmer, 1987a, Rinnooy-Kan and Timmer, 1987b, Locatelli and Schoen, 1996, Schoen, 1998, Schoen, 1999, Locatelli and Schoen, 1999, Schoen, 2002, Kucherenko and Sytsko, 2005]. We test each of these algorithms on instances of varying sizes generated with the two generating methods, one taken from the literature [Moré and Wu, 1997] and the other new [Lavor].

Our computational results show that, in terms of user CPU time, VNS is the most efficient of the methods we tested. As the size of the instance grows, however, the performance difference between VNS and MLSL decreases. Whilst VNS and MLSL are stochastic algorithms, sBB is a deterministic algorithm. As such, it provides a guarantee of ε-global optimality, but at a practically high computational cost on most global optimization problems. With MDGP instances, however, sBB was found to be competitive with VNS and MLSL at least for small and medium-sized instances.

It is worth mentioning explicitly that, somewhat unusually for this type of problems, we included no smoothing techniques in our algorithms, as the aim of this test was to verify the applicability of general-purpose global optimization algorithms to the problems in original form. Similar tests, but with smoothing techniques included, are currently work in progress.

The rest of this paper is organized as follows: Section 2 describes the global optimization algorithms used; Section 3 describes the two sets of instances used to generate the experiments and discusses the computational results.

2.   Global optimization methods

In this section, we shall briefly describe the three algorithms we used to solve the MDGP. All these methods are general-purpose, in the sense that they can be used without modification to solve all global optimization problems. In other words, they do not take into account the structure of the problem.

2.1   Spatial Branch-and-Bound

Spatial Branch-and-Bound (sBB) algorithms locate the global optimum by generating converging sequences of upper and lower bounds to the objective function. The upper bounds are obtained by locally solving the original (non-convex) problem. The lower bounds are obtained by locally solving a convex (in this case, linear) relaxation of the original problem. Since any local solution of a convex problem is also global, locally solving the linear relaxation yields a valid lower bound to the original problem. The algorithm, first proposed in [Smith, 1996, Smith and Pantelides, 1999], is shown in Fig. 1.1. The implementation details, as well as many refinements and improvements with respect to the original algorithm, are given in [Liberti, 2004].

The most outstanding feature of sBB algorithm described in this section is the automatic construction of the convex relaxation via symbolic reformulation. This involves identifying all the nonconvex terms in the problem and replacing them with the respective convex relaxations. The algorithm that carries out this task is symbolic in nature as it has to recognize the nonconvex operators in any given function. The relaxation is built in two stages: first the problem is reduced to a standard form where the nonlinear terms are linearized. This means that each nonlinear term is replaced by a linearizing variable, and a constraint of type "linearizing variable = nonlinear term" is added to the problem formulation. Such constraints are called *defining equations*, or *defining constraints*. In the second stage of the linear relaxation each nonlinear term is replaced by the corresponding linear under- and over-estimators. Note that this process is wholly automatic, and part of the implementation software.

2.2   Variable Neighbourhood Search

Variable Neighbourhood Search (VNS) is a relatively recent metaheuristic which relies on iteratively exploring neighbourhoods of growing size to identify better local optima [Hansen and Mladenović, 2001]. More precisely, VNS escapes from the current local minimum x^* by initiating other local searches from starting points sampled from a neighbourhood of x^* which increases its

1 (Initialization) Initialize a list of regions to a single region comprising the entire set of variable ranges. Set the convergence tolerance $\varepsilon > 0$, the best objective function value found up to the current step as $U := \infty$ and the corresponding solution point as $x^* := (\infty, \ldots, \infty)$. Optionally, perform optimization-based bounds tightening.

2 (Choice of Region) If the list of regions is empty, terminate the algorithm with solution x^* and objective function value U. Otherwise, choose a region R (the "current region") from the list. Delete R from the list. Optionally, perform feasibility-based bounds tightening on R.

3 (Lower Bound) Generate a convex relaxation of the original problem in the selected region R and solve it to obtain an underestimation l of the objective function with corresponding solution $\bar{x}$. If $l > U$ or the relaxed problem is infeasible, go back to step 2.

4 (Upper Bound) Attempt to solve the original (generally nonconvex) problem in the selected region to obtain a (locally optimal) solution $\tilde{x}$ with objective function value u. If this fails, set $u := +\infty$ and $\tilde{x} = (\infty, \ldots, \infty)$.

5 (Pruning) If $U > u$, set $x^* = \tilde{x}$ and $U := u$. Delete all regions in the list that have lower bounds bigger than U as they cannot possibly contain the global minimum.

6 (Check Region) If $u - l \leq \varepsilon$, accept u as the global minimum for this region and return to step 2. Otherwise, we may not yet have located the region global minimum, so we proceed to the next step.

7 (Branching) Apply a branching rule to the current region to split it into sub-regions. Add these to the list of regions, assigning to them an (initial) lower bound of l. Go back to step 2.

Figure 1.1. The spatial Branch-and-Bound algorithm.

size iteratively until a local minimum better than the current one is found. These steps are repeated until a given termination condition is met.

VNS has been applied to a wide variety of problems both from combinatorial and continuous optimization. Its early applications to continuous problems were based on a particular problem structure. In the continuous location-allocation problem the neighbourhoods are defined according to the meaning of problem variables (assignments of facilities to customers, positioning of yet unassigned facilities and so on) [Brimberg and Mladenović, 1996]. In the bilinearly constrained bilinear problem the neighbourhoods are defined in terms of the applicability of the successive linear programming approach, where the problem variables can be partitioned so that fixing the variables in either set yields a linear problem; more precisely, the neighbourhoods of size k are defined as the vertices of the LP polyhedra that are k pivots away from the current vertex [Hansen and Mladenović, 2001]. In summary, none of the early applications of VNS to continuous problems solved problems in general form.

The first VNS algorithm targeted at problems with fewer structural requirements, namely, box-constrained NLPs, was given in [Mladenović et al., 2003] (the paper focuses on a particular class of box-constrained NLPs, but the proposed approach is general). The very same code used in [Mladenović et al., 2003] (which is different from that used in this paper) has been already applied to another molecular conformation problem with considerable success [Drazić et al., 2004]. Since the problem is assumed to be box-constrained, the neighbourhoods arise naturally as hyperrectangles of growing size centered at the current local minimum x^*. In the pseudocode algorithm in Fig. 1.2, the termi-

---

1 Set $k \leftarrow 1$, pick random point $\tilde{x}$, perform local descent to find a local minimum x^*.

2 Until $k > k_{\text{max}}$ repeat the following steps:

    (a) define a neighbourhood $N_k(x^*)$;

    (b) sample a random point $\tilde{x}$ from $N_k(x^*)$;

    (c) perform local descent from $\tilde{x}$ to find a local minimum x';

    (d) if x' is better than x^* set $x^* \leftarrow x'$ and $k \leftarrow 1$; go to step 2;

    (e) set $k \leftarrow k + 1$

Figure 1.2. The VNS algorithm.

nation condition is taken to be $k > k_{\text{max}}$. This is the most common behaviour, but not the only one (the termination condition can be based on CPU time or other algorithmic parameters). The definition of the neighourhoods may vary. If $N_k(x)$ is taken to be a hyperrectangle $H_k(x)$ of "size" k centered at x, sampling becomes easy; there is a danger, though, that sampled points will actually be inside a smaller hyperrectangular neighbourhood. A way to deal with this problem is to take $N_k(x) = H_k(x) \backslash H_{k-1}(x)$, although this makes it harder to sample a point inside the neighbourhood. For each $k \leq k_{\max}$ we define $H_k(x^*)$ to be hyper-rectangles similar to $x^L \leq x \leq x^U$, all centered at x^*, whose sides have been scaled by $\frac{k}{k_{\max}}$. More formally, let $H_k(x^*)$ be the hyper-rectangle $y^L \leq x \leq y^U$ where, for all $i \leq n$:

$$
y_i^L = x_i^* - \frac{k}{k_{\max}}(x_i^* - x_i^L)
$$

$$
y_i^U = x_i^* + \frac{k}{k_{\max}}(x_i^U - x_i^*).
$$

This construction forms a set of hyper-rectangular "shells" centered at x^*. In our computational experiments, we used $N_k(x) = H_k(x)$ for simplicity.

2.3   Multi-Level Single Linkage

In this section we shall describe the main features of a Multi-Level Single Linkage (MLSL) stochastic algorithm for global optimization. The algorithm is called *SobolOpt* [Kucherenko and Sytsko, 2005]. Its main strength is that it employs certain Low-Discrepancy Sequences (LDSs) of sampling points called *Sobol' sequences* whose distributions in Euclidean space have very desirable uniformity properties. Uniform random distributions where each point is generated in a time interval (as is the case in practice when generating a sampling on a computer) are guaranteed to "fill the space" in infinite time with probability 1. In fact, these conditions are very far from the normal operating conditions. LDSs, and in particular Sobol' sequences, are guaranteed to fill the space as uniformly as possible even in finite time. In other words, for any integer $N > 0$, the first N terms of a Sobol' sequence does a very good job of filling the space evenly. One further very desirable property of Sobol' sequences is that any projection on any coordinate hyperplane of the Euclidean space $\mathbb{R}^n$ containing N n-dimensional points from a Sobol' sequence will still contain N projected $(n-1)$-dimensional Sobol' points. This clearly does not hold with the uniform grid distribution where each point is located at a coordinate lattice point (in this case the number of projected points on any coordinate hyperplanes is $O(N^{\frac{n-1}{n}})$, as shown in Fig. 1.3). The comparison between grid and Sobol' points in $\mathbb{R}^2$ is shown in Fig. 1.4.

The regularity and uniformity properties of Sobol' sequences are exploited in the following MLSL algorithm. Let Q be the set of pairs of sampled points q together with their evaluation $f(q)$ (where f is the objective function). Let S be the list of all local minima found up to now.

The algorithm terminates with a list S of all the local minima found. Finding the global minimum is then a trivial matter of identifying the minimum with lowest objective function value $f(y)$. Two of the most common termination

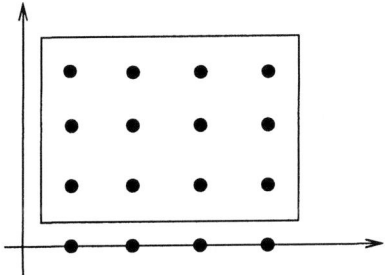

Figure 1.3. Projecting a grid distribution in $\mathbb{R}^2$ on the coordinate axes reduces the number of projected points. In this picture, $N = 12$ but the projected points are just 4.

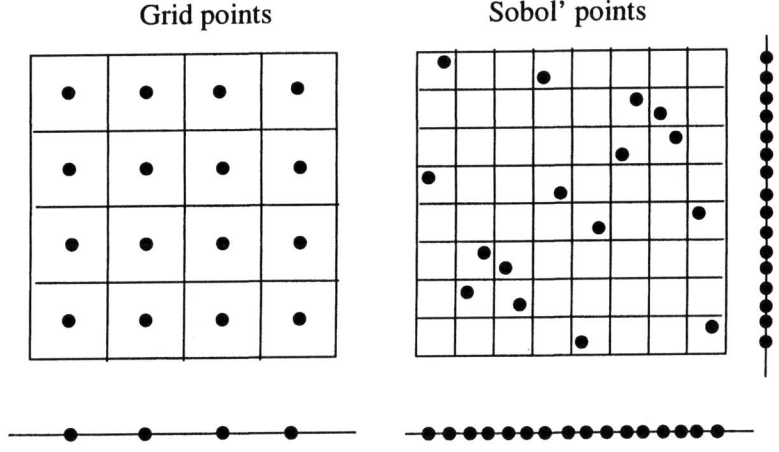

Figure 1.4. Comparison between projected distribution of grid points and Sobol' points in $\mathbb{R}^2$.

conditions are (a) maximum number of sampled points and (b) maximum time limit exceeded. In our tests we accepted a default termination condition based on the number of local searches not exceeding 320.

Sobol' sequences are generated as follows. Let $P(x)$ be a primitive polynomial of degree q in $GF(2)[x]$, say $P(x) = \sum_{i=0}^{q} a_{q-i} x^i$ where $a_0 = a_q = 1$. Now for all $i > q$ define Q_i recursively as the result of the bitwise XOR operation on the following set of numbers: $\{2^k a_k Q_{i-k} \mid 1 \le k \le q-1\} \cup \{2^q Q_{i-q}, Q_{i-q}\}$ (this is a q-term recurrence relation; the first q terms of the sequence can be chosen as arbitrary odd integers respectively less than $2, \ldots, 2^q$). Let $V_i = \frac{Q_i}{2^i}$ for all i. For each integer j, let $\Lambda_j = \{V_i \mid$ the i-th bit of j is nonzero$\}$. We define X_j, the j-th element of the Sobol' sequence, as the result of the bitwise XOR operation on Λ_j. Multidimensional

1 (Initialization) Let $Q = \emptyset$, $S = \emptyset$, $k = 1$ and set $\varepsilon > 0$.

2 (Termination) If a pre-determined termination condition is verified, stop.

3 (Sampling) Sample a point q_k from a Sobol' sequence; add $(q_k, f(q_k))$ to Q.

4 (Clustering distance) Compute a distance r_k (which is a function of k and n; there are various ways to compute this distance, so this is considered as an "implementation detail" — one possibility is $r_k = \beta k^{-\frac{1}{n}}$, where β is a known parameter).

5 (Local phase) If there is no previously sampled point $q_j \in Q$ (with $j < k$) such that $\|q_k - q_j\| < r_k$ and $f(q_j) \leq f(q_k) - \varepsilon$, solve the problem locally with q_k as a starting point to find a solution y with value $f(y)$. If $y \notin S$, add y to S. Set $k \leftarrow k + 1$ and repeat from step 2.

Figure 1.5.   The SobolOpt algorithm.

Sobol' sequences are obtained by building each vector out of a different primitive polynomial. For a full discussion on the implementation details, see [Press et al., 1997], p. 311.

3.   Computational experiments

In this section we compare the results of the three global optimization algorithms mentioned in Section 2 on two sets of instances.

3.1   Instance generation methods

The first set of instances (Moré instances) was generated using the first model proposed in [Moré and Wu, 1997]. The model is based on a molecule with s^3 atoms ($s = 1, 2, 3, \ldots$) located in the three-dimensional lattice defined by

$$\{(i_1, i_2, i_3) \in \mathbb{R}^3 : 0 \leq i_k \leq s - 1, \ k = 1, 2, 3\}.$$

An order is defined for the atoms of the lattice by letting atom i be the atom at position (i_1, i_2, i_3), where

$$i = 1 + i_1 + s i_2 + s^2 i_3,$$

and the set S is defined by

$$S = \{\{i,j\} : |i - j| \leq s^2\}.$$

For example, for $s = 2$, the atoms located at $(0,0,0)$, $(1,0,0)$, and $(0,1,0)$, are the first ($i = 1$), second ($i = 2$), and third ($i = 3$) atoms.

The second set of instances (Lavor instances) was generated according to another model. This model considers a molecule as a chain of n atoms with Cartesian coordinates given by $x_1, \ldots, x_n \in \mathbb{R}^3$. For every pair of consecutive atoms i, j, let r_{ij} be the bond length which is the Euclidean distance between them. For every three consecutive atoms i, j, k, let θ_{ik} be the bond angle corresponding to the relative position of the third bead with respect to the line containing the previous two. Likewise, for every four consecutive atoms i, j, k, l, let ω_{il} be the angle, called the torsion angle, between the normals through the planes determined by the atoms i, j, k and j, k, l. The sets M_1, M_2 are the sets of pairs of atoms separated by one and two covalent bonds, respectively. The bond lengths and bond angles are set to $r_{ij} = 1.526$Å (for all $(i, j) \in M_1$) and $\theta_{ij} = 109.5°$ (for all $(i, j) \in M_2$), respectively. Torsion angles are obtained by selecting first one value ω from the set $\{60°, 180°, 300°\}$ and another one from the set $\{\omega + i : i = -5°, \ldots, 5°\}$. Both of these selections are random. To generate distances, it is necessary to calculate Cartesian coordinates for each atom of the chain. This can be done, for example, using the procedure described in [Phillips et al., 1996]. For each molecule, described by the selection of the torsion angles, we define the set S using a cut-off value of 4Å. That is, $(i, j) \in S$ if and only if $d_{ij} < 4$. The pairs of atoms (i, j) selected, associated to the distances d_{ij}, constitute an instance for the molecular distance geometry problem. For a complete description of this set of instances, see [Lavor].

3.2   Numerical results

All computations were performed on an Intel Xeon 2.8GHz with 2GB RAM running Linux. The local NLP optimization code we used to perform the local descents is SNOPT v.5 [Gill, 1999]. The global optimization algorithms were implemented as global solvers in the $oo\mathcal{OPS}$ optimization framework [Liberti et al., 2001, Liberti]. As such, they are not specially fine-tuned to solve the MDGP — this, together with the choice not to employ smoothing methods, explains the relatively small size of the largest molecules we can tackle.

The results are reported in Tables 1.1 (for the Moré instances) and 1.2 (for the new Lavor instances). The global optimum (with value 0) was found in all of the tested instances but the Lavor instance with 40 atoms. Three general trends emerge:

1 the Lavor instances, on average, are harder to solve than the Moré instances. In particular, one of the randomly generated Lavor instance (the

Atoms	Variables	sBB		VNS		SobolOpt	
		OF Value	Time	OF Value	Time	OF Value	Time
8	24	0	0.22	0	1.21	0	13.56
27	81	0	30.39	0	34.01	0	300.285
64	192	0	2237.73	0	398.875	0	2765.13

Table 1.1. Computational results for the Moré instances. Timings are in seconds of user CPU time.

Atoms	Variables	sBB		VNS		SobolOpt	
		OF Value	Time	OF Value	Time	OF Value	Time
5	15	0	0.02	0	0.48	0	0.57
10	30	0	1.12	0	7.06	0	69.71
20	60	0	2.25	0	49.99	0	411.152
30	90	0	488.87	0	352.06	0	1634.09
40	120	-	-	0.09	1258.13	0.547	2376.01
50	150	-	-	0	673.48	0	3002.88

Table 1.2. Computational results for the Lavor instances. Missing values are due to excessive computational requirements. Timings are in seconds of user CPU time.

one with 40 atoms), was so hard to solve that we could not reach the global optimum with any of the proposed global optimization algorithms;

2 the deterministic sBB algorithm is the fastest method for solving small to medium-sized instances. This result is rather surprising, as sBB is usually slower than heuristic methods;

3 both VNS and SobolOpt usually manage to find the correct solution, but VNS is faster. However, as the size of the molecule grows, the performance difference decreases.

Here are some notes and remarks about these computational experiments.

- The results obtained by the SobolOpt solver might be improved by careful tuning of parameters. Our tests were run with all the default parameter values. On the other hand, we did spend some time tuning the parameters of the VNS solver. It appears that for very hard instances (like the Lavor with 40 atoms), we can get nearer the global optimum by setting a very high k_{max} parameter (possibly in the region of 10^3) and a number of trials in each neighbourhood (i.e. maximum number of local searches to carry out in each neighbourhood) to something between 5 and 15. This slows the search down considerably, but it does produce better results.

- We also conducted a number of tests using a different VNS neighbourhood structure. In practice, we focused the search on the corners of each hyper-rectangle $H_k(x^*)$: this made it possible to sample starting points from disjoint neighbourhoods, but it affected the convergence properties of the VNS (certain regions were not sampled extensively). Surprisingly, VNS managed to locate the global optimum for all instances but the Lavor with 40 atoms, where it succeeded in locating a point with extremely low (albeit clearly non-zero) objective function value. This constitutes numerical evidence that the best minima are to be found near the extreme points of the hyper-rectangle.

- The convergence speed of the sBB solver can be improved by relaxing the ϵ tolerance (set by default to 1×10^{-3}).

- One of the reasons why sBB is so effective on this problem is that it has a known globally optimal value (0), and that the automatic convexification of sBB provides a tight lower bound (namely, 0 itself). Since the lower bound is so tight, many regions are discarded very soon in the Branch-and-Bound tree.

4.   Conclusion

In this paper we described computational experiments performed in globally solving instances of the molecular distance geometry problem. We discussed three global optimization methods: a deterministic one (spatial Branch-and-Bound) and two heuristic ones (Variable Neighbourhood Search and Multi Level Single Linkage with deterministic low-discrepancy sampling based on Sobol' sequences). We solved instances from two different classes: one taken from the literature and the other new. Rather surprisingly, sBB is clearly the best choice for small-scale problems, both because it provides a guarantee of ε-global optimality and because it is faster than the other methods. SobolOpt and VNS perform rather well for medium to large-scale problems, with VNS being faster than SobolOpt.

Acknowledgments

We wish to acknowledge the invaluable help of Dr. Sergei Kucherenko who provided the SobolOpt global solver. We also would like to thank CNPq and FAPERJ for their support. One of the authors (LL) is grateful to Prof. Nelson Maculan for financial support.

References

Adjiman, C.S., Dallwig, S., Floudas, C.A., and Neumaier, A. (1998). A global optimization method, αBB, for general twice-differentiable constrained NLPs: I. Theoretical Advances. *Computers & Chemical Engineering*, 22(9):1137–1158.

An, L.T. Hoai (2003). Solving large scale molecular distance geometry problems by a smoothing technique via the Gaussian transform and d.c. programming. *Journal of Global Optimization*, 27:375–397.

Brimberg, J. and Mladenović, N. (1996). A variable neighbourhood algorithm for solving the continuous location-allocation problem. *Studies in Location Analysis*, 10:1–12.

Crippen, G.M. and Havel, T.F. (1988). *Distance Geometry and Molecular Conformation*. Wiley, New York.

Drazić, M., Lavor, C., Maculan, N., and Mladenović, N. (2004). A continuous vns heuristic for finding the tridimensional structure of a molecule. *Le Cahiers du GERAD*, G-2004-22.

Gill, P.E. (1999). *User's Guide for SNOPT 5.3*. Systems Optimization Laboratory, Department of EESOR, Stanford University, California.

Hansen, E. (1992). *Global Optimization Using Interval Analysis*. Marcel Dekker, Inc., New York.

Hansen, P. and Mladenović, N. (2001). Variable neighbourhood search: Principles and applications. *European Journal of Operations Research*, 130:449–467.

Hendrickson, B.A. (1995). The molecule problem: exploiting structure in global optimization. *SIAM Journal on Optimization*, 5:835–857.

Kucherenko, S. and Sytsko, Yu. (2005). Application of deterministic low-discrepancy sequences in global optimization. *Computational Optimization and Applications*, 30(3):297–318.

Lavor, C. (to appear). On generating instances for the molecular distance geometry problem. In [Liberti and Maculan].

Liberti, L. (2004). *Reformulation and Convex Relaxation Techniques for Global Optimization*. PhD thesis, Imperial College London, UK.

Liberti, L. (to appear). Writing global optimization software. In [Liberti and Maculan].

Liberti, L. and Maculan, N., editors (to appear). *Global Optimization: from Theory to Implementation*. Kluwer, Dordrecht.

Liberti, L., Tsiakis, P., Keeping, B., and Pantelides, C.C. (2001). *ooOPS*. Centre for Process Systems Engineering, Chemical Engineering Department, Imperial College, London, UK, 1.24 edition.

Locatelli, M. and Schoen, F. (1996). Simple linkage: Analysis of a threshold-accepting global optimization method. *Journal of Global Optimization*, 9:95–111.

Locatelli, M. and Schoen, F. (1999). Random linkage: a family of acceptance/rejection algorithms for global optimization. *Mathematical Programming*, 85(2):379–396.

Mladenović, N., Petrović, J., Kovačević-Vujčić, V., and Čangalović, M. (2003). Solving a spread-spectrum radar polyphase code design problem by tabu search and variable neighbourhood search. *European Journal of Operations Research*, 151:389–399.

Moré, J.J. and Wu, Z. (1997). Global continuation for distance geometry problems. *SIAM Journal on Optimization*, 7:814–836.

Moré, J.J. and Wu, Z. (1999). Distance geometry optimization for protein structures. *Journal of Global Optimization*, 15:219–234.

Pardalos, P.M. and Romeijn, H.E., editors (2002). *Handbook of Global Optimization*, volume 2. Kluwer Academic Publishers, Dordrecht.

Phillips, A.T., Rosen, J.B., and Walke, V.H. (1996). Molecular structure determination by convex underestimation of local energy minima. In Pardalos, P.M., Shalloway, D., and Xue, G., editors, *Global Minimization of Nonconvex Energy Functions: Molecular Conformation and Protein Folding*, volume 23, pages 181–198, Providence. American Mathematical Society.

Press, W.H., Teukolsky, S.A., Vetterling, W.T., and Flannery, B.P. (1992, reprinted 1997). *Numerical Recipes in C, Second Edition*. Cambridge University Press, Cambridge.

Rinnooy-Kan, A.H.G. and Timmer, G.T. (1987a). Stochastic global optimization methods; part I: Clustering methods. *Mathematical Programming*, 39:27–56.

Rinnooy-Kan, A.H.G. and Timmer, G.T. (1987b). Stochastic global optimization methods; part II: Multilevel methods. *Mathematical Programming*, 39:57–78.

Ryoo, H.S. and Sahinidis, N.V. (1995). Global optimization of nonconvex NLPs and MINLPs with applications in process design. *Computers & Chemical Engineering*, 19(5):551–566.

Schoen, F. (1998). Random and quasi-random linkage methods in global optimization. *Journal of Global Optimization*, 13:445–454.

Schoen, F. (1999). Global optimization methods for high-dimensional problems. *European Journal of Operations Research*, 119:345–352.

Schoen, F. (2002). Two-phase methods for global optimization. In [Pardalos and Romeijn, 2002], pages 151–177.

Smith, E.M.B. (1996). *On the Optimal Design of Continuous Processes*. PhD thesis, Imperial College of Science, Technology and Medicine, University of London.

Smith, E.M.B. and Pantelides, C.C. (1999). A symbolic reformulation/spatial branch-and-bound algorithm for the global optimisation of nonconvex MINLPs. *Computers & Chemical Engineering*, 23:457–478.

Tawarmalani, M. and Sahinidis, N.V. (2002). Exact algorithms for global optimization of mixed-integer nonlinear programs. In [Pardalos and Romeijn, 2002], pages 1–63.

Chapter 10

Non Linear Optimization Models in Water Resource Systems

Saverio Liberatore, Giovanni M. Sechi, and Paola Zuddas[1]

Departement of Land Engineering, University of Cagliari (Italy) zuddas@unica.it

Summary. The progress made in past years in large-scale optimization algorithms led to a general interest in the possibility of applying mathematical optimization to real Water Resources Systems (WRS). As it is well known, this kind of problem typically generates computationally expensive models involving a large number of variables and constraints. Planning aspects can be represented by linear optimization models by introducing simplifications and approximations, even if linear assumptions are not strictly adherent to real WRS. In order to reach a more adequate level of adherence to the physical system more detailed models are resolved by taking into account non-linearity in objective function and constraints. An expansion technique interacting between primal and dual mathematical optimization models is proposed. This kind of approach is very useful to formulate trade-off between the dimension of water works, the reliability of the system and the prediction of short falls severity in demands. Moreover, the necessity to introduce system-vulnerability leads to solve a quadratic programming model taking into account additional non-linear costs due to the requirement of well operating during periods of drought. An adequate approach for the planning and maintenance optimization of pipes networks for water supply distribution, would consider the non-linear relations between head-loss in each pipe, its diameter, length and hydraulic property.

Standard non-linear optimization procedures frequently identify only local optima for this kind of problem. In recent years, a number of papers have demonstrated that optimization techniques, based on meta-heuristic algorithms, are particularly promising for solving problems related to water distribution systems. This new methodology may be considered as a useful alternative to traditional approaches, based on trial-and-error or mathematical programming methods. Some applications, results and perspectives are presented for the different approaches.

Key words: non-linear optimization, water resource systems

1 Introduction

As well known, to represent water resource systems adequately, we need to extend the analysis to a sufficiently wide time horizon in order to take into account the variability of its hydrological inflows and water demands. Moreover a large number of decision variables and constraints have to be considered for planning and management purposes. One of the main goals in this field is to reach a configuration that should guarantee an adequate level of reliability of water supply and provide management criteria to be adopted by the Water Authorities.

As a consequence, the Water Resource Systems (WRS) management problem generates computationally expensive mathematical models, involving a large number of variables and equations necessary to describe the physical components of the system, relative functional ties and operating modalities. In this connection a decisive contribution to the definition of optimum management criteria and to the scaling of the works that should guarantee adequate water supply reliability is given by mathematical programming techniques particularly referred to large scale problems [29].

In previous papers [30] it was stated that these techniques, did not allow a detailed modeling of the system, and that they had to be accompanied by a simulation testing process starting from the solution obtained by optimization phase. As will be shown in what follows, it is possible to set up an interactive process between the optimization phase and the simulation testing phase, that should limit recourse to this last burdensome computational procedure. In order to reduce the gap between the optimization solution and the simulation solution we need to resolve the WRS optimization model, adopting very efficient algorithms to reach a sufficiently good approximation of the solution of the problem. In this way the simulation phase can be reduced to few iterations. Moreover, in order to reduce the gap between the real physical system and its mathematical formalization we need to reach an adequate level of adherence to the physical system with a sufficiently detailed model taking into account non-linearity in objective function and constraints.

In this paper three kind of problems are studied, modeled and solved by different algorithm approaches. The first one is concerned with WRS planning problem in which shortage control is focused and a non-linear objective function is adopted. An expansion procedure interacting between primal and dual mathematical optimization models is proposed. This kind of approach is very useful to formulate trade-off between the dimension of water works, the reliability of the system and the prediction of severity in demand short falls. The second is concerned with WRS management problem in which the vulnerability of the system is considered dealing with water resource shortage risk in planning studies. The third model is concerned with the plan and maintenance optimization problem of pipe-networks for water supply in which new plan, rehabilitation and maintenance are taken into account.

2 Water Resource Planning and Management

The first two models are related to system planning and management. These exhibit some common features from a mathematical structure point of view, even if each of them describes different aspects that arise when analyzing extended water-systems. In a general frame we can refer to a model representing a WRS problem taking into account different variables that, as usual, are classified into planning variables (e.g. the storage volume of reservoirs, the transfer-work capacity, the extension of irrigation sites, etc.) as well as operating (or flow) variables that represent the water transfer to meet the system requirements.

In a general model, flows are linked to planning variables by relations to guarantee the fulfillment of request at demand centers, as well as to bound allowed flows by work dimensions. Flows are submitted to mass balance constraints and, if required, to some problem oriented constraints. In the following we describe the main features of this general model and analyze specific models arising when real problems are examined. The general model can then be expressed as (*Model 1*):

$$\min \ f\left(\mathbf{y} + c\mathbf{x}\right) \tag{1a}$$
$$F(\mathbf{x}, \mathbf{y}) \leq \mathbf{0} \tag{1b}$$
$$\mathbf{x} \in X \tag{1c}$$
$$\mathbf{y} \in Y \tag{1d}$$

where $\mathbf{x}$ represents the set of operating variables and $\mathbf{y}$ the set of planning variables. An important general feature is that it is possible to establish a correspondence between the model formal structure and the classification of variables and constraints. Referring, at first, to a single-period situation, we can consider a static point of view. In *Model 1* we identify an operative aspect relating to the network operative elements in constraints (1c) and in the objective function concerning $\mathbf{x}$ variables. In the same way we can identify planning aspects in constraints (1d), in the linking constraints (1b) and in the objective function concerning $\mathbf{y}$ variables.

In WRS analysis we also need to examine the evolution of flow-values in time, therefore, from a dynamic point of view we can give a complementary classification of the components of the general model. The planning variables $\mathbf{y}$ have a time independent character as we consider a fixed (but generally unknown) situation in work dimensions, while the operating variables $\mathbf{x}$ have a time dependant character, as they can be different in the different periods of time horizon. It has proved particularly useful to adopt a graph structure as a topological support to the model that allows the use of such highly efficient data structures to reach a significant reduction in computer storage and computational time during data input and processing [1] [2].

As pointed out in some previous studies [29] [20], the dynamic evolution of a water resource system can be represented by a multi-period network, in which the physical components of the system and the related spatial and dynamic interconnections are adequately represented. Modularity also allows the automatic construction of the multi-period network and of data generators that characterize the initialization phase of the resolutive algorithms. The generating module is the basic-graph of the river basin scheme with dummy nodes and arcs. Therefore we can obtain further information from the analysis and to prevent the risk of infeasibility.

A multi-period network $R=(N,A)$ is then generated by reproducing the module as many times as time-periods are [29]. N represents the set of nodes and A represents the set of arcs. In this kind of model a close correspondence between operating variables and flows of the multi-period network is observed. Operating variables are immediately confirmed in the multiperiod graph and correspond biunivocally to the flows on the arcs. Among the nodes of the network representing the allocation of the planning variable for each single period, the nodes referring to the reservoirs are typically connected together by interperiod arcs. The flow on these transfer arcs represents the volume input at the end of each period. This feature allows to identify a network kernel related to operating variables and to consider non-network constraints as "complicating constraints". This is presented in more detail later in this paper. Several WRS management problems can be represented by pure network models for which very efficient algorithms have been tested [2]. Some planning aspects, such as size and requirements of the works in WRS, are usually represented by linear optimization models introducing simplifications and approximations, even if linear assumptions are not strictly adherent to real WRS.

2.1 WRS Planning with Shortage Control

The formal representation of the physical problem by an optimization model, must take into consideration the different operating and project aspects that are present in the problem. Having located the time-horizon of reference of the studied basin, we determine the number of subperiods by a time-step corresponding to the hydrological and hydraulic characteristics of the physical model. The formulation of the model is characterized by the usual operative constraints and by the determination of the minimum scaling of supplementary works that allows reduction in the system shortages as much as possible. The shortage for each single period corresponds to the flows on the arcs representing the recourse to external and expensive resources by demand centers. The objective function is made up of a function $f(\mathbf{y})$ fitted to represent the set of construction, maintenance and operating costs in a way that is satisfactory for this kind of problem. Usually $f(\mathbf{y})$ is assumed as a polynomial convex function [30]. A significant part of the constraints is represented by the flow continuity equations at the nodes, that are not seats of resource accumulation and lower and upper bounds on the flows.

The constraints that describe the links between planning constraints and some operative variables are also present. These constraints are generally non-linear and, as in this model, they are referred to the link between the flows that represent the volume input in the reservoirs at the end of each period and the variables that represent the filling capacity of the same reservoirs. The planning variables are submitted to constraints of lower and upper bounds. Moreover the constraint that represents the control on the total shortage is a characterizing element. It requires that the sum of the flows on shortage arcs should not exceed the prefixed value that the manager of the system can accept. The optimization model minimizing construction, maintenance, operating costs of plants can be expressed as follows (*Model 2*):

$$\min f(\mathbf{y}) \tag{2a}$$

$$\mathbf{x_k} \leq F(\mathbf{y}) \tag{2b}$$

$$\mathbf{s} \leq \mathbf{y} \leq \mathbf{t} \tag{2c}$$

$$\mathbf{Ex} = \mathbf{b} \tag{2d}$$

$$\mathbf{r} \leq \mathbf{x} \leq \mathbf{u} \tag{2e}$$

$$\mathbf{1} \cdot \mathbf{x_d} \leq \mathbf{x_0} \tag{2f}$$

where (2b) and (2c) represent planning constraints. More precisely, constraints (2b) require that flow $\mathbf{x_k}$ of a subset of arcs $K \subseteq A$ are linked to planning variables; we assume that flows on arcs belonging to set K must adopt the appropriate planning variable y_k as an upper bound; (2c) represents bounds on planning variables $\mathbf{y}$; (2d) and (2e) are the pure network operative constraints on multi-period network $R=(N,A)$; (2f) represents the shortage control on set of shortage-arcs $D \subseteq A$.

Problem formulation shows that a configuration of flows feasible for network constraints and for the shortage constraint, corresponds to each prefixed configuration $\mathbf{y}^*$ of $\mathbf{y}$. The determination of such a configuration of flows corresponds to the solution of the pure network flows problem (*Model 3*):

$$g(\mathbf{y}^*) = \min \mathbf{1} \cdot \mathbf{x_d} \tag{3a}$$

$$\mathbf{Ex} = \mathbf{b} \tag{3b}$$

$$\mathbf{r} \leq \mathbf{x} \leq \mathbf{u}^* \tag{3c}$$

where the vector $\mathbf{u}^*$ depends on the current value of the vector $\mathbf{y}^*$ corresponding to the arcs of K. More precisely, upper bounds on flow-variables $x_{i,j}$ for an arc $(i, j) \subseteq K$, connecting nodes $i, j \in N$, are:

$$\begin{aligned} u_{i,j}^* &= u_{i,j} & (i,j) \notin K \\ u_{i,j}^* &= \min(u_{i,j}, y_s) & (i,j) \in K \end{aligned} \tag{4}$$

To solve the *Model 3* we can determine the direction of expansion improving the objective function, by converting the variation analysis of the minimum shortage in terms of variation of the planning variables. The variation of the optimum $g(\mathbf{y}^*) = \min \mathbf{1} \cdot \mathbf{x_d}$ with respect to y_s^* can be studied by the dual of network flow problem (*Model 3*):

$$
\begin{array}{ll}
\underline{\text{primal}} & \underline{\text{dual}} \\
g(\mathbf{y}^*) = \min \mathbf{1} \cdot \mathbf{x_d} & \max \mathbf{wb} - \boldsymbol{\pi}\mathbf{u} + \mathbf{hr} \\
\mathbf{Ex} = \mathbf{b} & \mathbf{wE} - \boldsymbol{\pi} + \mathbf{h} = \mathbf{c} \\
\mathbf{r} \le \mathbf{x} \le \mathbf{u}^* & \boldsymbol{\pi}, \mathbf{h} \ge \mathbf{0} \quad \forall \mathbf{w}
\end{array}
\tag{5}
$$

where the dual variables $\boldsymbol{\pi}$, $\mathbf{h}$ correspond to the constraints $\mathbf{r} \le \mathbf{x} \le \mathbf{u}^*$ in which $\mathbf{y}_s$ appears as an upper bound and represent the variation of the objective function with respect to right hand side y_s^*. The dual variables $\mathbf{w}$ correspond to continuity constraints. The R.H.S. of the dual is the vector $\mathbf{c} = (\mathbf{1}, \mathbf{0})$ that is the cost vector of primal. We can state the Kharush-Khun-Tucker conditions (KKT) [5]:

$$
\pi_{i,j} = w_i - w_j - c_{ij} \qquad (u_{ij} - x_{ij})\pi_{ij} = 0
\tag{6}
$$

where non null π_{ij}'s are related to the flow-variables that have not reached their upper bound u_{ij}. All the arcs $(i,j) \in K_s$, have y_s^* as upper bound and the only non zero and equal to one components of the vector $\mathbf{c}$ correspond to shortage flows. The term containing the upper bounds of these arcs in the current objective function of the problem is given by:

$$
g_u(\mathbf{y}) = -\sum_{s=1,S} \sum_{(i,j)\in K_S} (w_i - w_j - c_{ij})y_s
$$
$$
= -\sum_{s=1,S} y_s \sum_{(i,j)\in K_S} (w_i - w_j - c_{ij})
\tag{7}
$$

as $c_{ij} = 0$ for $(i,j) \in K_s$, the direction $\mathbf{q}^*$ of shortage reduction on varying the configuration $\mathbf{y}^*$ of $\mathbf{y}$ will therefore have components:

$$
\frac{\partial g(\mathbf{y}^*)}{\partial y_s} = - \sum_{(i,j)\in K_S} (w_i - w_j) = \sum_{(i,j)\in K_S} \pi_{ij} = \pi_s
\tag{8}
$$

The direction of shortage reduction can therefore be determined through the solution of a network problem. A few sample basins in which the most significant elements were included by the manager have been considered, and subsequently a drinkable water real supply scheme [31]. During the last decade two critical shortages occurred. Because of this event several demand centers were cut-off from all water supplies for some consecutive days and the scarce water had to be rationed.

In all the cases examined, our objective was to test the level of shortage in critical periods and we obtained the best scaling of the supplementary works that are necessary to guarantee adequate reliability of the system. Moreover we stated the rules of management of the system itself. Thanks to a minimum cost flow algorithm applied to the multi-period network, it has been possible to assess the minimum shortage values that an ideal manager would have obtained, if he had had a priori knowledge of the sequence of hydrological inflows and demands.

The obtained results point out the computational competitiveness of this technique with respect to the classical mathematical programming technique and the possibility of changing scenario rapidly thanks to the structure of the graph supporting the model [30].

2.2 WRS Management with Vulnerability Criteria

The problem of the optimal dimension of the water resource system and the related optimal configuration should take into account additional costs given by the criteria to operate satisfactorily during periods of drought. Particularly the vulnerability of the system should be considered dealing with water resource shortage risk in planning studies. The vulnerability express the severity of drought in terms of its consequences. The consequences of drought are generally expressed by a loss (cost) function and measure to estimate the severity of a drought is given considering cost functions weighting more the shortage flows as the severity of the drought event increases. Vulnerability trend can be examined simulating a simple system that considers a generalized expression using the standardized shortage to define the expected losses L [17]:

$$L = \left(\frac{T\text{-}R}{T}\right)^{\beta} \tag{9}$$

where R is the effective release and T is the target. A simulation procedure allowed to evaluate the vulnerability of the system for different values of exponent β. Vulnerability achieves its maximum at $\beta = 0$ and decrease with increasing β finding a minimum for $\beta = 2$. We assume $\beta = 2$ to minimize the system vulnerability in planning situations and introduce a multiplicative parameter k to represent the time extension of losses. Replacing releases with deficit values x^d, since cT is the expected benefit from the irrigation site, the cost term in the objectify function (OF) related to draught losses can be written:

$$c\, T\, k \left(\frac{x^d}{x^d_{max}}\right)^2 = c'(x^d)^2 \tag{10}$$

The OF of the optimization problem can then be written:

$$\min z = \sum_k f_k y_k + \sum_i c_i x_i + \sum_j c'(x^d_j)^2 \tag{11}$$

where the first two terms represent, as usual, costs on planning and operating variables (see the general model: Model 1). In this way the problem can be expressed as a quadratic programming model (*Model 4*) [5]:

$$\min z = \mathbf{f}_y^T + \mathbf{c}^T \mathbf{x} + \frac{1}{2}\mathbf{x}^T \mathbf{H}\mathbf{x} \tag{12a}$$

$$\mathbf{x}_k \leq \mathbf{F}(\mathbf{y}) \tag{12b}$$

$$\mathbf{s} \leq \mathbf{y} \leq \mathbf{t} \tag{12c}$$

$$\mathbf{E}\mathbf{x} = \mathbf{b} \tag{12d}$$

$$\mathbf{r} \leq \mathbf{x} \leq \mathbf{u} \tag{12e}$$

Some of the state-of the art LP codes allow to resolve quadratic problems by the resolution of the associated linear complementary problem. We use the public domain code LOQO based on the resolution of the reduced KKT system by the interior point method and it is developed to solve linear and quadratic programming problems. Referring to real water resource system, the LOQO code has been used to solve the model when a quadratic function cost has been used for the deficit flows in the multi-period problem. In this way it has been possible to solve a set of problems containing up to 29,000 variables and 14,400 constraints. In Figure 1 it is evident the computational time explosion in solving QP model compared to LP ones and the increase of CPU time when QP is used to solve problems with five planning variables and from 120 to 600 time period extension [31].

3 The Pipe Network Optimization Problem

In this paragraph we refer to the classical pipe distribution network planning problem with the following main features: network demands are known and configured as node-outflows, continuity of flow must be maintained at all nodes in the pipe-network, the head loss in each pipe-arc is a known function of the flow in the pipe, its diameter, length and hydraulic properties of the pipe, at each node minimum and maximum pressure head limitation must be satisfied and diameter constraint may be applied to pipes. In the network, existing pipes (with known diameters) as well as new pipes are taken into account. For each pipe-arc different possible states are examined: the possibility to leave exactly the same pipe-arc (leave), the possibility to clean the existing pipe (clean), the possibility to add a new pipe-arc to the existing one (duplicate) and the possibility to put a new pipe-arc (new). A notable number of recent works [32], [7], [28], [33], [22], [6] described applications of metaheuristic optimization procedures when solving problems concerning the design of new pipes and duplication and maintenance rehabilitation in water distribution networks. For these problems, metaheuristic algorithms afford several benefits compared with classical mathematical programming techniques, as they can be implemented without heavy a-priori model requirements, such as convexity or differentiability in objective function and constraints. Thanks to their ability to manage discrete variables, metaheuristic optimization procedures can deal directly with the alternatives available (commercial diameters, cleaning and duplication alternatives, etc.). Each alternative consists of a set

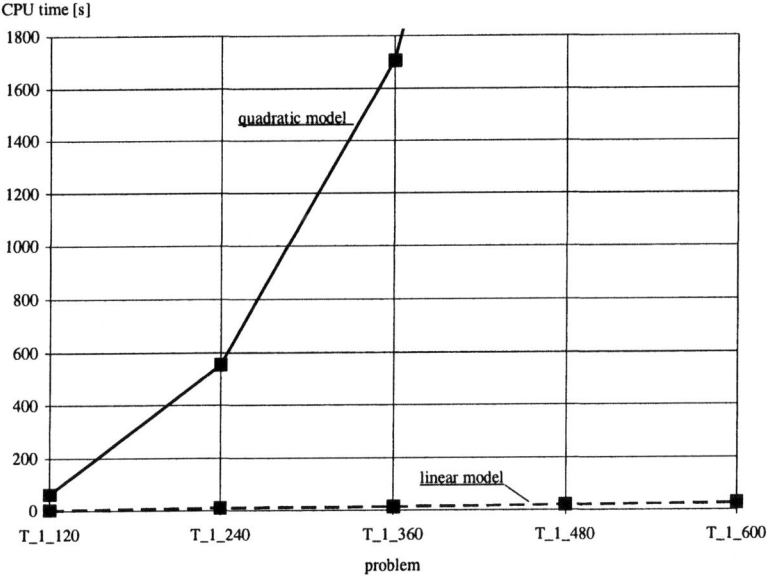

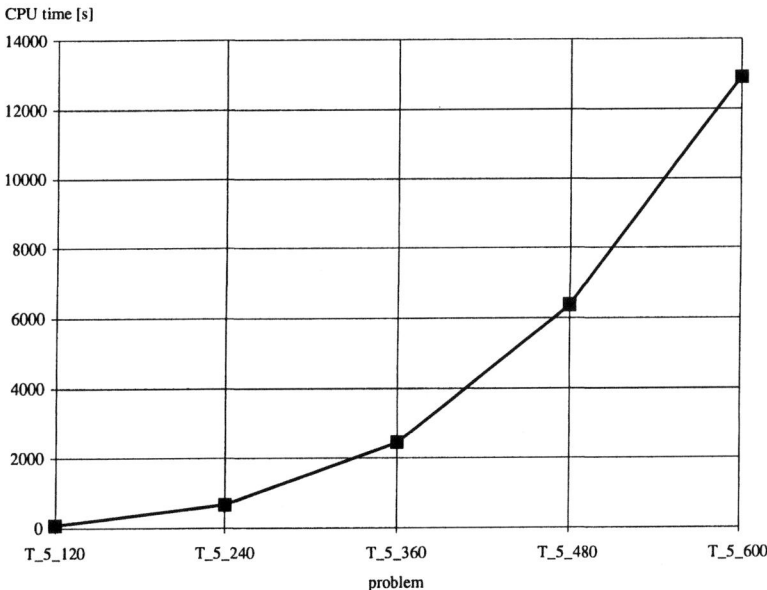

Fig. 1. CPU time comparison solving LP and QP model

of discrete, organized strings that are usually coded using predefined rules. Recently, metaheuristic approaches have also been used with the aim of optimizing the number of valves, their location and calibration. In this paper we refer to a simplified problem of pipe network design for a pressure system (without pumping-stations) with the following main features:

- the network demands are known and configured as node-outflows;
- different demand patterns are considered;
- the continuity of flow must be maintained at all nodes in the pipe-network;
- the head loss in each pipe-arc is a known function of the flow in the pipe, its diameter, length, and hydraulic properties;
- at each node minimum and maximum pressure head limitations must be satisfied;
- at each arc minimum and maximum velocity limitations can be imposed;
- diameter constraints may be applied to the pipes;
- different possible pipe-arc states and design options can be considered.

In the network $G = (N, R)$ the existing pipes (that have known diameters) as well as new pipes can be taken into account. For each pipe-arc, different options are examined, i.e. leaving exactly the same pipe-arc (leave), cleaning the existing pipe (clean), adding a new pipe-arc to the existing one (duplicate), and installing a new pipe-arc (new). The general constraint equations considered for a given demand pattern are as follows: Continuity at each node:

$$\sum_{j \in R_j} Q_j + q_i = 0 \qquad (13)$$

for each $i \in N$, where Q_j represents the flow in each of the set of pipes $R_i \in R$ connected to node i , and q_i is the demand at node i. Head-loss equation:

$$H_{i_1} - H_{i_2} = \frac{kL_j Q_j^{\beta_j}}{C_j^\alpha D_j^{n_j}} \qquad (14)$$

for $j = (i_1, i_2 \in R)$ where H_i is the node-head, L_j the length of pipe j from node i_1 to node i_2, C_j the roughness coefficient and D_j the pipe diameter. Minimum pressure head constraint:

$$H_i \geq H_i^* \qquad (15)$$

where H_i^* is the node hydraulic-head that must be guaranteed. Bounds on water velocity in the pipe:

$$V_{j,min} \leq V_j \leq V_{j,max} \quad j \in R \qquad (16)$$

Bounds on pipe diameters:

$$D_{j,min} \leq D_j \leq D_{j,max} \quad i \in R \qquad (17)$$

where the minimum diameter refers to the existing diameter in the event of duplication. Considering s different demand patterns $(s = 1, S)$, the purpose is to optimize a non-linear objective function, i.e. the total cost needed to construct new pipes, or clean or duplicate the existing ones. For the last two cases, we used the "equivalent diameter" approach [32].

Moreover, implementing the optimization procedures, equations (15) are relaxed on the OF as penalty components depending on whether the network satisfies the minimum pressure constraints at the nodes, and equations (16) can be treated as flow bounds for defined diameters. After generating an initial network configuration, the procedure performs a hydraulic analysis of the pipe network, resolving the non-linear system given by equations (13), (14), and (16). Node pressure differences from target values are then used in the OF to compute penalty costs. The OF assumes the general form:

$$
\min \sum_{j \in R_1} C_{1j} L_j + \sum_{j \in R_2} C_{2j} L_j + \sum_{j \in R_3} C_{3j} L_j +
$$

$$
+ \sum_{j \in R_4} C_{4j} L_j + \sum_{s=1,S} \left(\sum_{i \in N^*} C_{5i} (H_i^* - H_i)^\gamma \right)^{\lambda_s} \tag{18}
$$

where the first term refers to the maintenance of old pipes, the second to cleaned pipes, the third to the duplicate set, the fourth to new pipes and the fifth to hydraulic head differences.

3.1 The Metaheuristic Approach

Thanks to their ability to manage discrete variables, metaheuristic approach can deal directly with the alternatives available (commercial diameters, cleaning, duplication, etc.). Each alternative consists of a set of discrete, organized strings that are usually coded using predefined rules. Starting from initial pipe-network configurations, the proposed metaheuristic procedures only use OF cost values or other fitness information, (i.e.: the hydraulic-head constraint violation to be penalized) to allow the algorithm to reach a feasible solution as a final optimum. Over the past years metaheuristic approaches, mainly based on the Genetic Algorithm (GA) and Tabu Search (TS) technique methodology has been developed for pipe-network optimization [27] [8] [32] [26] [7] [21] [10] [23]

In this paper, comparisons between three different metaheuristic optimization approaches based on GA, reactive TS, and a combined Scatter Search (SS) - tabu search technique, have been considered. These techniques have been implemented and tested on a well-known test-problem given by [12], while introducing some extensions.

As extensively referred in [21] [22] the utilization of metaheuristic optimization procedures to the problem has been summarized in the following steps:

1. Initialization procedure: the sets of possible network element configurations (i.e. diameters, pipe-states, etc.) are proposed to the algorithm; an initial set of values is thus adopted.

2. Hydraulic verification procedure: continuity, head-loss, and velocity constraints equations are solved retrieving pressure-heads at the nodes.

3. OF evaluation: for the chosen system configuration, the economic OF evaluations are added to the penalty evaluation caused by target node pressure violation.

4. Design variables values replacement: using the fitness information, and thanks also to suitable strategic decisions, a new set is generated with the metaheuristic optimizer.

5. Optimization cycle closure: the cycle is closed and the procedure is returned to step 2.

The stop criteria can be related to the number of cycles and to improvements in a fixed number of iterations. Even though the tests indicate that optimal or near-optimal solutions are almost always obtained using well-calibrated procedure parameters related to the design variables replacement, metaheuristics do not guarantee that the global optimum will be reached. Moreover, the computational time needed to reach near-optimal configurations should be tested to check the ability of these approaches to fit real problems. In the following some general remarks on GA, TS, and SS will be given, as well as information on the applied codes. GA algorithms are search algorithms based on the mechanism of natural selection and natural genetics. The primary monograph on the topic is by Holland (1975), and extended applications of this approach have been made by Goldberg (1989). As stated in the opening paragraph, many applications of this technique are available in the recent literature on GA application on these problems. This study uses the PGA-Pack Library optimization module [3]. One of the main advantages of the genetic algorithm is that situations that cannot be adequately described with only one numeric parameter can be represented synthetically. This is possible thanks to the fact that the symbolic code system used by the GA is related to each variable configuration. The tests carried out on the GA algorithm have shown that a correct calibration of the model parameters must be reached. Using the GA, the following aspects should be kept under control:

- initial population size;
- definition procedure and string initialization;
- population replacement parameters.

The OF evaluation number needed to reach optimality remains the most limiting problem when applying GAs to real water systems. This is mainly due to the large number of system configurations in each GA population. The tabu search approach main concepts are collected in [13], and [14]. As a matter of fact, TS is usually defined as a meta-strategy in [14] [18] that guides several subordinated heuristics to produce solutions beyond those normally generated by the search for a local optimum. The employed methodologies were essentially two: an adaptive memory and a "sensitive" exploration, both of which typify the method. The system practically exploits its memories in an attempt to avoid being trapped in attraction basins, or better, in order to direct the search towards domain areas believed to be more promising. Limitations on search space generally operate by direct exclusion of search alternatives that are classified as prohibited, hence tabu. We implemented a so-called reactive kind of TS scheme: RTS [4]. An extended illustration of the RTS variant to WDND problems has been reported in [10]. In this variant, the RTS memory structure of the past solutions is dynamical, and its optimal value is estimated automatically from the algorithm by means of a retroactive assessment of the search history. The reaction mechanisms are related to the transition frequency of the current solution. The RTS algorithm has been implemented using a general purpose tabu search tool called *Universal Tabu Search* (UTS) [11]. Scatter search metaheuristics, as well as GAs, is designed to operate on a set of solutions maintained from iteration to iteration, while TS typically maintains only one solution by applying specific mechanisms to update solutions from one iteration to the next. A description of the SS can be found in [14]. In the present paper, we have developed an interface for the problems using the *OptQuest* general-purpose optimizer as a resolution module. *OptQuest* was developed by [15] using scatter search methodology. This optimizer uses the SS framework associated with tabu search strategies to obtain enhanced solutions for problems defined using complex settings. The optimization process is organized in such a way as to utilize auxiliary solutions in evaluating the combination obtained from the previous solutions, and in generating new solution vectors actively. A significant difference between classical GA implementation and SS is that, while the former heavily relies on randomization and somewhat limiting operations to create new solutions, the latter employs strategic choices and memory along with a combination of solutions to generate new solutions. Moreover *OptQuest* exploits a neural network accelerator trained on the historical data collected during search. Even though *OptQuest* can be used to take into account the linear and non-linear constraints of the model, in the interface we have implemented the *OptQuest* module as the GA and TS above, only in design variables replacement using fitness information.

3.2 Test Case

The Gessler Problem [12] can be used to compare solutions obtained using different approaches. The Gessler problem considers eight different pipe sizes (commercial diameters) available for new pipes, while existing pipes may be left as they are, cleaned, or duplicated with new pipes. The search space is extended to include, among the options for the existing arcs, replacing the pipeline completely, and doubling the alternatives for the available pipes. The total search space is $3.436 \; 10^{10}$ possible configurations (32 alternatives for existing arcs and 16 for new arcs). Using commercial NL optimization software only near-optimal solutions were been obtained and there is a problem to approximate the obtained pipe-size up or down to the nearest commercial available diameter. Results obtained applying metaheuristic approaches to the benchmark case, and comparing them. In Table 1, we report statistics over 100 run results obtained using the previously described resolution modules implementing the GA, TS, and SS approaches. Each run starts from randomly generated initial configurations. The presented results report the following indications:

- the average of obtained minimum cost functions;
- the relative occurrence frequency of the optimum;
- the average number of iterations needed to assess the corresponding optimum;
- the number of OF evaluations needed to reach the optimum.

As can be observed in Table 1, TS and SS find absolute optima in 100% of the cases, with a relatively low number of function calculations. The performance of the applied techniques is remarkably better than any other previously described method. For example, the cost of the optimum solution supplied by Gessler (1985) with selective enumeration is 4.8% higher than the absolute optimum.

Table 1. Average results (100 runs) obtained from modified Gessler Problem

Optim. technique	Average cost ($)	Success ($)	Average iter.	Average of eval.
GA	1,818,756	80%	205	20,790
TS	1,750,300	100%	172	2632
SS	1,750,300	100%	124	1110

Though according to correct guidelines, the search space may drop to the 2632 checks by TS (negligible, considering the total search space) and even further using SS, on average 1110 evaluations. Moreover, a comparison of the results shows that the success percentage rises from 80% to 100% using TS and SS instead of GA. Besides this clear difference in success percentage, it should be observed that the function calculations are remarkably more numerous in GA.

4 Conclusions

With respect to WRS models, in Sect. 2 and 3, the illustrated resolution techniques exploiting the pure network kernel, allow to reach a high computational efficiency with respect to the classical mathematical programming technique. This gives to the manager of the system the possibility of an easy examination of the changing scenario, aided by the structure of the graph supporting the model. With respect to the pipe network problem in Sect. 3, the illustrated resolution technique allows to analyze objective function without having to do an a priori study on convexity and differentiability and prevents by stalling in local optima. To improve solution time at each iteration, efficient resolution techniques exploiting the peculiarity of the matrix constraints, in the global optimization model, can be adopted in the resolution of a non-linear system, during a metaheuristic iteration. The proposed general modeling approach in water resource systems and in pipe network analysis can be applied for a wide variety of problem in his field and can give to Water Authority an easy and efficient support in taking decisions in critical conditions.

References

1. Aho, A.V., Hopcroft E., Ullman, J.D.: Data Structure and Algorithms, Addison-Wesley (1983)
2. Ahujia, R.K., Magnanti, T.L., Orlin, J.B.: Network Flows. Prentice-hall, New York (1993)
3. Argonne National Laboratory: PGA - Pack Library. Parallel Genetic Algorithm. Argonne(IL) (1996)
4. Battiti, R., Tecchiolli, G.: The Reactive Tabu Search, ORSA Journal on Computing, 6(2), 126-140 (1994)
5. Bertsekas, D.P.: Convex Analisys and Optimization. Athena Scientific (2003)
6. Broad, D.R., Dandy, G. C., Mayer, H.R.: Water distribution system optimization using metamodels. J. of Water Resource Planning and Management, ASCE, (131)3, 172-180 (2005)
7. Dandy G.C., Simpson A.R., Murphy L.J.: An improved genetic algorithm for pipe network optimization. Water Resour. Res. (32)2, 49-458 (1996)
8. Dandy, G.C., Simpson, A.R., Murphy, L.J.: A review of pipe network optimization techniques. Proceedings of Watercomp '93, Melbourne (1993)
9. DIMACS. The first DIMACS international algorithm implementation challenge: The benchmark experiments. Technical report. DIMACS, New Brunswick, N.J (1991)
10. Fanni, A., Liberatore, S., Sechi, G.M., Soro, M., Zuddas, P.: Optimization of Water Distribution Systems by a Tabu Search Metaheuristic. In: Laguna, J.L., Gonzales-Velarde (eds.) Computing Tools for Optimization and Simulation, Kluwer Academic Publ. Boston (1999)
11. Fanni, A., Bibbo', S., Giua, A., Matta, A.: A general purpose Tabu Search code: an application to Digital filters design, IEEE Int. Conf. on Systems, Man, and Cybernetics. San Diego (1998)
12. Gessler, J.: Pipe network optimization by enumeration, Proc. Computer applications for water resources: 572-581 (1985)
13. Glover, F.: Tabu Search fundamentals and uses. University of Colorado, Boulder (1994)
14. Glover, F., Laguna, M.:Tabu Search. Boston: Kluwer Academic Publishers (1997)
15. Glover, F., Kelly, J.P., Laguna, M.: New Advances and Applications of Combining Simulation and Optimization. In J.M. Charnes, D.J. Morrice, D.T.Brunner and J.J. Swain (eds), Proceedings of the 1996 Winter Simulation Conference: 144-152 (1996)

16. Goldberg, D.E.: Genetic Algorithms in Search. Optimization and Machine Learning. Reading: Addison-Wesley Publ.Co.Inc. (1989)
17. Hashimoto, T., Stedinger, J.R. and Loucks, D.P.: Reliability, Resiliency, and Vulnerability Criteria for Water Resource System Performance Evaluation". Water Resources Research, 18, 14-20. (1982)
18. Hertz, A., De Werra, D.: The Tabu Search metaheuristic: how we used it. Annals of Mathematics and Artificial Intelligence, 1, 111-121 (1990)
19. Holland, J.H.: Adaptation in natural and artificial systems. The University of Michigan Press, Ann Arbor (1975)
20. Kuczera, G.: Network Linear Programming Codes For Water-Supply Headworks Modelling, ASCE, J. of Water Resources Planning and Management, 3, 412-417 (1993)
21. Liberatore, S., Sechi, G.M., Zuddas, P.: A Genetic Algorithm approach for water system optimization. AICE Annual Conference, Milano (1998)
22. Liberatore S., Sechi G. M., Zuddas P.: Water Distribution Systems Optimization by Metaheuristic Approach. Advances in supply management, Lisse, 265-272, London (2003)
23. Lippai, I., Heaney, J.P., Laguna, M.: Robust water system design with commercial intelligent search optimizers. Accepted by J.Computing in Civil Eng. (1999)
24. Loucks, D.P., Stedinger, J.R., Haith, D.A.: Water Resource System Planning and analysis. Water resource research, 18, 14-20 (1981)
25. Lasdon, L.S.: Optimization Theory for Large Systems. Courier Dover Publications (2002)
26. Murphy, L.J., Simpson, A.R., Dandy, G.: Design of a network using genetic algorithms. Water, 20 40-42 (1993)
27. Murphy, L.J., Simpson, A.R.: Pipe optimization using genetic algorithms. Res. Rep. 93:95, Dept. of Civ.Eng., Univ. of Adelaide (1992)
28. Savic D.A., Walters G.A.: Genetic algorithms for least-cost design of water distribution networks. J. of Water Resource Planning and Management, ASCE, (123)2, 67-77 (1997)
29. Sechi, G.M., Zuddas, P.: Data management for extended multiperiod analisys of water resource systems. Prc. 11th Conf. on Operation Research, Buenos Aires (1987)
30. Sechi, G.M., Zuddas, P.: A Large-Scale Water Resource Network Optimization Algorithm. ICOTA '95, Chengdu-China, 378-385 (1995)
31. Sechi, G.M., Zuddas, P.: Algorithms for Large Scale Water Resource Structured Models with network flow kernels. Internal Report, Univ. of Cagliari (Italy) (1996)
32. Simpson, A.R., Dandy, G. C., Murphy, L.J.: Genetic algorithms compared to other techniques for pipe optimization. J. of Water Resource Planning and Management, ASCE, (120)4 (1994)
33. Walters, G.A., Halhal, D., Savic, D., Ouazar, D.: Improved design of Anytown distribution network using structured messy genetic algorithms. Urban Water, 1(1), 23-38 (1999)

Chapter 11

SOLVING THE PHASE UNWRAPPING PROBLEM BY A PARAMETRIZED NETWORK OPTIMIZATION APPROACH[*]

Pierluigi Maponi

Dipartimento di Matematica e Informatica, Università di Camerino,
62032 Camerino, Italy

pierluigi.maponi@unicam.it

Francesco Zirilli

Dipartimento di Matematica "G. Castelnuovo", Università di Roma "La Sapienza",
00185 Roma, Italy

f.zirilli@caspur.it

Abstract   The phase unwrapping problem consists in recovering a real function U defined on a discrete set (i.e. a rectangular grid) from the knowledge of its values modulus 2π. The phase unwrapping problem is the key problem in interferometry, for simplicity we restrict our attention to the SAR (Synthetic Aperture Radar) interferometry problem. The phase unwrapping problem is not well defined in fact it has infinitely many solutions, so that it must be "regularized" to be satisfactorily solvable. We propose a formulation of the phase unwrapping problem based on a network optimization problem depending on a parameter. We study the behaviour of the solution obtained as a function of this parameter. Numerical algorithms to solve the network optimization problems obtained are proposed. We report some numerical experience comparing the results obtained with the algorithm proposed here with the results obtained with another algorithm proposed in the scientific literature. The numerical experience proposed is relative to synthetic data and to

---

[*]This research has been partially supported by ASI-Agenzia Spaziale Italiana under grant ASI-ARS-99-50

real SAR interferometry data. The real data are taken from the ERS
missions of the European Space Agency (ESA).

Keywords: SAR interferometry problem, phase unwrapping problem, minimum
cost flow problem.

1. Introduction

We begin introducing some notations. Let $I\!N$, $Z\!\!\!Z$, $I\!R$ be the sets of
natural, integer and real numbers respectively. Let $N \in I\!N$, we denote
with $Z\!\!\!Z^N$ the set of N-tuples of integers and with $I\!R^N$ the N-dimensional
real Euclidean space. Let $\underline{x} = (x_1, x_2, \ldots, x_N)^t \in I\!R^N$ be a generic
vector, where the superscript t denotes the transposition operation, for
$\underline{x}, \underline{y} \in I\!R^N$ we denote with $\underline{x}^t \underline{y}$ the Euclidean scalar product of $\underline{x}$ and
$\underline{y}$. Let $1 \leq p \leq \infty$, we denote with $\|\underline{x}\|_p$ the usual p-norm in $I\!R^N$. We
denote with $\underline{0}_N$ the vector of $I\!R^N$ having all the components equal to
zero and with $\underline{1}_N$ the vector of $I\!R^N$ having all the components equal to
one.

Let U be a real valued function defined on a discrete set (i.e. in
the simplest case a rectangular grid). The phase unwrapping problem
consists in the reconstruction of the function U from the knowledge of
its values modulus 2π. More precisely we consider:

PROBLEM 1 (phase unwrapping) *Let $I = \{(i,j) \mid i = 1, 2, \ldots, N_1, j = 1, 2, \ldots, N_2\}$, let $U = U_{i,j}, (i,j) \in I$ be a real function defined on I. Let $W = W_{i,j}, (i,j) \in I$ be a real function defined on I, such that for every $(i,j) \in I$ we have $W_{i,j} = [U_{i,j}]_{2\pi}$, that is $W_{i,j}$ is equal to $U_{i,j}$ modulus 2π. The problem considered is: from the knowledge of W recover U.*

We use $[\cdot]_{2\pi} \in [-\pi, \pi)$ that is we take the value of $\cdot$ modulus 2π in
the interval $[-\pi, \pi)$ instead than in the more usual interval $[0, 2\pi)$. In
Figure 1.3 we show an illustrative picture of a one dimensional version
of the phase unwrapping problem. We note that usually W is called the
wrapped phase function and U is called the *(unwrapped) phase function*.

We note that the phase of an electric field is the argument of the electric field as a complex number. We want to use the property that every
complete rotation in the phase, i.e. every increment or decrement of
2π in the phase angle, the (time harmonic) electromagnetic field travels
a distance equal to the wavelength. Thus in principle we can recover
distances from the phase measurements. Problem 1 rises naturally from
the fact that phase measurements can be done only modulus 2π, in fact
the argument of a complex number is defined modulus 2π.

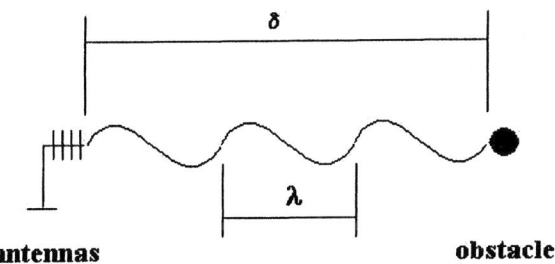

Figure 1.1. The example.

Problem 1 occurs in many application fields, for a review see [1, 2]. For example it arises naturally when from backscattering measurements one has to compute the distance δ between the scatterer and the location where the datum is collected, that is the position where both the transmitting antenna and the receiving antenna are located, see Figure 1.1. The transmitting antenna generates an electromagnetic wave, when the scatterer is hit by this wave generates a scattered electromagnetic wave that is measured by the receiving antenna. Let us suppose that all the electromagnetic waves involved in this experiment have wavelength λ. The distance between the antennas and the scatterer is equal to half of the path traveled by the electromagnetic wave from the transmitting antenna to the scatterer and back from the scatterer to the receiving antenna. This distance denoted with δ can be computed from the knowledge of λ and the knowledge of the difference u of the phase of the electromagnetic wave received by the receiving antenna and the phase of the electromagnetic wave generated by the transmitting antenna, in fact we have:

$$\delta = \frac{1}{2}\lambda\frac{u}{2\pi}. \tag{1.1}$$

However note that we can only measure the phase of an electromagnetic wave modulus 2π, as a consequence of that also the difference of phase u previously considered can be evaluated only modulus 2π.

We note that the reconstruction of the distance δ in the previous example can be obtained in many other ways, for example the distance δ can be obtained measuring the travel time of the electromagnetic waves involved in the experiment, that is the time necessary to the electromagnetic wave generated by the transmitting antenna to reach the scatterer and to come back to the receiver. However in some practical situations the measurement of a phase function can be the more convenient way to solve remote sensing problems. In section 2 an interesting application of the phase unwrapping problem will be presented, that is a remote sensing application aimed at recovering the digital elevation map of the Earth surface using interferometric data measured by two SAR (Synthetic Aperture Radar) systems travelling on board of an airplane or of a satellite, for details see [2–4]. This problem is called SAR interferometry problem and, due to its important applications, several different methods have been proposed for its solution, see [2, 4–10] for details.

We note that in Problem 1 from the knowledge of $W = W_{i,j}$, $(i, j) \in I$ we can recover $U = U_{i,j}$, $(i, j) \in I$ computing the integer function $\kappa^* = \kappa^*_{i,j}$, $(i, j) \in I$ such that $U_{i,j} = W_{i,j} + 2\pi\kappa^*_{i,j}$, $(i, j) \in I$. However any integer function $\kappa = \kappa_{i,j}$, $(i, j) \in I$ gives a legitimate reconstruction of U through the formula $U = W + 2\pi\kappa$. This means that Problem 1, the phase unwrapping problem is not well defined. A more sophisticated mathematical formulation of the phase unwrapping problem than the simple definition given in Problem 1 is needed to try to characterize κ^*. In this paper we reformulate the phase unwrapping problem as a network optimization problem, where the objective function is a measure of the "magnitude" of a function related to the function $\kappa = \kappa_{i,j}$, $(i, j) \in I$. Of course this can be done in many different ways. We choose to model the phase unwrapping problem using the "gradient" of the phase function as an integer minimum cost flow problem on a network. This choice generalizes the choice made in [5], where the objective function considered is the 1-norm of an unknown function related to κ. This problem is a very special one in fact it is equivalent to a linear minimum cost flow problem on a network. We note that the minimum cost flow problems considered in [5] and here are linear programming problems that have special properties, the most important one is that the integral constraint on the independent variables does not increase the difficulty of the problem, in fact these problems can be solved (or approximated) as linear programming problems with real variables, for details see [11] page 10 or [12] page 94. When we consider the phase unwrapping problem in the context of SAR interferometry this is a very important feature of minimum cost flow problems on a network. In fact this problem usually involves a large number of variables, so that in this case very efficient algorithms are needed to solve Problem 1. Here "efficient algorithm" is refered both to the computational cost as well as to memory requirements. The special properties of minimum cost problems on a network can be exploited to build very efficient algorithms, for details see [11] page 10 or [12] page 94. A first attempt to reconsider the choices made in [5] was done in [7] where the objective function is chosen to be the p-norm of the unknown function considered in [5] with $p > 1$ or $p = \infty$ and the solution of the corresponding integer nonlinear programming problem is computed through the solution of a minimum cost flow problem. This feature makes possible the construction of very efficient algorithms.

In this paper the objective function is chosen to be the ∞-norm of the unknown function. The corresponding network optimization problem is highly degenerate that is it has many different solutions, so that we propose a method to select one of these solutions. This selection mechanism is the main improvement on [5] and [7] made in this paper. The selection mechanism is based on a pair of minimum cost flow problems that depends on a parameter w; this parameter controls the magnitude of the 1-norm of the solution of the ∞-norm problem chosen. The parameter w should be chosen depending on the particular instance of Problem 1. Remember that the ∞-norm problem has a highly degenerate set of solutions. We show with some numerical examples based on synthetic data that the solution of Problem 1 obtained using the mathematical model proposed here is strongly dependent from the parameter w; moreover when the "right" value of w is chosen the solution of Problem 1 obtained with the method proposed here is usually substantially better than the one obtained with the methods proposed in [5, 7]. Based on statistical considerations we propose a simple way to choose the value of the parameter w. The numerical experience obtained on a different set of simulated data shows that this strategy to choose the value of the parameter w is only moderately effective. A more efficient strategy to perform this choice will be investigated in a future paper.

In section 2 we present the SAR interferometry problem. In section 3 we formulate the phase unwrapping problem as a network optimization problem depending on a parameter and we propose a numerical method to solve this problem. In section 4 we show some numerical experience based on simulated and real SAR interferometry data comparing the results obtained using the method proposed with the results obtained with the method proposed in [5, 7]. The real data are taken from the ERS missions of the European Space Agency (ESA). In section 5 we present some final remarks.

2.   The SAR interferometry problem

Let us consider the Earth surface, that we suppose flat, and at a distance h above it we consider the origin of a Cartesian coordinate system having the z-axis oriented downward, see Figure 1.2(a). Let $(x, y, z)^t$ be the corresponding Cartesian coordinates and $(r, y, \theta)^t$ be the cylindrical coordinates such that:

$$x = r \cos \theta, \tag{1.2}$$
$$z = r \sin \theta. \tag{1.3}$$

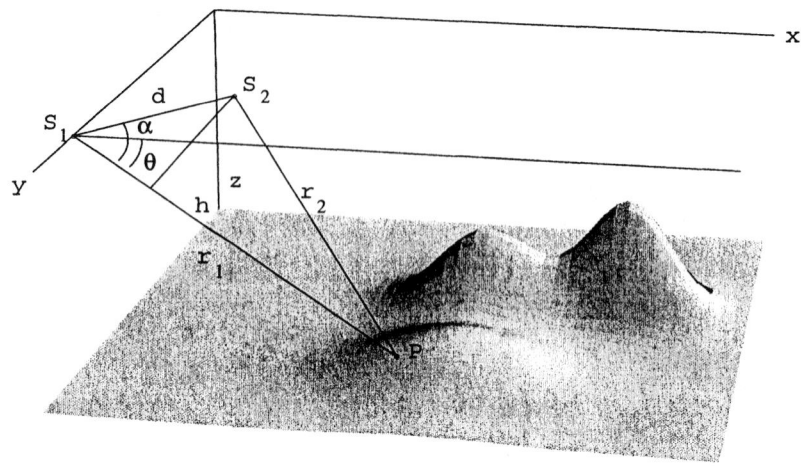

Figure 1.2. The SAR interferometry system.

In the radar jargon the coordinate r is called *slant range coordinate* and the coordinate y is called *azimut.*

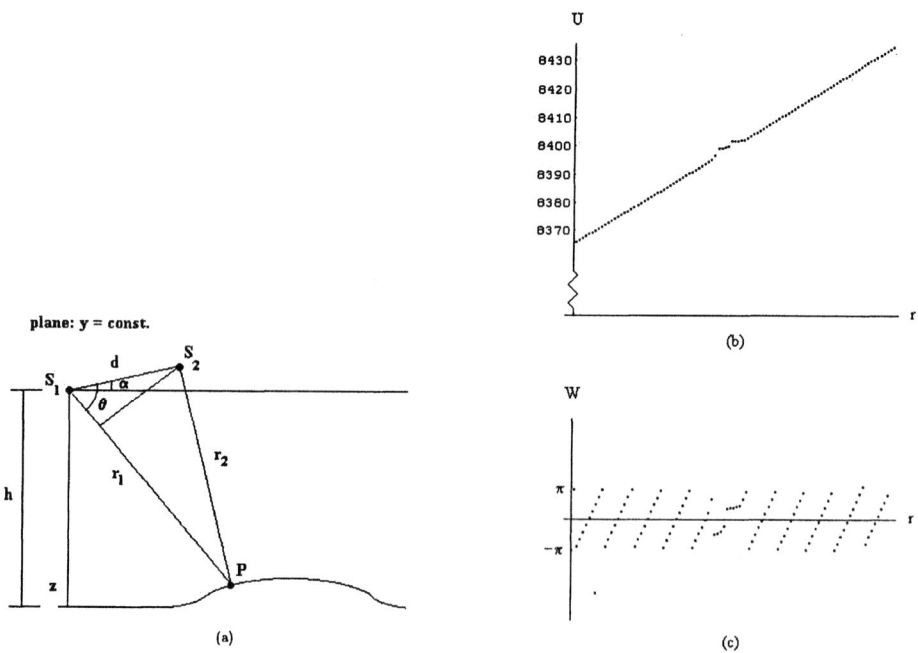

Figure 1.3. The SAR interferometry system in the plane $y = const$: (a) the SAR interferometry system, (b) the (unwrapped) phase function, (c) the wrapped phase function.

A SAR system usually travels on board of satellites or airplanes, and is able to emit an electromagnetic radiation to illuminate the Earth surface moreover it is able to measure the electromagnetic radiation scattered by the Earth surface when hit by the emitted electromagnetic radiation. Let λ be the wavelength of all the electromagnetic radiations involved in this experiment. The measurements of the backscattered electromagnetic radiation are processed taking advantage of the fact that the SAR system is travelling to obtain new synthetic data that can be interpreted as the measurements obtainable with an antenna having a larger aperture (i.e. synthetic aperture) than the aperture of the physical antenna. This procedure increases the resolution power of the instrument. For more details about SAR see [3].

We consider two SAR systems S_1, S_2 travelling along two parallel trajectories moreover S_1 is travelling along the y-axis and we suppose that S_1 and S_2 have the same y coordinate, see Figure 1.2. We note that under the above assumptions the relative position of S_1 and S_2 can be determined by the distance d between S_1, S_2 and the angle α, see Figure 1.2. Every SAR system measures the phase modulus 2π of the backscattered electromagnetic radiation as a function of the $(x, y)^t$ coordinates of the position P on the observed scene. We can assume that P has the same y coordinate of S_1, S_2. The modulus 2π of the difference of the phase measured by the two SAR systems is called *wrapped interferometric phase*, the difference of the whole phases of the signals arrived to the SAR systems is called *(unwrapped) interferometric phase*. We note that from the knowledge of the unwrapped interferometric phase we can compute the elevation map of the scene that is the z coordinate of a generic point P on the scene as a function of its $(x, y)^t$ coordinates, see equation (1.4) and see [2] for details. We consider the following problem:

PROBLEM 2 (SAR interferometry) *From the knowledge of the wrapped interferometric phase compute the corresponding unwrapped interferometric phase.*

In Figure 1.3(a),(b),(c) we report an illustrative picture of a one dimensional version of SAR interferometry problem, where the characteristic parameters h, d, α, λ are similar to those used in the ERS1 mission of ESA, that is: $h = 782563$m, $d = 143.13$m, $\alpha = -1.44$radiants, $\lambda = 0.056415$m. The scale of U in Figure 1.3(b) is different from the scale of W in Figure 1.3(c). We note that this problem is a phase unwrapping problem, that is this problem is a particular case of Problem 1. So that it can be solved with the method proposed in section 3. However the SAR interferometry problem has some special features, such as the shadow and the layover phenomena that are due to the special experimental setting. In practical situations these phenomena make impossible a complete knowledge of the wrapped interferometric phase function. Let us explain briefly these phenomena.

Let $\Omega_1 \subset \{(x,y)^t \in I\!\!R^2 : x \geq 0\}$ be a bounded open set, let $Z : \Omega_1 \to I\!\!R$ be the elevation map of the observed scene, that is for $(x,y)^t \in \Omega_1$ the surface of the observed scene is given by: $z = h - Z(x,y)$. Moreover in the cylindrical coordinates (1.2), (1.3) we suppose that the surface $z = h - Z(x,y)$ is represented by the function $\theta = \Theta(r,y)$, $(r,y)^t \in \Omega_2$, where $\Omega_2 \subset I\!\!R^2$ is a suitable bounded open set that depends on Ω_1 and the elevation map Z through the change of variables (1.2), (1.3). When the observed scene is such that the function Θ is single valued and injective the unwrapped interferometric phase is given by:

$$u_e(r,y) = \frac{4\pi}{\lambda} d\cos(\Theta(r,y) - \alpha) , \quad (r,y)^t \in \Omega_2, \qquad (1.4)$$

see Figure 1.2. In our numerical experiments the unwrapped interferometric phase is replaced with the following function:

$$u(r,y) = u_e(r,y) + \epsilon(r,y) , \quad (r,y)^t \in \Omega_2, \qquad (1.5)$$

where $\epsilon(r,y)$ is a random term that takes in account all the errors that will affect the wrapped phase function. For example in (1.5) we can consider the errors due to the inhomogeneities of the atmosphere, or the measurement errors that take place in the wrapping process. All these errors are grouped together in the term $\epsilon(r,y)$. Note that from the knowledge of the unwrapped phase $u_e(r, y)$ using equation (1.4) is possible to obtain informations about $\Theta(r, y)$ that is the digital elevation map of the scene. The wrapped interferometric phase is defined as follows:

$$w(r,y) = [u(r,y)]_{2\pi} , \quad (r,y)^t \in \Omega_2. \qquad (1.6)$$

We note that the measurements of w are usually made by a sampling operation, so that we define:

$$W_{i,j} = w(r_0 + j\delta_2, y_0 + i\delta_1) , \quad (i,j) \in I, \qquad (1.7)$$

where $r_0, y_0, \delta_1, \delta_2 \in I\!\!R$ with $r_0, \delta_1, \delta_2 > 0$ are parameters that characterize the measurement operations. As in the general phase unwrapping problem from the measurements of $W_{i,j}, (i,j) \in I$, we have to compute $U_{e,i,j}, (i,j) \in I$, where for $(i,j) \in I$ the values $U_{e,i,j}$ are the approximation of $u_e(r_0 + j\delta_2, y_0 + i\delta_1)$ computed by the reconstruction procedure. We note that formula (1.4) does not hold when the observed scene is such that Θ is a multivalued function and/or a non injective function with respect to the r variable. More precisely when for some value of the y coordinate the function Θ is not injective with respect to the r variable we have the so called *shadow phenomenon*. When for some value of the y coordinate the function Θ is a multivalued function with respect to the r variable we have the so called *layover phenomenon*. We must keep in mind that the data of a SAR interferometry experiment can be affected by shadow and/or layover phenomena. We note that the shadow and layover phenomena deteriorate the information content of the data, so that also the corresponding ability to reconstruct the unwrapped phase function is deteriorated, see [6] for a detailed discussion.

3. The Network Optimization Problem

Let us consider Problem 1. This problem is not well defined, in fact the set $\mathcal{N} = \{u : I \to \mathbb{R}|\ u = u_{i,j} = 2\pi n_{i,j},\ n_{i,j} \in \mathbb{Z},\ i = 1, 2, \ldots, N_1,\ j = 1, 2, \ldots, N_2\}$ can be seen as the "kernel" of the modulus 2π operation, that is for every $u \in \mathcal{N}$ we have $[u]_{2\pi} = 0$. Moreover the solution U of Problem 1 can be written as $U = W + u^*$ for a suitable function $u^* \in \mathcal{N}$, in fact $[U]_{2\pi} = [W + u^*]_{2\pi} = W$. So that without some a priori information on U from the knowledge of W we cannot distinguish the required solution U from all the other possible solutions that is the functions of the form $W + u$, $u \in \mathcal{N}$. That is in the set $\{W + u, u \in \mathcal{N}\}$ of all possible solutions of Problem 1 we must select a function which we consider the solution of Problem 1. This is done formulating the phase unwrapping problem as an optimization problem. The optimizers of this optimization problem will be the proposed solutions of Problem 1. When several optimizers corresponding to the same value of the objective function are present one of them, for example the one obtained by the numerical algorithm used, will be privileged.

Let $I_1 = \{(i,j), i = 1, 2, \ldots, N_1 - 1, j = 1, 2, \ldots, N_2\}$, and $I_2 = \{(i,j), i = 1, 2, \ldots, N_1, j = 1, 2, \ldots, N_2 - 1\}$. Let $F : I \to \mathbb{R}$ be a generic function, we define the *discrete partial derivatives* operators Δ_1, Δ_2 acting on F in the following way:

$$(\Delta_1 F)_{i,j} = F_{i+1,j} - F_{i,j}, \quad (i,j) \in I_1, \tag{1.8}$$

$$(\Delta_2 F)_{i,j} = F_{i,j+1} - F_{i,j}, \quad (i,j) \in I_2. \tag{1.9}$$

We note that equations (1.8), (1.9) define two functions which are usually denoted as follows: $F_\nu = F_{\nu,i,j} = (\Delta_\nu F)_{i,j}$, $(i,j) \in I_\nu$, $\nu = 1, 2$. The vector field $(F_1, F_2)^t = (\Delta_1 F, \Delta_2 F)^t$, defined on $I_1 \cap I_2$, is called the *discrete gradient vector field* of the function F. We note that a generic vector field $(F_1, F_2)^t$ defined on $I_1 \cap I_2$ is the gradient of a function F

defined on I when it satisfies the *irrotational property*, that is:

$$(\Delta_2 F_1)_{i,j} = (\Delta_1 F_2)_{i,j} , \quad (i,j) \in I_1 \cap I_2. \tag{1.10}$$

The function F can be computed from the knowledge of its value in a point, for example the knowledge of $F_{1,1}$, and the knowledge of $(F_1, F_2)^t$, that is the discrete gradient vector field of F, in fact we have:

$$F_{i,j} = F_{1,1} + \sum_{l=1}^{i-1} F_{1,l,1} + \sum_{l=1}^{j-1} F_{2,i,l} , \quad (i,j) \in I, \tag{1.11}$$

where the sums having upper index equal to zero must be considered equal to zero.

Let U, W be the functions defined in Problem 1. We define the following functions:

$$G_{\nu,i,j} = [\Delta_\nu W]_{2\pi\ i,j} + 2\pi k_{\nu,i,j} , \quad (i,j) \in I_\nu , \quad \nu = 1, 2, \tag{1.12}$$

where $k_{\nu,i,j}$, $(i,j) \in I_\nu$, $\nu = 1, 2$, are integer variables that must be determined. We note that:

$$[\Delta_\nu W]_{2\pi\ i,j} = \Delta_\nu W_{i,j} + 2\pi b_{\nu,i,j} , \quad (i,j) \in I_\nu , \quad \nu = 1, 2, \tag{1.13}$$

for a suitable choice of $b_{\nu,i,j} \in \{-1, 0, 1\}$, $(i,j) \in I_\nu$, $\nu = 1, 2$. We note that the values of the variables $b_{\nu,i,j}$, $(i,j) \in I_\nu$, $\nu = 1, 2$ are uniquely determined by equation (1.13) since $W : I \rightarrow [-\pi, \pi)$.

Let us impose that $(G_1, G_2)^t$ is a discrete gradient vector field, that is we require that $(G_1, G_2)^t$ verifies the irrotational property (1.10); from (1.12), (1.13) we can rewrite the irrotational property for $(G_1, G_2)^t$ as follows:

$$k_{1,i,j+1} - k_{1,i,j} - k_{2,i+1,j} + k_{2,i,j} = -(b_{1,i,j+1} - b_{1,i,j} - b_{2,i+1,j} + b_{2,i,j}),$$
$$(i,j) \in I_1 \cap I_2. \tag{1.14}$$

We note that equation (1.14) gives $M = (N_1 - 1)(N_2 - 1)$ constraints for the $N = N_1(N_2 - 1) + N_2(N_1 - 1)$ variables $k_{\nu,i,j}$, $(i,j) \in I_\nu$, $\nu = 1, 2$.

Let $\underline{k} \in \mathbb{Z}^N$ be the vector containing the variables $k_{\nu,i,j}$, $(i,j) \in I_\nu$, $\nu = 1, 2$, $\underline{\beta} \in \mathbb{Z}^M$ be the vector containing the terms on the right hand side of equations (1.14), A be the matrix of the coefficients of the linear system made of equations (1.14) having the rows and the columns arranged appropriately and finally let $A\underline{k} = \underline{\beta}$ be the linear system (1.14) in the matrix-vector notation.

Every integer solution $\underline{k}$ of the linear system $A\underline{k} = \underline{\beta}$ defines, via the equations (1.12), (1.13), a vector field $(G_1, G_2)^t$, that is a possible choice for the discrete gradient vector field of U. We note that $A\underline{k} = \underline{\beta}$ is always compatible in $\mathbb{Z}^N$, in fact the vector $\underline{k}_b$ whose components $k_{\nu,i,j}$ are defined as follows $k_{\nu,i,j} = -b_{\nu,i,j}$, $(i,j) \in I_\nu$, $\nu = 1,2$ verifies $A\underline{k}_b = \underline{\beta}$. So that we introduce a merit function to select a particular solution $\underline{k}^*$ of $A\underline{k} = \underline{\beta}$. That is given p with $1 \leq p \leq \infty$, we can compute $\underline{k}^*$ as a minimizer of the following problem:

$$\min f_p(\underline{k})$$
$$\text{s.t.} : A\underline{k} = \underline{\beta},$$
$$\underline{k} \in \mathbb{Z}^N, \tag{1.15}$$

where for $\underline{k} \in \mathbb{Z}^N$ we define:

$$f_p(\underline{k}) = \begin{cases} \|\underline{k}\|_p^p, & 1 \leq p < \infty, \\ \|\underline{k}\|_\infty, & p = \infty. \end{cases} \tag{1.16}$$

When in (1.13) we have $b_{\nu,i,j} = 0$, $(i,j) \in I_\nu$, $\nu = 1,2$ the solution of the optimization problem (1.15), (1.16) is $\underline{k}^* = \underline{0}_N$. Later we restrict our attention to the more interesting case when $b_{\nu,i,j} \neq 0$ for some $(i,j) \in I_\nu$, $\nu = 1,2$.

The optimization problem (1.15), (1.16) has been investigated in [7]. The constraints $A\underline{k} = \underline{\beta}$ of problem (1.15) can be seen as flow conservation conditions on the nodes of a grid graph and the vector $\underline{\beta}$ gives the exogenous supplies at the nodes of the graph, for a detailed discussion on network problems see [11] page 10 or [12] page 37. However to consider an equivalent genuine network problem we have to add four "ground nodes" and an artificial node P, see Figure 1.4 where is reported the graph $\mathcal{G}$ which is constituted by the original grid graph, by the four "ground nodes" and by the node P. Note that the node P is joined with the four ground nodes and in P we require an exogenous supply equal to $-\sigma$, where $\sigma = \underline{\beta}^t \underline{1}_M$. Thus problem (1.15), (1.16) is equivalent to the following network problem:

$$\min f_p(\underline{k})$$
$$\text{s.t.} : A_1 \begin{pmatrix} \underline{k} \\ \underline{h} \end{pmatrix} = \underline{\beta}_1, \tag{1.17}$$
$$\underline{k} \in \mathbb{Z}^N,$$
$$\underline{h} \in \mathbb{Z}^4,$$

where $\underline{h} = (h_1, h_2, h_3, h_4)^t$, $\underline{\beta}_1 = (\underline{\beta}^t, 0,0, 0,0, -\sigma)^t \in \mathbb{Z}^{M+5}$, $A_1 = \begin{pmatrix} A & Z \\ B & C \end{pmatrix}$, and Z is the null matrix having M rows and 4 columns. We

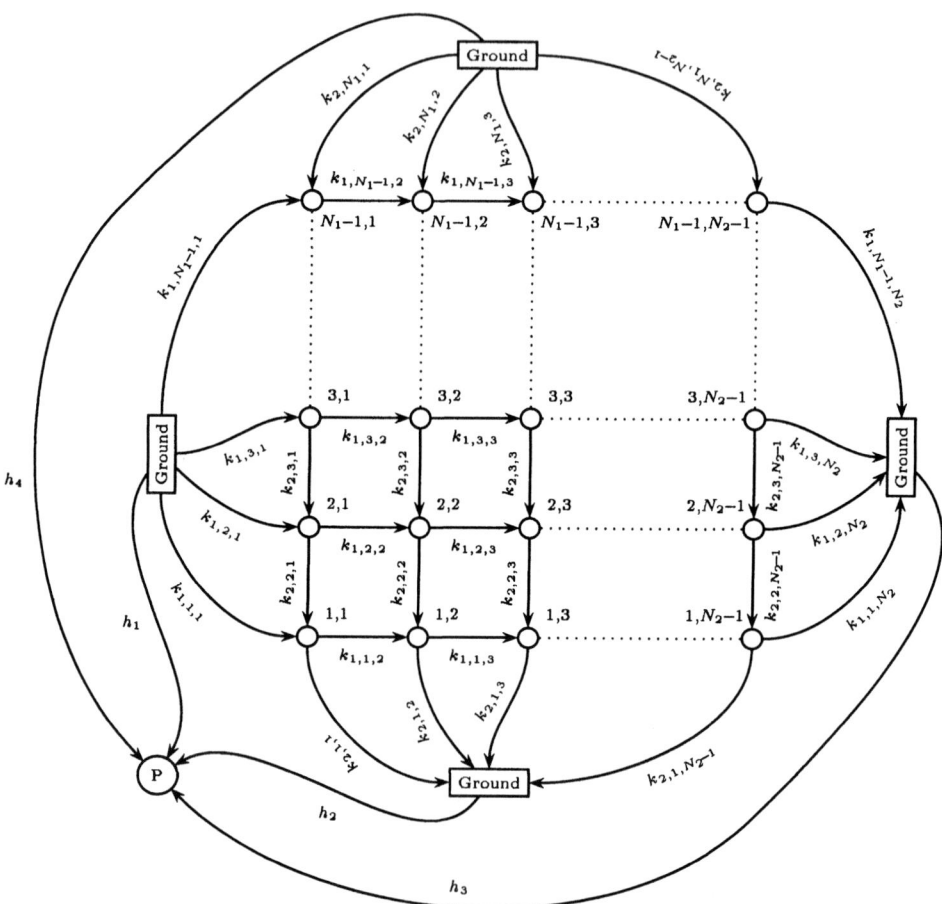

Figure 1.4. The graph $\mathcal{G}$.

note that the matrix B has 5 rows and N columns and the matrix C has 5 rows and 4 columns. Moreover the first four rows of the matrix (B, C) give the left hand side of the flow conservation conditions at the ground nodes and the last row of the matrix (B, C) gives the left hand side of the flow conservation condition at node P. So that when $p = 1$ problem (1.15), (1.16) can be seen as a minimum cost flow problem, for details see [5]. We note that in a minimum cost flow problem the integral constraint on the variables, that is the fact that $\underline{k} \in \mathbb{Z}^N$, $\underline{h} \in \mathbb{Z}^4$, does not increase the difficulty of the problem, for a detailed discussion see [11] page 10 or [12] page 94. When $1 < p \leq \infty$ problem (1.15), (1.16) cannot be reformulated as a linear optimization problem, in [7] a method to solve this problem is proposed, which is based on the approximation of the problem considered using minimum cost flow problems. So that when $p = 1$ we can consider the following simple algorithm to solve the phase unwrapping problem.

ALGORITHM 1 (proposed in [5], $p = 1$) *Read the parameters N_1, N_2 and the data $W = W_{i,j}$, $(i,j) \in I$; perform the following steps:*

1 compose the matrix A_1 and the vector $\underline{\beta}_1$ of problem (1.17);

2 compute $\underline{k}^*$, $\underline{h}^*$ as a minimizer of problem (1.17) with $p = 1$;

3 compute the discrete gradient vector field $(G_1, G_2)^t$ using formula (1.12), with $k_{\nu,i,j} = k^*_{\nu,i,j}$ $(i,j) \in I_\nu$, $\nu = 1,2$, where $k^*_{\nu,i,j}$ $(i,j) \in I_\nu$, $\nu = 1,2$ are the components of vector $\underline{k}^*$ and compute $U = U_{i,j}$, $(i,j) \in I$ up to an arbitrary constant using formula (1.11);

4 stop.

We note that step 2 consists in the solution of a linear programming problem with real variables and nicely structured constraint matrix.

We note that the minimizer of problem (1.15), (1.16) with $p = 1$ found determines the correction introduced in formula (1.12), so that problem (1.15), (1.16) with $p = 1$ can give a good solution of the phase unwrapping problem when the functions $[\Delta_\nu W]_{2\pi}$, $\nu = 1,2$, need a correction $\underline{k}^*$ of small 1-norm to define a reasonable approximation of G_ν, $\nu = 1,2$, that is the discrete partial derivatives of U. However in general there is no reason to believe that the required correction should have small 1-norm. For example considering the SAR interferometry problem we have that the correction $\underline{k} \in \mathbb{Z}^N$ introduced in formula (1.12) necessary to have a reasonable approximation of the discrete partial derivatives of U depends on the character of the observed scene and the 1-norm of this correction tends to increase when the observed scene becomes craggier.

It is easy to see that in general problem (1.15), (1.16) is degenerate, that is its solution is not unique. This degeneracy is particularly relevant when $p = \infty$, that is problem (1.15), (1.16) for $p = \infty$ is highly degenerate, in fact it has a large number of solutions. We want to take advantage on this property of problem (1.15), (1.16), with $p = \infty$, in particular among all the possible solutions of problem (1.15), (1.16) when $p = \infty$ we want to select the appropriate correction $\underline{k}^*$ to compute the solution of Problem 1. Thus we want to introduce a class of optimization problems to control the magnitude of the 1-norm of the correction $\underline{k}^*$ used to solve Problem 1.

Let $\mathcal{S}_\infty$ be the set of the solutions of problem (1.15), (1.16) with $p = \infty$, we select a solution in $\mathcal{S}_\infty$ using a couple of auxiliary optimization problems. We note that $\mathcal{S}_\infty$ is the set of vectors $\underline{k} \in I\!\!R^N$ satisfying $A\underline{k} = \underline{\beta}$ and $\|\underline{k}\|_\infty = 1$, in fact $\underline{k} = \underline{k}_b$ defines an element of $\mathcal{S}_\infty$. Moreover from the assumption that $b_{\nu,i,j}$, $(i,j) \in I_\nu$, $\nu = 1, 2$ are not identically zero we have $|b_{\nu,i,j}| = 1$ for some (i,j), that is $\|\underline{k}_b\|_\infty = 1$ and the fact that $\underline{0} \notin \mathcal{S}_\infty$.

Let us consider a graph $\mathcal{H}$ given by the graph $\mathcal{G}$ plus two artificial nodes S^+, S^-, such that the node S^+ is joined with each node of $\mathcal{G}$ (except P) having positive exogenous supply, the node S^- is joined with each node of $\mathcal{G}$ (except P) having negative exogenous supply, see Figure 1.5. Let $N^+ \in I\!\!N$ be the number of arcs between S^+ and $\mathcal{G}$ and $\underline{l} \in I\!\!N^{N^+}$ be the vector containing the flows on these arcs; let $N^- \in I\!\!N$ be the number of arcs between S^- and $\mathcal{G}$ and $\underline{m} \in I\!\!N^{N^-}$ be the vector containing the flows on these arcs; finally let S^+ be joined with S^- and let $\phi \in I\!\!N$ be the flow on this last arc. We note that when $N^+ = 0$ ($N^- = 0$) no arcs between S^+ and $\mathcal{G}$ (S^- and $\mathcal{G}$) are considered. In the following we assume $N^+ \neq 0$ and $N^- \neq 0$, however with the proper modifications the discussion that follows holds also when $N^+ = 0$ and/or when $N^- = 0$.

Let $\underline{\beta}_2 = (\underline{\beta}_1^t, 0, 0)^t$ and

$$A_2 = \begin{pmatrix} A_1 & B_1^+ & B_1^- & \underline{0}_M \\ \underline{0}_N^t & -\underline{1}_{N^+}^t & \underline{0}_{N^-}^t & 1 \\ \underline{0}_N^t & \underline{0}_{N^+}^t & \underline{1}_{N^-}^t & -1 \end{pmatrix}. \tag{1.18}$$

We note that the matrices B_1^+, B_1^- of dimensions $M \times N^+$ and $M \times N^-$ respectively take care of the flow conservation conditions on the arcs joining the nodes S^+, S^- to the nodes of $\mathcal{G}$, while the last two rows of A_2 take care of the flow conservation conditions on the nodes S^+, S^- respectively. Let $\Omega = \left\{ \underline{x} = (\underline{k}^t, \underline{h}^t, \underline{l}^t, \underline{m}^t, \phi)^t, \underline{k} \in \mathbb{Z}^N, \underline{h} \in \mathbb{Z}^4, \underline{l} \in I\!\!N^{N^+}, \underline{m} \in I\!\!N^{N^-}, \phi \in I\!\!N \right\}$. Let $\underline{w}_1 \in I\!\!N^{N^+}$, $\underline{w}_2 \in I\!\!N^{N^-}$, $w \in I\!\!N$ be

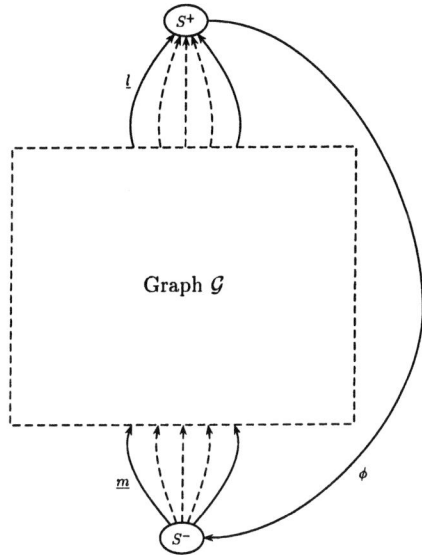

Figure 1.5.   The graph $\mathcal{H}$.

given weights. We consider the following problem:

$$
\begin{aligned}
&\min\{\underline{w}_1^t \underline{l} + \underline{w}_2^t \underline{m} - w\,\phi\} \\
&\text{s.t.} : A_2 \underline{x} = \underline{\beta}_2, \\
&\qquad -\underline{1}_N \le \underline{k} \le \underline{1}_N, \\
&\qquad \underline{x}' \in \Omega.
\end{aligned}
\tag{1.19}
$$

We note that here and in the following inequalities between vectors should be understood componentwise. Let $\underline{h}_b \in \mathbf{Z}^4$ be a vector such that $(\underline{k}_b^t, \underline{h}_b^t)^t$ is a feasible point of problem (1.17) (such a vector always exists since $A\underline{k}_b = \underline{\beta}$). We note that $\underline{x}_b$ defined by $\underline{k} = \underline{k}_b$, $\underline{h} = \underline{h}_b$, $\underline{l} = \underline{0}$, $\underline{m} = \underline{0}$, $\phi = 0$ is always a feasible point for problem (1.19). Let $\tilde{\underline{x}} = (\tilde{\underline{k}}^t, \tilde{\underline{h}}^t, \tilde{\underline{l}}^t, \tilde{\underline{m}}^t, \tilde{\phi})^t$ be a solution of problem (1.19).

It is easy to see that given the weights $\underline{w}_1$, $\underline{w}_2$ the number of non vanishing components of $\tilde{\underline{k}}$ tends to increase when the weight w increases. In fact in order to minimize the objective function in (1.19) we would like to take ϕ as large as possible, and as a consequence also the flows $\underline{l}$, $\underline{m}$ and the flows on $\mathcal{G}$ tend to increase componentwise in order to satisfy the conservation conditions given in (1.19). Note that the vector $\tilde{\underline{k}}$ does not verify the constraints of problem (1.15), (1.16) or equivalently $\tilde{\underline{k}}$, $\tilde{\underline{h}}$ do not verify the constraints of problem (1.17). In fact we have

$A_1 \left(\begin{array}{c} \tilde{\underline{k}} \\ \tilde{\underline{h}} \end{array} \right) = \underline{\beta}_1 - B_1^+ \tilde{\underline{l}} - B_1^- \tilde{\underline{m}}.$ Let $\tilde{\tilde{\underline{k}}}, \tilde{\tilde{\underline{h}}}$ be a solution of the following problem:

$$\min \|\underline{k}\|_1$$

$$\text{s.t.} : A_1 \left(\begin{array}{c} \underline{k} \\ \underline{h} \end{array} \right) = B_1^+ \tilde{\underline{l}} + B_1^- \tilde{\underline{m}}, \tag{1.20}$$

$$\tilde{\underline{k}} - \underline{1}_N \leq \underline{k} \leq \tilde{\underline{k}} + \underline{1}_N,$$

$$\underline{k} \in \mathbf{Z}^N,$$

$$\underline{h} \in \mathbf{Z}^4.$$

We note that problem (1.20) is always feasible. Let $(\underline{k}_b^t, \underline{h}_b^t)^t$ be the feasible point of problem (1.17) defined above. Note that due to the assumptions made on $b_{\nu,i,j}$, $(i,j) \in I_\nu$, $\nu = 1,2$ we have $\|\underline{k}_b\|_\infty = 1$. Then it is easy to see that $\left((\underline{k}_b - \tilde{\underline{k}})^t, (\underline{h}_b - \tilde{\underline{h}})^t \right)^t$ is a feasible point of problem (1.20).

We define $\underline{k}^* = \tilde{\tilde{\underline{k}}} - \tilde{\underline{k}}$. We note that as a simple consequence of the properties of the vectors $\tilde{\underline{k}}, \tilde{\tilde{\underline{k}}}$ we have that $\underline{k}^* \in S_\infty$. In other words the functions G_1, G_2 defined by (1.12) choosing $\underline{k} = \underline{k}^*$ verify the irrotational property (1.10), thus G_1, G_2 can be seen as the discrete partial derivatives of U and by (1.11), from the knowledge of G_1, G_2, we can compute the function U up to an arbitrary constant.

Note that the number of non zero components of $\underline{k}^*$ depends on the value of the parameter w and tends to increase when w increases. In fact this parameter controls the number of non zero components of the vector $\tilde{\underline{k}}$, however this vector in general is not a feasible point for problem (1.15), (1.16). The vector $\tilde{\underline{k}}$ should be corrected; the required correction is made by the vector $\tilde{\tilde{\underline{k}}}$ solution of problem (1.20). So that $\underline{k}^* = \tilde{\tilde{\underline{k}}} - \tilde{\underline{k}}$ is a solution of problem (1.15), (1.16) with $p = \infty$. Finally we note that problems (1.19), (1.20) are linear optimization problems, they have a global minimum corresponding eventually to multiple global minimizers.

Summarizing we have the following simple algorithm:

ALGORITHM 2 *Read the parameters* N_1, N_2, *the data* $W = W_{i,j}$, $(i,j) \in I$ *and the weights* $\underline{w}_1$, $\underline{w}_2$, w; *perform the following steps:*

1 *compose the matrix* A_2 *and the vector* $\underline{\beta}_2$ *of problem (1.19);*

2 *compute* $\tilde{\underline{k}}, \tilde{\underline{h}}, \tilde{\underline{l}}, \tilde{\underline{m}}, \tilde{\phi}$ *as a minimizer of problem (1.19);*

3 *compose the matrix* A_1 *and the vectors* $B_1^+ \tilde{\underline{l}} + B_1^- \tilde{\underline{m}}$, $\tilde{\underline{k}} - \underline{1}_N$, $\tilde{\underline{k}} + \underline{1}_N$ *of problem (1.20);*

4 compute $\tilde{\tilde{k}}$, $\tilde{\tilde{h}}$ *as a minimizer of problem (1.20)*

5 compute the vector $\underline{k}^* = \tilde{\underline{k}} - \tilde{\tilde{k}}$;

6 compute the discrete gradient vector field $(G_1, G_2)^t$ *using formula (1.12), with* $k_{\nu,i,j} = k^*_{\nu,i,j}$ $(i,j) \in I_\nu$, $\nu = 1, 2$, *where* $k^*_{\nu,i,j}$ $(i,j) \in I_\nu$, $\nu = 1, 2$ *are the components of vector* $\underline{k}^*$ *and compute* $U_{i,j}$, $(i,j) \in I$ *up to an arbitrary constant using formula (1.11);*

7 stop.

The steps 2 and 4 consist in the solution of a linear programming problem with real variables and a nicely structured constraint matrix. So that having some a priori information about the number of non vanishing components of the vector $\underline{k}^*$ that defines the approximation G_1, G_2 of the discrete partial derivatives of U through formulas (1.12), (1.13) we can choose the value of the weight w such that the solution $\underline{k}^*$ of problems (1.19), (1.20) will have approximately that number of nonzero components. In the next section we propose a method to choose the value of the weight w without the knowledge of a priori information on $\underline{k}^*$, that is of the solution of Problem 1.

4.   The Numerical Experience

Let us consider Problem 2. We solve this problem using synthetic and real SAR interferometry data. The real data are taken from the ERS mission of the European Space Agency (ESA).

We begin with synthetic SAR interferometry data generated using an implementation of formulas (1.4), (1.5), (1.6), (1.7) in a FORTRAN program. Let m denote meters. The numerical experiment is performed with values of the parameters h, d, α, λ similar to those used in the ERS1 mission of ESA, that is: $h = 782563$m, $d = 143.13$m, $\alpha = -1.44$radiants, $\lambda = 0.056415$m. In the ERS1 mission the measurements of the two SAR systems are obtained by two successive passages of the same instrument over the scene. In the following examples we consider a grid of points contained in Ω_2, whose indices belong to the set I, and from relations (1.2), (1.3), (1.4), (1.5), (1.6), (1.7) we compute the quantities $U_e = U_{e,i,j}$, $(i,j) \in I$ and $W = W_{i,j}$, $(i,j) \in I$ corresponding to a given elevation map Z. Moreover, with the notation of section 2, the grid on Ω_2 is defined by the following parameters: $N_1 = 100$, $N_2 = 100$, $r_0 = 846645$m, $y_0 = -797$m, $\delta_1 = 15.944$m, $\delta_2 = 7.905$m. Moreover in formula (1.5) we have chosen $\epsilon(r, y) = 0$.

We define a point $(i, j) \in I_1 \cap I_2$ to be "non-irrotational" when the vector field $([\Delta_1 W]_{2\pi}, [\Delta_2 W]_{2\pi})^t$ defined in (1.12), (1.13) does not verify the irrotational property (1.10) at the point (i, j).

In particular in Algorithm 2 we fix $\underline{w}_1 = \underline{1}_{N^+}$, $\underline{w}_2 = \underline{1}_{N^-}$, and we propose a method to choose the remaining weight w appearing in problem (1.19) according to some statistics obtained considering the weight $w = \hat{w}$ that gives the "best" reconstructed phase function as a function of the rate of non-irrotational points in the wrapped interferometric phase function in a class of simulated scenes. In particular we have considered one hundred scenes having different rates of non-irrotational points. In (1.12) we expect a large number of nonzero components in the correction needed $\underline{k} = \underline{k}^*$ when the rate of non-irrotational points is large and we expect a small number of nonzero components in the correction $\underline{k} = \underline{k}^*$ when the rate of non-irrotational points is small. These scenes are generated randomly in the way shown below in the following family of scenes:

$$
Z(x, y) = \max \left\{ 0, b - b \left(\sqrt{\sin^2 \left(\frac{cx + dy}{4a} \right) + \cos^2 \left(\frac{dx - cy}{4a} \right)} + \right. \right.
$$

$$
\left. \left. \frac{1}{10} \sqrt{\cos^2 \left(\frac{cx + dy}{a} \right) + \sin^2 \left(\frac{dx - cy}{a} \right)} \right) + \frac{y + 1000}{e} \right\},
$$

$$
(x, y)^t \in \Omega_1, \tag{1.21}
$$

where $a, b, c, d, e \in I\!R$ are parameters that characterize the scene.

The wrapped interferometric phase of each scene is processed using Algorithm 1 proposed in [5] and using Algorithm 2 with several choices of the value for the parameter w, that is $w = 1, 2, \ldots, 30$. Let U_e be the exact interferometric phase, let U be the unwrapped interferometric phase computed with Algorithm 1 or Algorithm 2 using in both cases $U_{1,1} = U_{e,1,1}$, then we define the following performance index:

$$
E_{l^2}(U, U_e) = \left(\frac{\sum_{(i,j) \in I} (U_{e,i,j} - U_{i,j})^2}{\sum_{(i,j) \in I} (U_{e,i,j} - U_{e,1,1})^2} \right)^{1/2}. \tag{1.22}
$$

The following results are obtained considering the ten classes of scenes (1.21) corresponding to all the possible choices of the three parameters a, b, e in the set: $a = 50, 30, b = 150, 170, 190, 210, 230, e = 3$. In each class of scenes we have selected randomly ten scenes sampling the parameter c from a random variable uniformly distributed in the interval $[0.15, 0.85]$ and defining $d = 1 - c$. For $i = 1, 2, \ldots, 100$ we denote with

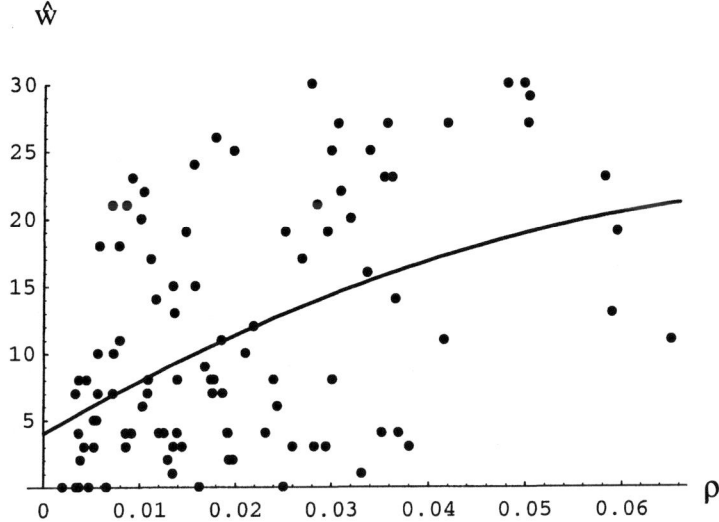

Figure 1.6.   The values of $\hat{w}$ versus ρ and the least square parabola fitting them.

$\hat{w}^i$ the value of weight w such that Algorithm 2 gives the best result, that is the value of w such that the performance index (1.22) is minimized. In Figure 1.6 the results of this experiment are reported, that is we have plotted $\hat{w}$ as a function of rate ρ of non-irrotational points for the one hundred randomly generated scenes considered. For $i = 1, 2, \ldots, 100$ we denote with $\tilde{U}_e^i$ the exact interferometric phase for the i-th scene, we denote with $\tilde{U}^i$ the solution obtained with Algorithm 1 for the i-th scene, and we denote with $\hat{U}^i$ the solution obtained with Algorithm 2 using the weight $w = \hat{w}^i$ for the i-th scene. We have:

$$\frac{1}{100} \sum_{i=1}^{100} E_{l^2}(\tilde{U}^i, U_e^i) \approx 0.24, \qquad (1.23)$$

$$\frac{1}{100} \sum_{i=1}^{100} E_{l^2}(\hat{U}^i, U_e^i) \approx 0.16. \qquad (1.24)$$

That is Algorithm 2 gives significant improvements on the performance of Algorithm 1 when the correct choice of the weight w is made.

Now we turn our attention to the problem of choosing the weight w. We have computed the least squares parabola fitting the data reported in Figure 1.6, and rounded its coefficients to the closest integer. We obtain the following parabola:

$$\hat{w} = -2398\rho^2 + 418\rho + 4. \qquad (1.25)$$

The parabola (1.25) is used to choose the value w when Algorithm 2 is used on a new set of data, that is for example on the set of one hundred scenes generated randomly in the following family of scenes:

$$Z(x,y) = \max\left\{ h_j e^{-\frac{(x-x_j)^2+(y-y_j)^2}{\sigma_j^2}}, j = 1, 2, \ldots, n_g \right\} + \frac{y+1000}{a_j},$$

$$(x,y)^t \in \Omega_1, \qquad (1.26)$$

where $n_g \in I\!N$ and h_j, x_j, y_j, $\sigma_j \in I\!R$, a_j, $j = 1, 2, \ldots, n_g$ are parameters that characterize the scenes. We choose randomly $n_g \in \{1, 2\ldots, 20\}$, and for $j = 1, 2, \ldots, n_g$ we choose randomly $h_j \in [50m, 300m]$, $x_j \in [-1000m, 1000m]$, $y_j \in [-500m, 500m]$, $\sigma_j \in [50m, 200m]$, $a_j \in [0.5, 5]$ sampling a random variable uniformly distributed in the above mentioned intervals. Moreover in formula (1.5) for every $(r, y)^t \in \Omega_2$ we choose $\epsilon(r, y)$ as the samples of a random variable uniformly distributed in $[-0.1, 0.1]$ multiplied by $u_e(r, y)$. For $i = 1, 2, \ldots, 100$ we denote with $\tilde{U}_e^i$ the exact interferometric phase for the i-th scene, we denote with $\tilde{U}^i$ the solution obtained with Algorithm 1 for the i-th scene, and we denote with $\hat{U}^i$ the solution obtained with Algorithm 2 using the weight $w = \hat{w}^i$, obtained rounding to the closest integer the value of w provided by formula (1.25), for the i-th scene. We consider rounded values for w since w is a coefficient of the objective function of the minimum cost flow problem (1.19) and this problem can be solved more efficiently when the cost function has integer coefficients. We have the following results:

$$\frac{1}{100} \sum_{i=1}^{100} E_{l^2}(\tilde{U}^i, U_e^i) \approx 0.12, \qquad (1.27)$$

$$\frac{1}{100} \sum_{i=1}^{100} E_{l^2}(\hat{U}^i, U_e^i) \approx 0.11. \qquad (1.28)$$

That is Algorithm 2 with the choice of w given by (1.25) and the rounding described above outperforms Algorithm 1 of about 10% in the test case considered.

For the real SAR interferometry data we report a simple numerical experiment using Algorithm 1 and Algorithm 2. We note that in this case we do not know the exact phase function, thus we cannot evaluate the performance indices (1.27), (1.28) as done with the synthetic data. We have considered the SAR interferometry data provided by the ERS mission of ESA that corresponds to a particular region of Sardinia, Italy. We note that Sardinia is a craggy island, so that we can hope to compute with Algorithm 2 a better solution than the one computed with

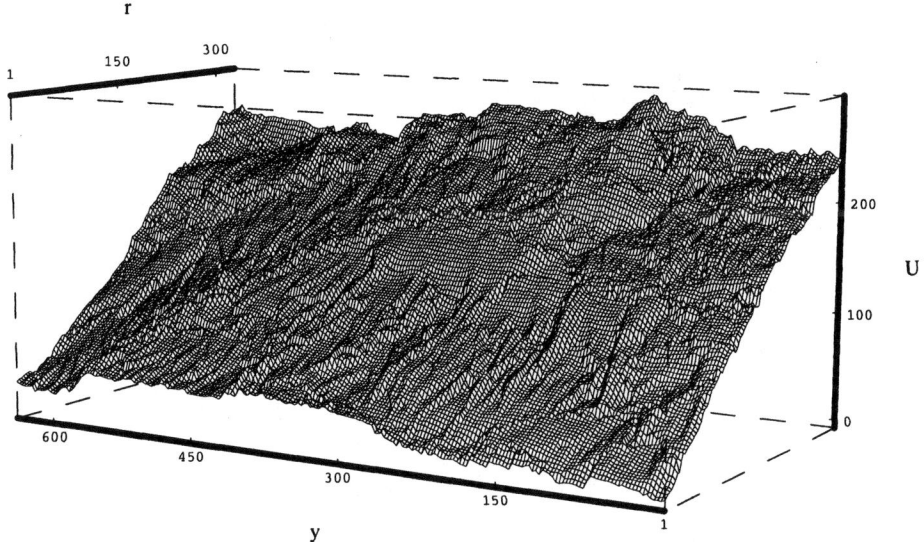

Figure 1.7. The phase U reconstructed using Algorithm 1.

Algorithm 1. The experiment of the data coming from the ERS mission is performed using the same values of the parameters h, d, α, λ than those mentioned above for the synthetic data. Moreover for the size of the interferogram we have: $N_1 = 641$, $N_2 = 329$. We note that the rate of non-irrotational points of these data is equal to 0.053.

The numerical experiment is performed as follows: we solve Problem 2 using Algorithm 1 and using Algorithm 2 with several values of weight w, that is with $w = 1, 2, \ldots, 20$. In Table 1.1 we report the 1-norm of the solution $\underline{k}^*$ obtained with Algorithm 1 and obtained with Algorithm 2 using several values for the weights w.

Finally in Figure 1.7 we report the unwrapped interferometric phase computed with Algorithm 1 and in Figure 1.8 we report the unwrapped interferometric phase computed with Algorithm 2 using $w = 19$, that

Algorithm 1	Algorithm 2							
$\|\underline{k}^*\|_1$	w	$\|\underline{k}^*\|_1$	w	$\|\underline{k}^*\|_1$	w	$\|\underline{k}^*\|_1$	w	$\|\underline{k}^*\|_1$
12810	1	15468	2	15531	3	17290	4	21270
	5	22538	6	24364	7	26786	8	28761
	9	30195	10	32245	11	33708	12	34641
	13	35825	14	37077	15	37774	16	38706
	17	39462	18	39932	19	40552	20	41279

Table 1.1. The value of $\|\underline{k}^*\|_1$ for several values of the parameter w.

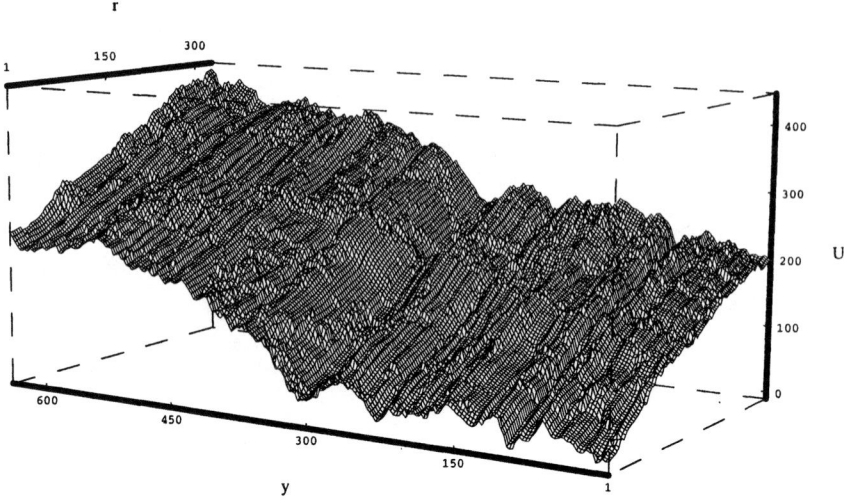

Figure 1.8. The phase U reconstructed using Algorithm 2 with $w = 19$.

is the rounded weight obtained using formula (1.25). We note that in the numerical experience proposed in this paper the minimum cost flow problems appearing in Algorithm 1 and in Algorithm 2 are solved with RELAX4[1]. This choice seems to be a good one, in fact RELAX4 has solved in a few minutes all the examples considered. However different codes can be used in place of RELAX4, see [13] page 53 for a complete review of the available minimum cost flow problems solvers. Moreover the code implementing Algorithm 2 is available and can be obtained free of charge contacting the authors by electronic mail.

5.　　Conclusions and Remarks

We note that the results shown in (1.23), (1.24) imply that on the set of data considered the solution of Problem 1, or equivalently of Problem 2, can be computed more accurately using Algorithm 2 than using Algorithm 1 when the weight w in Algorithm 2 is chosen appropriately. Moreover Table 1.1 shows that the results obtained using Algorithm 1 or using Algorithm 2 are quite different and that the results obtained with Algorithm 2 are highly dependent on the choice of the parameter w. Thus, according to the results obtained with the synthetic data considered, we can suppose that Algorithm 2 with a suitable choice of

---

[1]RELAX4 is a minimum cost flow problems solver designed by D.P. Bertsekas and P. Tseng, for details see [11] page 279. This software package is available free of charge in the Web site http://www.mit.edu/people/dimitrib/RELAX4.txt

the weight w gives also in the case of real data a substantially better result than Algorithm 1. The simple way of choosing the weight given by (1.25) is not entirely satisfactory as can be seen from (1.23), (1.24), (1.27), (1.28). Further investigation is needed to refine the choice of the weight parameter w.

We remark that the results of this paper are relevant for the phase unwrapping problem. The contribution presented is a first step of a promising research to study adaptive techniques to solve the phase unwrapping problem, that is techniques adapted to the particular instances of Problem 1.

References

[1] A. V. Oppenheim and J. S. Lim: The Importance of Phase in Signals, *Proceedings of the IEEE*, 69:529–541, 1981.

[2] H. A. Zebker and R. M. Goldstein: Topographic Mapping from Interferometric Synthetic Aperture Radar Observations, *Journal of Geophysical Research*, 91:4993-4999, 1986.

[3] C. Oliver and S. Quegan: *Understanding Synthetic Aperture Radar Images*, Artech House, Boston, 1998.

[4] R. M. Goldstein, H. A. Zebker and C. L. Werner: Satellite Radar Interferometry: Two-Dimensional Phase Unwrapping, *Radio Science*, 23:713-720, 1988.

[5] M. Costantini: A Novel Phase Unwrapping Method Based on Network Programming, *IEEE Transactions on Geoscience and Remote Sensing*, 36:813-821, 1998.

[6] M. Costantini, A. Farina and F. Zirilli: A Fast Phase Unwrapping Algorithm for SAR Interferometry, *IEEE Transactions on Geoscience and Remote Sensing*, 37:452-460, 1999.

[7] P. Maponi, F. Zirilli: A Class of Global Optimization Problems as Models of the Phase Unwrapping Problem, *Journal of Global Optimization*, 21:289-316, 2001.

[8] N. Egidi, P. Maponi: A Comparative Study of Two Fast Phase Unwrapping Algorithms, *Applied Mathematics and Computation*, 148:599-629, 2004.

[9] C.W. Chen, H.A. Zebker: Network approaches to two-dimensional phase unwrapping: intractability and two new algorithms, *Journal of the Optical Society of America A*, 17:401-414, 2000.

[10] C.W. Chen, H.A. Zebker: Two-dimensional phase unwrapping with use of statistical models for cost functions in nonlinear optimization, *Journal of the Optical Society of America A*, 18:338-351, 2001.

[11] D. P. Bertsekas: *Linear Network Optimization*, MIT Press, Cambridge (Ma), 1991.

[12] W. J. Cook, W. H. Cunningham, W. R. Pulleyblank and A. Schrijver: *Combinatorial Optimization*, John Wiley & Sons, New York, 1998.

[13] J. J. Moré, S. J. Wright. *Optimization Software Guide*. SIAM, 1993.

Chapter 12

Evolutionary Algorithms for Global Optimization

Osyczka Andrzej and Krenich Stanislaw
Department of Mechanical Engineering, Cracow University of Technology,
Al. Jana Pawla II 37, 31-864 Krakow, Poland.
email: osyczka@mech.pk.edu.pl, krenich@mech.pk.edu.pl

Abstract: The Chapter presents some evolutionary algorithm based methods used for solving global optimization problems. Firstly, the methods, which use proportional selection, are briefly described. Secondly, a bicriterion approach is applied for solving the constrained function minimization problem. This method transforms the constrained function minimization problem to the bicriteria optimization problem in the way that the first objective function is the sum of violated constraints and the second objective function is the one that should be minimized. Thirdly, the constraint tournament selection method is presented. The aim of the method is to reduce the number of function evaluations in the optimization process. The method is very effective while solving highly constrained single criterion optimization problems as well as the problems with the computationally expensive objective function. The method may also reduce the computation time for ordinary nonlinear programming problems as well as produce better results. Finally, the application of evolutionary algorithms to two design optimization problems is presented. The first problem deals with optimum design of concentric springs, whereas the second problem deals with the optimum design of a robot gripper. Both of these problems are considered as discrete programming problems. For problems like these the use of conventional optimization methods is very limited or even impossible. As it is shown in this Chapter evolutionary algorithm based methods can easily handle such problems.

Keywords: evolutionary algorithms, genetic algorithms, constrained nonlinear programming, discrete programming, computationally expensive objective functions, proportional selection, constraint tournament selection, Pareto set distribution method, concentric springs design, robot gripper design.

1. INTRODUCTION

Evolutionary and genetic algorithms are powerful and widely used wise stochastic optimization techniques which rely on analogies to natural processes. They can often outperform conventional optimization methods when applied to difficult real-word optimization problems. Many different evolutionary algorithm based strategies have been developed recently to find the global minimum for nonlinear programming problems (see review papers by Gen & Chang, 1996 and Michalewicz & Schoenauee, 1996). In contrast to conventional optimization methods evolutionary algorithm methods for nonlinear programming problems have the following two advantages:

(i)   they impose no restriction on the optimization problem; the objective function can be multimodal and noncontinuous, the feasible domain can be incoherent etc.,

(ii) they can be used to solve any optimization models i.e., models with continuous, integer, discrete and mixed continuous − integer and continuous − discrete decision variables.

A general nonlinear programming problem is formulated as follows: Find $\mathbf{x}^* = \left[x_1{}^*, x_2{}^*, \ldots x_N^* \right]^T$ which will satisfy the K inequality constraints

$$g_k(\mathbf{x}) \geq 0 \quad k = 1, 2, \ldots K \tag{1}$$

and the M equality constraints

$$h_m(\mathbf{x}) = 0 \quad m = 1, 2, \ldots, M < N \tag{2}$$

and minimize the objective function $f(\mathbf{x})$

$$f(\mathbf{x}^*) = min\ f(\mathbf{x}) \tag{3}$$

where $\mathbf{x} = \left[x_1, x_2, \ldots, x_N \right]^T$ is the vector of decision variables.

This Chapter presents some evolutionary algorithm based methods developed so far, which are used for finding the global minimum of the function (3) under the constraints (1) and (2). Firstly, the methods, which use proportional selection, are presented. Secondly, a bicriterion approach is applied for solving the constrained function minimization problem. Thirdly the constraint tournament selection method is presented. Finally, applications of evolutionary algorithms to two design optimization problems are presented. From the results obtained while solving these problems it is clear that these methods are very effective in solving some global optimization problems.

2. EVOLUTIONARY ALGORITHMS WITH PROPORTIONAL SELECTION

2.1. Scaling the objective function

In proportional selection, also called roulette wheel selection, individuals are selected according to their relative fitness value. This often leads to the situation that in early generations a few super chromosomes dominate the selection process and in later generations the competition among chromosomes is less strong and an evolutionary algorithm method works like an ordinary random search method. The main question, which arises here is: how to increase the selection pressure to find the maximum in a short computation time and at the same time avoid convergence to the local maximum.

Another problem is that most common evolutionary algorithm methods based on proportional selection require positive values of the fitness function for the process of evaluation. For function maximization problems it is, in many cases, easy to satisfy this requirement. Assuming that $f(\mathbf{x}) > 0$ for all $\mathbf{x}$, the objective function $f(\mathbf{x})$ can be treated as the fitness function. For function minimization problems, which occur more often in design optimization, the question arises: how to transform these problems so that evolutionary algorithms can solve them effectively. One of the simplest ways is to use some positive constant C and then the fitness function is evaluated from the following formula:

$$f'(\mathbf{x}) = C - f(\mathbf{x}) \tag{4}$$

where C is a constant, which should satisfy the following formula:

$$C > f(\mathbf{x}) \quad \text{for all } \mathbf{x} \tag{5}$$

If formula (5) is not satisfied the value of the fitness function is set to zero.

Value of constant C can be chosen in many different ways. However, the problem is that this choice has a great influence on the effectiveness of the algorithm. The constant C might be entered by the user of the evolutionary algorithm, it might be equal to the maximum of $f(\mathbf{x})$ found so far or it might be evaluated for each generation using the following formula:

$$C^t = \max_{j \in J}\{ f^j(\mathbf{x}) \} \tag{6}$$

where J is the population size. This means that in the t-th generation the value of C^t is equal to the maximal value of the function obtained in this generation. This seems to be a very universal approach since for each problem the zero fitness function is assigned only to one chromosome.

Michalewicz, 1996 in his genetic algorithm system GENESIS uses the so-called scaling window method. This method updates the constant C while running GENESIS in the following way: If the scaling window $W > 0$, the system sets C to the greatest value of $f(\mathbf{x})$ which has occurred in the last W generations. A value W indicates the infinite window, i.e., $C = max\{f(\mathbf{x})\}$ over all evolutions. If $W < 0$, the sigma truncation method can be used (see Michalewicz, 1996).

Some other methods of scaling the objective function are presented in Gen & Chang, 1997 and Osyczka, 2000.

2.2. Handling constraints

The main problem in applying evolutionary algorithms to solving a constrained problem is how to deal with constraints because evolutionary operators used for manipulating chromosomes may yield infeasible solutions. Quite a large number of methods have been developed recently to handle constraints when evolutionary algorithms are used (see Kim & Myung, 1996; Michalewicz, 1995; Myung & Kim, 1996; Orvosh & Davis, 1994). These methods can be roughly classified as follows:

(i)   Rejecting strategy
(ii) Repairing strategy
(iii) Modifying genetic operator strategy
(iv) Penalty function strategy

The last strategy has a universal character and thus it is the most often used strategy. Many different forms of this strategy have been developed recently and some of these forms will be presented below.

Penalty function strategy
This strategy is based on the strategies developed for conventional optimization methods in which solutions, which are out of the feasible domain, are penalized using a penalty coefficient. In other words a constrained optimization problem is transformed to an unconstrained optimization problem in which the function to be minimized has the form:

$$\phi(\mathbf{x},r) = f(\mathbf{x}) + r \sum_{m=1}^{M} [h_m(\mathbf{x})]^2 + r \sum_{k=1}^{K} G_k [g_k(\mathbf{x})]^2 \qquad (7)$$

where: G_k is the Heaviside operator such that $G_k = 0$ for $g_k(\mathbf{x}) \geq 0$ and $G_k = 1$ for $g_k(\mathbf{x}) < 0$, and r is a positive multiplier which controls the magnitude of the penalty terms.

It may seem logical to choose a very high value for r to ensure that no constraints are violated. However, this approach leads to difficulties. As r is increased the minimum of $\phi(\mathbf{x},r)$ moves closer to the constraint boundary (see Fig.1), but at the same time, the selective pressure for infeasible solutions decreases. This means that solutions, which differ significantly in their infeasibility, produce similar fitness values and in proportional selection all infeasible solutions have the same chance to be chosen for the next generation.

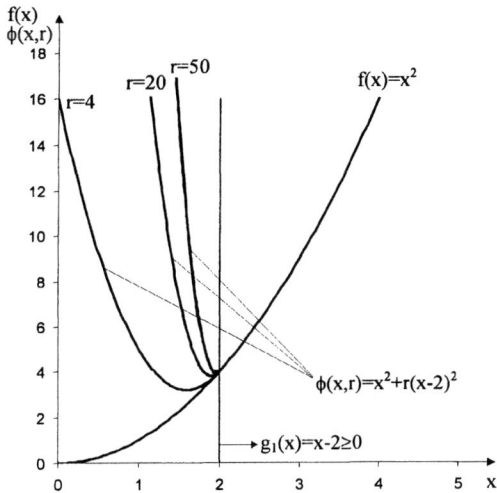

Fig.1. Illustration of the penalty function method
for the problem min f(x)=x^2 with the constraint g$_1$=x-2≥0.

Penalty function methods for both conventional and evolutionary algorithm methods can be classified as follows:
(i)   constant penalty method,
(ii)   variable penalty methods.
The constant penalty approach is less effective and most penalty function methods, which have been proposed in the area of evolutionary algorithms, are of a variable penalty character. It should be mentioned here that the concept of variable penalty methods is taken from conventional optimization methods where several approaches have been developed such as the penalty trajectory method (see Murray, 1969), the recursive quadratic programming method (see Biggs, 1975) and the sequential quadratic penalty function (see Broyden & Attia, 1983).

Homaifar, Lai and Qi's Method. Homaifar et al., 1994 proposed the method in which the penalty function has the form:

$$\phi(\mathbf{x},r) = f(\mathbf{x}) + \sum_{k=1}^{K} r_{kl} G_k \big[g_k(\mathbf{x})\big]^2 \tag{8}$$

where r_{kl} is a variable penalty parameter for the k-th constraint on the l-th level of violation, where $l = 1,2,...L$. It is assumed that higher levels of violation require larger values of the parameter r_{kl}.

The disadvantage of this method is that for K constraints it is required that K coefficients are established on the L different levels and the results of the optimization process depend significantly on the choice of r_{kl}. It is quite likely that for a given problem there exists an optimal set of coefficients for which the method gives a good solution but it might be difficult to find them. This means that the method is a problem orientated method rather than a universal one.

Joines and Houck's Method. Joines & Houck, 1994 proposed a method with dynamic penalties in which the penalty coefficient increases with the increasing number of generations. At each t-th generation the evaluation function has the form:

$$\phi(\mathbf{x},t) = f(\mathbf{x}) + (C \times t)^{\alpha} \left(\sum_{m=1}^{M} [h_m(\mathbf{x})]^{\beta} + \sum_{k=1}^{K} G_k [g_k(\mathbf{x})]^{\beta} \right) \qquad (9)$$

where C, α and β are constant. The method requires fewer coefficients than the previous one. Still, as indicated by Joines & Houck the quality of the solution is very sensitive to the values of these coefficients. Joines & Houck proposed $C = 0.5$ and $\alpha = \beta = 2$ as a reasonable choice of these constants. In this method the value of the penalty term $(C \times t)^{\alpha}$ increases constantly along with the number of generations. This leads to the situation in which in the last few generations infeasible chromosomes obtain death penalty and thus the method tends to converge in early generations.

Michalewicz and Attia's Method. Michalewicz & Attia, 1994 used the concept of the sequential quadratic penalty method in which the function, which is minimized, has the form:

$$\varphi(\mathbf{x},t) = f(\mathbf{x}) + \frac{1}{2r(t)} \mathbf{A}^T \mathbf{A} \qquad (10)$$

where $r(t) > 0$ and $\mathbf{A}$ is a vector of all active constraints $a_1, a_2, \ldots, a_l$.

The variable penalty coefficient $r(t)$, beginning with the starting value $r(0)$, decreases in each generation to reach the final assumed value $r(T)$ in the last generation. The method is very sensitive to the change of the coefficient $r(t)$ through the generations and how to settle these changes for a particular problem remains an open question.

Using this method Michalewicz and his co-workers have developed an optimization system GENECOP. The results they obtained using this system while testing several examples are very promising (see Michalewicz & Schoenauer, 1996).

Yokota, Gen, Ida and Taguchi's Method. Yakota, et al., 1995 considered the following nonlinear programming problem:

$$max\ f(\mathbf{x})$$

such that

$$g_m(\mathbf{x}) \le b_m, \quad m=1,2,...,M \tag{11}$$

and took the multiplication form of the penalty function:

$$\phi(\mathbf{x}) = f(\mathbf{x})p(\mathbf{x}) \tag{12}$$

The penalty term $p(\mathbf{x})$ is constructed as follows:

$$p(\mathbf{x}) = 1 - \frac{1}{M}\sum_{m=1}^{M}\left(\frac{\Delta b_m(\mathbf{x})}{b_m}\right)^{\alpha} \tag{13}$$

$$\Delta b_m(\mathbf{x}) = max\{0, g_m(\mathbf{x}) - b_m\} \tag{14}$$

where $\Delta b_m(\mathbf{x})$ is the value of violation of the m-th constraint. In this method the penalty function is designed with the non-parameterized approach and is problem-independent.

3. A BICRITERION APPROACH TO CONSTRAINED OPTIMIZATION PROBLEMS

This method transforms the constrained function minimization problem to the bicriteria optimization problem in the way that the first objective function is the sum of violated constraints and the second objective function is the one that should be minimized.

3.1. Problem formulation and transformation

Let us formulate for convenience once more the nonlinear programming problem for single criterion optimization
Find $\mathbf{x}^* = [x_1^*, x_2^*, ..., x_N^*]$ which will satisfy the K inequality constraints

$$g_k(\mathbf{x}) \ge 0 \quad \text{for } k = 1,2,...,K \tag{15}$$

and the M equality constraints

$$h_m(\mathbf{x}) = 0 \quad \text{for } m = 1,2,...,M \tag{16}$$

and minimize the objective function $f(\mathbf{x})$

$$f\left(\mathbf{x}^*\right) = \min f(\mathbf{x}) \tag{17}$$

The main idea of the proposed method consists in transforming the single criterion optimization problem into the bicriteria optimization problem with the following objective functions:

$$f_1(\mathbf{x}) = \sum_{m=1}^{M} |h_m(\mathbf{x})| + \sum_{k=1}^{K} G_k \times g_k(\mathbf{x}) \tag{18}$$

$$f_2(\mathbf{x}) = f(\mathbf{x}) \tag{19}$$

where:

G_k is the Heaviside operator such that $G_k = -1$ for $g_k(\mathbf{x}) < 0$ and $G_k = 0$ for $g_k(\mathbf{x}) \geq 0$,

$f(\mathbf{x})$ - the objective function that is to be minimized.

The minimum of $f_1(\mathbf{x})$ is known and equals zero. The function $f_1(\mathbf{x})$ will achieve its minimum for any solution that is in the feasible region. Assuming that the weak Pareto solutions are not taken under consideration, after solving the above bicriteria problem, the Pareto set can be obtained. From this set the first solution gives the global minimum of the minimized function (17) and satisfies constraints (15) and (16). The remaining Pareto solutions are slightly violated (see the numerical example below, section 3.3) which means they are very near to the feasible region. These solutions might be very important in the optimization process, especially when sensitivity analysis (see Olhoff & Lund, 1995) is to be performed.

3.2. Method of Solution

To solve the bicriterion problem formulated above, Pareto set distribution method (see Osyczka & Tamura, 1996) was used. The main idea of this method is as follows: Within each new generation a set of Pareto solutions is found on the basis of two sets: the set of Pareto solutions from a previous generation and the set of solutions created by genetic algorithm operations within the considered generation. The new set of Pareto solutions, thus created is distributed randomly to the next generation for a half of the population. There are two possible ways of breeding the remaining half of the population:

Approach 1. By randomly chosen strings from the previous generation,
Approach 2. By randomly generated strings.
The graphical illustration of this method is presented in Fig.2.

The method updates the set of Pareto solutions and then orders this set according to the increasing values of the first objective function. This ordering plays an important role because in our case the first objective function refers to constraints violation. Thus in the Pareto set the first solution will be the least violated solution, whereas, the last solution the most violated one. When the real set of Pareto optimal solutions is found, the first solution from the set gives zero violation.

For general applications of the Pareto set distribution method the whole set of Pareto optimal solutions has to be found.

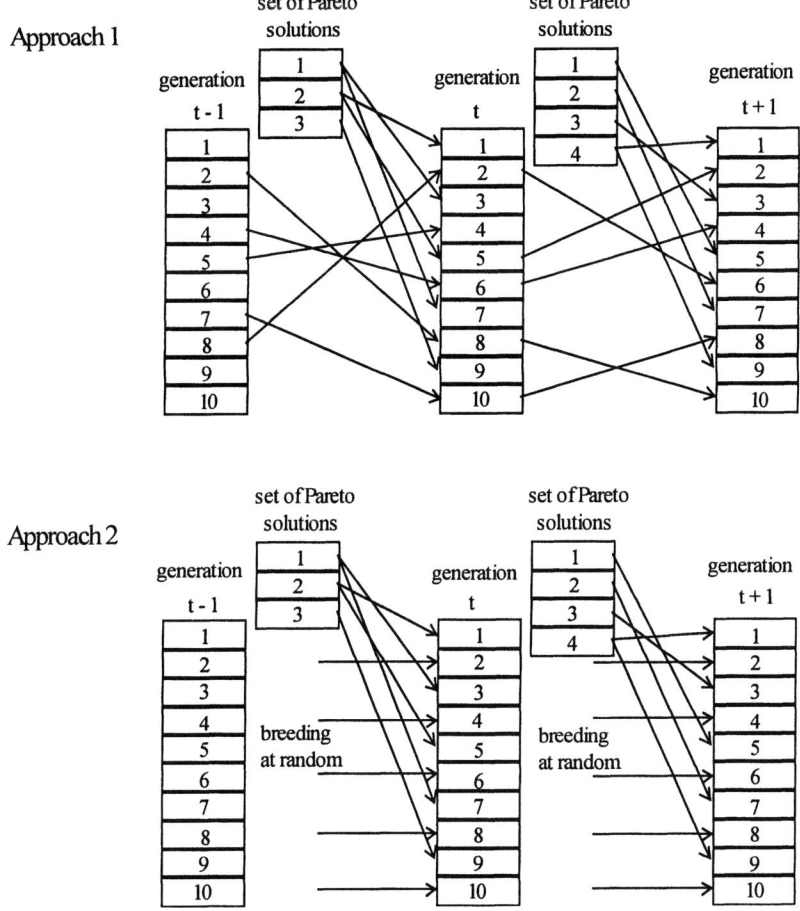

Fig.2. Graphical illustration of the Pareto set distribution method

But for the bicriterion approach good results can be obtained when the set of Pareto solutions is limited to the solutions that are close to the minimum of the first objective function, i.e., the solutions which are close to the feasible region. Thus after running each generation the set of Pareto solutions, which is remembered and distributed to the next generation, is limited to the assumed number of solutions. The best results were obtained when this set is between 10 and 20 solutions.

As it was mentioned before to solve a bicriteria optimization problem as formulated in (18) and (19), the Pareto set distribution method as described above was used. Using this method for the above problem the best results were obtained with the following options:

- The Approach 2 for breeding the remaining half of the population.
- Real number representation of the chromosomes.
- Variable point crossover (see Osyczka, 2000).
- Non-uniform mutation (see Michalewicz, 1996).

3.3. Numerical example

In order to show the way the method works the following numerical example will be considered (Floudas & Pardalos, 1992):
Minimize

$$f(\mathbf{x}) = 5x_1 + 5x_2 + 5x_3 + 5x_4 - 5\sum_{i=1}^{4} x_i^2 - \sum_{i=5}^{13} x_i$$

subject to

$$2x_1 + 2x_2 + x_{10} + x_{11} \le 10, \quad 2x_1 + 2x_3 + x_{10} + x_{12} \le 10,$$
$$2x_1 + 2x_3 + x_{11} + x_{12} \le 10, \quad -8x_1 + x_{10} \le 0, \quad -8x_2 + x_{11} \le 0,$$
$$-8x_3 + x_{12} \le 0, \quad -2x_4 - x_5 + x_{10} \le 0, \quad -2x_6 - x_7 + x_{11} \le 0,$$
$$-2x_8 - x_9 + x_{12} \le 0, \quad 0 \le x_i \le 1, \quad i = 1,2,...,9,$$
$$0 \le x_i \le 100, \quad i = 10,11,12, \quad 0 \le x_{13} \le 1,$$

The problem has 9 linear constraints; the function $f(\mathbf{x})$ is quadratic with its global minimum at

$$\mathbf{x}^* = (1,1,1,1,1,1,1,1,1,3,3,3,1,)$$

where: $f(\mathbf{x}^*) = -15$.

Six out of nine constraints are active at the global optimum (all except the following three: $-8x_1 + x_{10} \le 0, -8x_2 + x_{11} \le 0, -8x_3 + x_{12} \le 0$).

Using the bicriteria method the sets of Pareto solutions obtained after running 30, 50, 100 and 2000 generations with the population size equal 100 each are shown in Table 1 and illustrated graphically in Figure 3. The example was run under assumption that the set of Pareto solutions stored between generations is less or equal to 15 solutions.

All are ordered according to the increasing value of the first objective function. After 30 generations the first solution is in the feasible region but the function, which is minimized, is far from the global minimum. Similar results are obtained after running 50 and 100 generations, however, the minimized function is closer to the global minimum.

Table 1. Results of the numerical example.

	generation = 30		generation = 50		generation = 100		generation = 2000	
	$f_1(x)$	$f_2(x)$	$f_1(x)$	$f_2(x)$	$f_1(x)$	$f_2(x)$	$f_1(x)$	$f_2(x)$
1	0.0	-8.3343	0.0	-10.5220	0.0	-13.8610	0.0	-14.9990
2	0.0025	-9.0168	0.0158	-10.9049	0.0980	-13.9361	0.00004	-14.9991
3	0.3857	-9.2298	0.0464	-11.2853	0.1771	-14.3953	0.00009	-14.9993
4	0.4049	-9.3399	0.0920	-12.0309	0.6554	-14.7565	0.00012	-14.9993
5	0.5126	-9.5632	0.0930	-12.4564	1.0417	-14.7660	0.00043	-14.9994
6	0.8292	-9.7139	0.6844	-13.6639	1.2937	-14.8542	0.00049	-14.9996
7	0.9233	-10.3139	1.5650	-13.7253	1.8433	-14.8840	0.00068	-14.9997
8	1.3143	-10.9543	2.0770	-13.7517	2.2659	-14.9873	0.00081	-14.9997
9	2.2454	-11.0049	3.9748	-13.9124	2.6537	-15.1261	0.00099	-14.9997
10	2.4665	-11.4977	4.0369	-14.0490	4.3450	-15.34.06	0.00099	-14.9998
11	3.0314	-11.7973	4.4263	-14.2431	10.1634	-16.5952	0.00110	-14.9998
12	3.0404	-11.8809	5.4202	-14.7359	14.4600	-17.8367	0.00110	-14.9998
13	3.5870	-12.2400	6.9159	-14.9687	16.5506	-18.1385	0.00160	-15.0000
14	4.4061	-12.3043	7.0647	-15.8519	18.9173	-19.7414	0.00170	-15.0001
15	4.4797	-12.6996	10.3906	-16.0907	19.3740	-20.2950	0.00170	-15.0001

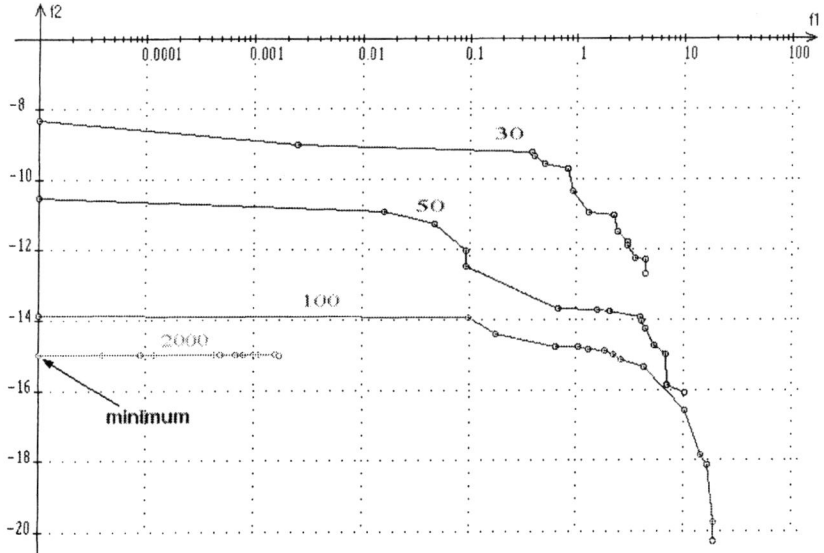

Fig. 3. Sets of Pareto solutions for numerical example

After 2000 generations the first Pareto solution is feasible and gives the global minimum of the minimized function. The remaining Pareto solutions are slightly violated but they might be very useful in the optimization process. For these solutions the rate of violation of each constraint can be analyzed. Note that using the proposed method the set of Pareto solutions can be presented to the decision maker after running for example every 50 generations. The solutions from the generations, which are close to the last generation, might be very interesting for the decision maker. He will know what the influence of the constraint's relaxation on the objective function value is.

3.4. Test Cases

In order to check the efficiency of the bicriteria method four test cases were performed. Michalewicz & Schoenauer, 1996 using different GA based methods, compared the first four tests. Test cases 2, 3 and 4 were originally formulated by Hock & Schitkowski, 1981.

Test Case #1

The numerical example from the previous section 3.3 will be considered as the Test Case #1.

Test Case #2

The problem is to minimize the function:

$$f(\mathbf{x}) = x_1 + x_2 + x_3$$

where

$$1 - 0.0025 \cdot (x_4 + x_6) \geq 0, \ 1 - 0.0025 \cdot (x_5 + x_7 - x_4) \geq 0,$$
$$1 - 0.01 \cdot (x_8 - x_5) \geq 0, \ x_1 x_6 - 833.33252 x_4 - 100 x_1 + 83333.33 \geq 0,$$
$$x_2 x_7 - 1250 x_5 - x_2 x_4 + 1250 x_4 \geq 0,$$
$$100 \leq x_1 \leq 10000, \ 1000 \leq x_i \leq 10000, \ i = 2,3, \ 10 \leq x_i \leq 1000, \ i = 4,5 ..., 8.$$

The problem has 3 linear and 2 nonlinear constraints; the function $f(\mathbf{x})$ is linear and has its global minimum at

$$x^* = (579.3167, 1359.943, 5110.071, 182.0174, 295.5985, 217.9799, 286.4162,$$
$$395.5979)$$

where $f(\mathbf{x}^*) = 7049.330923$

All six constraints are active at the global optimum.

Test case #3

The problem is to minimize the function:

$$f(\mathbf{x}) = (x_1 - 10)^2 + 5 \cdot (x_2 - 12)^2 + x_3^4 + 3 \cdot (x_4 - 11)^2 + 10 x_5^6 + 7 x_6^2 +$$
$$+ x_7^4 - 4 x_6 x_7 - 10 x_6 - 8 x_7$$

where:

$$127 - 2x_1^2 - 3x_2^4 - x_3 - 4x_4^2 - 5x_5 \geq 0$$
$$282 - 2x_1 - 3x_2 - 10x_3^2 - x_4 + x_5 \geq 0,$$
$$196 - 23x_1 - x_2^2 - 6x_6^2 - 8x_7 \geq 0$$
$$-10 \leq x_i \leq 10, \ i = 1,2,...,7$$

The problem has 3 nonlinear constraints; the function $f(\mathbf{x})$ is nonlinear and has its global minimum at

$$x^* = (2.330499, 1.951372, -0.4775414, 4.365726, -0.6244870, 1.038131,$$
$$1.594227)$$

where $f(x^*) = 680.6300573$.

Two out of four constraints, the first and the fourth, are active at the global optimum.

Test case #4

The problem is to minimize the function:

$$f(\mathbf{x}) = x_1^2 + x_2^2 + x_1 x_2 - 14x - 16x_2 + (x_3 - 10)^2 + 4 \cdot (x_4 - 5)^2 +$$
$$+ (x_5 - 3)^2 + 2 \cdot (x_6 - 1)^2 + 5x_7^2 + 7 \cdot (x_8 - 11)^2 + 2 \cdot (x_9 - 10)^2 +$$
$$+ (x_{10} - 7)^2 + 45$$

where:

$$105 - 4x_1 - 5x_2 + 3x_7 - 9x_8 \geq 0,$$
$$-10x_1 + 8x_2 + 17x_7 - 2x_8 \geq 0,$$
$$8x_1 - 2x_2 - 5x_9 + 2x_{10} + 12 \geq 0,$$
$$-3 * (x_1 - 2)^2 - 4 \cdot (x_2 - 3)^2 - 2x_3^2 + 7x_4 120 \geq 0,$$
$$-5x_1^2 - 8x_2 - (x_3 - 6)^2 + 2x_4 + 40 \geq 0,$$
$$-x_1^2 - 2 \cdot (x_2 - 2)^2 + 2x_1 x_2 - 14x_5 + 6x_6 \geq 0,$$
$$-0.5 \cdot (x_1 - 8)^2 - 2 \cdot (x_2 - 4)^2 - 3x_5^2 + x_6 + 30 \geq 0,$$
$$3x_1 - 6x_2 - 12 \cdot (x_9 - 8)^2 + 7x_{10} \geq 0,$$
$$-10 \leq x_i \leq 10, \ i = 1,2,...,10.$$

The problem has 3 linear and 5 nonlinear constraints; the function $f(\mathbf{x})$ is quadratic and has its global minimum at

$$x^* = (2.171996, 2.363683, 8.773926, 5.095984, 0.9906548, 1.430574, 1.321644,$$
$$9.828726, 8.280092),$$

where $f(\mathbf{x}^*) = 24.3062091$.

Six out of eight constraints, except the last two, are active at the global optimum.

These test cases were run for the following data:

variant I: number of generations = 3500, population size = 100,

variant II: number of generations = 350, population size = 1000,

The remaining data for the evolutionary algorithm method are the same for both variants:

- length of chromosome = N × computer precision for a double number,
- crossover rate = 0.6,
- mutation rate = 0.2,
- range of Pareto solutions = 15.

Ten experiments were carried out for each test case and the results of these experiments are presented in the last two columns of Table 2. These results were compared with the best results obtained by different methods tested by Michalewicz & Schoenauer, 1996. For the test case #1 the same results were obtained. For the test cases #2 and #4 much better results were obtained for the best, worst and median solutions. For the test case #3 slightly worse results were obtained using the bicriterion method for the best, worst and median results.

Table 2. Results of the first four test cases

	Exact optimum		Best results from Michalewicz & Schonauer (1996)	Bicriterion method Variant I	Bicriterion method Variant II
test #1	−15.000	best	− 15.000	-15.000	− 15.000
		worst	− 15.000	-15.000	− 15.000
		median	− 15.000	-15.000	− 15.000
			Method in Section 3.2.3[*]		
test #2	7049.331	best	7377.979	7191.430	7079.875
		worst	9652.901	9140.260	7800.337
		median	8206.151	7764.712	7343.7737
			Method in Section 3.2.3[*]		
test #3	680.640	best	680.642	680.681	680.688
		worst	680.955	681.333	681.318
		median	680.718	680.938	680.883
			Method in Section 3.2.3[*]		
test #4	24.306	best	25.486	24.611	24.749
		worst	42.358	27.425	26.392
		median	26.905	25.610	25.212
			Method in Section 3.2.2[*]		

*The methods describe in Michalewicz and Schoenauer, 1996

It should be mentioned here that the real advantage of the discussed Pareto constrained optimizer is that search direction is unconstrained. All penalty methods have preferred search directions, which may or may not be amenable to evolutionary algorithm search. Pareto schemes take their solutions from any direction regardless of their feasibility or other considerations.

4. TOURNAMENT SELECTION IN CONSTRAINED OPTIMIZATION

The tournament selection method seems to be a more effective method while solving constrained nonlinear programming problems. This selection picks up the better solution for the next generation. Thus this selection does not depend on the way of scaling the function as well as the kind of the problem what is considered. Both maximization and minimization problems are treated in the same way because tournament selection will choose a better solution in both cases. For a minimization problem, the solution with the smaller fitness value is selected and kept in an intermediate population, whereas for a maximization problem, the solution with a higher fitness is selected. It is also possible to transform a maximization problem to a minimization problem using the identity

$$max\{f(\mathbf{x})\} = -min\{-f(\mathbf{x})\} \tag{20}$$

For the constrained optimization the tournament selection method is less sensitive as to the choice of the penalty function. In most cases the constant penalty method with a great value of the penalty coefficient will produce good results. Kundu & Osyczka, 1996 showed that tournament selection gives better results while solving multicriteria constrained optimization problems using the distance method. Deb, 1997 used tournament selection in his Genetic Adaptive Search method and applied it to solve several mechanical design problems, which are also highly constrained problems. Recent works by Osyczka et al., 1999 and Zhang & Kim, 1999 indicate that tournament selection and ranking selection can give much better results than proportional selection. Osyczka et. al., 1999 studied nonlinear programming problems and show that tournament selection produces better results than proportional selection. Zhang & Kim, 1999 study refers to the machine layout design problem and from this study it is clear that ranking selection and tournament selection are a better choice than proportional selection. Osyczka et. al., 1999 considered the example as presented below.

Example 1

Let us consider a fairly complicated nonlinear optimization problem provided by Himmelblau, 1972 which was then solved by Gen & Cheng, 1997 using the genetic algorithm method with proportional selection. The optimization problem is:

The vector of decision variables is $\mathbf{x} = [\,x_1, x_2, x_3, x_4, x_5\,]^T$
The objective function is:

$$f(\mathbf{x}) = 5.3578547x_3^2 + 0.8356891x_1x_5 + 37.293239x_1 - 40792.141$$

The constraints are:

$$g_1(\mathbf{x}) \equiv 85.334407 + 0.0056858x_2x_5 + 0.0006262x_1x_4 - 0.0022053x_3x_5 \geq 0$$

$$g_2(\mathbf{x}) \equiv 92.0 - (85.334407 + 0.0056858x_2x_5 + 0.0006262x_1x_4 - \\ - 0.0022053x_3x_5) \geq 0$$

$$g_3(\mathbf{x}) \equiv 80.51249 + 0.0071317x_2x_5 + 0.0029955x_1x_2 + 0.0021813x_3^2 - \\ - 90.0 \geq 0$$

$$g_4(\mathbf{x}) \equiv 110.0 - (80.51249 + 0.0071317x_2x_5 + 0.0029955x_1x_2 + \\ + 0.0021813x_3^2) \geq 0$$

$$g_5(\mathbf{x}) \equiv 9.300961 + 0.0047026x_3x_5 + 0.0012547x_1x_3 0.0019085x_3x_4 - 20 \geq 0$$

$$g_6(\mathbf{x}) \equiv 25 - (9.300961 + 0.0047026x_3x_5 + 0.0012547x_1x_3 + \\ + 0.0019085x_3x_4) \geq 0$$

Lower and upper bounds on the decision variables are:

$$78 \leq x_1 \leq 102, \quad 33 \leq x_2 \leq 45, \quad 27 \leq x_3 \leq 45, \quad 27 \leq x_4 \leq 45, \quad 27 \leq x_5 \leq 45$$

For the above problem the tournament selection method was used with the same parameters for the evolutionary algorithm as in Gen & Cheng, 1997. These parameters are:
  - Length of the string for every decision variable - 19 bits,
  - Crossover rate p_c = 0.8,
  - Mutation rate p_m = 0.088,
  - Penalty parameter r =1000,
  - Population size J = 400,
  - Number of generations T = 200.

The comparison of the results obtained using different methods, including the evolutionary algorithm proportional selection method and the tournament selection method is shown in Table 3.

In this table the first four columns are taken from Gen & Cheng, 1997 and the fifth column presents the best results obtained using the tournament selection method. This table shows that the tournament selection method produces much better results than the other methods. The experiment was carried out for ten different initial seeds and in each case a better solution than the other solutions in Table 3 were obtained. The worst solution was - 30412.03 and the median was -30502.42. Similarly, better results were obtained for other test examples.

Table 3. Results of the experiment for the numerical nonlinear programming problem.

Items	Reference Solution	General Reduced Gradient Method	GA Solution by Gen & Cheng, 1997 with proportional selection	The best GA solution with tournament selection
F(x)	-30665.5	-30373.950	-30182.269	-30573.244
x_1	78.00	78.62	81.49	78.30
x_2	33.00	33.44	34.09	33.20
x_3	29.995	31.07	31.24	30.44
x_4	45.00	44.18	42.20	44.91
x_5	36.776	35.22	34.37	35.69

5. CONSTRAINT TOURNAMENT SELECTION METHOD

5.1. Constraints and computationally expensive functions.

Many optimization problems are highly constrained problems with a very limited feasible domain. Thus, while running an evolutionary algorithm, quite often the first several generations contain only nonfeasible solutions. As it was presented in the previous section, the tournament selection method may produce better results than the proportional selection method using the simple penalty function method. Using most of evolutionary methods we waste computing time for evaluation of the objective functions for both feasible and nonfeasible chromosomes. Moreover, in quite a large number of optimization problems objective functions are computationally expensive. Several conventional optimization methods have been developed to solve such problems using mainly the experimental design theory (see Kardis & Turns, 1984, Schoofs, 1988 and Osyczka et al., 1994). The aim of these methods is to reduce the number of function evaluations in the optimization process. In this section a tournament selection method very recently developed by Osyczka & Krenich, 1999 is discussed. This method is very effective while solving highly constrained single criterion optimization problems as well as the problems with the computationally expensive objective function. The method may also reduce the computation time for ordinary nonlinear programming problems as well as produce better results.

5.2. Description of the method.

In this method the tournament between two chromosomes is carried out in the following way:
(i)   If both chromosomes are not in the feasible region the one which is closer to the feasible region is taken to the next generation. The

values of the objective function are not calculated for either of chromosomes.

(ii) If one chromosome is in the feasible region and the other one is out of the feasible region the one, which is in the feasible region, is taken to the next generation. The values of the objective function are not calculated for either chromosomes.

(iii) If both chromosomes are in the feasible region, the values of the objective function are calculated for both chromosomes and the one, which has a better value of the objective function is taken to the next generation.

The constraint violation function can be evaluated as follows:

$$\Psi(\mathbf{x}) = \sum_{m=1}^{M} [h_m(\mathbf{x})]^2 + \sum_{k=1}^{K} G_k [g_k(\mathbf{x})]^2 \qquad (21)$$

where: G_k is the Heaveside operator such that $G_k = 0$ for $g_k(\mathbf{x}) \geq 0$ and $G_k = 1$ for $g_k(\mathbf{x}) < 0$.

It is clear that for the solutions, which are in the feasible region, the value of function (21) equals zero and for those, which are out of feasible region, the value of $\psi(\mathbf{x})$ indicates how far the solutions are from the feasible region.

In the tournament selection method a comparison between violated solutions is made and the one, which is less violated, i.e., which is closer to the feasible region, will be chosen to the next generation.

The steps of the method are as follows:

Step 1.   Set $t = 1$, where t is the index of the generation.

Step 2.   Set $j = 1$, where j is the index of the chromosome in each generation.

Step 3.   Generate an initial population of chromosomes at random $\mathbf{x}^{j,1} = \left[x_1, x_2, x_3, ..., x_N \right]$ for j=1, 2, ..., J, where J is assumed number of chromosomes in each population.

For the first chromosome from the population calculate $f\left(\mathbf{x}^{1,1}\right)$ and $\Psi\left(\mathbf{x}^{1,1}\right)$ and substitute $f(\mathbf{x}^*) = f\left(\mathbf{x}^{1,1}\right) + \Psi\left(\mathbf{x}^{1,1}\right)$ (the first chromosome is treated as the optimal solution).

Step 4.   Select at random chromosomes r and s from the t-th generation.

Step 5.   Calculate $\Psi\left(\mathbf{x}^{r,t}\right)$ and $\Psi\left(\mathbf{x}^{r,t}\right)$.

Step 6.   If $\Psi\left(\mathbf{x}^{r,t}\right) > 0$ and $\Psi\left(\mathbf{x}^{s,t}\right) > 0$ go to 7, otherwise go to 8.

Step 7.   If $\Psi\left(\mathbf{x}^{r,t}\right) < \Psi\left(\mathbf{x}^{s,t}\right)$ go to 15, otherwise go to 16.

Step 8.   If $\Psi\left(\mathbf{x}^{r,t}\right) > 0$ and $\Psi\left(\mathbf{x}^{s,t}\right) = 0$ go to 16, otherwise go to 9.

Step 9.   If $\Psi\left(\mathbf{x}^{r,t}\right) = 0$ and $\Psi\left(\mathbf{x}^{s,t}\right) > 0$ go to 15, otherwise go to 10.

Step 10. Calculate $f\left(\mathbf{x}^{r,t}\right)$ and $f\left(\mathbf{x}^{s,t}\right)$.

Step 11. If $f\left(\mathbf{x}^{r,t}\right) < f\left(\mathbf{x}^{s,t}\right)$ go to 12, otherwise go to 13.

Step 12. If $f\left(\mathbf{x}^{r,t}\right) < f(\mathbf{x}*)$ substitute $\mathbf{x}* = \mathbf{x}^{r,t}$ and $f(\mathbf{x}*) = f\left(\mathbf{x}^{r,t}\right)$ and go to 14, otherwise go straight to 14.

Step 13. If $f\left(\mathbf{x}^{s,t}\right) < f(\mathbf{x}*)$ substitute $\mathbf{x}* = \mathbf{x}^{s,t}$ and $f(\mathbf{x}*) = f\left(\mathbf{x}^{s,t}\right)$ and go to 15, otherwise go straight to 15.

Step 14. Take the r-th chromosome and place it as the j-th string in the $t+1$ generation, i.e., substitute $\mathbf{x}^{j,t+1} = \mathbf{x}^{r,t}$, then go to 16.

Step 15. Take the s-th chromosome and place it as the j-th string in the $t+1$ generation $\mathbf{x}^{j,t+1} = \mathbf{x}^{s,t}$, then go to 16.

Step 16. Set $j = j+1$ and if $j < J$ go to 4, otherwise go to 17. J is the number of the chromosomes in the population.

Step 17. Perform the evolutionary algorithm operations (crossover and mutation) on the $t+1$ generation.

Step 18. Set $t = t+1$ and if $t < T$ substitute $r = 1$ and go to 4, otherwise terminate calculations. T is the assumed number of generations.

The main idea of the constraint tournament selection method is illustrated graphically in Fig.4.

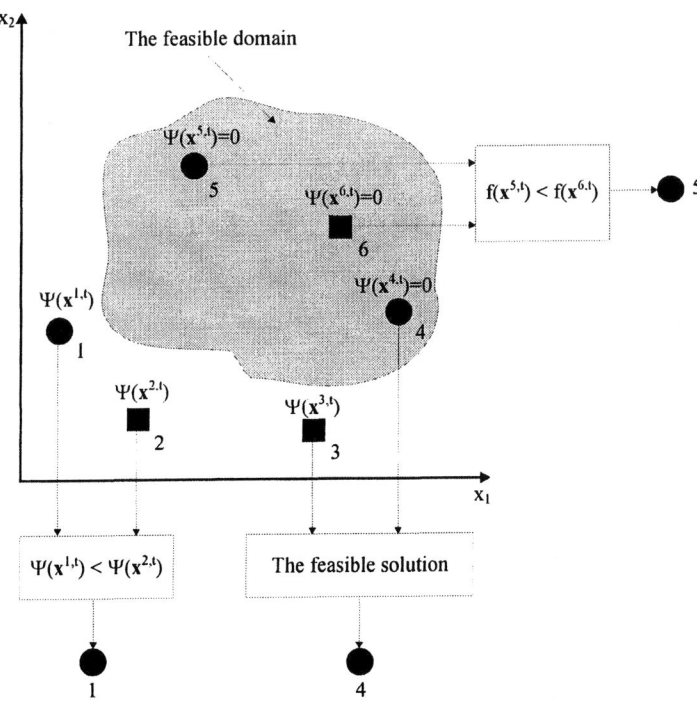

Fig. 4. The idea of the constraint tournament method.

Example 2.

Let us consider once more the problem from Example 1. The experiments were carried out ten times for the constraint tournament selection method, simple tournament selection method and proportional selection method. The data for this experiment were the same as in Example 1. The results of the experiments are shown in Table 4. This table shows clearly that the constraint tournament selection method produces best results in the shortest computation time.

Table 4. Comparison of the methods for the numerical nonlinear programming problem

Method of solution	Results (best, worst, average)	Average time of solving [s]	Average number of the objective function callings	Last improvement on the best solution
Constraint Tournament Method	-30588,42 -30423,52 -30503,82	26	109.102	180
Simple Tournament Method	-30573.244 -30412.03 -30502.42	38	160.000	159
Proportional Method	-30475,54 -30301,09 -30384,76	52	160.000	309

6. APPLICATION OF EVOLUTIONARY ALGORITHMS TO DESIGN OPTIMIZATION

Some evolutionary algorithm methods presented above were applied to solve real-life design optimization problems two of which are presented below. The first problem deals with optimum design of concentric springs, whereas the second problem deals with the optimum design of a robot gripper. Both of these problems are considered as discrete programming problems. For problems like these the use of conventional optimization methods is very limited or even impossible.

6.1 Optimum design of concentric springs

The problem is to find dimensions of concentric springs, which satisfy constraints and minimize the volume of the springs. The scheme of the springs is presented in Fig.5. The optimization model was built on the basis of formulas from Polish Standard PN-85/M-80701-3, which is based on

British Standard BS 1726, German Standard DIN 2989 and Japanese Standard JIS B 2704-1978.

The concentric springs are under repeated loading P within the range 0 and P_{max}. Both springs are made of the same material. The basic equations for these springs are:

The load on the outer spring

$$P_z = \frac{c_z}{c_z + c_w} \cdot P \tag{22}$$

The load on the inner spring

$$P_w = \frac{c_w}{c_z + c_w} \cdot P \tag{23}$$

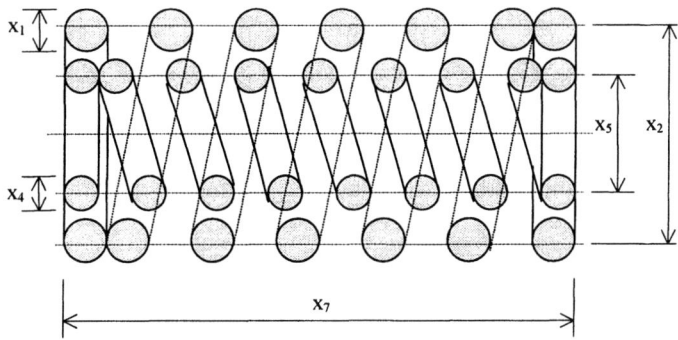

Fig. 5. Scheme of the concentric springs.

The stiffness of the outer spring

$$c_z = G \cdot \frac{x_1^4}{8 \cdot x_2^3 \cdot x_3} \tag{24}$$

where: G [N/mm^2] – shear modulus.
The stiffness of the inner spring

$$c_w = G \cdot \frac{x_4^4}{8 \cdot x_5^3 \cdot x_6} \tag{25}$$

The shear stress in the outer spring:

$$\tau_{mz} = k \cdot \frac{8 \cdot x_2 \cdot P_z}{\pi \cdot x_1^3} \tag{26}$$

where: k – Wahl factor that can be approximated very closely by the relation:

$$k = 1 + \frac{5}{4} \cdot \left(\frac{1}{w}\right) + \frac{7}{8} \cdot \left(\frac{1}{w}\right)^2 + \left(\frac{1}{w}\right)^3 \tag{27}$$

where: w – index of the spring expressed by $w = \dfrac{x_2}{x_1}$,

The shear stress in the inner spring:

$$\tau_{mw} = k \cdot \frac{8 \cdot x_5 \cdot P_w}{\pi \cdot x_4^3} \tag{28}$$

where k is calculated using formula (7.30) for $w = \dfrac{x_5}{x_4}$

The variable shear stress amplitudes for the outer and the inner spring are

$$\tau_{az} = \frac{\tau_{mz}}{2} \text{ and } \tau_{aw} = \frac{\tau_{mw}}{2} \tag{29}$$

Optimization model

Decision variables:

$$\mathbf{x} = [\, x_1, x_2, x_3, x_4, x_5, x_6, x_7\,]$$

where:

x_1 - wire diameter of the outer spring	[mm],
x_2 - meancoil diameter of the outer spring	[mm],
x_3 - number of active coils of the outer spring	[],
x_4 - wire diameter of the inner spring	[mm],
x_5 - meancoil diameter of the inner spring	[mm],
x_6 - number of active coils of the inner spring	[],
x_7 - length of the spring system	[mm].

Objective function:

Volume of both springs

$$f(\mathbf{x}) = \frac{\pi^2}{2}\left(x_1^2 \sqrt{x_2^2 x_3^2 + (x_7 - x_1)^2} + x_4^2 \sqrt{x_5^2 x_6^2 + (x_7 - x_4)^2} \right) + \\ + \frac{\pi}{4}\left(x_1^2 x_2 + x_4^2 x_5 \right) \tag{30}$$

Constraints:

(i)   Shear stress constraint for the outer spring

$$g_1(\mathbf{x}) \equiv s_f \cdot \tau_{dop} - k \cdot \frac{8 \cdot x_2 \cdot P}{\pi \cdot x_1^3} \geq 0 \qquad (31)$$

where: τ_{dop} [N/mm^2] – allowable shear stress,

s_f – coefficient of allowable changes in shear stress,

k - is calculated using formula (7.30) for $w = \frac{x_2}{x_1}$.

(ii)   Shear stress constraint for the inner spring

$$g_2(\mathbf{x}) \equiv s_f \cdot \tau_{dop} - k \cdot \frac{8 \cdot x_5 \cdot P}{\pi \cdot x_4^3} \geq 0 \qquad (32)$$

where: k - is calculated using formula (7.30) for $w = \frac{x_5}{x_4}$.

(iii) Total stiffness of the spring

$$g_3(\mathbf{x}) \equiv \Delta c \cdot c - \left| c - \left(\frac{G \cdot x_1^4}{8 \cdot x_2^3 \cdot x_3} + \frac{G \cdot x_4^4}{8 \cdot x_5^3 \cdot x_6} \right) \right| \geq 0 \qquad (33)$$

where:
Δc[N/mm] – allowable deviation of the stiffness of the springs,
c[N/mm] – required total stiffness of the springs.

(iv) Clearance between the coils for the outer spring

$$g_4(\mathbf{x}) \equiv x_7 - \frac{8 \cdot x_2^3 \cdot x_3 \cdot P}{G \cdot x_1^4} - x_1 \cdot x_3 \cdot (1 + \alpha) \geq 0 \qquad (34)$$

where:
α - coefficient of the clearance, which can be approximated as follows:
for $x_1 < 0.8$:

$$\alpha = f(w) = -0.00003 \cdot w^3 + 0.0025 \cdot w^2 - 0.027 \cdot w + 0.1390 \quad (35a)$$

for $x_1 \geq 0.8$:

$$\alpha = f(w) = -0.00002 \cdot w^3 + 0.002 \cdot w^2 + 0.0018 \cdot w + 0.0627 \quad (35b)$$

where: $w = \frac{x_2}{x_1}$

(v)   Clearance between the coils for the inner spring

$$g_5(\mathbf{x}) \equiv x_7 - \frac{8 \cdot x_5^3 \cdot x_6 \cdot P}{G \cdot x_4^4} - x_4 \cdot x_6 \cdot (1+\alpha) \geq 0 \qquad (36)$$

where:

α is calculated from the formula (7.45a) for $x_4 < 0.8$ or from

the formula (7.45b) for $x_4 \geq 0.8$ and $w = \dfrac{x_5}{x_4}$

(vi) Buckling constraint of the outer spring

$$g_6(\mathbf{x}) \equiv \eta_{dop} - \frac{100 \cdot 8 \cdot x_2^3 \cdot x_3 \cdot P}{G \cdot x_7 \cdot x_1^4} \geq 0 \qquad (37)$$

where:

$\lambda = \dfrac{x_7}{x_2}$ - spring slenderness ratio,

$\eta_{dop} = f(\lambda)$ - spring allowable flexibility ratio, which can be approximated as follows:

$$\eta_{dop} = f(\lambda) = -0.0122 \cdot \lambda^6 + 0.2562 \cdot \lambda^5 - 1.977 \cdot \lambda^4 + \qquad (38)$$
$$+ 6.448 \cdot \lambda^3 - 12.807 \cdot \lambda^2 + 9.897 \cdot \lambda + 52.444$$

(vii) Buckling constraint of the inner spring

$$g_7(\mathbf{x}) \equiv \eta_{dop} - \frac{100 \cdot 8 \cdot x_5^3 \cdot x_6 \cdot P}{G \cdot x_7 \cdot x_4^4} \geq 0 \qquad (39)$$

where:

$\lambda = \dfrac{x_7}{x_5}$ - spring slenderness ratio,

$\eta_{dop} = f(\lambda)$ is calculated using the formula (7.48) for $\lambda = \dfrac{x_7}{x_5}$

(viii) Minimal ratio between the meancoil diameter and the wire diameter of the inner spring

$$g_8(\mathbf{x}) \equiv x_5 - 3 \cdot x_4 \geq 0 \qquad (40)$$

(ix) Maximal ratio between the meancoil diameter and the wire diameter of the inner spring

$$g_9(\mathbf{x}) \equiv 20 \cdot x_4 - x_5 \geq 0 \qquad (41)$$

(x) Dependencies between diameters of both springs

$$g_{10}(\mathbf{x}) \equiv x_2 - x_1 - (x_5 + x_4) \geq 0 \qquad (42)$$

(xi) Maximal ratio between meancoil diameter and wire diameter of the outer spring

$$g_{11}(\mathbf{x}) \equiv 20 \cdot x_1 - x_2 \geq 0 \qquad (43)$$

(xii) Maximal outer diameter of the outer spring

$$g_{12}(\mathbf{x}) \equiv D_{max} - (x_1 + x_2) \geq 0 \qquad (44)$$

(xiii) Minimal inner diameter of the outer spring

$$g_{13}(\mathbf{x}) \equiv x_5 - x_4 - D_{min} \geq 0 \qquad (45)$$

(xiv) Maximal length of both springs

$$g_{14}(\mathbf{x}) \equiv L_{max} - x_7 \geq 0 \qquad (46)$$

(xv) Minimal length of both springs

$$g_{15}(\mathbf{x}) \equiv x_7 - L_{min} \geq 0 \qquad (47)$$

(xvi) Minimal safety factors of strength fatigue for the outer spring

$$g_{16}(\mathbf{x}) \equiv \frac{z_{so}}{\beta_z \cdot \gamma_z \cdot \tau_{az} + \tau_{mz} \cdot \left(\dfrac{2 \cdot z_{so}}{z_{sj}} - 1 \right)} - 1{,}33 \geq 0 \qquad (48)$$

(xvii)   Minimal safety factors of strength fatigue for the inner spring

$$g_{17}(\mathbf{x}) \equiv \frac{z_{so}}{\beta_w \cdot \gamma_w \cdot \tau_{aw} + \tau_{mw} \cdot \left(\dfrac{2 \cdot z_{so}}{z_{sj}} - 1 \right)} - 1{,}33 \geq 0 \qquad (49)$$

(xvii)   Difference between the safety factors of strength fatigue of both springs

$$g_{18}(\mathbf{x}) = SF_{\lim it} - \left| \frac{z_{so}}{\beta_z \cdot \gamma_z \cdot \tau_{az} + \tau_{mz} \cdot \left(\dfrac{2 \cdot z_{so}}{z_{sj}} - 1 \right)} - \frac{z_{so}}{\beta_w \cdot \gamma_w \cdot \tau_{aw} + \tau_{mw} \cdot \left(\dfrac{2 \cdot z_{so}}{z_{sj}} - 1 \right)} \right| \geq 0 \quad (50)$$

where:

SF_{limit} – assumed upper limit of the difference between the safety factors of strength fatigue of both springs,

z_{so} - alternating tortional fatigue strength,

z_{sj} - fluctuating tortional fatigue strength,

$\beta_z = [1 + \eta_z \cdot (\alpha_{k\_z} - 1)] \cdot \beta_{ps\_z}$ - fatigue factor of strength concentration for the outer spring,

$\beta_w = [1 + \eta_w \cdot (\alpha_{k\_w} - 1)] \cdot \beta_{ps\_w}$ - fatigue factor of strength concentration for the inner spring,

where:

β_{ps_z}, β_{ps_w} - surface state factors of the outer and inner springs, respectively, which can be approximated using the formula:

$$\beta_{ps} = f(R_r) = 1.154 \cdot 10^{-4} + 1.01$$

where: R_r - tensile strength of the spring material.

α_{k_z}, α_{k_w} - shape factors of the outer and inner springs, respectively, which can be approximated using the formula:

$$\alpha_k = f(w) = 0.574 \cdot 10^{-4} \cdot w^4 - 0.285 \cdot 10^{-3} +$$
$$0.05208 \cdot w^2 - 0.424 \cdot w + 2.452 \qquad (51)$$

where: $w = \dfrac{x_2}{x_1}$ for the outer spring,

$w = \dfrac{x_5}{x_4}$ for the inner spring.

η_z, η_w – stress concentration factors of the outer and inner springs respectively, which can be approximated using the formula:

$$\eta = f(Z_{go}) = 5.0 \cdot 10^4 + 0.75 \qquad (52)$$

γ_z, γ_z – quantity factors of the outer and inner springs, respectively, which can be evaluated as follows:
- for the spring with diameter $d < 10$ $\gamma = 1$,
- for the spring with diameter $d \geq 10$ using the following formula:

$$\gamma = f(\alpha_k, Z_{go}, d) = B \cdot log(d) + A \qquad (53)$$

where:

$$A = (1 - \gamma_1) \cdot \frac{ln2}{2 \cdot ln5} - \frac{\gamma_1 - 3}{2}, \qquad (54)$$

$$B = (\gamma_1 - 1) \cdot \frac{ln2}{2 \cdot ln5} + \frac{\gamma_1 - 1}{2}, \qquad (55)$$

$$\gamma_1 = f(Z_{go}, \alpha_k) = 4{,}2860 \cdot 10^{-4} \cdot (Z_{go} - 500{,}0) +$$
$$0{,}1038 \cdot \alpha_k^3 - 0{,}9265 \cdot \alpha_k^2 + 2{,}6931 \cdot \alpha_k - 0{,}5870 \tag{56}$$

Results of optimization

The optimization problem formulated above was considered as a discrete programming problem with the following arbitrary chosen sets of discrete values of the decision variables:

$X_1 = \{1.0, 1.5, 2.0,...,15.0\}$,
$X_2 = \{35.0, 40.0, 45.0, ..., 105.0\}$,
$X_3 = \{5.0, 5.5, 6.0, ...,15.0\}$,
$X_4 = \{1.0, 1.5, 2.0,...,15.0\}$,
$X_5 = \{10.0, 15.0, 20.0, ...,80.0\}$,
$X_6 = \{5.0, 5.5, 6.0, ...,15.0\}$,
$X_7 = \{50.0, 60.0, 70.0,..., 200\}$.

For the exhaustive search method, which is able to find the global minimum of the volume of the springs, these sets require calculations in 1 959 655 824 points and no one would recommend this method for finding the optimum. Other conventional optimization methods may probably fail in searching minimum or may be not able to find this minimum in a reasonable computation time.

Data for the optimization process:

(i) Data for the spring
- Load $P = 1500$ [N],
- Spring constant $c = 150$ [N/mm],
- Allowable deviation of the spring constant $\Delta c = 3$ [%],
- Material of the spring – toughened spring steel 5HG for which we have:
  - modulus of rigidity $G = 82000$ [N/mm^2],
  - tensile strength $R_r = 1300$ [N/mm^2].

  From the given tensile strength R_r the remaining strength factors are calculated as follows:

  $Z_{go} = 0.6 \cdot R_r = 780$ [N/mm^2],
  $Z_{so} = 0.6 \cdot Z_{go} = 468$ [N/mm^2],
  $Z_{sj} = 1.1 \cdot Z_{go} = 858$ [N/mm^2],
  $\tau_{dop} = 0.5 \cdot R_r = 650$ [N/mm^2].

- Constraints on the dimensions of the concentric springs:

  $D_{max} = 100$ [mm],
  $D_{min} = 10$ [mm],

$$L_{max} = 50 \text{ [mm]},$$
$$L_{min} = 200 \text{ [mm]}.$$

- Assumed upper limit of the difference between the safety factors of strength fatigue of both springs $SF_{limit}=1$,

(ii) Data for the evolutionary algorithm

- population size $J = 400$,
- number of generations $T = 400$,
- crossover rate $p_c = 0.4$,
- mutation rate $p_m = 0.08$.

Three different evolutionary algorithm methods: the constraint tournament method, the simple tournament method and the proportional method were used for solving this problem. The experiments were carried out several times for different initial populations (different initial seeds) with the same parameters in each method. The results of experiments are presented in Table 5 and Table 6.

Table 5. The best solution obtained for the concentric springs.

Items	Constraint Tournament Method	Simple Tournament Method	Proportional Selection Method
f(x)	$10.583 \cdot 10^4$	$10.961 \cdot 10^4$	$11.022 \cdot 10^4$
x_1	6.5	6.5	6.5
x_2	45	45	45
x_3	7.0	8.5	7.5
x_4	6.5	5.5	6.5
x_5	20	20	20
x_6	5.0	5.0	5.0
x_7	130	130	130

Table 6. Comparison of the methods for the concentric springs.

Method of solution	Average time of solving [s]	Average number of the objective function callings	Last improvement on the best solution
Constraint Tournament Method	22	101.102	254
Simple Tournament Method	28	160.000	359
Proportional Method	34	160.000	317

From this example it is clear that recently developed evolutionary algorithm based constraint tournament selection method is a very effective tool in finding the global minimum for highly constrained optimization problem.

6.2. Optimum design of robot gripper

Let us consider an example of a robot griper the scheme of which is presented in Fig. 6.

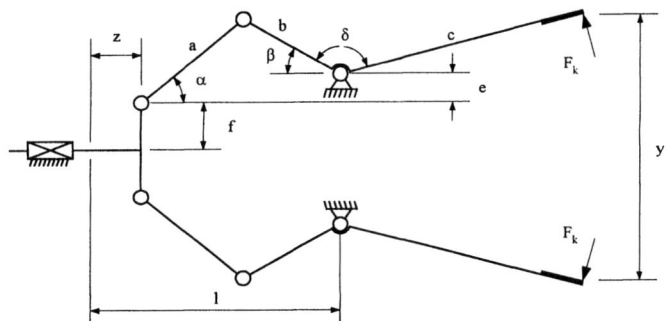

Fig.6. The scheme of robot griper mechanism.

The geometrical dependences of the griper mechanism are (see Fig.7):

$$g^2 = (l\text{-}z)^2 + e^2, \quad g = \sqrt{(l-z)^2 + e^2} \, ,$$

$$a^2 = b^2 + g^2 - 2 \cdot b \cdot g \cdot cos(\beta + \phi),$$

$$b^2 = a^2 + g^2 - 2 \cdot a \cdot g \cdot cos(\alpha - \phi),$$

$$\phi = a\,tan\left(\frac{e}{l-z}\right), \alpha = arccos\left(\frac{a^2 + g^2 - b^2}{2 \cdot a \cdot g}\right) + \phi \, ,$$

$$\beta = arccos\left(\frac{b^2 + g^2 - a^2}{2 \cdot b \cdot g}\right) - \phi \, ,$$

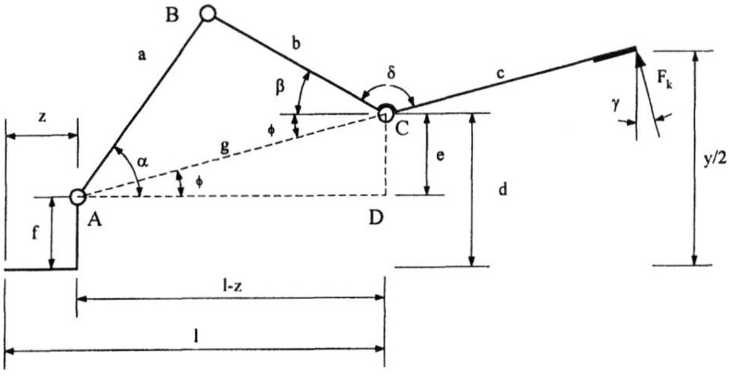

Fig.7. The geometrical dependences of the griper mechanism.

The distribution of the forces is presented in Fig.8 and from this figure we have:

$$R \cdot sin(\alpha+\beta) \cdot b = F_k \cdot c, \quad R = \frac{P}{2} \cdot cos(\alpha), \quad F_k = \frac{P \cdot b}{2 \cdot c} \cdot sin(\alpha+\beta) \cdot cos(\alpha)$$

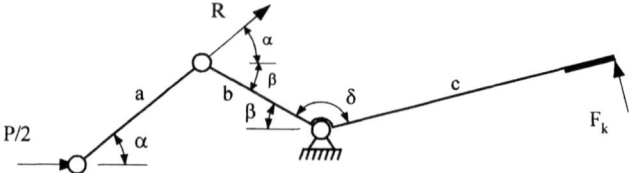

Fig.8. The distribution of the forces in the mechanism of the griper.

Using the above formulas the optimization model can be evaluated as follows:

Vector of decision variables

$\mathbf{x} = [\ a,\ b,\ c,\ e,\ f,\ l,\ \delta\ ]^T$, where $a,\ b,\ c,\ e,\ f,\ l$ are dimensions of the griper and δ is the angle between b and c elements of the griper.

Objective function: difference between the maximum and the minimum griping forces for the assumed range of griper ends displacement:

$$f(\mathbf{x}) = \max_{z} F_k(\mathbf{x},z) - \min_{z} F_k(\mathbf{x},z)$$

This objective function depends on the vector of decision variables and displacement z, thus we have to use a procedure, which makes these function computationally expensive.

From the geometry of the griper the following constraints can be derived:

(I) $g_1(x) = Y_{min} - y(x, Z_{max}) \geq 0,$
(II) $g_2(x) = y(x, Z_{max}) \geq 0,$
(III) $g_3(x) = y(x, 0) - Y_{max} \geq 0,$
(IV) $g_4(x) = Y_G - y(x, 0) \geq 0,$
(V) $g_5(x) = (a+b)^2 - l^2 - e^2 \geq 0,$
(VI) $g_6(x) = (l-Z_{max})^2 + (a-e)^2 - b^2 \geq 0,$
(VII) $g_7(x) = l-Z_{max} \geq 0,$

where: $y(x,z) = 2 \cdot [e + f + c \cdot sin(\beta+\delta)]$ displacement of the griper ends
        Y_{min} – minimal dimension of griping object
        Y_{max} – maximal dimension of griping object
        Y_G – maximal range of the griper ends displacement
        Z_{max} – maximal displacement of the griper actuator

The optimization process was carried out using the following data:
1. The arbitrary chosen discrete values of decision variables:
        a={10, 50, 90, 130, 170, 210, 250},
        b={10, 50, 90, 130, 170, 210, 250},

c={100, 140, 180, 220, 260, 300},
e={0, 10, 20, 30, 40, 50},
f={10, 50, 90, 130, 170, 210, 250},
l={100, 140, 180, 220, 260, 300},

$$\delta=\{\frac{\pi}{3}, \frac{\pi}{2}, \frac{2}{3}\cdot\pi, \frac{5}{6}\cdot\pi, \pi\}$$

$Y_{min} = 50$, $Y_{max} = 100$, $Y_G = 150$, $Z_{max} = 50$, $P = 100$

2. The parameters for GA:
- Number of generation $T=400$
- Population size $J = 400$
- Crossover rate $R_c = 0,6$
- Mutation rate $R_m = 0,08$
- Penalty rate $r = 10^3$

For the above problem the exhaustive search method requires 1810 seconds of calculations. The results obtained using the evolutionary algorithm methods i.e. constraint tournament selection method, simple tournament selection method and proportional selection method are presented in Table 7 and Table 8. As for the previous example the constraint tournament selection method produces the best results considering the computation time and the obtained minimum.

Table 7. The best solution obtained for the robot gripper design problem.

Items	Constraint Tournament Selection Method	Simple Tournament Selection Method	Proportional Selection Method
f(x)	0.568	0.568	1.358
A	250	250	250
B	210	210	210
C	300	300	220
E	30	30	10
F	50	50	50
L	220	220	260
δ	2.094	2.094	2.094

Table 8. Comparison of the methods for the robot gripper design problem.

Method of solution	Average time of solving [s]	Average number of the objective function callings	Last improvement on the best solution
Constraint Tournament Method	24	30.234	372
Simple Tournament Method	117	160.000	279
Proportional Method	134	160.000	301

7. CONCLUSIONS

In this Chapter the evolutionary algorithm based methods for solving global optimization problems are presented. Special attention is devoted to the two methods recently developed by the authors: the bicriterion method for solving constrained minimization problems and the constraint tournament method. The evolutionary algorithm based methods are a new tool for solving a wide range of nonlinear programming problems including those with integer and discrete decision variables. They simulate natural evolutionary processes of living organisms and they can often outperform conventional optimization methods used for seeking the global minimum. It is difficult to prove that these methods are better than conventional global optimization methods, but from the examples presented in this Chapter it is clear that these methods are very effective in solving some global optimization problems mainly those which are discrete nonlinear programming problems.

REFERENCES

1. Biggs M.C., (1975): Constrained Minimization Using Recursive Quadratic Programming: Some Alternative Subproblem Formulation. In: *Towards Global Optimization*, Dixon L.C.W. & Szego G.P., Eds., North Holland.

2. Broyden C.G. & Attia N.F., (1983): A Smooth Sequential Penalty Function Method for Solving Nonlinear Programming Problem. In: *System Modelling and Optimization, Proceeding of the 11-th IFIP Conference,* Lecture Notes in Control and Information Sciences, Vol. 59, Springer-Verlag.

3. Deb K., (1997): A Robust Optimal Design Technique for Mechanical Component Design. In: *Evolutionary Algorithms in Engineering Applications*. D. Dasgupta and Z. Michalewicz, Eds. Springer Verlag, pp. 497 – 514.

4. Floudas,C.A. & Pardalos,P.M. (1992): *Recent advances in global optimization*, Princeton series in Computer Science, Princeton University Press, Princeton, NJ.

5. Gen M. & Chang R. (1996): A survey of penalty techniques in genetic algorithms, In: *Proc. of IEEE International Conference on Evolutionary Computation*, pp. 804-809.

6. Gen M. P., Cheng R. (1997*): Genetic Algorithms and Engineering Design*. John Wiley and Sons, Inc. New York.

7. Himmbelblau M., (1972): *Applied Nonlinear Programming,* McGraw-Hill, New York.

8. Hock W., Schittkowski K., (1981): *Test Examples for Nonlinear Programming Codes,* Springer-Verlag, New York.

9. Homaifar A., Qi C. & Lai S., (1994): Constrained optimization via genetic algorithms, *Simulation.* pp.242 – 254.

10. Joines J. & Houck C. (1994): On the use of non-stationary penalty functions to solve nonlinear constrained optimization problems with GAs. In: *Proceedings of the First IEEE Conference on Evolutionary Computation,* Fogel D. Ed. IEEE Press Orlando, FL, pp. 579 – 584.

11. Kardis, J.P. & Turns, S.R. (1984) Efficient Optimization of Computationally Expensive Functions. *Proc. ASME 18-th Mechanisms Conf.,* pp. 218-226.

12. Kim, J. H. & Myung, H., (1996): A Two Phase Evolutionary Programming for general constrained optimization problem. In: *Proc. Of the Fifth Annual Conference on Evolutionary Programming,* San Diego.

13. Kundu S. & Osyczka A. (1996a): Genetic Multicriteria Optimization of Structural Systems, In: *Proceedings of the 19th International Congress on Theoretical and Applied Mechanics* (ICTAM) Kyoto, Japan, IUTAM Volume of Abstracts 272.

14. Kundu S. & Osyczka A., (1996b): The Effect of Genetic Algorithm Selection Mechanism on Multicriteria Optimization Using the Distance Method. In: *Proceedings of the Fifth International Conference on Intelligent Systems,* Reno, Nevada, pp.164-168.

15. Michalewicz Z. & Attia N. (1994): Evolutionary optimization of constrained problems. In: *Proceedings of Third Annual Conference on Evolutionary Programming,* Sebald A. V. & Fogel L. J. Eds., World Scientific, River Edge, NJ, pp. 98 – 108.

16. Michalewicz Z. & Schoenauer M. (1996): Evolutionary Algorithms for Constraints Parameter Optimization Problems, *Evolutionary Computation,* **4**(1) pp. 1-32.

17. Michalewicz Z. (1996): *Genetic Algorithms + Data Structures = Evolution Programs* (third edition) Springer-Verlag.

18. Michalewicz, Z. (1995) Genetic Algorithms, Numerical Optimization, and Constraints. In: *Proceedings of the Sixth International Conference on Genetic Algorithms.* Echelman, L. J. Ed., pp. 151-158.

19. Myung, H. & Kim, J. H., (1996): Hybrid Evolutionary Programming for Heavily Constrained Problems, *Bio-Systems,* 38, pp. 29-43.

20. Olhoff N. & Lund E. (1995) Finite element engineering design sensitivity analysis and optimization, In: *Advances in Structural Optimization,* Herskovits J., Ed., Kluwer Academic Publishers, pp. 1-45.

21. Orvosh, D. Davis, L. (1995): Using a Genetic Algorithm to Optimize Problems with Feasibility Constraints. In: *Proc. of the Sixth*

International Conference on Genetic Algorithms. Echelman, L. J., Ed., pp. 548-552.

22. Osyczka A, & Krenich S. (1999) : A New Method of Solution of Nonlinear Programming Problems using Genetic Algorithms. (in Polish) *3rd Polish Conference on Evolutionary Algorithms and Global Optimization*, Potok Zloty 25-28 Maj.pp. 253-260.

23. Osyczka A. & Krenich S. (2000): A New Constraint Tournament Selection Method for Multicriteria Optimization Using Genetic Algorithm. Accepted for *Congress of Evolutionary Computing*, San Diego, USA, 16-19 July 2000.

24. Osyczka A. & Montusiewicz J. (1994): A Random Search Approach to Multicriterion Discrete Optimization. In: *IUTAM Symposium on Discrete Structural Optimization, Springer-Verlag*, pp. 71-79.

25. Osyczka A. & Tamura H. (1996): Pareto Set Distribution Method for Multicriteria Optimization Using Genetic Algorithm, In: *Proceedings of Genetic Algorithms '96, (MENDEL - 96) - International Conference*, Brno, Czech Republic. pp. 135-143.

26. Osyczka A., Krenich S. & Kundu S. (1999): Proportional and tournament selections for constrained optimization problems using Genetic Algorithms, *Evolutionary Optimization an International Journal on the Internet*, 1(1) pp. 89-92.

27. Osyczka A., Krenich S., Karas J., (1999): *Optimum design of robot gripers using Genetic Algorithms, In: The 3rd World Congress on Structural and Multidisciplinary Optimization*, Buffalo 17-22 May, USA.

28. Osyczka A., Kuchta W. & Czula W. (1994): Computer Aided Multicriterion Optimization System for Computationally Expensive Functions, *Structural Optimization*, **8,** pp. 37-41.

29. Schoofs, A.J.G. (1988) Experimental Design and Structural Optimization. In: *Structural Optimization*, Rozwany, G.I.N. & Karihaloo, B.L. (Eds.) Kluwer Academic Publishers, pp. 307-314.

30. Yokota T., Gen M., Ida K. & Taguchi T. (1995): Optimal design of system reliability by an approved genetic algorithm, *Transactions of Institute of Electronics, Information and Communication Engineers,* vol. J78A, no.6, pp. 702 – 709.

31. Zhang, B-T. & Kim, J-J. (2000): Comparison of Selection Methods for Evolutionary Optimization, *Evolutionary Optimization, an International Journal on the Internet.* (2)1, pp.54-69.

Chapter 13

DETERMINING 3-D STRUCTURE OF SPHERICAL VIRUSES BY GLOBAL OPTIMIZATION

Ozcan Ozturk

Peter C. Doerschuk

Saul B. Gelfand
School of Electrical and Computer Engineering, Purdue University, West Lafayette, IN USA

Abstract   State-of-the-art global optimization (GO) algorithms are applied to computation of 3-D virus reconstructions from solution x-ray scattering data by minimizing a weighted least squares error. Starting with random uniformly-distributed initial conditions, high quality reconstructions are achieved using GO algorithms. Similar quality reconstructions using classical multistart require initial conditions that incorporate more detailed *a priori* knowledge which is not always available. The behavior of the GO algorithms with respect to the parameterization of the error function and with respect to the accuracy of the quadrature rule embedded in the calculation of the regressor are described. Several future directions in terms of structural biology and GO algorithms are discussed.

1.   Introduction

Determination of three-dimensional structures of viruses is an important problem in biophysics. In this chapter we pose this reconstruction problem as an optimization problem by appropriate mathematical modeling of the virus and measurement process. We solve this problem by applying global optimization (GO) techniques to determine the values of

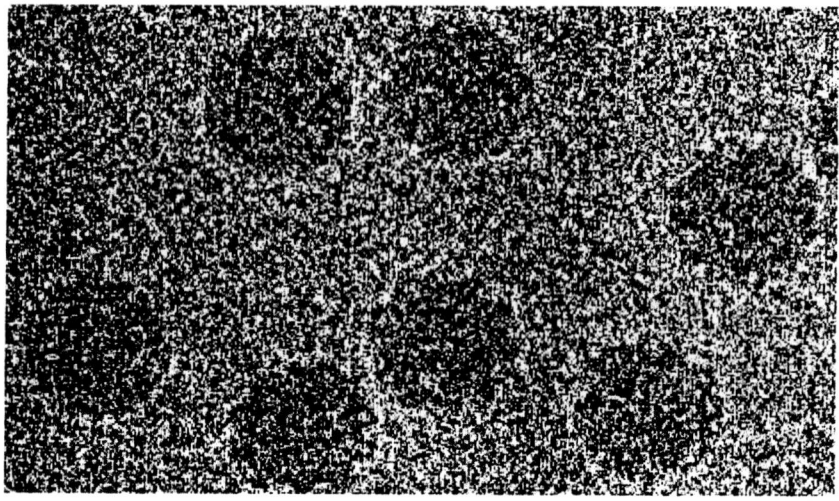

Figure 1.1. A small subimage from a cryo electron micrograph of the *Nudaurelia capensis ω* virus after histogram equalization showing seven particles.

the parameters in the model and thereby determine the virus structure in 3-D. Though other techniques exist, e.g., techniques based on interval mathematics [Corliss and Kearfott, 1999], we focus on stochastic methods.

Viruses are of great biological interest in and of themselves, as pathogens of plants and animals, and as relatively simple models of more complex organisms. There is a large range of virus structures [Chiu et al., 1997]. Our work considers the so-called "spherical viruses" which are viruses with a shell of protein (the so-called "capsid") surrounding an inner core of nucleic acid. The capsid is "crystalline" in the sense that it is constructed from many repetitions of the same polypeptides and the entire capsid is invariant under the rotational symmetries of the icosahedron. Typical radii for these viruses are in the range 10^2–10^3 Å. A cryo electron micrograph[1] showing several virus particles is shown in Figure 1.1.

The structure of viruses can be determined from x-ray crystal diffraction data, cryo electron microscope images, or x-ray solution scattering data. In this chapter we focus on x-ray solution scattering data which is the least informative type of data (see below) but which is the only type of data that is recorded while the virus particle is in its natural aqueous environment and is the type of data that can be recorded most quickly (e.g., 50 ms using a synchrotron x-ray source). Therefore this type of data has the potential of tracking, in real time, the changes in 3-D virus structure that occur during maturation and infection.

The reason that solution x-ray scattering data is uninformative is that each virus particle is constantly rotating in the aqueous solution essentially independently of all of the other virus particles and therefore the data is the spherical average of the magnitude-squared of the Fourier transform of the electron density in the virus. This is 1-D data even though we seek a 3-D reconstruction. Furthermore the resolution range of the data is limited, e.g., less than 0.05Å^{-1}. Therefore, exploiting known properties of the virus particle (such as its icosahedral symmetry) and using a simple model with a small number of parameters is crucial and we describe results based on 9 or 10 parameters. Because of the presence of the icosahedral symmetry and Fourier transforms, we make extensive use of so-called "icosahedral harmonics" in our mathematical representations of these viruses. Such harmonics are a complete orthonormal basis for the subspace of square-integrable functions on the sphere.

Given a mathematical model of the virus particle and data collection process, we can compute predicted data for every possible set of parameter values. Our approach to computing a reconstruction is to determine that set of parameter values which minimizes the difference between the measured data and the predicted data. We measure this difference by the least squares criteria. We regularize the problem indirectly by controlling the number of parameters in the model, in particular, the value of L in Eq. 1.7. Because the predicted data is a nonlinear function of the parameters that describe the virus, this is a nonlinear least squares problem. There are also constraints to be considered, e.g., the shell of capsid protein must lie outside of the core of nucleic acid. In addition to those constraints, we use *a priori* knowledge of the possible parameter values and solve the optimization problem in the corresponding hypercube. Our numerical experiments show that the weighted least squares error function[2] has many local minima with long and narrow valleys. Therefore a GO method is necessary to find the global minimum.

We have done similar calculations in the past [Zheng et al., 1995] using simple multistart ideas. What is attractive to a virologist about the GO approach is the possibility of automatic location of the solution and the impartial nature of the initial condition algorithms used by GO algorithms. These features will be increasingly important when the problems are scaled up to handle, for instance, multiple x-ray solution scattering curves where different curves were measured with different x-ray wavelength and/or originate from solutions of chemically modified virus particles where the modification is to introduce a strong point scatterer.

The cost function and its gradient can be expressed analytically (see Section 2) and requires the computation of multidimensional quadratures which is expensive. For example, evaluation of the cost with a 10×10 (5×5) point Gaussian quadrature takes approximately 2 (0.5) minutes on one processor of a Sun Microsystems HPC-3500. We also examine the effect of the quadrature rule order on the quality of the solution. This type of nonrandom error has not been much studied in GO.

The remainder of this chapter is organized in the following fashion. After a description of the virus model and the cost function (Section 2), we describe several state-of-the-art optimization algorithms that are tested (Section 3) and then the numerical results (Section 4). Finally, we close with some discussion in Section 5.

2.   Virus Model and the Cost

The electron density in most virus particles exhibits some sort of symmetry [Chiu et al., 1997]. This has the effect of reducing the amount of information that must be stored in the virus genome in order to describe the virus particle. It is also important in the reconstruction problems considered in this chapter because it reduces the number of unknown degrees of freedom. We focus on spherical viruses where the global symmetries (i.e., those symmetries that apply to the entire particle) are the 60 operations of the icosahedral group which are described in, for instance, Zheng and Doerschuk, 1995. Because these operations are all rotations and the rotation axes share a common point, we describe the electron density in spherical coordinates with the origin placed at the intersection of the rotation axes.

Let f be a square-integrable function from $\mathbb{R}^3$ to $\mathbb{R}$. Use spherical coordinates (r, θ, ϕ) in $\mathbb{R}^3$. Then f can be expressed as a linear superposition of spherical harmonics [Jackson, 1975, Eq. 3.53], denoted by $Y_{l,m}(\theta, \phi)$ for $l \in \{0, 1, \ldots\}$ and $m \in \{-l, \ldots, +l\}$, where the weights in the superposition, denoted by $A_{l,m}(r)$, are functions of radius and each harmonic is a function of the two spherical coordinate angles:

$$ f(r, \theta, \phi) = \sum_{l=0}^{\infty} \sum_{m=-l}^{+l} A_{l,m}(r) Y_{l,m}(\theta, \phi). $$

For practical computation, the l sum is terminated at some value L. The disadvantages of this approach are that there are a large number of spherical harmonics ($2l + 1$ for each value of l) and, for most choices of weights, the resulting linear superposition will have no particular symmetry. Therefore, if symmetry is required, as it is in our problem, then the symmetry must be described as constraints on the weights.

Both disadvantages of the approach described in the previous paragraph can be remedied by replacing the spherical harmonics with a basis for a subspace of the space spanned by the spherical harmonics, in particular, the subspace of suitably symmetric functions. We employ this approach and use the so-called icosahedral harmonics [Zheng and Doerschuk, 1996a, Zheng and Doerschuk, 2000] (denoted by $T_{l,n}(\theta, \phi)$ for $l \in \{0, 1, \ldots\}$ and $n \in \{0, 1, \ldots, N_l - 1\}$) as the basis functions for the subspace of icosahedrally-symmetric functions. (See the citations for the value of N_l, which is far smaller than the $2l + 1$ occurring with spherical harmonics). There are a variety of ways to define the icosahedral harmonics. As described in the references, we define $T_{l,n}$ to be a real-valued linear combination of $Y_{l,m}$ with $m \in \{-l, \ldots, l\}$:

$$T_{l,n}(\theta, \phi) = \sum_{m=-l}^{+l} b_{l,n,m} Y_{l,m}(\theta, \phi),$$

where recursions for the weights $b_{l,n,m}$ are given in the references. The number of icosahedral harmonics of a particular l (i.e., N_l) is 0 for

$$l \in \{1, 2, 3, 4, 5, 7, 8, 9, 11, 13, 14, 17, 19, 23, 29\},$$

since there are no linear combinations of $Y_{l,m}(\theta, \phi)$ that have icosahedral symmetry, and 1 for

$$l \in \{0, 6, 10, 12, 15, 16, 18, 20, 21, 22, 24, 25, 26, 27, 28\}.$$

For $l \geq 30$ the number of icosahedral harmonics of a particular l (i.e., N_l) is at least 1 (i.e., $N_l \geq 1$). The lowest order [i.e., $(l, n) = (0, 0)$] icosahedral harmonic is a constant: $T_{0,0}(\theta, \phi) = 1/\sqrt{4\pi}$. Plots of the next three lowest order icosahedral harmonics [i.e., $(l, n) = (6, 0)$, $(10, 0)$, or $(12, 0)$] are shown in Figure 1.2.

The most general icosahedrally-symmetric object, written in spherical coordinates, is

$$\rho(\mathbf{x}) = \sum_{l=0}^{\infty} \sum_{n=0}^{N_l-1} A_{l,n}(r) T_{l,n}(\theta, \phi), \tag{1.1}$$

where $A_{l,n}(r)$ are unspecified coefficient functions. We use the most simple standard physical model: the scattered field, denoted by $P(\mathbf{k})$, is the Fourier transform of $\rho(\mathbf{x})$ and the solution scattering, denoted by $I(k)$, is the spherical average of the magnitude-squared of $P(\mathbf{k})$:

$$P(\mathbf{k}) = \frac{1}{(2\pi)^{3/2}} \int \rho(\mathbf{x}) \exp(-i\mathbf{k}^T \mathbf{x}) d^3 \mathbf{x} \tag{1.2}$$

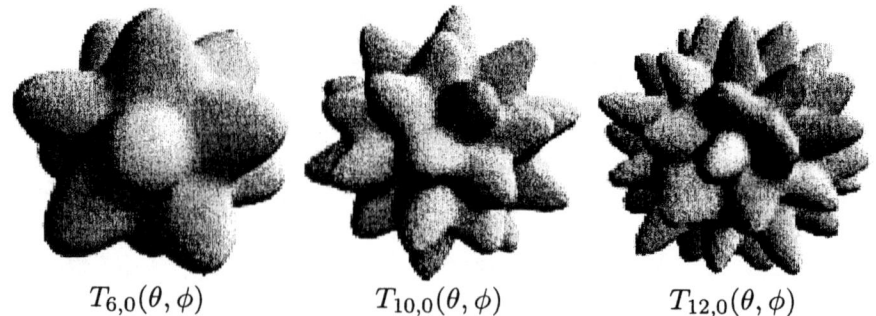

$$T_{6,0}(\theta, \phi) \qquad\qquad T_{10,0}(\theta, \phi) \qquad\qquad T_{12,0}(\theta, \phi)$$

Figure 1.2. Icosahedral harmonics. For each value of θ and ϕ the distance of the surface from the origin is $c_{l,n} + T_{l,n}(\theta, \phi)$ where $c_{l,n} = 2\max_{\theta,\phi}(|T_{l,n}(\theta,\phi)|)$. The icosahedral symmetry axes are located as shown in Figure 1.4(a).

$$I(k) = \frac{1}{4\pi} \int |P(\mathbf{k})|^2 d\Omega' \tag{1.3}$$

where $\int d\Omega'$ denotes integration over solid angles in Fourier space [$\mathbf{k} = (k, \theta', \phi')$] which, in spherical coordinates, is $\int_{\phi'=0}^{2\pi} \int_{\theta'=0}^{\pi} \sin\theta' d\theta' d\phi'$. Using Eq. 1.1 in Eqs. 1.2 and 1.3 we find that

$$I(k) = \frac{1}{4\pi} \sum_{l=0}^{\infty} \sum_{n=0}^{N_l-1} a_{l,n}^2(k) \tag{1.4}$$

where $a_{l,m}(k)$ is the spherical Hankel transform of $A_{l,m}(r)$, i.e.,

$$a_{l,m}(k) = \sqrt{\frac{2}{\pi}} \int_0^{\infty} A_{l,m}(r) j_l(kr) r^2 dr. \tag{1.5}$$

The importance of the 3-D Fourier transform is the reason we emphasize harmonic functions.

It would be desirable if $I(k)$ uniquely determined $\rho(\mathbf{x})$. However, this is not true. In particular, if ρ is an electron density with solution scattering intensity I then $\rho'(\mathbf{x}) = \rho(-\mathbf{x})$ is a second electron density which is not related to ρ by an operation from the icosahedral group and which also has solution scattering intensity I.

We desire the mathematical model of the viral electron density to incorporate the information that is known about the viral particle independent of the solution scattering data. The information we have focused on is the icosahedral symmetry of the particle; the finite size of the particle (typically $\rho(\mathbf{x}) = 0$ for $|\mathbf{x}| > R_+$ or, for empty particles, $\rho(\mathbf{x}) = 0$ for $|\mathbf{x}| < R_-$ and $|\mathbf{x}| > R_+$), which is often called a support constraint; the fact that the electron density is real; and the fact that the electron density is positive (i.e., $\rho(\mathbf{x}) \geq 0$ for all $\mathbf{x}$), which is often called a positivity constraint. For the model described in this chapter, all four properties can be exactly enforced in a practical way.

We have considered a variety of mathematical models for the electron density of spherical virus particles [Zheng and Doerschuk, 1998, Zheng and Doerschuk, 1996b, Zheng et al., 1995] but here we focus on generalizations of the piecewise constant "Envelope" model [Zheng et al., 1995], because of the strong nonlinearities present in that model. In the Envelope model there are I concentric shells with different electron densities which are constant throughout a shell and the mathematical model focuses on the boundaries (called "envelopes") between the shells. The interpretation of a simple two-shell model is that the inner shell (which extends into the origin) is the nucleic acid coding the genes of the virus and the outer shell is the protein capsid that protects the nucleic acid, recognizes a new host cell, and achieves appropriate entry of the nucleic acid into the new host cell thereby infecting the new host cell.

The shells are restricted to sets that are star-shaped at the origin [Gardner, 1995, Section 0.7, pp. 18–20] and so the boundaries can be described by the distance of the boundary from the origin at a particular (θ, ϕ). Therefore overhanging regions on the surface of the virus particle and voids within the virus particle are not allowed. These restrictions imply that this model is only appropriate for low spatial resolution studies, which is exactly the situation that occurs with solution x-ray scattering. These angularly-dependent distances are denoted by the functions $\gamma^{(i)}(\theta, \phi)$ ($i \in \{0, \ldots, I\}$), which must have icosahedral symmetry if the particle is to have icosahedral symmetry, and the electron densities are denoted by ρ_i ($i \in \{0, \ldots, I\}$). The electron density is

$$\rho(\mathbf{x}) = \begin{cases} \rho_i, & \gamma^{(i-1)}(\theta, \phi) \leq r < \gamma^{(i)}(\theta, \phi) \\ 0, & \gamma^{(I)}(\theta, \phi) \leq r \end{cases} . \tag{1.6}$$

While the $\gamma^{(i)}(\theta, \phi)$ functions could be arbitrary icosahedrally-symmetric functions, for numerical computation we describe them as finite sums of icosahedral harmonics:

$$\gamma^{(i)}(\theta, \phi) = \begin{cases} \sum_{l=0}^{L} \sum_{n=0}^{N_l-1} \gamma_{l,n}^{(i)} T_{l,n}(\theta, \phi) & i \in \{1, \ldots, I\} \\ 0 & i = 0 \end{cases} . \tag{1.7}$$

When using this model, computing a 3-D structure for the virus particle based on the data is the same as estimating the real-valued parameters ρ_i and $\gamma_{l,n}^{(i)}$ from the data.

Let ω be a vector containing all of the parameters. For any mathematical model and value of ω, Eqs. 1.1–1.5 imply a predicted scattering that we denote by $\hat{I}(k; \omega)$. Let $I(k)$ denote the measured scattering.

A general approach to determining the parameters is to minimize, perhaps with constraints, a cost function J, which is a function of ω and which is the sum of two terms denoted by J_1 and J_2. The first term measures the difference between $I(k)$ and $\hat{I}(k;\omega)$ for a proposed value of ω. The second term, often called a "regularizer", measures the desirability of a proposed value of ω. In a maximum *a posteriori* estimation interpretation, the first term would be the logarithm of the probability density function (pdf) of the data conditional on the parameters and the second term would be the logarithm of the *a priori* pdf of the parameters.

We take a weighted least squares approach so that the first term is

$$J_1(\omega) = \sum_k [I(k) - \hat{I}(k;\omega)]^2 w(k) \qquad (1.8)$$

where $w(k) = k^\alpha$ with α typically near 8. This choice of weight $w(k)$ with $\alpha = 8$ is the inverse of the asymptotic behavior of $[\hat{I}(k;\omega)]^2$ for large k and so leads to equal weighing across the entire range of scattering angles k. While least squares could be interpreted as a Gaussian measurement model, this is probably not realistic. For instance, the standard deviation in a Gaussian model should probably be proportional to the data, which (see, e.g., Fig. 1.4) varies much more rapidly than $1/\sqrt{w(k)}$.

We use an extensive set of constraints. In particular, we require

$$\gamma^{(i-1)}(\theta_q, \phi_q) < \gamma^{(i)}(\theta_q, \phi_q) \qquad (1.9)$$

for $i \in \{1, \ldots, I\}$ where (θ_q, ϕ_q) are all of the abscissas in the quadrature rule required to approximately evaluate the integration hidden in the calculation of $\hat{I}(k;\omega)$ (see Eqs. A.4–A.5). In addition, we require that $\rho_i \geq 0$ for $i \in \{1, \ldots, I\}$.

We note that there are many aspects of biological knowledge that are not incorporated into our model, cost, or constraints. Such knowledge could be incorporated into the regularizer J_2, but here we use $J_2 = 0$. One reason for this choice is the great difficulty in selecting and weighting such knowledge, especially given the highly nonlinear nature of the optimization problem. Another reason is that we seek a compromise between automatic and expert-aided analysis. Hence, instead of computing the global minimum of J and declaring that value of ω as "the answer", we compute the global minimum and a set of local minima with low cost function values, and present the set of ω values (or actually various images of $\rho(\mathbf{x})$) to virologist collaborators and ask them to make a selection.

Eqs. 1.1–1.8 completely define the cost function. However there are two pairs of parameterizations that merit discussion. The first pair concerns the method used to describe the piecewise constant regions. Define

$s^{(i)}(\mathbf{x}) = 1$ for $\gamma^{(i-1)}(\theta, \phi) \leq r < \gamma^{(i)}(\theta, \phi)$ and $= 0$ otherwise. Then, $\rho(\mathbf{x}) = \sum_{i=1}^{I} \rho_i s^{(i)}(\mathbf{x})$. Define $t^{(i)}(\mathbf{x}) = 1$ for $0 \leq r < \gamma^{(i)}(\theta, \phi)$ and $= 0$ otherwise. Assume that $\gamma^{(1)}(\theta, \phi) \leq \gamma^{(2)}(\theta, \phi) \leq \cdots \leq \gamma^{(I)}(\theta, \phi)$ for all (θ, ϕ), which is what we enforce, at least at the quadrature abscissas, by our constraint. Then,

$$\rho(\mathbf{x}) = \sum_{i=1}^{I} \left(\rho_i - \sum_{i'=i+1}^{I} \rho_{i'} \right) t^{(i)}(\mathbf{x}).$$

Therefore we could use either ρ_i or $\delta_i \doteq \rho_i - \sum_{i'=i+1}^{I} \rho_{i'}$ as parameters along with $\gamma_{l,n}^{(i)}$. Use of ρ_i requires a 1-D numerical quadrature from $k\gamma^{(i'-1)}(\theta, \phi)$ to $k\gamma^{(i')}(\theta, \phi)$ (see Eqs. A.4 A.5, and A.6) while use of δ_i requires a similar quadrature from 0 to $k\gamma^{(i')}(\theta, \phi)$. Directly computing the quadratures required by δ_i is an unfavorable option since the region of integration is larger than the region required for ρ_i. However, since the larger quadrature region required by δ_i can be assembled from the smaller regions required by ρ_i at the cost of some bookkeeping, the two approaches can be made equivalent and we have used ρ_i in order to minimize the bookkeeping while at the same time minimizing the size of the total quadrature region.

The second pair of parameterizations concerns the method used to describe the electron density. We focus on ρ_i rather than δ_i since they are essentially equivalent. Instead of using ρ_i we could select a particular shell, number κ, and define g and Ω_i by $\rho_i = g\Omega_i$ ($i \in \{1, \ldots, I\}$) and $\Omega_\kappa = 1$. Then the parameter vector would be $\omega = (g, \Omega_1, \ldots, \Omega_{\kappa-1}, \Omega_{\kappa+1}, \ldots, \Omega_I, \gamma_{l,n}^{(i)})$. Because g linearly scales $\rho(\mathbf{x})$ it follows that g linearly scales $P(\mathbf{k})$ and g^2 linearly scales $\hat{I}(k; \omega)$. Therefore, because we minimize a weighted least squares cost, we can symbolically determine the optimal value of g^2 and substitute that value back into the cost to get a reduced cost that is a function only of $\omega' = (\Omega_1, \ldots, \Omega_{\kappa-1}, \Omega_{\kappa+1}, \ldots, \Omega_I, \gamma_{l,n}^{(i)})$. This is a special case of separable nonlinear least squares [Golub and Pereyra, 1973, Kaufman, 1975]. Two attractive aspects of using ω_i as parameters are that the dimension of the parameter space is reduced by one and at every step of an iterative GO algorithm, the prediction $\hat{I}(k; \omega')$ is reasonably scaled to the data $I(k)$. An unattractive aspect is that the cost function becomes even more nonlinear (compare the more complicated Eqs. A.7 and A.8 with Eqs. A.1 and A.2). In particular, the cost function is no longer of a least-squares type which implies that the large body of specialized algorithms for least squares (e.g., [Björck, 1996]) cannot be applied. In our numerical experiments, we have considered both $\omega = (\rho_i, \gamma_{l,n}^{(i)})$

and $\omega' = (\Omega_1, \ldots, \Omega_{\kappa-1}, \Omega_{\kappa+1}, \ldots, \Omega_I, \gamma_{l,n}^{(i)})$. We refer to the cost as the "scaled cost" when using ω' and the "unscaled cost" when using ω.

The nonintersection constraint on the envelopes (Relation 1.9) does not imply that the parameter space is compact. Therefore the parameter space is further restricted to a hypercube. The dimensions of the hypercube depend on the specific virus and are described in Section 4.

3.   Optimization Algorithms

We employ four state-of-the-art GO algorithms which represent a range of different approaches to the problem, along with the classical multistart technique. This set of algorithms, however, does not include all techniques of current interest. For instance, we have not applied algorithms based on interval mathematics [Corliss and Kearfott, 1999]. The algorithms are

1 ASA 18.2 by L. Ingber, a simulated annealing algorithm, which uses adaptive temperature schedules (both for cost and each parameter) [Ingber, 1993][3];

2 DIFFEVOL 3.6, a heuristic genetic type of algorithm, which creates new points by adding the difference of two points to a third point [Storn and Price, 1995][4];

3 GENESIS 5.0, a genetic algorithm by John J. Grefenstette [Grefenstette, 1986][5];

4 GLOBAL, a derivative free implementation by T. Csendes of the clustering algorithm of Boender, Rinnooy Kan, Stougie, and Timmer [Boender et al., 1982, Csendes, 1985][6]; and

5 MULTISTART [Zhigljavsky, 1991].

For the first four algorithms we used the suggested or default values for all parameters and generation of the initial point or population was performed by the algorithm using its standard approach.

For MULTISTART we used Levenberg-Marquardt [Marquardt, 1963] local search applied to pseudo-random initial points uniformly distributed over the feasible set determined by the nonintersection constraint (Relation 1.9) and the hypercube constraint (Relation 1.10 or 1.11). We considered 50 initial points. The algorithm for selection of these points is based on rejection. Let f be the cost and $x_n \in \mathbb{R}^p$ be the nth parameter vector. A local search was stopped when any of the following occur.

1 The change in the function value is small: $|f(x_n) - f(x_{n-1})| < 10^{-5} f(x_n)$.

2 The l_1 norm of the gradient is small: $\sum_{i=1}^{p} |[\nabla f(x_n)]_i| < 10^{-5} f(x_n)$.

3 x_n exits the feasible set which can occur since the Levenberg-Marquardt algorithm solves the unconstrained optimization problem. By terminating the search when x_n exits the feasible set we are doing a random projection back inside of the feasible set.

4 The number of iterations exceeds a limit n_0 which we take as $n_0 = 100$ and which, for our cost functions, is only infrequently reached.

The impartial nature of the initial conditions is important to the biological interpretation of the results and so we enumerate the methods that are used in each of the algorithms:

1 ASA: the initial condition is a random sample that comes from a uniform distribution over the feasible set.

2 DIFFEVOL and GLOBAL: the initial population is a set of independent random samples that come from a uniform distribution over the hypercube (Relation 1.10 or 1.11). Therefore, some of the initial population will violate the nonintersection constraint (Relation 1.9) but, as described in Section 4, we have chosen the hypercube in a way that less than 10% of the points will violate the nonintersection constraint.

3 GENESIS: the initial population is a set of independent random samples which are vectors of 32 independent random bits, each taking values 0 and 1 with equal probability. These bit strings code uniform scalar quantizations of each component of the parameter over the allowed range of the parameter which is an interval since only the hypercube constraint (Relation 1.10 or 1.11) is considered. This implies, ignoring the quantization, that the samples are uniformly distributed over the hypercube. Therefore, as with DIFFEVOL and GLOBAL, some of the initial population will violate the nonintersection constraint (Relation 1.9).

4 MULTISTART: the situation was described above.

The sequence of points or populations of points may or may not remain in the feasible set for different algorithms. The feasible set is the intersection of the nonintersection constraint (Relation 1.9) and the hypercube (Relation 1.10 or 1.11) and we consider these two sets separately. With regard to the hypercube, ASA, GENESIS, and MULTISTART guarantee that all points or populations of points remain within the hypercube but GLOBAL and DIFFEVOL allow points outside of the hypercube (see, e.g., the negative ρ_i values in Table 1.1). With regard to the nonintersection constraint, ASA provides a Boolean function for implementing a constraint, i.e., if the proposed point is feasible then the function returns TRUE and if infeasible then the function returns FALSE. This was used to ensure that all points satisfy the non-intersection constraint. Similarly, in MULTISTART all points satisfy the nonintersection constraint. In the other programs, which lacked a feature analogous to the Boolean function in ASA, the nonintersection constraint was implemented by returning a cost that was 1 unit greater than the maximum cost value found so far. Therefore, while some members of the population might violate the nonintersection constraint, it is highly likely that at least the minimum cost member will satisfy the nonintersection constraint.

Only GLOBAL has a statistical stopping condition, which stops when global sampling does not produce new clusters or local searches from un-clustered points do not generate a new local minimum. For the other algorithms, it was necessary to fix a reasonable maximum number of function evaluations, and for consistency all algorithms were treated in the same way. Also, MULTISTART runs its local searches to convergence and provides a precise location of the local minimum and evaluation of its cost. For the other algorithms, we therefore follow GO with a local search using the Levenberg-Marquardt algorithm starting from the best member of the population.

4.   Numerical Experiments

All of our numerical experiments concern models with two shells (i.e., $I = 2$) where the inner shell represents the nucleic acid core and the outer shell represents the protein capsid. The data is an experimental solution x-ray scattering curve from empty capsids of cowpea mosaic virus (CpMV) [Chen et al., 1990], whose electron density has been solved at near atomic resolution based on x-ray crystal diffraction studies and fit with an atomic model [Chen et al., 1990]. The range of the data is from $k = 0.004279\text{Å}^{-1}$ to $k = 0.020853\text{Å}^{-1}$. The number of samples over this range is 156 and the data is subsampled by a factor of 3 to speed the computation. Because the data is low resolution, only a low resolution reconstruction can be computed. In particular, the expansion of the envelopes (i.e., $\gamma^{(i)}(\theta, \phi)$) uses only the icosahedral harmonics $T_{0,0}$, $T_{6,0}$, $T_{10,0}$, and $T_{12,0}$ which are the four lowest order harmonics. (Note that there are no harmonics of order $l \in \{1\text{-}5, 7\text{-}9, 11, 13, 14, 17, 19, 23, 29\}$ and there is at most one harmonic, that is, $N_l = 1$, for all orders

$l < 30$). Therefore there are eight $\gamma_{l,n}^{(i)}$ parameters (since there are two envelopes) plus two ρ_i parameters (since there are two shells) for a total of ten parameters for the first parameterization and nine parameters for the second parameterization. In all calculations, the least-squares weighting function $w(k)$ is $w(k) = k^6$.

The nonintersection constraint on the envelopes (Relation 1.9) does not imply that the parameter space is compact. Therefore the parameter space is further restricted to a hypercube. For the first parameterization the hypercube is

$$300 \leq \gamma_{0,0}^{(1)} \leq 450, \quad 400 \leq \gamma_{0,0}^{(2)} \leq 600$$
$$-100 \leq \gamma_{l,0}^{(1)} \leq 100, \quad -100 \leq \gamma_{l,0}^{(2)} \leq 100 \quad \text{for } l \in \{6, 10, 12\} \cdot \qquad (1.10)$$
$$0 \leq \rho_1 \leq 10, \quad 0 \leq \rho_2 \leq 10$$

For the second parameterization we have always used $\kappa = 2$ and a hypercube that is the same as Relation 1.10 in the $\gamma_{l,n}^{(i)}$ coordinates but with the ρ_i coordinates replaced by

$$0 \leq \Omega_1 \leq 100. \qquad (1.11)$$

The dimensions of the hypercube have the following motivation. Since $\int Y_{l,m}(\theta, \phi)d\Omega = \delta_{l,0}\delta_{m,0}\sqrt{4\pi}$ it follows that $\int T_{l,n}(\theta, \phi)d\Omega = \delta_{l,0}\delta_{n,0}\sqrt{4\pi}$ and therefore that $\int \gamma^{(i)}(\theta, \phi)d\Omega = \gamma_{0,0}^{(i)}\sqrt{4\pi}$. Therefore, $\gamma_{0,0}^{(i)}$ is $1/\sqrt{4\pi} \approx$ .28 times the average radius of the ith envelope and these radii can be estimated from other data, e.g., electron microscopy images. Because $\gamma_{l,n}^{(i)}$ control the average radii of the envelopes, they have a strong interaction with the nonintersection constraint and a poor choice of the hypercube dimensions in these variables, e.g., increased overlap in the allowed range of $\gamma_{0,0}^{(1)}$ and $\gamma_{0,0}^{(2)}$, leads to rejection of most uniformly sampled points in the hypercube by the nonintersection constraint. Efficiency of all of the genetic algorithms depended on whether the initial population contained enough feasible points. This was ensured by careful choice of the hypercube such that at least 90% of the initial points in the hypercube (Relation 1.10 or 1.11) satisfied the nonintersection constraint (Relation 1.9).

Each panel of Fig. 1.3 shows a plane through the scaled cost function demonstrating the nonconvexity of the cost function (the unscaled cost function is also nonconvex). The unscaled cost function has a very large dynamic range (around $10^{-2} - 10^{14}$) and is very sensitive to the values ρ_1 and ρ_2. The scaled cost has a much smaller dynamic range (about $10^{-2} - 10^2$). It can be shown that the scaled cost function has at least as many local minima as the unscaled cost function, but also has one less

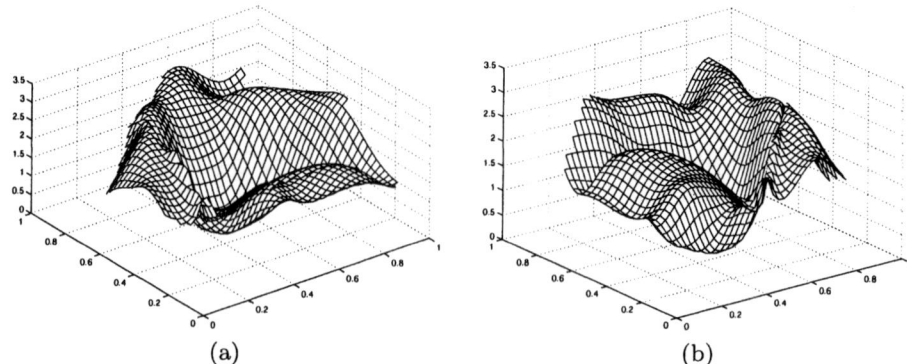

(a)   (b)

Figure 1.3. The scaled cost function, computed using a 10×10 quadrature rule, evaluated on two planes in parameter space demonstrating the nonconvexity of the cost function. For Panel (a) the plane is defined by the local minima found by local search from a starting point provided by GLOBAL (scaled cost, 10×10 quadrature rule, automatic termination around 16000 function evaluations), ASA (scaled cost, 5×5 quadrature rule, 5000 function evaluations), and GLOBAL (scaled cost, 10×10 quadrature rule, 5000 function evaluations) while for Panel (b) the same first two points were used but the third was replace by ASA (scaled cost, 10×10 quadrature rule, 5000 function evaluations).

parameter. To aid in interpreting our results, Fig. 1.4 shows data and models computed from atomic resolution 3-D structure based on x-ray crystal diffraction data. All 3-D figures which appear in our results are shown with the virus positioned so that the icosahedral symmetry axes have the locations shown in Fig. 1.4(a). In Fig. 1.4 (c)-(d) we show envelope models computed from the atomic resolution 3-D structure. The envelopes have the same number of $\gamma_{l,n}^{(i)}$ coefficients as are used in all of the other calculations in this chapter and therefore have resolution far less than atomic resolution. The underlying atomic resolution structure is regarded as the "gold standard" and therefore these envelopes are as close as we can come to knowing the truth for this reconstruction problem. Fig. 1.4(b) shows the predicted x-ray solution scattering computed from the envelopes shown in Fig. 1.4(c)-(d) and the experimental curve. Note that the two are somewhat different. The difference is attributed to the preprocessing of the experimental curve, unmodeled physics in the scattering process and the x-ray detector, and failures of the envelope model to precisely represent the complexity of the virus structure (as described in Section 2). Note also that the x-ray scattering curve and the envelopes associated with the (unknown) global minimum of the cost function need not coincide with the curves and envelopes in Fig. 1.4(b)-(d).

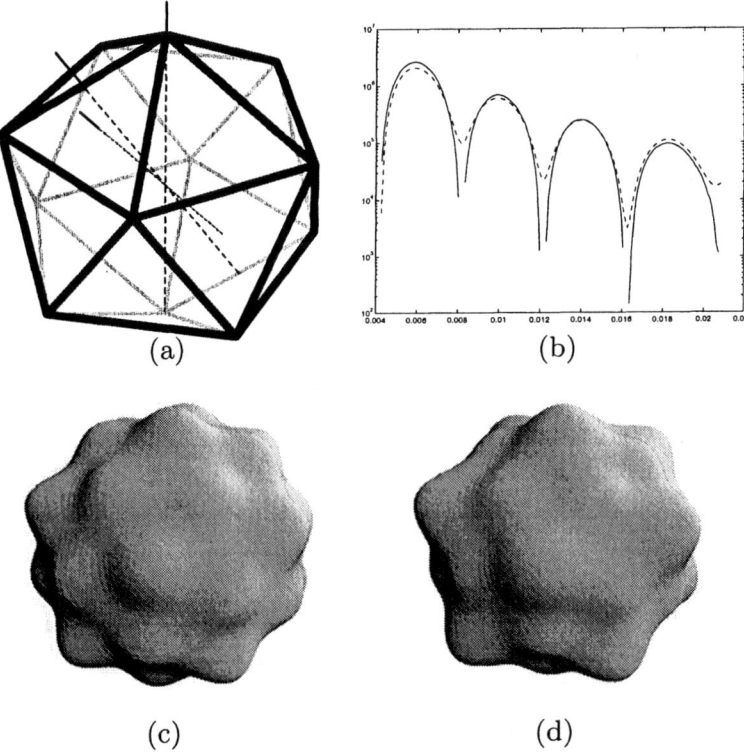

(a)   (b)

(c)   (d)

Figure 1.4. Results based on the atomic resolution model of CpMV. Panel (a): An icosahedron in the standard orientation used in all images in this chapter. One symmetry axis of each type—2-, 3-, and 5-fold—is shown. Panel (b): The experimental and computed solution scattering curves. Panels (c) and (d): Inner [(c)] and outer [(d)] envelopes. The computed solution scattering curves and both envelopes are based on envelopes computed from the atomic resolution structure. Each envelope is individually scaled to fit the available space which makes it appear that the inner envelope is too large to fit inside of the outer envelope.

The five GO optimization algorithms were applied to the unscaled and scaled cost functions, with high (10×10) and low (5×5) order quadrature rules. The low order quadrature results in less expensive cost function evaluation than the high order quadrature (about 0.5 minute compared with 2 minutes on a Sun HPC-3500), but introduces more approximation error. The algorithms were compared after 5000 function evaluations followed by a local search performed using the Levenberg-Marquardt algorithm from the best member of the terminal population. When performing the local search we always used a 10×10 quadrature rule (even if local search in the GO algorithm used a 5×5 quadrature rule). None of the algorithms' performance improved significantly with more than 5000 function evaluations, and the subsequent local search almost never required more than 200 function and gradient evaluations to converge to a local minimum (only $10 - 20$ function evaluations were usually needed for the latter). MULTISTART with 50 initial search points was roughly equivalent in computation to 5000 function evaluations. Table 1.1 shows the local minimum locations and cost values (after local search). From the table it is seen that for the unscaled cost ASA performed the best with a function value of 0.02076, while for the scaled cost GLOBAL was best with a function value of 0.01051. Generally, the performance of the algorithms was better on the scaled cost, and with the higher order quadrature rule (in either case the algorithms found different local minima).

Additional information about the GO algorithms can be obtained from a learning curve. The learning curve $l(n)$ is defined to be the smallest cost found at all iterations less than or equal to n, and over all population members (in some algorithms, such as ASA, there is only one population member). For MULTISTART this is not a natural definition because the local searches from different initial search points have no natural order. Therefore in this case we concatenate the local searches for a random permutation of initial search points, and then average over the permutations. In our numerical results there are 50 initial search points, so we were unable to compute the average over all permutations and instead we replaced it by an average over 10,000 randomly chosen permutations. Fig. 1.5 shows some learning curves where time is measured in units of function evaluations rather than iterations so that the different algorithms can be fairly compared. The algorithms were run for a total of 5000 function evaluations, with local searches performed using the Levenberg-Marquardt algorithm from the best member of the population at 500 and multiples of 1000 function evaluations (the terminal point of these learning curves is the data in Table 1.1). When performing the local searches we always used a 10×10 quadrature rule

Table 1.1. Local minimum locations and cost values. The alorithms are "XY" where the values of "X" and their meanings are A=ASA, D=DIFFEVOL, Gl=GLOBAL, Ge=GENESIS, and M=MULTISTART and the values of "Y" are the order of the quadrature rules (10×10 or 5×5).

Unscaled cost	Alg.	$\gamma^{(1)}_{0,0}$ $\gamma^{(2)}_{0,0}$	$\gamma^{(1)}_{6,0}$ $\gamma^{(2)}_{6,0}$	$\gamma^{(1)}_{10,0}$ $\gamma^{(2)}_{10,0}$	$\gamma^{(1)}_{12,0}$ $\gamma^{(2)}_{12,0}$	ρ_1 ρ_2	Cost
	A10	372.5 484.1	15.80 26.13	-11.70 -3.245	10.61 -4.791	0.000232 0.004937	0.02286
	A5	372.7 486.5	17.82 20.54	-16.39 -3.741	6.611 -3.386	0.000205 0.004883	0.02076
	D10	117.5 480.3	-14.59 -26.95	30.13 70.78	-22.50 -6.976	0.02735 0.003110	2.873
	D5	185.9 539.8	-10.22 -71.50	17.14 -48.36	8.531 35.99	-0.02068 0.003991	33.49
	Ge10	384.0 575.8	-6.621 21.41	-17.61 -40.93	6.701 -39.24	0.006731 0.004905	2.133
	Ge5	310.8 549.9	-7.672 6.789	-2.747 7.178	-15.29 35.82	0.01030 0.000239	5.765
	Gl10	354.9 535.3	-40.60 83.98	-39.09 30.19	-36.03 79.13	0.000671 -0.003497	0.7052
	Gl5	444.5 458.8	65.58 68.67	2.958 8.521	63.58 61.91	-0.002269 -0.02666	2.850
	M10	360.2 558.2	-29.32 76.10	-25.22 24.59	-30.24 69.07	0.0002 0.0036	0.8157
	M5	378.5 514.3	-36.54 3.617	-15.76 -20.69	-31.57 4.189	0.002060 0.004990	2.304

Scaled cost	Alg.	$\gamma^{(1)}_{0,0}$ $\gamma^{(2)}_{0,0}$	$\gamma^{(1)}_{6,0}$ $\gamma^{(2)}_{6,0}$	$\gamma^{(1)}_{10,0}$ $\gamma^{(2)}_{10,0}$	$\gamma^{(1)}_{12,0}$ $\gamma^{(2)}_{12,0}$	Ω_1	Cost
	A10	374.2 487.1	-14.90 -24.91	7.534 5.916	10.96 -2.637	9.993	0.01087
	A5	377.5 485.9	12.12 19.82	-16.54 -8.463	3.165 -0.4666	9.948	0.01994
	D10	374.8 486.5	-6.894 -16.19	16.68 10.74	-15.45 -7.794	12.94	0.02739
	D5	380.2 486.0	-15.08 -20.29	-3.125 11.34	-10.15 1.427	11.94	0.04351
	Ge10	379.8 487.4	-14.11 12.07	0.2384 7.372	-5.559 13.12	8.764	0.08300
	Ge5	379.8 487.6	-14.19 12.07	0.2804 7.429	-5.696 13.05	8.754	0.08373
	Gl10	376.7 488.0	-2.938 -13.54	11.48 15.88	3.965 0.9524	12.72	0.01051
	Gl5	374.2 487.8	-13.54 -22.28	10.90 9.711	-0.3923 -4.936	13.78	0.01485
	M10	322.2 589.4	-60.50 -48.41	39.72 73.63	-32.94 -53.85	56.55	0.9184
	M5	310.7 508.1	13.59 84.95	-90.24 51.57	16.72 87.38	76.01	3.699

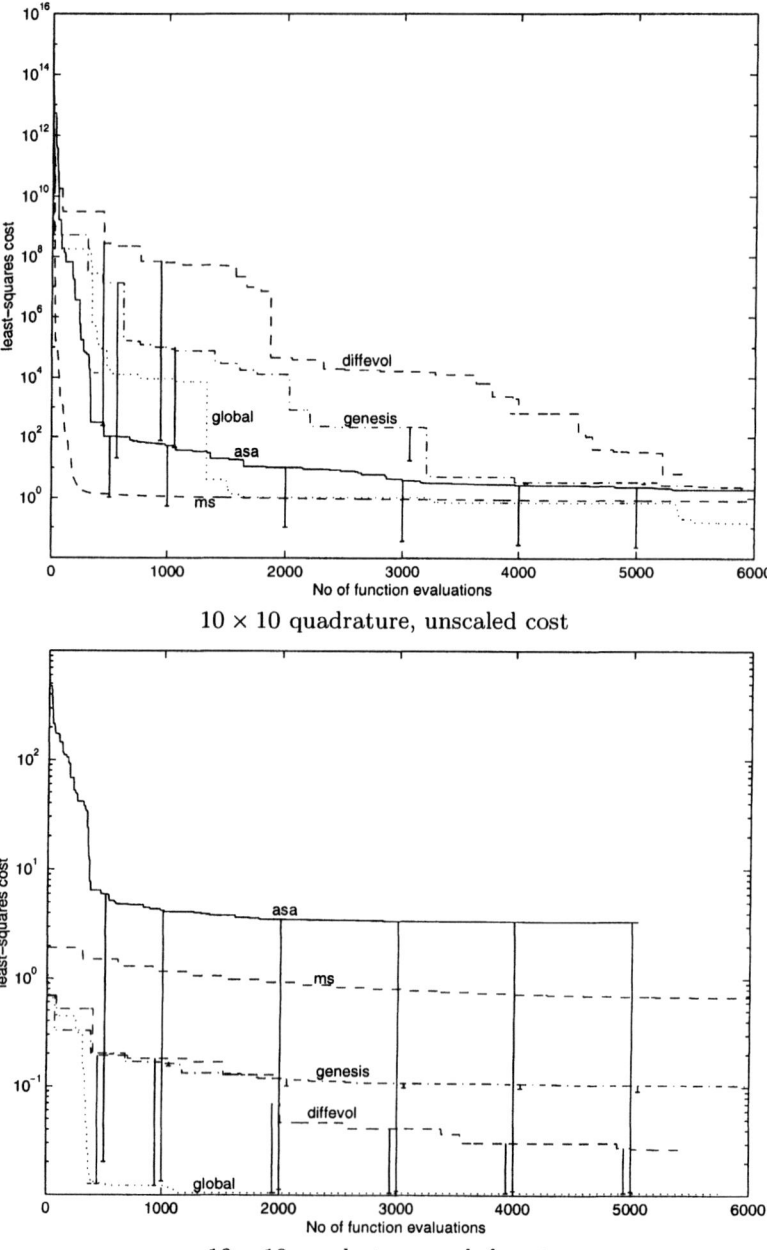

10×10 quadrature, unscaled cost

10×10 quadrature, scaled cost

Figure 1.5. Learning curves, i.e., cost as a function of function evaluation number, for each of the five algorithms. Curves are given only for the 10×10 quadrature rule since the performance with the 5×5 rule was poor. Separate curves are given for the two parameterizations. As described in the text, the vertical lines show the decrease in cost achieved with a local search at that iteration of the algorithm.

(even if local search in the GO algorithm used a 5 × 5 quadrature rule). The cost at the local minimum is often substantially reduced from the cost at the starting vector, and is shown in the learning curve plots as a vertical line segment where the bottom (top) is the cost at the local minimum (starting vector). Since all five learning curves are plotted together we offset these line segments in function evaluation count ("time") so that they do not overlap: for ASA, DIFFEVOL, GENESIS, and GLOBAL the line segment trails by 0, -60, +60, and -120 function evaluations (in the case of MULTISTART there is no line segment since additional local search is not performed as it runs its own local search to convergence). The apparent absence of a line segment is due to the negligible difference between the cost before and after the local search was performed. From the figures it is seen that local searches improved ASA performance significantly more than GLOBAL, which already performs some local search. Also, note that although MULTISTART has amongst the steepest learning curves, all the algorithms outperformed MULTISTART when local searches were applied.

As discussed in Section 1, our cost criteria do not nearly capture all of a virologist's knowledge concerning the structure of viruses. In particular, recall that we have no regularizing term but only a term measuring the fit of the predicted data to the experimental data in the least squares sense. Therefore some local minima, even minima with low cost values, will be biologically unacceptable. In particular, there are two common problems. One is that the curvature of the virus surface is to great. The second is that quasi-symmetry [Chiu et al., 1997] is violated. Unlike the icosahedral symmetry, which applies exactly and globally (i.e., to the entire particle), quasi symmetries apply approximately and locally. Based on information such as the number of protein subunits that are contained in a single virus capsid, it is expected that CpMV would have so-called $T = 3$ quasi symmetry which means that each 3 fold axis of the icosahedral symmetry is approximately a 6 fold axis. Clearly both of these biological ideas could be added to the cost as a regularizer (though the weight to apply to regularization versus fitting the experimental data is not clear). However, instead of further complexifying the cost function, we prefer to find these local minima and reject them manually in concert with an expert virologist. In addition, for some local minima, the predicted and experimental scattering curves are very different which implies that such a local minimum is biological unacceptable. Unacceptability in this sense is generally reflected in a high least squares cost. Figs. 1.6–1.8 display some results of Table 1.1 in a form (scattering data and envelopes) which an expert virologist can assess. The biological acceptability of each of the local minima found

Table 1.2. Biological acceptability of local minima. "+", "±", and "-" indicate acceptable, marginal, and unacceptable characteristics, respectively. The algorithms are "XY" where the values of "X" and their meanings are A=ASA, D=DIFFEVOL, Gl=GLOBAL, Ge=GENESIS, and M=MULTISTART and the values of "Y" are the order of the quadrature rules (10 × 10 or 5 × 5). The abbreviated column headings are *C*, *Q*, and *D* for *Curvature*, *Quasisymmetry*, and *Data fit*.

Alg.	Unscaled Cost				Scaled Cost			
	C	Q	D	Cost	C	Q	D	Cost
A10	+	-	+	0.02286	+	-	+	0.01087
A5	+	-	+	0.02076	+	-	+	0.01994
D10	-	-	-	2.873	+	-	+	0.02739
D5	-	-	-	33.49	+	-	+	0.04351
Ge10	+	-	+	2.133	+	-	+	0.08300
Ge5	±	-	-	5.765	+	-	+	0.08373
Gl10	-	-	-	0.7052	+	+	+	0.01051
Gl5	-	-	-	2.850	+	-	+	0.01485
M10	-	-	-	0.8157	-	-	-	0.9184
M5	-	-	-	2.304	-	-	-	3.699

at 5000 function evaluations is listed in Table 1.2. It is seen that the only completely satisfactory virus structure results from GLOBAL with scaled cost and high quadrature rule.

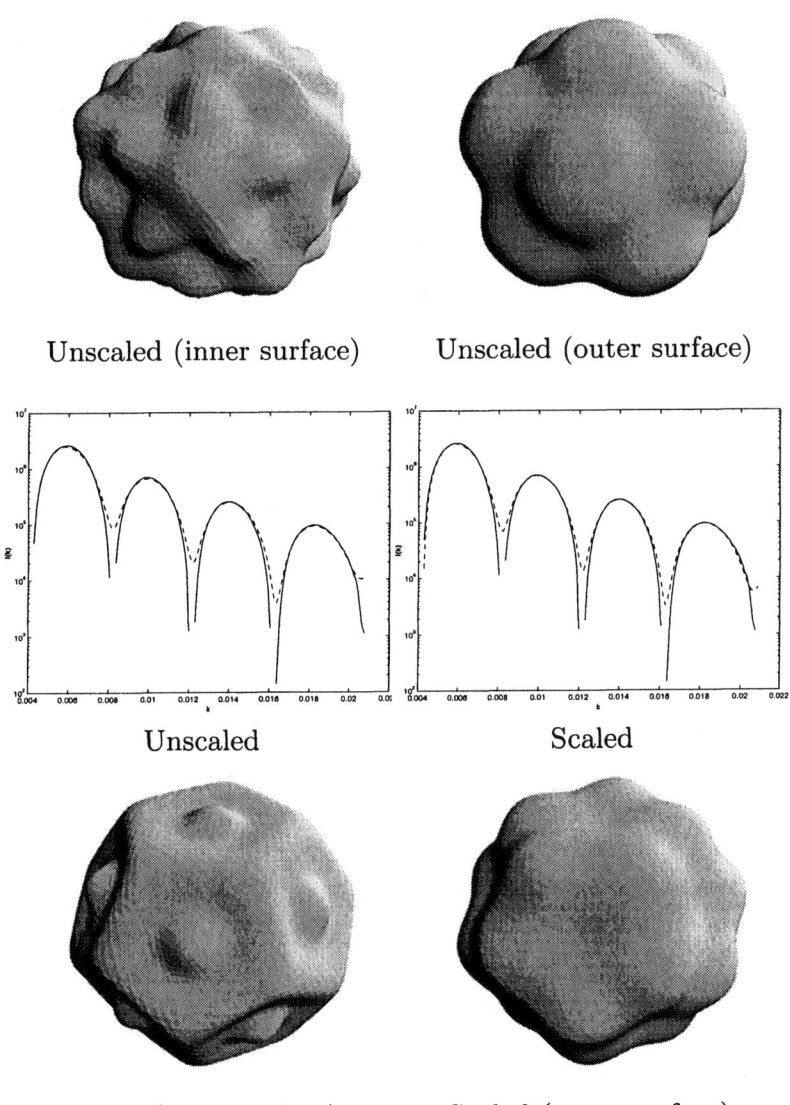

Figure 1.6. Results for ASA for the scaled and unscaled cost functions using 10×10 quadrature. Each envelope is individually scaled to fit the available space which makes it appear that the inner envelope is too large to fit inside of the outer envelope.

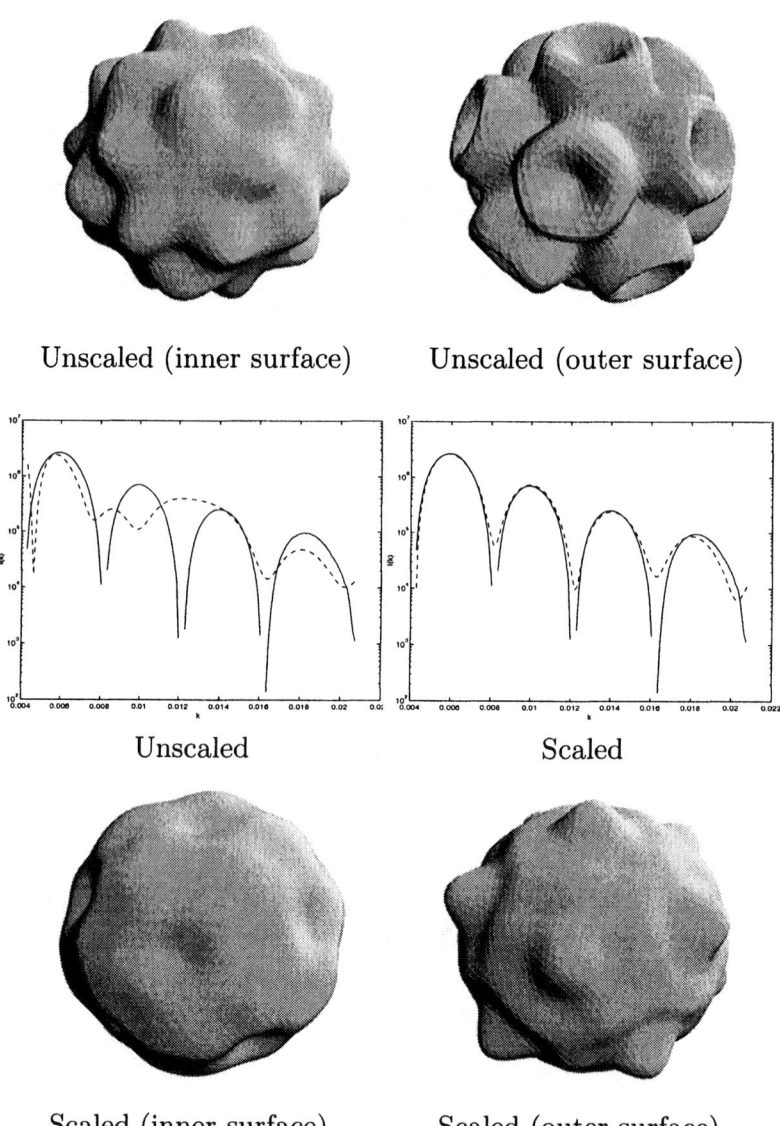

Figure 1.7. Results for GENESIS for the scaled and unscaled cost functions using 10×10 quadrature. Each envelope is individually scaled to fit the available space which makes it appear that the inner envelope is too large to fit inside of the outer envelope.

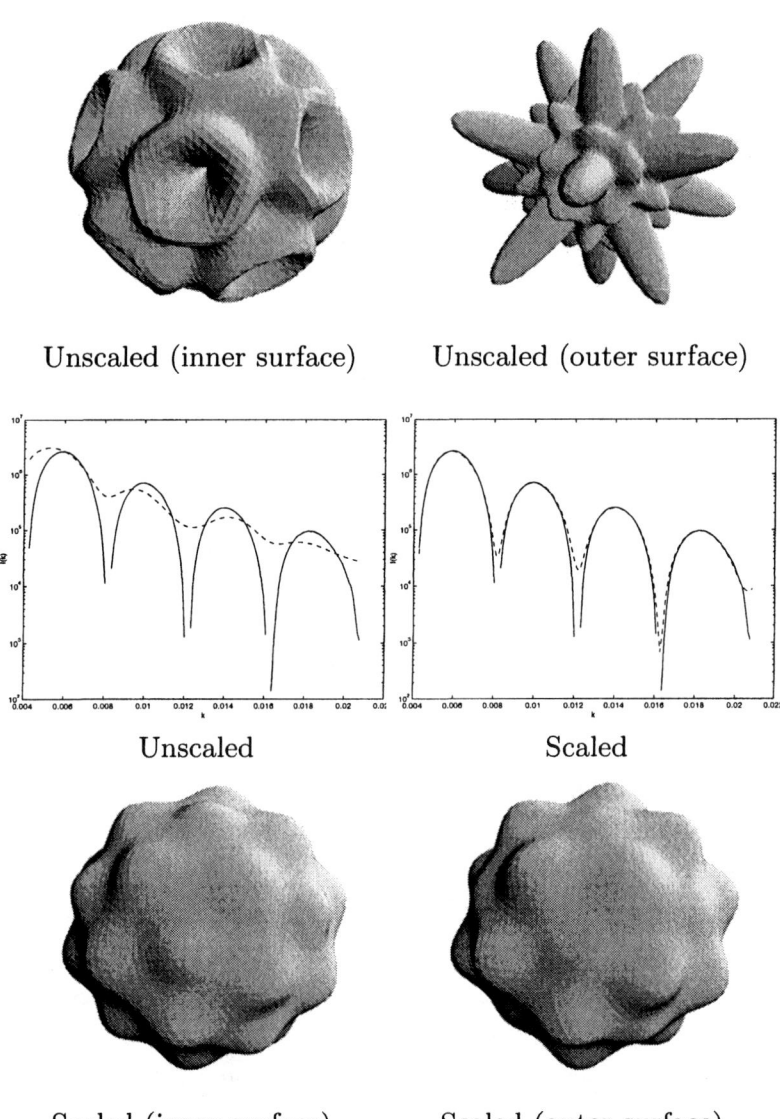

Unscaled (inner surface)   Unscaled (outer surface)

Unscaled   Scaled

Scaled (inner surface)   Scaled (outer surface)

Figure 1.8. Results for GLOBAL for the scaled and unscaled cost functions using 10 × 10 quadrature. Each envelope is individually scaled to fit the available space which makes it appear that the inner envelope is too large to fit inside of the outer envelope.

5.   Discussion and Conclusions

In this chapter we studied the application of GO to 3-D reconstruction of spherical viruses using experimental solution x-ray scattering data. Starting from previous work on modeling and estimation of the electron density for spherical viruses [Zheng et al., 1995], we formulated the 3-D reconstruction problem as a nonlinear weighted least squares problem based on a two shell envelope model. Two versions of the cost function were considered, the original unscaled cost, and a reduced scaled cost based on a linear scaling of the electron density (with this scaling, the original cost function is partly separable and the optimal scaling can be computed analytically which results in one less parameter and smaller dynamic range, but the reduced cost function is no longer of the least squares type). There are also constraints based on the structure and size of the virus. The evaluation of the cost function and its gradient require expensive multidimensional quadratures, whose expense can be controlled. The cost function is observed to have many local minima in long deep valleys. For these reasons, it was felt that advanced GO techniques were of potential use for this problem.

We applied several state-of-the-art optimization algorithms and also the classical multistart approach for comparison. Our results showed that certain algorithms were far more effective for finding low-lying local minima. In particular, amongst the algorithms we considered, GLOBAL and ASA worked best, followed by genetic algorithms and multistart, and the performance gap between GLOBAL and ASA and the others was significant. Generally, it was found that the algorithms required the scaled cost and high order quadrature rules to obtain the best solutions. Although the algorithms found many local minima, and there was clearly a positive correlation between the depth of a local minimum and an expert virologist's assessment of biological correctness of the 3-D virus reconstruction, only one algorithm (GLOBAL) was able to find a completely satisfactory reconstruction according to the expert. This suggests two approaches: (i) regularize the cost function by adding additional information which hopefully would remove spurious local minima and produce more realistic virus structures, or (ii) generate many low-lying local minima and have then evaluated by an expert. In view of the difficulty of identifying and incorporating the additional information, and also the additional complexity in the cost function (and in particular the possibility of introducing as well as removing spurious local minima), we favor the second approach.

Our results suggest that 3-D virus reconstruction is a rich area for both modeling and GO. The quality of reconstructions presented here have only been obtained previously using a simple multistart approach when extensive knowledge of a good initial condition is available. Using the methodology described here, we can get these results from an arbitrary initial condition, followed by screening of solutions by an expert. There are also many interesting areas for further work. One such area involves developing cost functions for other virus models and data, with different cost criteria and/or constraints. For instance, the virus may have a different (typically lower order) global symmetry which would imply a different model or multiple x-ray solution scattering curves may be available (as described in Section 1) which would imply a different cost. Another area for future work involves development and experimentation with other GO algorithms which are particularly suited to the 3-D virus reconstruction problem. Here new algorithms should be developed which can adaptively control the quadrature error, allowing longer runs on more complex problems. There is very little work on GO in the presence of cost function errors, and what work exists is primarily theoretical analysis with random errors (c.f. [Gelfand and Mitter, 1991, Yakowitz, 1993]). Also new algorithms should be investigated which seek to identify many sufficiently distinct (according to some metric) low-lying local minima which can then be presented to an expert for further evaluation.

Acknowledgments

We thank Professor John E. Johnson (The Scripps Research Institute) for the data and for many helpful discussions. This work was supported by the National Science Foundation under Grants DBI-9630427 and MCB-9873139 and by the Army Research Office under Grant DAAG55-98-1-0360. The computations were carried out using facilities supported by the Army Research Office under Contract DAAD19-99-1-0015.

Appendix: The cost and its gradient

Let $K \subset \mathbb{R}$ be the finite set of scattering angles at which measurements are made. The weighted least squares cost and its gradient are

$$J(\omega) \quad = \quad \sum_{k \in K} [I(k) - \hat{I}(k; \omega)]^2 w(k) \tag{A.1}$$

$$\nabla J(\omega) \quad = \quad 2 \sum_{k \in K} [I(k) - \hat{I}(k; \omega)] w(k) \nabla \hat{I}(k; \omega). \tag{A.2}$$

The first parameterization is $\omega = (\rho_i, \gamma_{l,n}^{(i)})$. After some symbolic computation, the cost and the gradient of the cost can be computed by using

$$\hat{I}(k; \omega) \quad = \quad \sum_{l=0}^{\infty} \sum_{n=0}^{N_l - 1} a_{l,n}^2 (k; \omega) \tag{A.3}$$

$$\frac{\partial \hat{I}}{\partial \gamma_{l',n'}^{(i')}}(k; \omega) \quad = \quad 2 \sum_{l=0}^{\infty} \sum_{n=0}^{N_l - 1} a_{l,n}(k; \omega) \frac{\partial a_{l,n}}{\partial \gamma_{l',n'}^{(i')}}(k; \omega)$$

$$\frac{\partial \hat{I}}{\partial \rho_{i'}}(k; \omega) \quad = \quad 2 \sum_{l=0}^{\infty} \sum_{n=0}^{N_l - 1} a_{l,n}(k; \omega) \frac{\partial a_{l,n}}{\partial \rho_{i'}}(k; \omega)$$

$$a_{l,n}(k; \omega) \quad = \quad \frac{1}{k^3} \sum_{i=1}^{I} \rho_i \int T_{l,n}(\theta, \phi) \mu_l(k\gamma^{(i-1)}(\theta, \phi), k\gamma^{(i)}(\theta, \phi)) \mathrm{d}\Omega \tag{A.4}$$

$$\frac{\partial a_{l,n}}{\partial \gamma_{l',n'}^{(i')}}(k; \omega) \quad = \quad [\rho_{i'} - \rho_{i'+1}(1 - \delta_{i',I})] \times$$

$$\times \int T_{l,n}(\theta, \phi) \left[\gamma^{(i')}(\theta, \phi) \right]^2 j_l(k\gamma^{(i')}(\theta, \phi)) T_{l',n'}(\theta, \phi) \mathrm{d}\Omega$$

$$\frac{\partial a_{l,n}}{\partial \rho_{i'}}(k; \omega) \quad = \quad \frac{1}{k^3} \int T_{l,n}(\theta, \phi) \mu_l(k\gamma^{(i'-1)}(\theta, \phi), k\gamma^{(i')}(\theta, \phi)) \mathrm{d}\Omega \tag{A.5}$$

$$\mu_l(\eta_1, \eta_2) \quad \doteq \quad \int_{\eta_1}^{\eta_2} \eta^2 j_l(\eta) \mathrm{d}\eta \tag{A.6}$$

where j_l are spherical Bessel functions of the first kind [Jackson, 1975, Eq. 16.9] for which sophisticated numerical evaluation methods exist [Press et al., 1992, Section 6.7, pp. 240–252].

The second parameterization is

$$\omega' = (\Omega_1, \ldots, \Omega_{\kappa-1}, \Omega_{\kappa+1}, \ldots, \Omega_I, \gamma_{l,n}^{(i)}).$$

We can write the prediction using the second parameterization, denoted by $\hat{I}_2(k; \omega')$, in terms of the prediction using the first parameterization, denoted by $\hat{I}_1(k; \omega)$, for which formulas are given in Eqs. A.3–A.6. In particular, define

$$\hat{I}_2(k; \omega') = I_1(k; (\Omega_1, \ldots, \Omega_{\kappa-1}, \Omega_\kappa = 1, \Omega_{\kappa+1}, \ldots, \Omega_I, \gamma_{l,n}^{(i)}))$$

and then $\hat{I}_2(k;\omega') = g_0^2 \hat{\mathcal{I}}_2(k;\omega')$ where g_0 is the optimal gain which is

$$g_0^2 = \left[\sum_{k \in K} I(k)\hat{\mathcal{I}}_2(k;\omega')w(k) \right] / \left[\sum_{k \in K} [\hat{\mathcal{I}}_2(k;\omega')]^2 w(k) \right].$$

After some symbolic computation, the cost evaluated at the optimal g is

$$J'(\omega') = \frac{1}{2} \left\{ \sum_{k \in K} I^2(k)w(k) - \frac{\left[\sum_{k \in K} I(k)\hat{\mathcal{I}}_2(k;\omega')w(k) \right]^2}{\sum_{k \in K} [\hat{\mathcal{I}}_2(k;\omega')]^2 w(k)} \right\} \qquad (A.7)$$

and the derivative of the cost with respect to the pth component of ω' is

$$\frac{\partial J'}{\partial \omega_p'}(\omega') = g_0^4 \sum_{k \in K} \hat{\mathcal{I}}_2(k;\omega') \frac{\partial \hat{\mathcal{I}}_2}{\partial \omega_p'}(k;\omega')w(k)$$

$$- g_0^2 \sum_{k \in K} I(k) \frac{\partial \hat{\mathcal{I}}_2}{\partial \omega_p'}(k;\omega')w(k). \qquad (A.8)$$

Use of Eqs. A.3–A.6 in Eqs. A.7 and A.8 allows the computation of the cost and its gradient at the optimal g. Our current code for evaluating the cost and its gradient is in the C programming language.

Notes

1. "Cryo" refers to the fact that the specimen is frozen in ice when the image is made.
2. In the remainder of this chapter, we refer to the weighted least squares error function as the "cost".
3. http://www.ingber.com/ASA-CODE
4. http://http.icsi.berkeley.edu/~storn/code.html
5. http://www.aic.nrl.navy.mil/galist/src/genesis.tar.Z
6. ftp://ftp.jate.u-szeged.hu/pub/math/optimization/c/global.tar.gz

References

Björck, Åke (1996). *Numerical Methods for Least Squares Problems.* SIAM.

Boender, G., Rinnooy Kan, A., Stougie, L., and Timmer, G. (1982). A stochastic method for global optimization. *Mathematical Programming*, 22:125–140.

Chen, Zhongguo, Stauffacher, Cynthia V., and Johnson, John E. (1990). Capsid structure and RNA packaging in comoviruses. *Seminars in Virology*, 1:453–466.

Chiu, Wah, Burnett, Roger M., and Garcea, Robert L., editors (1997). *Structural Biology of Viruses.* Oxford Univ Press.

Corliss, George F. and Kearfott, R. Baker (1999). Rigorous global search: Industrial applications. In Csendes, T., editor, *Developments in Reliable Computing*, pages 1–16. Kluwer Academic Publishers.

Csendes, T. (1985). Two non-derivative implementations of Boender et al's global optimization method: numerical performance. Technical report, József Attila University, Szeged, Hungary.

Gardner, Richard J. (1995). *Geometric Tomography*. Cambridge University Press.

Gelfand, S. B. and Mitter, S. K. (1991). Recursive stochastic algorithms for global optimization in R^d. *SIAM J. Control Opt.*, vol. 29:999–1018.

Golub, G. H. and Pereyra, V. (1973). The differentiation of pseudo-inverses and nonlinear least squares problems whose variables separate. *SIAM J. Numer. Anal.*, 10(2):413–432.

Grefenstette, J. (1986). Genesis: a system for using genetic search procedures. *Proc. Conf. Intelligent Systems and Machines*, pages 161–165.

Ingber, Lester (1993). Simulated annealing: Practice versus theory. *J. Math. Comput. Modelling*, 18(11):29–57.

Jackson, John David (1975). *Classical Electrodynamics*. John Wiley, New York, 2nd edition.

Kaufman, Linda (1975). A variable projection method for solving separable nonlinear least squares problems. *BIT*, 15:49–57.

Marquardt, D.W. (1963). An algorithm for least-squares estimation of nonlinear parameters. *J. Soc. Ind. Appl. Math*, 11:431–441.

Press, William H., Flannery, Brian P., Teukolsky, Saul A., and Vetterling, William T. (1992). *Numerical Recipes in C: The Art of Scientific Computing*. Cambridge Univ. Press, Cambridge, 2nd edition.

Storn, David and Price, Kennetth (1995). Differential evolution—a simple and efficient adaptive scheme for global optimization over continuous spaces. Technical Report TR-95-012, International Computer Science Institute.

Yakowitz, S. (1993). A globally convergent stochastic approximation. *SIAM J. Control Opt.*, 31:30–40.

Zheng, Yibin and Doerschuk, Peter C. (1995). Symbolic symmetry verification for harmonic functions invariant under polyhedral symmetries. *Comput. in Phys.*, 9(4):433–437.

Zheng, Yibin and Doerschuk, Peter C. (1996a). Explicit orthonormal fixed bases for spaces of functions that are totally symmetric under the rotational symmetries of a Platonic solid. *Acta Cryst.*, A52:221–235.

Zheng, Yibin and Doerschuk, Peter C. (1996b). Iterative reconstruction of three-dimensional objects from averaged Fourier-transform magnitude: solution and fiber x-ray scattering problems. *J. Opt. Soc. Am. A*, 13(7):1483–1494.

Zheng, Yibin and Doerschuk, Peter C. (1998). 3D image reconstruction from averaged Fourier transform magnitude by parameter estimation. *IEEE Trans. Image Proc.*, 7(11):1561–1570.

Zheng, Yibin and Doerschuk, Peter C. (2000). Explicit computation of orthonormal symmetrized harmonics with application to the identity representation of the icosahedral group. *SIAM Journal on Mathematical Analysis*, 32(3):538–554.

Zheng, Yibin, Doerschuk, Peter C., and Johnson, John E. (1995). Determination of three-dimensional low-resolution viral structure from solution x-ray scattering data. *Biophys. J.*, 69(2):619–639.

Zhigljavsky, Anatoly A. (1991). *Theory of Global Random Search*. Kluwer Academic Publishers. [Janos Pinter, ed.].

Chapter 14

A COLLABORATIVE SOLUTION METHODOLOGY FOR INVERSE POSITION PROBLEM

Chandra Sekhar Pedamallu & Linet Ozdamar*
*Nanyang Technological University, School of Mechanical and Production Engineering, Systems and Engineering Management Division, 50 Nanyang Avenue, Singapore 639798 *:Corresponding author: e-mail: lozdamar@hotmail.com, First author e-mail: pcs_murali@hotmail.com*

Abstract: In this study, a difficult advanced kinematics problem is presented and solved with a collaborative methodology that integrates an effective symbolic inference scheme and a local search method into a global interval partitioning algorithm. The resulting methodology proves to be very effective in discarding infeasible sub-spaces, because with some exceptions, the symbolic inference scheme guarantees to reduce infeasibility in each re-partitioning iteration. Thus, the local search method is called much less frequently as compared to a subdivision scheme without symbolic inference. Empirical results are obtained on two applications, the 6R inverse position and modified kinematics problems. The first test problem is quite difficult to solve as compared to the second one, however, our available commercial solvers were not able to solve any of them. The proposed collaborative methodology is generic and can handle any Constraint Satisfaction Problem where the goal might be to cover all solutions or identify a first feasible solution.

Key words: Inverse Kinematics; Interval Analysis; Symbolic-Interval cooperation; Sequential Quadratic Programming.

1.   INTRODUCTION

A manipulator can be defined as a group of rigid bodies, or links, connected together by joints that are either revolute or prismatic. The relative motion associated with each joint can be controlled such that the free-end (the hand) is positioned in a desired manner. The revolute joint allows only rotational motion along the joint axis whereas prismatic joint allows only linear motion.

The kinematic analysis of manipulators can be represented either as a direct position problem or inverse position problem. In direct position problem all the relative joint displacements are given and the positions of every link including the free end are to be found. This type of problem can be solved by the matrix method of analysis (Hartenberg and Denavit, 1964). On the other hand, in inverse position problem the hand position and orientation are given and joint displacements are to be found. The inverse position problem is more difficult to solve because the governing equations are highly nonlinear. However, the solution to the inverse problem is easier to use by designers.

Inverse position problems are solved by identifying the closed-form solution to algebraic equations relating a given position and orientation of the hand to an unknown joint displacement. All possible solutions can be found using this approach (Tsai and Morgan, 1985).

In Pieper and Roth (1969), it is mentioned that the analysis of an open-loop manipulator is related to the displacement analysis of a closed –loop spatial mechanism. The authors identify the sufficient conditions for a closed-form solution of a manipulator where three adjacent joint axes intersect at a common point. Pieper (1968) also develop closed form solutions to six-revolute-joint (6R) manipulators, which result in a total degree of 64,000. Since the analysis of a six-degree-of-freedom 6-revolute-joint (6R) manipulator is equivalent to that of the single–loop, 7- revolute-joint (7R) spatial mechanism, all the methods used for spatial mechanisms, can be applied to manipulators. Examples of such approaches are screw algebra (Kohi and Somi, 1975), dual numbers (Yang and Freudenstein, 1964), vector methods (Chase,1963).

In an effort to solve the 6R problem, Roth et al. (1973) show that it has at most 32 solutions. Duffy and Crane (1980) derive the corresponding 32 degree polynomial. A lower degree polynomial for 6R problem with prismatic or cylindrical joints on consecutive parallel axes is developed by Duffy (1980). Tsai and Morgan (1985) convert the problem into a system of **eight second-degree equations** (with a total degree 256), which they solved numerically using polynomial continuation method. The 6R problem is reformulated of into different degree equations in Morgan and Sommese (1987a, 1987b). The problem is then reduced into a 16th degree polynomial (Lee and Liang, 1988). Computations with both formulations support that there are always 16 (sometimes complex) finite solutions.

The solution methodologies developed for solving these equations are in general numerical (Uicker et al.,1964; Yuan and Freudenstei, 1971; Roth et al., 1973; Albala and Angeles, 1979). The success of these methods depend on the initial solutions and if the desired position of the hand is far away from a known position of the manipulator or when there is no knowledge of the state of the manipulator, they fail to solve the inverse position problem.

A well-known numerical technique is the Continuation method (Drexler, 1978; Chow et al., 1979; Garcia and Zangwill, 1979; Watson, 1979; Allgower and Georg, 1980; Morgan, 1983a, 1983b) designed to solve n equations with n unknowns. The method is based on the concept of continuation paths which consist of a sequence of solutions obtained by solving a system of equations $K(x)$ that gradually approach the original set of equations $F(x)$ of the 6R problem. $K(x)$ is a weighted combination of a simplified solvable set of equations $I(x)$ and the original system $F(x)$. With each update of the continuation weight the similarity between $K(x)$ and $F(x)$ increases. By starting at different solutions of $I(x)$, the method hopes to find all solutions to $F(x)$. The resulting continuation paths are computed numerically by defining a differential equation. In the literature, there are reliable techniques of polynomial path tracking (e.g., Morgan and Sommese, 1989).

Some problems with the Continuation methods are related to particular values of the continuation weight. At some values, there might not be a solution to $K(x)$ or the solution may diverge to infinity. Further, $I(x)$ might not have enough solutions to start with in order to trace all solutions of $F(x)$. Though remedies are sought by adding random perturbations to parameters, the stability of these methods is parameter-dependent.

More recent efforts to solve the 6R kinematics problem are made in Wampler and Morgan (1991) who propose coefficient-parameter polynomial continuation, in Recio and Gonzalex-Lopez (1994) who use symbolic simplification and in Manocha and Canny (1994) who describe matrix operations and reduce the problem into an eigenvalue problem rather than a root finding one. Sommese et al. (2002) review the progress in continuation methods. Leykin et al. (2004) provide the most recent software contribution, *PHC*maple, for solving general polynomial equations.

Here, we propose a reliable and non-parametric solution methodology that integrates symbolic computing elements with an interval partitioning algorithm. The approach is exhaustive in the sense that it guarantees to identify all solutions. Symbolic computing elements involve a symbolic interval inference procedure that guides the search in selecting the most influential variables for subdivision in the next partitioning iteration. The convergence rate of the interval partitioning algorithm improves significantly because this approach chops off infeasible sub-spaces very fast. The proposed symbolic scheme applies parallel subdivision of all selected variables. A local search strategy (Sequential Quadratic Programming

(Panier and Tits, 1993; Zhou and Tits,1996)) is also incorporated and activated whenever discarded infeasible area does not increase significantly in a number of consecutive iterations. Hence, in this collaborative methodology, symbolic inferencing, interval partitioning and local search work in cooperation to determine all solutions to 6R inverse position system or maximize the number of solutions identified within a given computation time. The latter is important in real time robotic applications.

In Section 2, we introduce the model for six-revolute-joint (*6R*) manipulator. In Section 3, we describe the generic collaborative approach. The final section provides the experimental results for the manipulator model.

2.   MODEL FORMULATION

A 6R manipulator has six moving links, numbered sequentially from 2 to 7, as shown in Fig. 1-1(Tsai and Morgan, 1985). Link 1 is designated as the base (fixed to ground) and link7 as the hand or the manipulator. Every two neighboring links are connected by a joint that is associated with a joint axis Z_i, i=1to 6. Let Z_i, and Z_{i+1} be two adjacent joint axes and H_iO_{i+1} be the directed common normal between Z_i, and Z_{i+1}. H_i is the intersection of H_iO_{i+1} and Z_i, and O_{i+1} is the intersection of H_iO_{i+1} and Z_{i+1}. Then one can define the following link parameters shown in Fig. 1-2 (Hartenberg and Denavit, 1964).

a_i = the offset distance from the common normal H_iO_{i+1}.

α_i = the angle to rotate the axis Z_i about the common normal H_iO_{i+1} so that Z_i is parallel to Z_{i+1}. The sign of rotation is given by the right hand screw rule with the screw taken along normal H_iO_{i+1}.

d_i = the distance between the two normals $H_{i-1}O_i$ and H_iO_{i+1} measured from Z_i. The sign of d_i is positive if O_iH_i points to the positive Z_i direction. Otherwise, d_i is negative.

θ_i = the angle to rotate extended line of $H_{i-1}O_i$ about Z_i so that the extended line $H_{i-1}O_i$ is parallel to H_iO_{i+1}. The sign of rotation is given by the right hand screw with the screw pointing along the positive Z_i –axis.

If the ith joint is revolute, then a_i, d_i, and α_i are constant while θ_i is variable. If the ith joint is prismatic, then a_i, α_i, and θ_i are constant while d_i is a variable.

A coordinate system (X_i, Y_i, Z_i) is attached to each link of the manipulator as shown in Fig. 1-2. In each coordinate system, the Z_i – axis is defined to align with the ith joint axis, the X_i-axis is the one along the extended line of $H_{i-1}O_i$; and the Y_i- axis is defined according to the right-hand screw rule. The first coordinate system is fixed to ground. Since the common normal H_0O_1 does not exist, the X_1-axis is chosen perpendicular to Z_1, in an arbitrary manner. Also, a seventh coordinate system is attached to the free-end to specify the position of the hand. Z_7-axis lies in the direction from which the hand would approach an object as shown in Fig. 1-1. X_7-axis is defined by the common normal between Z6 and Z_7 axes, and Y_7-axis is defined according to the right-hand screw rule.

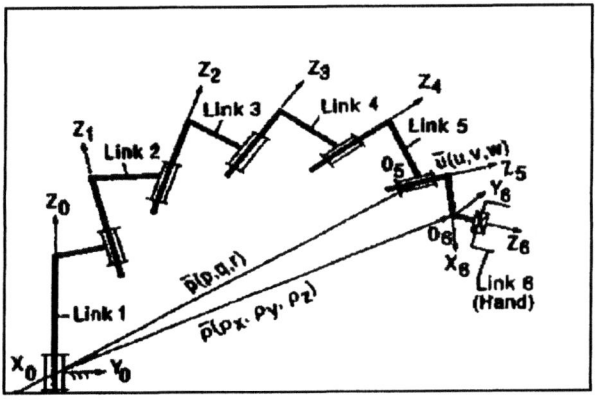

Figure 14-1. A general 6-R Manipulator

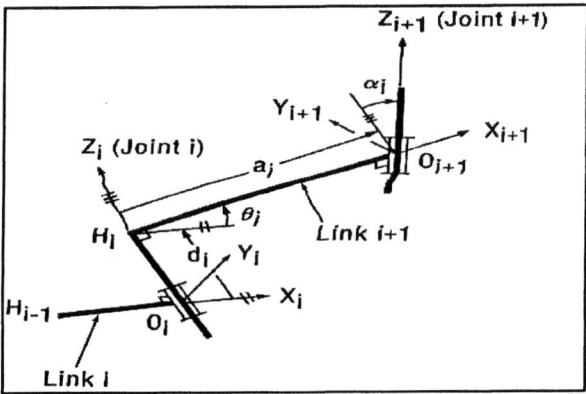

Figure 14-2. The basic notation

Let the coordinates of a point P in the ith and $(i+1)^{st}$ coordinate systems be expressed as (p_{xi}, p_{yi}, p_{zi}) and $(p_{xi+1}, p_{yi+1}, p_{zi+1})$, respectively. Then, vectors p_i and p_{i+1} can be written in the (4x1) matrix form as follows:

$$\mathbf{p}_i = \begin{bmatrix} p_{xi} \\ p_{yi} \\ p_{zi} \\ 1 \end{bmatrix} \quad \text{and} \quad \mathbf{p}_{i+1} = \begin{bmatrix} p_{xi+1} \\ p_{yi+1} \\ p_{zi+1} \\ 1 \end{bmatrix} \tag{1}$$

These vectors, are related to the hand position and orientation vectors by the following two equations:

The transformation of coordinates from the $(i+1)^{st}$ system to ith system is achieved by Eq. (2).

$$\mathbf{p}_i = A_i\,\mathbf{p}_{i+1} \tag{2}$$

where A_i is a (4x4) matrix defined as follows:

$$A_i = \begin{bmatrix} c_i & -s_i\lambda_i & s_i\mu_i & a_ic_i \\ s_i & c_i\lambda_i & -c_i\mu_i & a_is_i \\ 0 & \mu_i & \lambda_i & d_i \\ 0 & 0 & 0 & 1 \end{bmatrix} \tag{3}$$

where $c_i=\cos\theta_i$, $s_i=\sin\theta_i$, $\lambda_I=\cos\alpha_i$ and $\mu_i=\sin\alpha_i$.
The inverse also transformation exists:

$$\mathbf{p}_{i+1} = A_i^{-1}\mathbf{p}_i \tag{4}$$

$$A_i^{-1} = \begin{bmatrix} c_i & s_i & 0 & -a_i \\ -s_i\lambda_i & c_i\lambda_i & \mu_i & d_i\mu_i \\ s_i\mu_i & -c_i\mu_i & \lambda_i & d_i\lambda_i \\ 0 & 0 & 0 & 1 \end{bmatrix} \tag{5}$$

Similarly, let the components of a unit vector u in the ith coordinate and in the $(i+1)^{st}$ coordinate systems be expressed as $(u_{xi,}\,u_{yi,}\,u_{zi})$ and $(u_{xi+1,}\,u_{yi+1,}\,u_{zi+1})$. Then the vectors $u_{i,}$ and u_{i+1} can be written as follows.

$$\mathbf{u}_i = \begin{bmatrix} u_{xi} \\ u_{yi} \\ u_{zi} \\ 0 \end{bmatrix} \text{ and } \mathbf{u}_{i+1} = \begin{bmatrix} u_{xi+1} \\ u_{yi+1} \\ u_{zi+1} \\ 0 \end{bmatrix} \tag{6}$$

Coordinate transformation is obtained as follows.

$$\mathbf{u}_i = A_i \mathbf{u}_{i+1} \tag{7}$$

$$\mathbf{u}_{i+1} = A_i^{-1} \mathbf{u}_i \tag{8}$$

By applying the matrix transformation to each pair of coordinate systems between two successive links and proceeding from link 7 to link 1, Eq. (9) is obtained.

$$\mathbf{p}_1 = A_1\, A_2\, A_3\, A_4\, A_5 A_6\ \mathbf{p}_7 \tag{9}$$

Pieper and Roth (1969) define the following equivalent transformation matrix

$$A_{eq} = A_1\, A_2\, A_3\, A_4\, A_5 A_6 \tag{10}$$

Therefore, Eq. (9) can be re-written as

$$\mathbf{p}_1 = A_{eq}\,\mathbf{p}_7 \tag{11}$$

Similarly, the transformation of the unit vector can be written as

$$\mathbf{u}_1 = A_{eq}\,\mathbf{u}_7 \tag{12}$$

Since the equivalent transformation matrix defines the relationship between the coordinates of any point in the seventh system $\mathbf{p}_7$, and that of the same point expressed in the first system, $\mathbf{p}_1$, the matrix A_{eq} is known when the position and orientation of the hand is specified. Let $\boldsymbol{\rho}$ $(\rho_x,\ \rho_y,\ \rho_z)$ be the position vector from the origin of the first system to the origin of the seventh system as shown in Fig. 1-1; and $\mathbf{l}$ $(l_x,\ l_y,\ l_z)$, $\mathbf{m}$ $(m_x,\ m_y, m_z)$ and $\mathbf{n}$ (n_x, n_y, n_z) be three mutually perpendicular unit vectors aligned with X_7, Y_7, and

Z_7 axes, respectively. Then, when ρ, **l**, **m** and **n** are given in the first system, the equivalent matrix is given by:

$$A_{eq} = \begin{bmatrix} l_x & m_x & n_x & \rho_x \\ l_y & m_y & n_y & \rho_y \\ l_z & m_z & n_z & \rho_z \\ 0 & 0 & 0 & 1 \end{bmatrix} \tag{13}$$

By applying coordinate transformation and variable elimination, one can arrive at a system of eight nonlinear equations with eight unknowns expressed in the system of equations given in Eq. (14) (Tsai and Morgan, 1985).

The following model describes the dense constraint system ($1 \le i \le 4$) for a *6R* problem.

$$x^2_i + x^2_{i+1} - 1 = 0$$

$$a_{1i} x_1 x_3 + a_{2i} x_1 x_4 + a_{3i} x_2 x_3 + a_{4i} x_2 x_4 + a_{5i} x_5 x_7$$

$$+ a_{6i} x_5 x_8 + a_{7i} x_6 x_7 + a_{8i} x_6 x_8 + a_{9i} x_1 + a_{10i} x_2$$

$$+ a_{11i} x_3 + a_{12i} x_4 + a_{13i} x_5 + a_{14i} x_6 + a_{15i} x_7$$

$$+ a_{16i} x_8 + a_{17i} x_8 = 0$$

$$-1 \le x_i \le 1$$

$$\tag{14}$$

Where x_1, x_2, x_3, x_4, x_5, x_6, x_7 and x_8 represent c_1, s_1, c_2, s_2, c_4, s_4, c_5 and, s_5.

In Eq. (14) coefficients a_{ki} (These are listed in Appendix A for a particular instance of the problem used for testing) are defined as the manipulator parameters.

3.   COLLABORATIVE SOLUTION METHODOLOGY

The 6R inverse position problem has not received much attention from the interval community. An exception is the work described in Van-Hentenryck et al. (1997a, 1997b) who use the interval partitioning method, Numerica, that is based on box consistency and interval Newton method. In the following sections, we provide a general background for interval partitioning algorithms and describe the symbolic inferencing method for subdivision direction selection.

3.1   Background on Interval Partitioning

Interval Partitioning methods (*IP*) produce reliable results for Constrained Optimization (COP) and Constraint Satisfaction Problems (CSP) (overviews on interval methods can be found in (Ratschek and Rokne, 1988; Hansen, 1992; Neumaier, 1990; Ratschek and Rokne, 1995)). CSP consists of a system of nonlinear equalities and inequalities that have to be satisfied simultaneously and it is expressed as:

Find $x^* \in X \subseteq \mathbb{R}^n$ such that:

$$g_i(x^*) \leq 0 \quad i=1,.....k \tag{15}$$

$$h_j(x^*) = 0 \quad j=k+1,....r \tag{16}$$

where X is the search box within which the union of feasible sub-boxes X^* exist with arguments x^* satisfying (1.1) and (1.2). In vector notation, constraints can be expressed as $g(x) \leq 0$ and $h(x) = 0$, where functions g_i and h_j are components of vector-valued functions $g: X \rightarrow \mathbb{R}^k$ and $h: X \rightarrow \mathbb{R}^{r-k}$. The search box is assumed to be a closed interval with bounds: $X=[\underline{X}, \overline{X}]$ where $\underline{X}_l = \min x_l$ and $\overline{X}_l = \max x_l$, for $l=1,2...n$. A subset of X, or sub-box, is denoted as $Y=[\underline{Y}, \overline{Y}] \subseteq X$.

6R inverse position problem given in the system of equations given in Eq. (14) is a CSP where the goal is to maximize the number of feasible solutions found. In this problem general gradient based solvers (e.g, CONOPT (Drud, 1995)) usually fail to identify any feasible solution, even if they do, they identify only one. In general, it is quite difficult to find all solutions to a CSP (this can be guaranteed only by complete solvers) and often impossible to know whether all solutions are found (verification phase for all complete solvers).

IP is a non-parametric exhaustive method that is ideal for identifying all solutions reliably. The basic idea of *IP* is to subdivide a given domain into smaller sub-spaces (boxes) and assess them according to the ranges of constraints calculated through an approximating inclusion function. In practice, even if all solutions cannot be identified within a reasonable computation time, approaches that improve the convergence of *IP* are welcome to obtain as many as possible within a short time span. Here, we improve *IP*'s convergence rate substantially with a symbolic variable selection rule that re-partitions the variable domains that are most responsible for infeasibilities. In each re-partitioning iteration, we guarantee a reduction in the infeasibility degree of each constraint, except for some instances. This definitely improves convergence because infeasible regions are identified in the early stages of the search and chopped off. Furthermore, the symbolic re-partitioning scheme involves parallel partitioning of selected variables. After a box is re-partitioned, the local search code CFSQP developed by Lawrence et al. (1997) is applied within the boundaries of each pending box if the total infeasible space chopped off does not increase significantly for a number of consecutive iterations. In this manner, feasible solutions are collected by CFSQP during the partitioning process, and this continues until either the user is satisfied with the number of feasible solutions obtained or there are no more pending boxes to be partitioned (all solutions are identified). To describe this collaborative procedure, we need to define inclusion functions and their properties.

Definition 1: Let $g(Y) = \{g(x): x \in Y\}$ be the range of g over $Y \in \mathbb{I}(X)$ where $\mathbb{I}$ is the set of *n*-dimensional compact intervals in X. A function $G: \mathbb{I}$ $(X) \rightarrow I$, is an <u>inclusion function</u> for g, if $g(Y) \subseteq G(Y)$ for any $Y \in \mathbb{I}(X)$. An inclusion function $H(Y)$ is defined similarly for $h(Y)$.

It is assumed that for all functions in the real domain, the natural interval extensions of g and h over Y are always defined. An important property for inclusion functions is to be inclusion isotone, which enables to reduce constraint range in sibling boxes as compared to their parent.

In *IP*, box assessment can end up with one of the following three results.
1. If $G_i(Y) \leq 0$, $\forall i$, and $H_j(Y) = 0$, $\forall j$, then box Y is a <u>feasible</u> box and it is stored.
2. If $G_i(Y) > 0$ for any i, or, $0 \notin H_j(Y)$ for any j, then box Y is called a <u>infeasible</u> box and it is pruned.

3. If $0 \in G_i(Y)$ for any i and $0 \in H_j(Y) \neq 0$ for any j, (implying that indeterminate constraints exist), then Y is called an <u>indeterminate</u> box and it holds the potential of containing both X^* (partially or completely) and $X \setminus X^*$.

Checking these three conditions to determine the status of a box is called box consistency check.

Definition 3: The degree of infeasibility, PG_Y (PH_Y) of an indeterminate inequality (equality) is defined by Eqs. (17)-(18), respectively.

$$PG_Y = \overline{G}(Y) \tag{17}$$

$$PH_Y = \overline{H}(Y) + |\underline{H}(Y)| \tag{18}$$

IP sub-divides indeterminate boxes so as to reduce PG_Y and PH_Y and drive them to zero in the limit by nested partitioning. The latter takes place due to inclusion monotonocity of H and G and also due to the contraction property that states that $w(Y) \to 0$, $w(G(Y)) \to 0$ and $w(H(Y)) \to 0$ (Ratchek and Rokne, 1995), where $w()$ is the width of the argument. These properties render convergence to *IP*.

IP implementation is scarce in CSP or COP literature. A generic *IP* algorithm is described in Byrne and Bogle (Byrne and Bogle, 1996), however numerical results consist of a few test problems. Markot et al. (Markot et al., to appear) present a multi-section *IP* algorithm where a new box selection rule based on feasibility index is described. The approach is fit for inequalities and numerous test instances are generated from bound constrained optimization problems. Kearfott (Kearfott, 1996) conducts tests on different approaches to verify existence of feasible solutions in problems with equality constraints. The branch and prune approach (Van-Hentenryck 1997b) uses *IP* in conjunction with interval Newton procedure to narrow down the domains defined by each box. This method applies box-consistency (Benhamou et al., 1994).

3.2 Symbolic Interval Inference Approach

In the literature, symbolic-interval cooperation has been formulated in terms of consistency techniques and symbolic expression transformation (simplification-factorization) (Granvilliers et al., 2001). Consistency based pruning that incorporates intervals is first proposed by Cleary (Cleary, 1987), followed by L'Homme (Lhomme et al., 1998), Faltings (Faltings, 1994), Benhamou and Older (Benhamou and Older, 1997). The latter techniques are called hull consistency techniques and they narrow the intervals of variables by using constraint inversion and interval substitution (Ceberio and Granvilliers, 2000; Granvilliers and Benhamou, 2001; Granvilliers, 2004; Lhomme et al., 1998; Sam-Haroud and Faltings, 1996). Hull consistency techniques propagate intervals over multiple constraints, so that, with the inversion of each constraint another variable domain is hopefully reduced, and finally, a feasible solution is identified.

While the above symbolic-interval cooperation is based on the full function expression, here, we propose a scheme (*Symbolic Interval Inference*) that concentrates on interpreting a constraint in terms of its hierarchically recursive sub-expressions and the corresponding interval propagation. This perspective is different from hull consistency techniques in the sense that it does not require any symbolic pre-processing nor inversion of constraints, which might prove to be quite difficult in complex and highly non-linear constraints. The goal here is to dissect the constraint, represent it as a binary tree, propagate sub-expression intervals on the tree and use chained symbolic interpretation to identify the pair of variables (source variables) that are most influential on the constraint's infeasibility degree. The variables that are identified as such have their ranges partitioned in the next iteration of *IP* with a guarantee of infeasibility degree reduction in three out of four sibling boxes generated for each constraint. Hence, rather than narrowing down variable domains implicitly through external constraint propagation, we narrow down variable domains explicitly using internal interval propagation through constraint sub-expressions, and hence diminish total infeasibility degree of boxes.

For the purpose of generality, we discuss the generic Symbolic Interval Inferencing approach (*SII*) for CSP with equalities and inequalities, though this application only involves equalities. In *SII*'s collaborative framework, we develop three basic components to enable a recursive symbolic propagation through sub-expressions: a parser, a tree builder, and a rule operator. The tree builder constructs a binary tree that represents a given function after parsing. The rule operator uses the binary tree for propagating intervals to make an inference on the source variables for each constraint. Next, source variables are collected into a single pool. Then, a reduced subset of the pool is generated as follows: The pool consists of an ordered set of variables based on constraint infeasibility degree ranking. Pairs of source variables are selected in this order, but if a selected source variable is also a source variable for a constraint in a more inferior rank, then, the remaining source variable of the inferior rank constraint is deleted from the pool. All source variables that are able to survive this pool screening procedure are subdivided in parallel in the next iteration.

The mechanics of *SII*'s three enabling components are as follows. The parser is activated once before *IP* is executed. It dissects each constraint's expression and passes the output to the tree builder. A binary tree that represents the constraint is then constructed. Next, *IP* is executed. At each box assessment, *SII* activates a tree traversal to identify the pair of variables to be re-partitioned for the given constraint. This is achieved by calling the *Interval Library* at each (molecular) level of the hierarchical binary tree so that the impact of all terms can be assessed in descending order of complexity until the first atomic element (variable) having the maximum impact on $\overline{G}(Y)$ is reached. Since PG_Y is a function of $\overline{G}(Y)$ in inequalities, *SII* targets $\overline{G}(Y)$ to reduce PG_Y. For equalities, the binay tree is traversed twice, the first with target $|\underline{H}(Y)|$ and the second, $\overline{H}(Y).PH_Y$ is reduced, because it is a function of both $|\underline{H}(Y)|$ and $\overline{H}(Y)$. Once the first source variable is identified, a backward traversal is activated to identify the coupling maximum impact (source) variable.

In the following, we briefly describe how *SII* works and prove that the degree of infeasibility of a constraint in the parent box is strictly reduced in three of the four siblings generated if both of the source variables are re-partitioned, and it is reduced in one of the two siblings generated if one source variable is re-partitioned. There exist exceptions to this guarantee for even power functions and absolute value function whose interval arguments include both positive and negative bounds, and for trigonometric functions.

3.2.1   Symbolic Interval Inference over a Binary Tree

Suppose a binary tree is constructed for a constraint and its source variables are to be identified by a tree traversal after intervals for sub-expressions have been propagated for a given box Y. We denote a parent at tree level k as D^k, and its immediate Left and Right sub-branches, as L^{k+1} and R^{k+1}. Let us also denote the interval bounds of parent node D^k by $[\underline{D}^k, \overline{D}^k]$, and those of the sub-branches as $[\underline{L}^{k+1}, \overline{L}^{k+1}]$ and $[\underline{R}^{k+1}, \overline{R}^{k+1}]$.

The box selection strategy in our *IP* implementation is based on the simple criterion of maximizing the sum of infeasibility degrees taken over all violated equalities and inequalities. This is in agreement with the goal of *SII* in the sense that it aims at chopping off infeasible boxes as early as possible. Hence, at the topmost level (level zero) of the tree, our interest is focused on

$\bar{D}^0$ or $\underline{D}^0$ for $g(x)$ or $h(x)$. For the topmost node, we determine which pair of interval bounds ($\{\underline{L}^1 \Theta \underline{R}^1\} \vee \{\underline{L}^1 \Theta \bar{R}^1\} \vee \{\bar{L}^1 \Theta \underline{R}^1\} \vee \{\bar{L}^1 \Theta \bar{R}^1\}$) result exactly in $\bar{D}^0$ or $\underline{D}^0$ when connected by their operator. Then, we compare the absolute values of individual bounds in the pair and take their maximum to choose and label the corresponding L or R branch. For instance, if $\underline{L}^1 \Theta \underline{R}^1 = \bar{D}^0$ or $\underline{L}^1 \Theta \underline{R}^1 = \underline{D}^0$, and if $|\underline{L}^1| = \max\{|\underline{L}^1|, |\underline{R}^1|\}$, then we take Left branch and label $|\underline{L}^1|$ for going to the next level down the tree (level 2). This procedure is applied recursively from top to bottom, each time searching for the bound pair resulting in the labeled bound at the upper level till a leaf (a variable) is hit. Once this forward tree traversal is over, all leaves in the tree corresponding to the variable selected are set to "Closed" status. The procedure then backtracks to the next higher level of the tree to identify the other leaf in the couple of variables that produce the labeled bound.

    An example is given in Fig. 1-3 for the function "$((x_1+x_2)*(x_3+x_4))+\mathrm{Sin}(x_1+x_3) = 0$". The domain intervals of the box are $x_1 = [-1.0, 10.0]$, $x_2 = [-10.0, 20.0]$, $x_3 = [1.0, 5.0]$, and $x_4 = [1.0, 10.0]$ and the expression interval is $[-166, 451]$. Suppose we target at $\overline{H}(Y)$. In Fig. 1-3, the branches selected by *SII* rule are indicated: arrows with dotted lines indicate the possible choices at each level while full arrows indicate the actual choice of *SII*. Bound calculations for labeled bounds are indicated in the figure for clarity, and bounds in bold are labeled bounds for the current level and the one immediately below. *SII* selects the Left branch two times consecutively, leading to source variable x_2, after which, backtracking leads to the second source variable x_1. Tree traversal is applied as follows.

Level 0: $[\underline{D}^0, \bar{D}^0] = [-166, \mathbf{451}]$

Select $\bar{D}^0$ ($\overline{H}(Y)$) at level zero.

a Θ b = $\{(-165 + 1)$ or $(450 + 1)$ or $(-165-1)$ or $(450-1)\}$ = 451.

Hence, a Θ b = $\overline{L}^1 + \overline{R}^1$.

$$I_{MAX} = \max \{| \overline{L}^1 |, | \overline{R}^1 |\} \Rightarrow \max \{|450|, |1|\}$$

$\Rightarrow$ select $\overline{L}^1$.

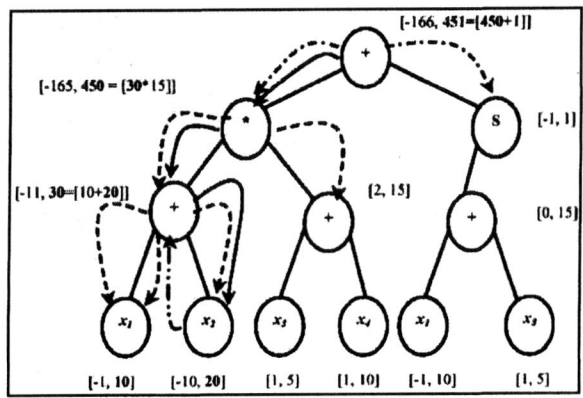

Figure 14-3. Implementation of SII "((x1+x2)*(x3+x4))+Sin(x1+x3)".
(Legend: S: Sine function)

Level 1: $[\underline{D}^1, \overline{D}^1] = [-165, 450]$

At level zero, 450 has been labeled as most contributing bound of level one
$a \odot b = \{(-11*2) \text{ or } (30*2) \text{ or } (-11*15) \text{ or } (30*15)\} = 450$.

$\Rightarrow a \odot b = \overline{L}^2 * \overline{R}^2$

$$I_{MAX} = \max \{| \overline{L}^2 |, | \overline{R}^2 |\} \Rightarrow \max \{|30|, |15|\}$$

$\Rightarrow$ select $\overline{L}^2$.

Level 2: $[\underline{D}^2, \overline{D}^2] = [-11, 30]$

At level one, 30 has been labeled as most contributing bound of level two.

a Θ b = {$(-1+-10)$ or $(-1+20)$ or **(10+20)** or $(10-10)$} = **30**.

$\Rightarrow$ a Θ b= $\overline{L}^3 + \overline{R}^3$

$I_{MAX} = \max \{|\overline{L}^3|, |\overline{R}^3|\} \Rightarrow \max \{|10|, |20|\}$

$\Rightarrow$ select $\overline{R}^3$.

This leads to $\overline{R}^3$, bound of leaf x_2.

The leaf pertaining to x_2 is "Closed" from here onwards, and the procedure backtracks to Level 2. Then, *SII* leads to the second source variable, x_1.

3.2.2   Convergence of SII

There are two exceptions where *SII* cannot identify source bounds for a given sub-expression.

LEMMA 1.
    At any level k of a binary tree, *SII* cannot identify source bounds for trigonometric (*trig*) sub-expressions, and for even power or absolute value (*abs*) type sub-expressions whose interval argument includes both negative and positive bounds.

PROOF.
    The proof is trivial since counterexamples can be easily constructed for each case. ∎

LEMMA 2.

For expressions excluding the exceptional sub-expressions stated in Lemma 1, *SII* tree traversal identifies the couple of variables that contribute most to $\overline{G}(Y)$ or $|\underline{H}(Y)|$, or $\overline{H}(Y)$ among all variables that can be re-partitioned in the next iteration.

PROOF.

By the monotonicity property of elementary interval operations and functions, *SII* tree traversal identifies the correct couple of bounds that result exactly in the bound labelled at the immediate upper level. By selecting the next level's labelled bound based upon the principle: *max* $\{|L|, |R|\}$, the maximum impact bound is identified at each level of the tree. A recursive repetition of this selection until an "Open" leaf is hit provides the maximum impact or first source variable for a given box. Similarly, recursive backtracking until a second "Open" leaf is hit provides the second source variable. ■

THEOREM 1.

Suppose a given constraint $g_i(x)$ or $h_j(x)$ does not contain exceptional sub-expression types indicated in Lemma 1. Let Y be a parent box and let S_1, S_2, S_3 and S_4 be its four siblings produced by the parallel bisection of the two source variables identified by *SII* for a given indeterminate inequality or equality constraint. Then, $PG_{S_i} < PG_Y$ or $PH_{S_i} < PH_Y$ for $i=1, 2, 3$. Therefore, as nested partitioning iterations (j) increase,

$$\lim_{j \to \infty} PG_{S_i} \to 0, \text{ or, } \lim_{j \to \infty} PH_{S_i} \to 0 \tag{19}$$

Hence, *SII* is a convergent method and it guarantees positive improvement in total infeasibility degree of a box in every partitioning iteration provided that the set of feasible boxes $X^* \neq \phi$.

PROOF.

Let x_r^Y, x_m^Y be the source variables of box Y identified by *SII* for a given constraint $g_i(x)$ or $h_j(x)$. We denote intervals of x_r^Y and x_m^Y in box Y as: $I_r^Y = [\underline{x}_r^Y, \overline{x}_r^Y]$ and $I_m^Y = [\underline{x}_m^Y, \overline{x}_m^Y]$, respectively. Variable domains in siblings are denoted by I_l^S, $l=1,2..n$.

Further, let $\overline{X}_r^{Y}$ and $\overline{X}_m^{Y}$ be identified as most contributing source bounds to $\overline{G}(Y)$ or $\left|\underline{H}(Y)\right| or \overline{H}(Y)$. We obtain four siblings S_1, S_2, S_3, S_4 from repartitioning I_r^{Y} and I_m^{Y} in parallel. Sibling domains are given in Table 1-1. We assume that $I_l^{S} = I_l^{Y}, \forall l \neq r,m$. Below, we show that either $PG_{S_1} < PG_Y$ or $PH_{S_1} < PH_Y$ for sibling S_1 depending on whether $g_i(x)$ or $h_j(x)$ is analyzed by SII.

Table 14-1. Domain boundaries of sibling boxes.

Sibling	I_r^{S}	I_m^{S}
S_1	$[\underline{X}_r^{Y}, \underline{X}_r^{Y} + w(I_r^{Y})/2]$	$[\underline{X}_m^{Y}, \underline{X}_m^{Y} + w(I_m^{Y})/2]$
S_2	$[\underline{X}_r^{Y} + w(I_r^{Y})/2, \overline{X}_r^{Y}]$	$[\underline{X}_m^{Y}, \underline{X}_m^{Y} + w(I_m^{Y})/2]$
S_3	$[\underline{X}_r^{Y}, \underline{X}_r^{Y} + w(I_r^{Y})/2]$	$[\underline{X}_m^{Y} + w(I_m^{Y})/2, \overline{X}_m^{Y}]$
S_4	$[\underline{X}_r^{Y} + w(I_r^{Y})/2, \overline{X}_r^{Y}]$	$[\underline{X}_m^{Y} + w(I_m^{Y})/2, \overline{X}_m^{Y}]$

Case of S_1: Based on sibling domains indicated in Table 1-1, $S_1 \subseteq Y$. Then, by inclusion isotonicity, $w(G(S_1)) \leq w(G(Y))$ and $\overline{G}(S_1) \leq \overline{G}(Y)$. Similarly, if the constraint analyzed is $h_j(x)$, then, either $\overline{H}(S_1) \leq \overline{H}(Y)$, or $\left|\underline{H}(S_1)\right| \leq \left|\underline{H}(Y)\right|$, whichever is targeted. Further, since $\overline{X}_r^{S_1} \neq \overline{X}_r^{Y}$ and $\overline{X}_m^{S_1} \neq \overline{X}_m^{Y}$, then, $\overline{G}(S_1) \neq \overline{G}(Y)$. The reasoning for $h_j(x)$ is similar.

From the above, $\overline{G}(S_1) < \overline{G}(Y)$ (for $h_j(x)$: $\overline{H}(S_1) < \overline{H}(Y)$, or $\left|\underline{H}(S_1)\right| < \left|\underline{H}(Y)\right|$), holds as strict inequality which leads to either $PG_{S_1} < PG_Y$ or $PH_{S_1} < PH_Y$.

One can show by similar reasoning that $PG_{S_2} < PG_Y$ and $PG_{S_3} < PG_Y$.

However, $PG_{S_4} = PG_Y$, because $\overline{X_r}^{S_4} = \overline{X_r}^Y$ and $\overline{X_m}^{S_4} = \overline{X_m}^Y$. The latter is also valid for $h_j(x)$.

The above proof is applicable to all combinations (4 bound combinations in total) of contributing source bounds other than $(\overline{X_r}^{-S}, \overline{X_m}^{-S})$ pair. In each case, three of the siblings result in reduced PG_S or PH_S. ∎

For dealing with the exceptions indicated in Lemma 1, we propose the heuristic rules described in the next section.

3.2.3   Heuristic rules for exceptional sub-expressions

- Heuristic rule for even powers and abs type sub-expressions: Consider a section of a binary tree containing a sub-expression at level k that is an even power. Suppose $\underline{L}^{k+1}$ and $\overline{L}^{k+1}$ are the lower and upper bounds for the sub-expression interval at level $k+1$, the immediate Left branch of the "Power" operator at level k. The heuristic rule for this situation is defined as follows.

Rule 1. If $\underline{L}^{k+1} < 0$ and $\overline{L}^{k+1} > 0$, then, select the stamped bound at level $k+1$ as $\underline{L}^{k+1}$. Else, apply *SII*.

The motivation behind Rule 1 is as follows: It is the negative bound that causes the problem indicated in Lemma 1. Hence, using this strategy, the heuristic aims to change $\underline{L}^{k+1}$ (in the next iteration) in order to eliminate the negative argument bound for the sub-expression at level k. The rule for *abs* type sub-expression is the same.

- Heuristic rule for trigonometric functions: The heuristic rule for *trig* operator is defined as follows.

Rule 2. If $(\overline{L}^{k+1} - \underline{L}^{k+1}) \geq \pi/2$, then select the labeled bound at level $k+1$ as $\max\{|\underline{L}^{k+1}|, |\overline{L}^{k+1}|\}$. Otherwise, apply *SII*.

This rule is motivated by the fact that trigonometric cycles are repetitive every $\pi/2$ units in a given interval. Therefore, the heuristic targets the portion of the interval having the maximum number of $\pi/2$ units, so that the bounds of the sub-expression soon become confined to $[-\pi/2, \pi/2]$ and the source bound identification problem is resolved.

3.3   Sequential Quadratic Programming (CFSQP)

C code for Feasible Sequential Quadratic Programming (CFSQP) finds a feasible solution the problem considering only nonlinear inequalities and linear equations. After the first feasible solution is obtained all subsequent solutions remain feasible for these constraints. Nonlinear equality constraints are then converted into inequality constraints and they are penalized in the objective function. The objective function is converted into a quadratic form consisting of second and first order derivatives of constraints and objective function. The model is solved to find an improving direction. Convergence of the method to a KKT solution is superlinear. The algorithms implemented are described in Lawrence et al. (1997). Here, we prefer to use finite derivative option of CFSQP rather than feeding in the gradient.

In our collaborative methodology, every time an indeterminate box is subdivided, a decision is taken on whether or not CFSQP should be applied to every indeterminate box in the pending list within their corresponding boundaries. The pending list is always sorted according to maximum total infeasibility. A decision is made on the number of times a partitioning iteration does not improve the total infeasible area chopped off. For instance, if total infeasible area is not 10% more than that of the last iteration, *IP* is assumed to fail. If failure occurs in 3 successive partitioning iterations, then all boxes in the pending list are subjected to CFSQP. Feasible solutions found during CFSQP applications are stored.

4.   NUMERICAL RESULTS

We compare the performance of the collaborative methodology with an IP algorithm that subdivides all variables in parallel once a box is selected to be re-partitioned. This parallel algorithm (*All_Vars*) is shown to be effective in unconstrained optimization (Casado et al., 2000). The algorithm calls CFSQP in the same manner. The only difference between the two algorithms is the utilization of *SII* in selecting the variables to be partitioned. Two test problems are used to assess performance. The first one is the *6R* inverse position problem defined in Eq. (14) (Tsai, and Morgan, 1985) and the second one is the Modified Kinematics problem, Kin1, (Coprin, 2004,

Shcherbina et al., 2002). Kin1 has trigonometric expressions. The model of Kin1 is given in Appendix B. The first test problem is used as a benchmark by Van-Hentenryck (1997a) and has 10 real solutions. The second one is used by Coprin (2004) and Shcherbina et al. (2002), and it has 16 real solutions.

Table 14-2. Comparison of results.

Function Name [D, NE, NI]	Collaborative			All_Vars		
	#CFSQ P Calls	# FE	T (min.)	#CFSQ P Calls	# FE	T (min.)
6R [8, 8, 0]	3511	130495	10.564	7602	459416	20.368
Kin1 Modified [6,6,0]	132	4686	0.3012	384	22631	1.2133

Table 1-2 shows numerical results. Properties of the two test problems are indicated as (D- Dimension of Problem, NE- Number of Equality Constraints, NI-Number of Inequality Constraints). Performance is measured in terms of CPU time (in minutes), T, Number of Function calls (# FE) and Number of CFSQP Calls. Table 1-3 and Table 1-4 summarize the percentage of infeasible area discarded from the search space at given function evaluations, as well as the number of real solutions found for 6R and Kin1 problems.

All solutions identified by the collaborative methodology and the corresponding constraint precisions are listed for both problems in Appendices C and D, respectively. The runs are executed on a PC with 256 MB RAM, 2.0 GHz P4 Intel CPU, under Windows OS system. All codes are developed with Visual C++ 6.0 interfaced with PROFIL interval arithmetic library (Knuppel, 1994) and CFSQP (Lawrence et al., 1997). It is observed that in the 6R problem both methods take a long time to identify the last and the last two solutions, respectively. *SII*'s impact on convergence can easily be seen in the percentage of infeasible area discarded from the search space as well as the number of solutions found at given *FE*. Hence, it is able to identify all solutions within 10 minutes whereas *All_Vars* has to run for about 20 minutes. It is also observed that Kin1 is much easier to solve than 6R problem. Still, *SII* improves convergence by cutting off 96% of the area during the first 1000 function calls.

Table 14-3. Percentage of infeasible area discarded and number of solutions found versus number of function calls in 6R Inverse position Problem.

Collaborative			All_Vars		
# FE	% Area Discarded	# of Solutions Found	# FE	% Area Discarded	# of Solutions Found
30000	24.011	9	30000	15.211	8
60000	28.023	9	60000	17.777	8
90000	28.71	9	90000	20.649	8
120000	46.846	9	120000	26.633	8
130495	51.745	10	130495	29.261	8
			150000	33.674	8
			180000	38.475	8
			210000	41.809	8
			240000	46.203	9
			459416	83.69	10

5.   CONCLUSION

A new collaborative methodology is developed here to solve Constraint Satisfaction Problems. The implementation is illustrated on a difficult advanced kinematics problem. The collaborative method is founded on global interval partitioning approach. The convergence of this basic approach is enhanced by a symbolic interval inference scheme that selects variables to be subdivided with a guarantee of reducing infeasibility degrees of constraints. Feasible points in boxes are identified by the Sequential Quadratic Programming (SQP) algorithm that is activated if a certain condition is satisfied.

Implementation on two test instances indicate that the collaborative method is able to discard a substantial area from the search space by classifying them as infeasible, hence, it economizes on both the number of function calls and SQP calls. Comparisons with the non-symbolic version of interval partitioning algorithm illustrates the enhancement obtained by the symbolic scheme. A final remark on the proposed methodology is that it has a wide applicability in both unconstrained and constrained optimization fields.

Table 14-4. Percentage of infeasible area discarded and number of solutions found versus number of function calls in Kin1 modified Problem.

	Collaborative			All_Vars	
# FE	% Area Discarded	# of Solutions Found	# FE	% Area Discarded	# of Solutions Found
1000	96.655	9	1000	86.67	0
2000	99.024	9	2000	90.1344	6
3000	99.143	9	3000	96.5134	9
4000	99.525	12	4000	97.966	9
4686	99.730	16	4686	97.95	9
			6000	98.0097	10
			22631	99.2125	16

ACKNOWLEDGEMENT:

We wish to thank Professor Andre Tits (Electrical Engineering and the Institute for Systems Research, University of Maryland, USA) for providing the software CFSQP.

APPENDIX A

The coefficients a_{ki} for a test instance.

Table 14-5. coefficients a_{ki} for a test instance.

a11= -0.24915068	a52 = -0.43241927	a134= 1.2499117	a112= 0.085383482
a12= 0.12501635	a53 = 0.29070203	a141= -1.0916287	a113= 0.0
a13= -0.63555007	a54 = 1.1651720	a142= 0.0	a114:= -0.69910317
a14= 1.4894773	a61 = 0.40026384	a74= 0.53816987	a143= -0.057131430
a21= 1.6091354	a62 = 0.0	a81= 0.0	a144= 1.4677360
a22= -0.68660736	a63 = 1.2587767	a82= -0.86483855	a151= 0.0
a23= -0.11571992	a64 = -0.26908494	a83= 0.58140406	a152= -0.43241927
a24= 0.23062341	a71 = -0.80052768	a84= 0.58258598	a153= -1.1628081
a31= 0.27942343	a72 = 0.0	a91= 0.074052388	a154= 1.1651720
a32= -0.11922812	a73 = -0.62938836	a92= -0.037157270	a161= 0.049207290
a33= -0.66640448	a121:= -0.75526603	a93= 0.19594662	a162= 0.0
a34= 1.3281073	a122:= 0.0	a94= -0.20816985	a163= 1.2587767
a41= 1.4348016	a123= -0.079034221	a101= -0.083050031	a164= 1.0763397
a42= -0.71994047	a124= 0.35744413	a102= 0.035436896	a171= 0.049207290
a43= 0.11036211	a131= 0.50420168	a103= -1.2280342	a172= 0.013873010
a44= -0.25864503	a132= -0.039251967	a104= 2.6868320	a173= 2.1625750
a51= 0.0	a133= 0.026387877	a111= -0.38615961	a174= -0.69686809

APPENDIX B

Model for the Modified Kinematics Problem (Shcherbina et al., 2002; Coprin, 2004)

$\sin(\theta_2)*\cos(\theta_5)*\sin(\theta_6)-\sin(\theta_3)*\cos(\theta_5)*\sin(\theta_6)-\sin(\theta_4)*\cos(\theta_5)*\sin(\theta_6)+$
$\cos(\theta_2)*\cos(\theta_6)+\cos(\theta_3)*\cos(\theta_6)+\cos(\theta_4)*\cos(\theta_6)-0.4077 =0$
$\cos(\theta_1)*\cos(\theta_2)*\sin(\theta_5)+\cos(\theta_1)*\cos(\theta_3)*\sin(\theta_5)+\cos(\theta_1)*\cos(\theta_4)*\sin(\theta_5)$
$+\sin(\theta_1)*\cos(\theta_5)-1.9115 =0$
$\sin(\theta_2)*\sin(\theta_5)+\sin(\theta_3)*\sin(\theta_5)+\sin(\theta_4)*\sin(\theta_5)-1.9791 =0$
$\cos(\theta_1)*\cos(\theta_2)+\cos(\theta_1)*\cos(\theta_3)+\cos(\theta_1)*\cos(\theta_4)+\cos(\theta_1)*\cos(\theta_2)+$
$\cos(\theta_1)*\cos(\theta_3)+\cos(\theta_1)*\cos(\theta_2)-4.0616 =0$
$\sin(\theta_1)*\cos(t2)+\sin(\theta_1)*\cos(\theta_3)+\sin(\theta_1)*\cos(\theta_4)+\sin(\theta_1)*\cos(\theta_2)+\sin(\theta_1)*$
$\cos(t3)+\sin(\theta_1)*\cos(\theta_2)-1.7172=0$
$\sin(\theta_2)+\sin(\theta_3)+\sin(\theta_4)+\sin(\theta_2)+\sin(\theta_3)+\sin(\theta_2)-3.9701 =0$
where $\theta_1, \theta_2, \theta_3, \theta_4, \theta_5, \theta_6,$ are the design variables which ranges between $[0, 2\Pi]$

APPENDIX C

Table 14-6. 10 real solutions for the 6R inverse position problem

SOLUTION $(x_1, x_2, x_3, x_4, x_5, x_6, x_7, x_8)$	CONSTRAINT PRECISION $(h_1, h_2, h_3, h_4, h_5, h_6, h_7, h_8)$	SOLUTION $(x_1, x_2, x_3, x_4, x_5, x_6, x_7, x_8)$	CONSTRAINT PRECISION $(h_1, h_2, h_3, h_4, h_5, h_6, h_7, h_8)$
-0.7675197725	-3.287300987e-016	-0.554195147	-5.5435257e-016
0.6410252716	-2.177620063e-016	0.8323867723	-5.2681383e-016
0.6964160419	-4.272840762e-016	0.5626360082	-5.2263965e-016
-0.717638277	-3.870601756e-017	0.826704737	-7.176334e-016
-0.9999792481	1.078781162e-017	0.8285714396	-4.1825809e-016
-0.0064423	-2.025842347e-016	-0.559883353	5.4830814e-016
0.8149506345	1.380026312e-016	-.6908546768	5.05990377e-016
0.5795303817	7.350348628e-016	-0.7229936484	-1.838264783e-016
0.9527580345	-1.009392223e-016	0.9779218583	-2.170572749e-016
0.3037303537	-1.488609583e-016	0.2089709049	-1.605703417e-016
-0.7905265443	-1.489693785e-016	0.02611329026	-2.481738773e-016
0.6124277775	-4.499981117e-016	0.9996589899	-4.277719672e-016
0.2415115238	-1.283966423e-016	-0.09786861	-1.844465065e-016
0.9703979513	-3.977937771e-016	0.9951993443	-5.133697287e-016
0.2373597961	3.739616581e-016	0.07707361041	3.075305581e-016
-0.971421807	5.7890975e-016	-0.9970254052	6.332824889e-016
-0.9749979045	-3.944327504e-016	-0.8332582975	-2.945235202e-016
-0.2222140551	-3.410357934e-016	-0.5528839025	-5.0404559e-016
-0.9938796286	-4.797052512e-016	-0.954937654	-1.374226254e-016
-0.1104684746	-2.108773225e-016	0.2968064639	-2.803746818e-016
-0.9999920795	3.056365924e-016	0.8646283146	-2.589074788e-016
0.00398006627	-3.338843074e-016	-0.5024120595	2.584466929e-016
0.9983398141	2.98228482e-016	-0.2334735726	2.67421854e-016
-0.05759874586	4.304282625e-016	-0.9723631476	-3.50359932e-016
-0.5419279206	-2.419939249e-016	0.9807938843	-1.604619215e-016
0.8404249692	-1.329231863e-016	0.1950470621	-3.279711572e-016
0.4670431944	-5.912696548e-016	-0.9941103362	-1.250085105e-016
0.884234502	-1.256590318e-016	-0.1083726874	-3.756760528e-016
-0.8367190935	3.775463015e-016	0.8844523118	-2.408555126e-016
-0.5476323206	-5.011453492e-016	0.4666305908	1.646632049e-016
0.8951440397	1.357285595e-016	0.2943403616	3.106950732e-016
0.4457770162	1.163565772e-015	-0.95570066	5.758197738e-016

SOLUTION $(x_1, x_2, x_3, x_4, x_5, x_6, x_7, x_8)$	CONSTRAINT PRECISION $(h_1, h_2, h_3, h_4, h_5, h_6, h_7, h_8)$	SOLUTION $(x_1, x_2, x_3, x_4, x_5, x_6, x_7, x_8)$	CONSTRAINT PRECISION $(h_1, h_2, h_3, h_4, h_5, h_6, h_7, h_8)$
0.9455046053	-1.667502941e-016	0.8476730058	7.641456912e-016
0.3256087244	-1.16768574e-016	0.5305190621	-2.364102837e-016
0.9385612474	-4.702726923e-016	0.9852621645	-5.577135975e-016
0.3451127134	-3.034681881e-016	-0.1710510662	-5.309338039e-016
-0.9391447018	4.602709273e-016	0.4900210554	-5.27776065e-016
-0.343521803	-4.522071736e-016	-0.8717105972	5.617793557e-016
0.8707744714	2.020664858e-016	-0.551930866	4.656987144e-017
0.4916826415	9.680299097e-016	-0.8338898723	1.354168513e-015

APPENDIX D

Table 14-7. 16 real solutions for the Modified Kinematics problem

SOLUTION $(x_1, x_2, x_3, x_4, x_5, x_6, x_7, x_8)$	CONSTRAINT PRECISION $(h_1, h_2, h_3, h_4, h_5, h_6, h_7, h_8)$	SOLUTION $(x_1, x_2, x_3, x_4, x_5, x_6, x_7, x_8)$	CONSTRAINT PRECISION $(h_1, h_2, h_3, h_4, h_5, h_6, h_7, h_8)$
3.54158911587688	1.43277e-016	0.399996462287087	-7.322701e-016
2.54163820909018	-1.37423e-015	0.676593554513104	-6.643990e-016
2.34152470623069	1.07585e-015	0.653413405776954	-2.024205e-016
2.14157870708576	-1.54347e-015	1.06725170617493	-6.683022e-016
1.94168714644523	-4.51028e-017	1.21680987054719	1.0516761e-016
4.3434014151202	8.88178e-016	1.22969809134691	7.77156117238e-016
0.399996462287087	9.4157 e-016	0.399996462287087	6.56213362e-017
0.676593554513105	-7.018 e-016	0.819005889921112	-9.477011e-016
0.653413405776952	-2.1684 e-017	0.524824446603842	-3.632077e-016
1.06725170617493	-5.819 e-016	0.889212794930671	-7.333543e-016
1.2168098705472	3.980 e-016	1.74096428607806	4.6403852e-016
4.76631936842528	1.11022 e-016	1.42521254487548	7.77156117238e-016
0.399996462287087	-1.47993 4271e-016	3.54158911587688	-2.9086433e-016
0.819005889921112	-3.5149e-016	2.32258676366868	-3.271037e-016
0.524824446603841	-2.8015e-016	2.61676820698595	-1.643650e-016
0.889212794930671	-4.9439 65339e-017	2.25237985865912	-6.310056e-016
1.74096428607806	5.06322 0731e-017	1.40062836751173	4.72278466 e-016
4.9429956645897	1.110223 2516e-016	1.80140301099991	6.6613381 e-016

SOLUTION $(x_1, x_2, x_3, x_4, x_5, x_6, x_7, x_8)$	CONSTRAINT PRECISION $(h_1, h_2, h_3, h_4, h_5, h_6, h_7, h_8)$	SOLUTION $(x_1, x_2, x_3, x_4, x_5, x_6, x_7, x_8)$	CONSTRAINT PRECISION $(h_1, h_2, h_3, h_4, h_5, h_6, h_7, h_8)$
3.54158911587688	1.43277e-016	0.399996462287087	-7.322701e-016
2.54163820909018	-1.37423e-015	0.676593554513104	-6.643990e-016
2.34152470623069	1.07585e-015	0.653413405776954	-2.024205e-016
2.14157870708576	-1.54347e-015	1.06725170617493	-6.683022e-016
1.94168714644523	-4.51028e-017	1.21680987054719	1.0516761e-016
4.3434014151202	8.88178e-016	1.22969809134691	7.7715611724e-16
3.54158911587688	-1.422 6e-015	0.399996462287087	5.7435610 e-017
2.46499909907669	-8.991 29e-016	0.599954444499607	-6.893357 e-016
2.48817924781284	-1.0217 034e-015	0.800067947359109	-2.0903417 e-016
2.07434094741487	-7.56773 4882e-016	1.00001394650403	-9.67975699 e-016
1.9247827830426	2.0199 5774e-015	1.19990550714456	-2.2096040 e-016
4.3712907449367	5.551 1 8e-016	1.20180876153041	-2.220446041e-016
0.399996462287087	-4.10018156 e-016	0.399996462287087	6.67868538e-017
0.599954444499609	-2.6096746 1e-016	0.612920767546544	-2.2096040e-016
0.800067947359107	8.1315162 e-017	0.941561364401486	2.6996634e-017
1.00001394650403	-7.2858385 e-016	0.678167384505091	-3.9248113e-016
1.19990550714456	-8.5326710 e-017	1.74943577165617	-7.31836415e-017
4.73538644590494	3.33066907 e-016	1.45345288025028	0
0.399996462287087	-9.61470486 e-016	3.54158911587688	-1.2633392e-015
0.612920767546544	-3.074797e-016	2.52867188604325	-2.060309e-015
0.941561364401486	2.699663 1e-017	2.20003128918831	1.6622987e-015
0.678167384505091	-3.924811 3e-016	2.4634252690847	-2.165802e-015
1.74943577165617	-7.31836466 5e-017	1.39215688193363	-3.266701e-016
4.96946138038386	0	1.82786872679406	9.9920072e-016
3.54158911587688	5.43456338 e-017	3.54158911587688	-2.027458e-017
2.52867188604325	-2.726443 e-015	2.54163820909018	-1.774838e-016
2.2000312891883	2.2120976 e-015	2.34152470623069	1.5460722e-016
2.4634252690847	-2.91693 e-015	2.14157870708576	-1.118896e-016
1.3921568819336	-6.78493 e-016	1.94168714644523	5.6020726e-016
4.59504553384007	8.8817845e-016	1.59379379231515	0

REFERENCE

Albala, H., and Angeles, J., 1979, Numerical Solution to the Input-Output displacement equation of the general 7R Spatial Mechanism, *Proceedings of 5th world congress on Theory of Machines and Mechanisms*, pp.1008-1011.

Allgower, E. L., and Georg, K., 1980, Simplicial and continuation methods for approximating fixed points and solutions to systems of equations. *SIAM Review.* **22**:28-85.

Benhamou, F., and McAllester, D., and Van Hentenryck, P., 1994, CLP(Intervals) Revisited, *Proceedings of ILPS'94, International Logic Programming Symposium*, pp. 124-138.

Benhamou, F., and Older, W. J., 1997, Applying Interval Arithmetic to Real, Integer and Boolean Constraints, *Journal of Logic Programming.* **32**:1-24.

Byrne, R. P., and Bogle, I. D. L., 1996, Solving Nonconvex Process optimization problems using Interval Subdivision Algorithms, *Global Optimization in Engineering Design*, I. E. Grossmann, ed., Kluwer Academic publisher, USA, pp. 155-173.

Casado, L.G., Garcia, I., and Csendes, T., 2000, A new multi-section technique in interval methods for global optimization, *Computing,* **65**:263-269.

Ceberio, M., and Granvilliers, L., 2000, Solving Nonlinear Systems by Constraint Inversion and Interval Arithmetic, *Lecture Notes in Artificial Intelligence.* **1930**:127-141.

Chase, M. A., 1963, Vector Analysis, *ASME Journal of Engineering for Industry.* 85:289-297.

Chow, S. N., Mallet-paret, J., and Yorke, J. A., 1979, A Homotopy Method for locating all zeros of a system of polynomials : *In Functional Differential Equations and Approximation of Fixed Points*, H.O. Peitgen, and H.O. Walther ed., Springer-Verlg Lecture Notes in Mathematics, Springer-Verlag, New York, pp. 228-237.

Cleary, J. G., 1987, Logical Arithmetic, *Future Computing Systems.* 2:125-149.

Coprin., 2004, http://www-sop.inria.fr/ coprin/ logiciels/ALIAS/Benches/.

Drexler, F. J., 1978, A homotopy method for the calculation of zeros of zero-dimensional polynomial ideals. In *Continuation Methods,* H.G. Wacker, ed., Academic Press, New York, pp. 69-93.

Drud, S., 1995, *CONOPT: A System for Large Scale Nonlinear Optimization, Tutorial for CONOPT Subroutine Library,* ARKI Consulting and Development A/S, Bagsvaerd, Denmark.

Duffy, J., and Crane, C., 1980, A displacement analysis of the general spatial 7R Mechanism, *Mechanism and Machine Theory.* **15**:153-169.

Duffy, J., 1980, *Analysis of Mechanisms and Robot Manipulators*, John Wiley Publishers, New York.

Faltings, B., 1994, Arc consistency for continuous variables, *Artificial Intelligence.* **65**:363-376.

Granvilliers, L., and Benhamou, F., 2001, Progress in the Solving of a Circuit Design Problem, *J. Global Optim.* **20**:155-168.

Granvilliers, L., Monfroy, E., and Benhamou, F., 2001, Symbolic-interval cooperation in constraint programming, *Proceedings of the 2001 International Symposium on Symbolic and Algebraic Computation*, London, Ontario, Canada.

Granvilliers, L., 2004, An Interval Component for Continuous Constraints, *J. Comput. Appl. Math.* **162** :79–92.

Garcia, C. B., and Zangwill, W. I., 1979, Finding all solutions to polynomial systems and other systems of equations. *Math. Program.* **16** :159-176.

Hartenberg, R. S., and Denavit, J., 1964, *Kinematic synthesis of linkages*, McGraw-Hill, New York.

Hansen, E., 1992, *Global Optimization Using Interval Analysis*, Marcel Dekker Inc, New York.

Hesse, R., 1973, A Heuristic Search Procedure for Estimating a Global Solution of Nonconvex Programming Problems, *Op. Res.* **21**:1267.

Kearfott, R. B., 1996, Test results for an interval branch and bound algorithm for equality constrained optimization, *State of the art in global optimization*, Kluwer, Dordrecht, pp. 181-199.

Kohi, D., and Somi, A. H., 1975, Kinematic analysis of spatial mechanisms via successive screw displacements, *ASME Journal of Engineering for Industry.* **97**:739-747.

Knuppel, O., 1994, PROFIL/BIAS – A Fast Interval Library, *Computing*, **53**:277-287.

Lawrence, C. T., Zhou, J. L., and Tits, A, L., 1997, *User's Guide for CFSQP version 2.5: A Code for Solving (Large Scale) Constrained Nonlinear (minimax) Optimization Problems, Generating Iterates Satisfying All Inequality Constraints*, Institute for Systems Research, University of Maryland, College Park, MD.

Lee, H.-Y., and Liang, C-G., 1988, Displacement analysis of the general spatial 7-link 7R mechanism, *Mechanism and Machine Theory.* **23**:219-226.

Leykin, A., and Verschelde, J., 2004, PHCmaple: A Maple Interface to the Numerical Homotopy Algorithms in PHCpack, In *Proceedings of the Tenth International Conference on Applications of Computer Algebra* (ACA'2004), pp.139-147.

Lhomme, O., Gotlieb, A., and Rueher, M., 1998, Dynamic Optimization of Interval Narrowing Algorithms, *Journal of Logic Programming.* **37**:165-183.

Manocha, D., and Canny, J. F., 1994, Efficient inverse kinematics for general 6R manipulators. *IEEE Journal on Robotics and Automation.* **10**:648-657.

Morgan, A., 1983a, A method for computing all solutions to Systems of polynomial equations, *ACM Transactions on Mathematical Software.* **2(1)**:1-17.

Morgan, A., 1983b, Solving Systems of Polynomial Equations Using Homogenous Coordinates, *GM Research Publication GMR-4220.*

Markót, M. Cs.., Fernández, J., Casado, L. G., and T. Csendes, T., New interval methods for constrained global optimization, Conditionally accepted for publication in *Mathematical Programming.*

Morgan, A., 1987a, *Solving Polynomial Systems Using Continuation for Scientific and Engineering Problems.* Prentice-Hall, Englewood Cliffs, New Jersey.

Morgan, A. P. 1987b, Computing All Solutions To Polynomial Systems Using Homotopy Continuation. *Appl. Math. Comput.* **24**:115-138.

Morgan, A. P., and Sommese, A. J., 1989, Coefficient-parameter polynomial continuation, *Appl. Math. Comput.* **29**:23-160.

Morgan, A. P., Sommese, A. J., and Watson, L. T., 1989a, Finding all isolated solutions to polynomial systems using HOMPACK, ACM Trans. Math. Softw. **15**:93-122.

Neumaier, A., 1990, *Interval Methods for Systems of Equations Encyclopedia of Mathematics and its Applications 37*, Cambridge University Press, Cambridge.

Panier, E. R., and Tits, A.L., 1993, On Combining Feasibility, Descent and Superlinear Convergence in Inequality Constrained Optimization, *Math. Programming.***59**:261-276.

Pieper, D. L., 1968, *The kinematics of Manipulators Under Computer Control.* Ph.D Thesis, Standford University, USA.

Pieper, D. L., and Roth, B., 1969, The kinematics of manipulators under computer control, *Proceedings II-International congress on the Theory of Machines and Mechanisms.* **2**:159-168.

Ratschek, H., and Rokne, J.,1988, *New computer Methods for Global Optimization*, John Wiley, New York.

Ratschek, H., and Rokne, J.,1995, Interval Methods, *Handbook of Global Optimization*, R. Horst and P.M. Pardalos, ed., Kluwer Academic publisher, Netherlands, pp. 751-828.

Recio, T., and Gonzalex-Lopez, M. J., 1994, On the symbolic insimplification of the general 6R-manipulator kinematic equations, In *Proceedings of the international symposium on Symbolic and algebraic computation.* pp. 354 - 358.

Roth, B., Rasteger, J., and Scheinman, V., 1973, On the design of computer controlled manipultors, *First CISM-IFToMM Symposium*, pp. 93-113.

Sam-Haroud, D., and Faltings, B., 1996, Consistency techniques for continuous constraints, Constraints, 1:85–118.

Shcherbina, O., Neumaier, A., Sam-Haroud, D., Vu, X.-H., and Nguyen, T.-V., 2002, Benchmarking Global Optimization and Constraint Satisfaction Codes, *Global Optimization and Constraint Satisfaction: First International Workshop on Global Constraint Optimization and Constraint Satisfaction, COCOS 2002,* Valbonne-Sophia Antipolis, France, 2002.

Sommese, A. J., Verschelde, J., and Wampler, C. W., 2002, Advances in Polynomial Continuation for solving problems in Kinematics, In *Proc. ASME Design Engineering Technical Conf. (CDROM),* Montreal, Quebec.

Tsai, L. W., and Morgan, A. P., 1985, Solving the Kinematics of the most general six and five-degree-of-freedom manipulators by continuation methods, *Journal of Mechanisms, Transmissions, and Automation in Design.* **107**:189-200.

Uicker Jr., Denavir J. J., and Hartenberg R.S., 1964, An iterative method for the displacement analysis of spatial mechanisms, *ASME Journal of Applied Mechanics.* **31**: 309-314.

Van-Hentenryck, P., Michel, L., and Deville, Y., 1997a, *Numerica: a Modeling Language for Global Optimization,* MIT press, London, England.

Van-Hentenryck, P., Mc Allester, D., and Kapur, D., 1997b, Solving polynomial systems using branch and prune approach, *SIAM Journal on Numerical Analysis.* **34**:797-827.

Verschelde, J., Verlinden, P., and Cools., R., 1994, Homotopies Exploiting Newton Polytopes For Solving Sparse Polynomial Systems, *SIAM Journal on Numerical Analysis.* **31**: 915-930.

Wampler, C. and Morgan, A., 1991, Solving the *6R* Inverse Position Problem using a Generic-Case Solution Methodology, *Mech. Mach. Theory.* **26**:91-106.

Watson, L. T., 1979, A Global Convergent Algorithm for Computing Fixed points of C^2 Maps, *Applied Math. Comput.* **5**:297-311.

Yang, A. T., and Freudenstein, F., 1964, Application of Dual number quaterian algebra to the analysis of spatial mechanisms, *ASME journal of Applied mechanics.* **86**:300-308.

Yuan, M.S.C., and Freudenstei, F., 1971, Kinematic Analysis of Spatial Mechanisms by means of Screw Coordinates (two parts), *ASME Journal of Engineering for Industry.* **93**:61-73.

Zhou, J. L., and Tits, A. L.,1996, An SQP Algorithm for Finely Discretized Continuous Minimax Problems and Other Minimax Problems with Many Objective Functions, *SIAM J. on Optimization .* **6**:461-487.

Zhou, J. L.,and Tits, A. L., 1993, Nonmonotone Line Search for Minimax Problems, *J. Optim.Theory Appl.* **76**:455-476.

Chapter 15

IMPROVED LEARNING OF NEURAL NETS THROUGH GLOBAL SEARCH

V.P. Plagianakos

Department of Mathematics, University of Patras, University of Patras Artificial Intelligence Research Center–UPAIRC, GR–26110 Patras, Greece

vpp@math.upatras.gr

G.D. Magoulas

School of Computer Science and Information Systems, Birkbeck College, University of London, Malet Street, London WC1E 7HX, UK

gmagoulas@dcs.bbk.ac.uk

M.N. Vrahatis

Department of Mathematics, University of Patras, University of Patras Artificial Intelligence Research Center–UPAIRC, GR–26110 Patras, Greece

vrahatis@math.upatras.gr

Abstract   Learning in artificial neural networks is usually based on local minimization methods which have no mechanism that allows them to escape the influence of an undesired local minimum. This chapter presents strategies for developing globally convergent modifications of local search methods and investigates the use of popular global search methods in neural network learning. The proposed methods tend to lead to desirable weight configurations and allow the network to learn the entire training set, and, in that sense, they improve the efficiency of the learning process. Simulation experiments on some notorious for their local minima learning problems are presented and an extensive comparison of several learning algorithms is provided.

Keywords: Global search, local minima, simulated annealing, genetic algorithms, evolutionary algorithms, neural networks, supervised training, swarm intelligence.

Introduction

Scientific interest in models of neuronal networks or artificial neural networks mainly arises from their potential ability to perform interesting computational tasks. Nodes, or artificial neurons, in neuronal network models are usually considered as simplified models of biological neurons, i.e. real nerve cells, and the connection weights between nodes resemble to synapses between neurons. In fact, artificial neurons are much simpler than biological neurons. But, for the time being, it is far from clear how much of this simplicity is justified because, as yet, we have only poor understanding of neuronal functions in complex biological networks. Artificial neural nets (ANNs) provide to computing an alternative algorithmic model, which is biologically motivated: the computation is massively distributed and parallel and the learning replaces a priori program development, i.e. ANNs develop their functionality based on training (sampled) data

In neural net learning the objective is usually to minimize a cost function defined as the multi–variable error function of the network. This perspective gives some advantage to the development of effective learning algorithms, because the problem of minimizing a function is well known in the field of numerical analysis. However, due to the special characteristics of the neural nets, learning algorithms can be trapped in an undesired local minimum of the error function: they are based on local search methods and have no mechanism that allows them to escape the influence of an undesired local minimum.

This chapter is focused on the use of Global Optimization (GO) methods for improved learning of neural nets and presents global search strategies that aim to alleviate the problem of occasional convergence to local minima in supervised training. Global search methods are expected to lead to "optimal" or "near-optimal" weight configurations by allowing the network to escape local minima during training.

In practical applications, GO methods can detect just *sub–optimal solutions* of the objective function. In many cases these sub–optimal solutions are acceptable but there are applications where the optimal solution is not only desirable but also indispensable. Therefore, the development of robust and efficient GO methods is a subject of considerable ongoing research.

It is worth noting that, in general, GO–based learning algorithms possess strong theoretical convergence properties, and, at least in principle, are straightforward to implement and apply. Issues related to their numerical efficiency are considered by equipping GO algorithms with a "traditional" local minimization phase. Global convergence, however, needs to be guaranteed by the global–scope algorithm component which, theoretically, should be used in a complete, "exhaustive" fashion. These remarks indicate the inherent computational demand of the GO algorithms, which increases non–polynomially, as a function of problem–size, even in the simplest cases.

The remaining of this chapter is organized as follows. Section 1 formulates the learning problem in the optimization context. In section 2, deterministic monotone and nonmonotone strategies for developing globally convergent modifications of learning algorithms are presented. Section 3 focuses on global search methods and error function transformations to alleviate convergence to undesired local minima. Section 4 presents simulations and comparisons with commonly used learning algorithms, and discusses the results.

1. Learning in neural nets

Let us consider an ANN whose l-th layer contains N_l neurons ($l = 1, \ldots, L$). The neurons of the first layer receive inputs from the external world and propagate them to the neurons of the second layer (also called hidden layer) for further processing. The operation of the neurons for $l = 2, \ldots, L$ is usually based on the following equations:

$$net_j^l = \sum_{i=1}^{N_{l-1}} w_{ij}^{l-1,l} y_i^{l-1}, \qquad y_j^l = f(net_j^l),$$

where net_j^l is for the j-th neuron in the l-th layer ($l = 2, \ldots, L; j = 1, \ldots, N_l$), the sum of its weighted inputs. The weights from the i-th neuron at the $(l-1)$ layer to the j-th neuron at the l-th layer are denoted by $w_{ij}^{l-1,l}$, y_j^l is the output of the j-th neuron that belongs to the l-th layer, and $f(net_j^l)$ is the j-th's neuron activation function.

If there is a fixed, finite set of input–output pairs, the square error over the training set, which contains P representative cases, is:

$$E(w) = \sum_{p=1}^{P} \sum_{j=1}^{N_L} (y_{j,p}^L - t_{j,p})^2 = \sum_{p=1}^{P} \sum_{j=1}^{N_L} [\sigma^L(net_j^L + \theta_j^L) - t_{j,p}]^2.$$

This equation formulates the error function to be minimized, in which $t_{j,p}$ specifies the desired response at the j–th neuron of the output layer at the input pattern p and $y_{j,p}^L$ is the output at the j–th neuron of the output layer L that depends on the weights of the network and σ is a nonlinear activation function, such as the well known sigmoid $\sigma(x) = (1 + e^{-x})^{-1}$. The weights of the network can be expressed in a vector notation:

$$w = \left(\ldots, w_{ij}^{l-1,l}, w_{i+1\ j}^{l-1,l}, \ldots, w_{N_{l-1}\ j}^{l-1,l}, \theta_j^l, w_{i\ j+1}^{l-1,l}, w_{i+1\ j+1}^{l-1,l}, \ldots \right)^{\top},$$

where θ_j^l denotes the bias of the j–th neuron ($j = 1, \ldots, N_l$) at the l–th layer ($l = 2, \ldots, L$). This formulation defines the weight vector as a point in the N–dimensional real Euclidean space $\mathbb{R}^N$, where N denotes the total number of weights and biases in the network.

Minimization of $E(w)$ is attempted by updating the weights using a learning algorithm. The weight update vector describes a direction in which the weight vector will move in order to reduce the network training error. The weights are modified using the iterative scheme:

$$w^{k+1} = w^k + \Delta w^k, \quad k = 0, 1, \ldots,$$

where w^{k+1} is the new weight vector, w^k is the current weight vector and Δw^k the weight update vector.

Various choices of the correction term Δw^k give rise to distinct learning algorithms, which are usually first–order or second–order methods depending on the derivative–related information they use to generate the correction term. Thus, first–order algorithms are based on the first derivative of the learning error with respect to the weights, while second–order algorithms on the second derivative (see [5] for a review on first–order and second–order training algorithms).

A broad class of first–order algorithms, which are considered much simpler to implement than second–order methods, uses the correction term $-\mu \nabla E(w^k)$; μ is a heuristically chosen constant that usually takes values in the interval $(0, 1)$ (the optimal value of the stepsize μ depends on the shape of the N–dimensional error function) and $\nabla E(w^k)$ defines the gradient vector of the ANN obtained by applying the chain rule on the layers of the network [50].

The most popular first–order algorithm is called Backpropagation (BP) and uses the steepest descent [36] with constant stepsize μ:

$$w^{k+1} = w^k - \mu \nabla E(w^k), \quad k = 0, 1, \ldots.$$

It is well known that the BP algorithm leads to slow network learning and often yields suboptimal solutions [16]. Attempts to speed up back–propagation training have been made by dynamically adapting the stepsize μ during training [29, 55], or by using second derivative related information [32, 34, 54]. Adaptive stepsize algorithms are more popular due to their simplicity. The stepsize adaptation strategies that are usually suggested are: (i) start with a small stepsize and increase it exponentially, if successive iterations reduce the error, or rapidly decrease it, if a significant error increase occurs [4, 55], (ii) start with a small small stepsize and increase it, if successive iterations keep gradient direction fairly constant, or rapidly decrease it, if the direction of the gradient varies greatly at each iteration [9], (iii) for each weight, an individual stepsize is given, which increases if the successive changes in the weights are in the same direction and decreases otherwise [21, 52], and (iv) use a closed formula to calculate a common stepsize for all the weights at each iteration [29, 46] or a different stepsize for each weight [14, 31]. Note that all the above–mentioned strategies employ heuristic parameters in an attempt to enforce the decrease of the learning error at each iteration and to secure the converge of the learning algorithm.

Methods of nonlinear optimization have also been studied extensively in the context of NNs [32, 54, 56]. Various Levenberg–Marquardt, quasi–Newton and trust–region algorithms have been proposed for small to medium size neural nets [18, 25]. Variations on the above methods, limited-memory quasi-Newton and double dogleg, have been also proposed in an attempt to reduce the memory requirements of these methods [1, 6]. Nevertheless, first–order methods, such as variants of gradient descent [27, 41] and conjugate-gradient algorithms [34] appear to be more efficient in training large size neural nets.

At this point it is worth mentioning an important consideration for adopting an iterative scheme in practical learning tasks is its susceptibility to ill–conditioning: the minimization of the network's learning error is often ill–conditioned, especially when there are many hidden units [51]. Although second–order methods are considered better for handling ill–conditioned problems [5, 32], it is not certain that the extra computational/memory cost these methods require leads to speed ups of the minimization process for nonconvex functions when far from a minimizer [35]; this is usually the case with the neural network training problems, [5], especially when the networks uses a large number of weights [27, 41].

Moreover, BP–like learning algorithms, as well as second–order algorithms, occasionally converge to undesired local minima which affect the efficiency of the learning process. Intuitively, the existence of local minima is due to the fact that the error function is the superposition of nonlinear activation functions that may have minima at different points, which sometimes results in a nonconvex error function [16]. The insufficient number of hidden nodes as well as improper initial weight settings can cause convergence to an undesired local minimum, which prevents the network from learning the entire training set and results in inferior network performance.

Several researchers have presented conditions on the network architecture, the training set and the initial weight vector that allow BP to reach the optimal solution [26, 60]. However, conditions such as the linear separability of the patterns and the pyramidal structure of the ANN [16] as well as the need for a great number of hidden neurons (as many neurons as patterns to learn) make these interesting results not easily interpretable in practical situations even for simple problems.

2.   Globally Convergent Variants of Local Search Methods

A local search learning algorithm can be made globally convergent by determining the stepsize in such a way that the error is exactly minimized along the current search direction at each iteration, i.e. $E(w^{k+1}) < E(w^k)$. To this end, an iterative search, which is often expensive in terms of error function evaluations, is required. It must be noted that the above simple condition does not guarantee global convergence for general functions, i.e. converges to a local minimizer from any initial condition (see [11] for a general discussion of globally convergent methods).

Monotone Learning Strategies

In adaptive stepsize algorithms, monotone reduction of the error function at each iteration can be achieved by searching a local minimum with small weight steps. These steps are usually constrained by problem-dependent heuristic learning parameters.

The use of heuristic strategies enforces the monotone decrease of the learning error and secures the convergence of the training algorithm to a minimizer of E. However, the use of inappropriate values for the heuristic learning parameters can considerably slow down the rate of training or even lead to divergence and to premature saturation [26, 49]; there is a trade-off between convergence speed and stability of the training algorithm. Additionally, the use of heuristics for bounding the stepsize prevents the development of efficient algorithms with the property that starting from any initial weight vector the weight updates will converge to a local minimum, i.e. globally convergent training algorithms.

A monotone learning strategy, which does not apply heuristics to bound the length of the minimization step, consists in accepting a positive stepsize η^k along the search direction $\varphi^k \neq 0$, if it satisfies the *Wolfe conditions*:

$$E(w^k + \eta^k \varphi^k) - E(w^k) \leqslant \sigma_1 \eta^k \left\langle \nabla E(w^k), \varphi^k \right\rangle, \qquad (1.1)$$

$$\left\langle \nabla E(w^k + \eta^k \varphi^k), \varphi^k \right\rangle \geqslant \sigma_2 \left\langle \nabla E(w^k), \varphi^k \right\rangle, \tag{1.2}$$

where $0 < \sigma_1 < \sigma_2 < 1$ and $\langle \cdot, \cdot \rangle$ stands for the usual inner product in $\mathbb{R}^n$. The first inequality ensures that the error is reduced sufficiently, and the second prevents the stepsize from being too small. It can be shown that if φ^k is a descent direction and E is continuously differentiable and bounded below along the ray $\{w^k + \eta \varphi^k \mid \eta > 0\}$, then there always exists a stepsize satisfying (1.1)–(1.2) [11, 35]. Relation (1.2) can be replaced by:

$$E(w^k + \eta^k \varphi^k) - E(w^k) \geqslant \sigma_2 \eta^k \left\langle \nabla E(w^k), \varphi^k \right\rangle, \tag{1.3}$$

where $\sigma_2 \in (\sigma_1, 1)$ (see [11]). The strategy based on Wolfe's conditions provides an efficient and effective way to ensure that the error function is globally reduced sufficiently. In practice, conditions (1.2) or (1.3) are generally not needed because the use of a backtracking strategy avoids very small learning rates [31, 57].

An alternative strategy has been proposed in [47]. It is applicable to any descent direction φ^k and uses two parameters $\alpha, \beta \in (0, 1)$. Following this approach the stepsize is $\eta^k = \beta^{m_k}$, where $m_k \in \mathbb{Z}$ is any integer such that:

$$E(w^k + \beta^{m_k} \varphi^k) - E(w^k) \leqslant \beta^{m_k} \alpha \left\langle \nabla E(w^k), \varphi^k \right\rangle, \tag{1.4}$$

$$E(w^k + \beta^{m_k - 1} \varphi^k) - E(w^k) > \beta^{m_k - 1} \alpha \left\langle \nabla E(w^k), \varphi^k \right\rangle. \tag{1.5}$$

To ensure global convergence, monotone strategies that employ conditions (1.1)–(1.2) or (1.4)–(1.5) must be combined with stepsize tuning subprocedures. For example, a simple subprocedure for tuning the length of the minimization step is to decrease the stepsize by a reduction factor q^{-1}, where $q > 1$ [36], so that it satisfies conditions (1.1)–(1.2) at each iteration. This *backtracking strategy* has the effect that the stepsize is decreased by the largest number in the sequence $\{q^{-m}\}_{m=1}^{\infty}$, so that condition (1.1) is satisfied. When seeking to satisfy (1.1) it is important to ensure that the stepsize is not reduced unnecessarily so that condition (1.2) is not satisfied. Since in training, the gradient vector is known only at the beginning of the iterative search for a new weight vector, condition (1.2) cannot be checked directly (this task requires additional gradient evaluations at each iteration), but is enforced simply by placing a lower bound on the acceptable values of the stepsize. This bound on the stepsize has the same theoretical effect as condition (1.2), and ensures global convergence [11].

Nonmonotone Learning Strategies

Although monotone learning strategies provide an efficient and effective way to ensure that the error function is reduced sufficiently, they have the disadvantage that no information, which might accelerate convergence, is stored and used [15]. To alleviate this situation we propose a nonmonotone learning strategy that exploits the accumulated information with regard to the M most recent values of the error function. The following condition is used to formulate the new approach and to define a criterion of acceptance of any weight iterate:

$$E\left(w^k - \eta^k \nabla E(w^k)\right) - \max_{0 \leqslant j \leqslant M} E(w^{k-j}) \leqslant \gamma \, \eta^k \left\langle \nabla E(w^k), \phi^k \right\rangle, \quad (1.6)$$

where M is a nonnegative integer, named *nonmonotone learning horizon*, $0 < \gamma < 1$, η^k indicates the learning rate and ϕ^k is the search direction at the kth iteration. The above condition allows for an increase in the function values without affecting the global convergence properties, as it has been proved theoretically in [17, 48].

Furthermore, it can be shown that the nonmonotone learning strategy generates a globally convergent sequence for any algorithm that follows a search direction $\varphi^k \neq 0$, provided that two positive numbers c_1, c_2 exist, such that:

$$\left\langle \nabla E(w^k), \varphi^k \right\rangle \leqslant -c_1 \|\nabla E(w^k)\|, \quad (1.7)$$

$$\|\varphi^k\| \leqslant c_2 \|\nabla E(w^k)\|. \quad (1.8)$$

This follows directly from the convergence theorem in [17].

Next, we summarize the basic steps of the nonmonotone learning strategy at the kth iteration:

1: Update the weights $w^{k+1} = w^k + \eta^k \varphi^k$.

2: If $E(w^{k+1}) - \max_{0 \leqslant j \leqslant M^k} E(w^{k-j}) \leqslant \gamma \, \eta^k \left\langle \nabla E(w^k), \phi^k \right\rangle$, store w^{k+1}, set $k = k + 1$ and go to Step 1; otherwise go to the next step.

3: Use a tuning technique for η^k and return to Step 2.

Experimental results indicate that the choice of the parameter M is critical for the implementation and depends on the nature of the problem [42, 46]. Therefore, instead of using a user–defined value for the nonmonotone learning horizon M, an adaptive procedure can be applied to dynamically evaluate M.

To this end, the following procedure, based on the notion of the Lipschitz constant, dynamically adapts the value of the nonmonotone learning horizon M at each iteration:

$$M^k = \begin{cases} M^{k-1} + 1, & \Lambda^k < \Lambda^{k-1} < \Lambda^{k-2}, \\ M^{k-1} - 1, & \Lambda^k > \Lambda^{k-1} > \Lambda^{k-2}, \\ M^{k-1} & , \quad \text{otherwise,} \end{cases} \qquad (1.9)$$

where Λ^k is the local estimation of the Lipschitz constant at the kth iteration [29]:

$$\Lambda^k = \frac{\left\| \nabla E(w^k) - \nabla E(w^{k-1}) \right\|}{\left\| w^k - w^{k-1} \right\|}, \qquad (1.10)$$

which can be obtained without additional error function or gradient evaluations. If Λ^k is increased for two consecutive iterations, the sequence of the weight vectors approaches a steep region and the value of M has to be decreased in order to avoid overshooting a possible minimum point. On the other hand, when Λ^k is decreased for two consecutive iterations, the method possibly enters a valley in the weight space, so the value of M has to be increased. This allows the method to accept larger stepsizes and move faster out of the flat region. Finally, when the value of Λ^k has a rather random behavior (increasing or decreasing for only one iteration), the value of M remains unchanged. It is evident that M has to be positive. Thus, if Relation (1.9) gives a non positive value in M, the nonmonotone learning horizon is set equal to 1 in order to ensure that the error function is sufficiently reduced at the current iteration.

At this point it is useful to remark that a simple technique to tune η^k at Step 3 is to decrease the stepsize by a reduction factor $1/q$, where $q > 1$, as mentioned in the previous subsection. The selection of q is not crucial for successful learning, however, it has an influence on the number of error function evaluations required to obtain an acceptable weight vector. Thus, some training problems respond well to one or two reductions in the stepsize by modest amounts (such as $1/2$), while others require many such reductions, but might respond well to a more aggressive stepsize reduction (for example by factors of $1/10$, or even $1/20$). On the other hand, reducing η^k too much can be costly since the total number of iterations will be increased. The value $q = 2$ is usually suggested in the literature [2] and, indeed, it was found to work effectively and efficiently in the experiments [41, 46]. The above procedure constitutes an efficient method of determining an appropriate stepsize without additional gradient evaluations. As a consequence, the number of gradient evaluations is, in general, less than the number of error function evaluations.

The nonmonotone learning strategy can be used as a subprocedure that secures and accelerates the convergence of a learning algorithm by providing the ability to handle arbitrary large stepsizes, and, in this way, learning by neural nets becomes feasible on a first–time basis for a given problem. Additionally, it alleviates problems generated by poor selection of the user–defined learning parameters, such as decreased rate of convergence, or even divergence and convergence to undesired local minima due to premature saturation [26]. It is worth noting that any stepsize adaptation strategy can be incorporated in Step 1 of the above algorithm model. For example, in [41, 46] the nonmonotone Backpropagation with variable stepsize (NMBPVS) and the nonmonotone Barzilai–Borwein Backpropagation (NMBBP) have been proposed.

The NMBPVS is the nonmonotone version of the Backpropagation with Variable Stepsize (BPVS) [29], which exploits the local shape of the error surface to obtain a local estimate the Lipschitz constant at each iteration and uses this estimate to adapt the stepsize η^k. The nonmonotone strategy helps to eliminate the possibility of using an unsuitable local estimation of the Lipschitz constant.

With regards to the NMBBP, the nonmonotone strategy helps to secure the convergence of the BBP method [42], even when the Barzilai–Borwein formula [3] gives an unsuitable stepsize. Experimental results show that the NMBBP retains the ability of BBP to escape from undesirable regions in the weight space, i.e. undesired local minima and flat valleys, whereas other methods are trapped within these regions [41, 46].

Furthermore, alternative weight adaptation rules can be used in Step 2 of the above algorithm model to develop their nonmonotone version. For example, in [21, 50] a simple, heuristic strategy for accelerating the BP algorithm has been proposed based on the use of a momentum term. The momentum term can been incorporated in the steepest descent method as follows:

$$w^{k+1} = w^k - (1 - m)\eta \nabla E(w^k) + m(w^k - w^{k-1}),$$

where m is the momentum constant. A drawback with the above scheme is that, if m is set to a comparatively large value, gradient information from previous iterations is more influential than the current gradient information in updating the weights. A solution is to increase the stepsize, however, in practice, this approach frequently proves ineffective and leads to instability or saturation. Thus, if m is increased, it may be necessary to make a compensatory reduction in η to maintain network stability. Combining the BP with Momentum (BPM) with the nonmonotone learning strategy (this is named NMBPM) helps to alleviate this problem.

3.   Learning Through Global Search Methods

In this section we focus on global search methods for neural network learning and we propose objective function transformation techniques that can be combined with any search method (either local or global) to alleviate the problem of occasional convergence to undesired local minima.

Adaptive stochastic search algorithms

Adaptive stochastic search algorithms include, simulated annealing [8, 24], genetic and evolutionary algorithms [33], as well as swarm intelligence [13, 22, 23]. Next, the fundamentals of those methods are reviewed.

The method of simulated annealing. *Simulated Annealing* (SA) refers to the process in which random noise in a system is systematically decreased at a constant rate so as to enhance the response of the system [24].

In the numerical optimization framework, SA is a procedure that has the capability to move out of regions near local minima [10]. SA is based on random evaluations of the objective function, in such a way that transitions out of a local minimum are possible. It does not guarantee, of course, to find the global minimum, but if the function has many good near–optimal solutions, it should find one. In particular, SA is able to discriminate between "gross behavior" of the function and finer "wrinkles". First, it reaches an area in the function domain space where a global minimizer should be present, following the gross behavior irrespectively of small local minima found on the way. It then develops finer details, finding a good, near–optimal local minimizer, if not the global minimum itself.

In the context of neural network learning the performance of the classical SA is not the appropriate one: the method needs a greater number of function evaluations than that usually required for a single run of first–order learning algorithms and does not exploit derivative related information. Notice that the problem with minimizing the neural network error function is not the well defined local minima but the broad regions that are nearly flat. In this case, the so–called Metropolis move is not strong enough to move the algorithm out of these regions [59].

In [8], it has been suggested to incorporate SA in the BP algorithm:

$$w^{k+1} = w^k - \mu \nabla E(w^k) + nc2^{-dk},$$

where n is a constant controlling the initial intensity of the noise, $c \in (-0.5, +0.5)$ is a random number and d is the noise decay constant. In the experiments reported below we have applied this technique for updating the weights from the beginning of the training as proposed by Burton *et al.* [8]. Alternatively, we update the weights using plain BP until convergence to an undesired local minimum is obtained, then we switch to SA. This combined BP with SA is named BPSA.

Genetic Algorithms. *Genetic Algorithms* (GA) are simple and robust search algorithms based on the mechanics of natural selection and natural genetics. The mathematical framework of GAs was developed in the 1960s and is presented in Holland's pioneering book [19]. GAs have been used primarily in optimization and machine learning problems and their operation is briefly described as follows. At each *generation* of a GA, a new set of approximations is created by the process of selecting individuals according to their level of fitness in the problem domain and breeding them together using operators borrowed from natural genetics. This process leads to the evolution of populations of individuals that are better suited to their environment than their progenitors, just as in natural adaptation. For a high level description of the simple GA see Figure 1.1.

More specifically, a simple GA processes a finite population of fixed length binary strings called *genes*. GAs have two basic operators, namely: *crossover* of genes and *mutation* for random change of genes. The crossover operator explores different structures by exchanging genes between two strings at a crossover position and the mutation operator is primarily used to escape the local minima in the weight space by altering a bit position of the selected string; thus introducing diversity in the population. The combined action of crossover and mutation is responsible for much of the effectiveness of GA's search. Another operator associated with each of these operators is the *selection* operator, which produces survival of the fittest in the GA.

The parallel noise–tolerant nature of GAs, as well as their hill–climbing capability, make GAs eminently suitable for training neural networks, as they seem to search the weight space efficiently. The "Genetic Algorithm for Optimization Toolbox (GAOT)" [20] has been used for the experiments reported here. GAOT's default crossover and mutation schemes, and a real–valued encoding of the ANN's weights have been employed.

Evolutionary Algorithms. *Evolutionary algorithms* (EA) are adaptive stochastic search methods which mimic the metaphor of natural

```
STANDARD GENETIC ALGORITHM MODEL
{
   //initialise the time counter
   t := 0;
   //initialise the population of individuals
   InitPopulation(P(t));
   //evaluate fitness of all individuals
   Evaluate(P(t));
   //test for termination criterion (time, fitness, etc.)
   while not done do
      t := t + 1;
      //select a sub-population for offspring production
      Q(t) := SelectParents(P(t));
      //recombine the "genes" of selected parents
      Recombine(Q(t));
      //perturb the mated population stochastically
      Mutate(Q(t));
      //evaluate the new fitness
      Evaluate(Q(t));
      //select the survivors for the next generation
      P(t + 1) := Survive(P(t), Q(t));
   end
}
```

Figure 1.1.   A high level description of the simple GA Algorithm

biological evolution. Differently from other adaptive stochastic search algorithms, evolutionary computation techniques operate on a set of potential solutions, which is called *population*, applying the principle of survival of the fittest to produce better and better approximations to a solution, and, through cooperation and competition among the potential solutions, they find the optimal one. This approach often helps finding optima in complicated optimization problems more quickly than traditional optimization methods.

To demonstrate the efficiency of the EA in alleviating the local minima problem, we have used the *Differential Evolution* (DE) strategies [53]. DE strategies have been designed as stochastic parallel direct search methods that can efficiently handle non differentiable, nonlinear and multimodal objective functions, and require few, easily chosen control parameters. Experimental results [28] have shown that DE algorithms have good convergence properties and outperform other evolutionary methods [44, 45]. To apply DE algorithms to neural network learning we start with a specific number (*NP*) of N–dimensional weight vectors, as an initial weight population, and evolve them over time. The number of individuals *NP* is kept fixed throughout the learning process and the weight vectors population is initialized randomly following a uniform probability distribution. As in GAs, at each iteration of the DE algorithm, called *generation*, new weight vectors are generated by the combination of weight vectors randomly chosen from the population, which is called *mutation*. The outcoming weight vectors are then mixed with another predetermined weight vector, the *target* weight vector. This operation is called *crossover* and it yields the so–called *trial* weight vector. This vector is accepted for the next generation if and only if it reduces the value of the error function E. This last operation is called *selection*.

Below, we briefly review the two basic DE operators used for ANN learning. The first DE operator, we consider, is mutation. Specifically, for each weight vector w_g^i, $i = 1, \ldots, NP$, where g denotes the current generation, a new vector v_{g+1}^i (mutant vector) is generated according to one of the following relations:

$$\text{Alg. DE}_1: \quad v_{g+1}^i = w_g^{r_1} + \xi \left(w_g^{r_1} - w_g^{r_2} \right), \tag{1.11}$$

$$\text{Alg. DE}_2: \quad v_{g+1}^i = w_g^{\text{best}} + \xi \left(w_g^{r_1} - w_g^{r_2} \right), \tag{1.12}$$

$$\text{Alg. DE}_3: \quad v_{g+1}^i = w_g^{r_1} + \xi \left(w_g^{r_2} - w_g^{r_3} \right), \tag{1.13}$$

$$\text{Alg. DE}_4: \quad v_{g+1}^i = w_g^i + \xi \left(w_g^{\text{best}} - w_g^i \right) + \xi \left(w_g^{r_1} - w_g^{r_2} \right), \tag{1.14}$$

$$\text{Alg. DE}_5: \quad v_{g+1}^i = w_g^{\text{best}} + \xi \left(w_g^{r_1} - w_g^{r_2} \right) + \xi \left(w_g^{r_3} - w_g^{r_4} \right), \tag{1.15}$$

$$\text{Alg. DE}_6: \quad v_{g+1}^i = w_g^{r_1} + \xi \left(w_g^{r_2} - w_g^{r_3} \right) + \xi \left(w_g^{r_4} - w_g^{r_5} \right), \tag{1.16}$$

where w_g^{best} is the best member of the previous generation, $\xi > 0$ is a real parameter, called mutation constant, which controls the amplification of the difference between two weight vectors, and

$$r_1, r_2, r_3, r_4, r_5 \in \{1, 2, \ldots, i-1, i+1, \ldots, NP\}$$

are random integers mutually different and different from the running index i.

Relation (1.11) has been introduced as crossover operator for GAs [33] and is similar to Relations (1.12) and (1.13). The remaining relations are modifications which can be obtained by the combination of (1.11), (1.12) and (1.13). It is clear that many more relations of this type can be generated using the above ones as building blocks. In recent works [44, 45], we have shown that the above relations can efficiently be used to train ANNs with arbitrary integer weights as well.

The second DE operator, i.e. the crossover, is applied to increase the diversity of the mutant weight vector. Specifically, for each component j ($j = 1, 2, \ldots, N$) of the mutant weight vector v_{g+1}^i, we randomly choose a real number r in the interval $[0, 1]$. Then, this number is compared with the crossover constant ρ; if $r \leqslant \rho$ we replace the j-th component of the trial vector u_{g+1}^i with the j-th component of the mutant vector v_{g+1}^i; otherwise, we pick the j-th component of the target vector w_g^i.

The particle swarm optimization method. In *Particle Swarm Optimization* (PSO) algorithm the population dynamics simulates a "bird flock's" behavior where social sharing of information takes place and individuals can profit from the discoveries and previous experience of all other companions during the search for food. Thus, each companion, called *particle*, in the population, which is now called *swarm*, is assumed to "fly" over the search space in order to find promising regions of the landscape. For example, in the minimization case, such regions possess lower functional values than other visited previously. In this context, each particle is treated as a point in a N–dimensional space which adjusts its own "flying" according to its flying experience as well as the flying experience of other particles (companions).

There are many variants of the PSO proposed so far, after Eberhart and Kennedy introduced this technique [13, 22]. In our experiments we have used a version of this algorithm, which is derived by adding a new inertia weight to the original PSO dynamics [12]. This version is described in the following paragraphs.

First let us define the notation used: the i-th particle of the swarm is represented by the N–dimensional vector $X_i = (x_{i1}, x_{i2}, \ldots, x_{iN})$ and the best particle in the swarm, i.e. the particle with the smallest function value, is denoted index g. The best previous position (the position giving the best function value) of the i-th particle is recorded and represented as $P_i = (p_{i1}, p_{i2}, \ldots, p_{iN})$, and the position change (velocity) of the i-th particle is $V_i = (v_{i1}, v_{i2}, \ldots, v_{iN})$.

The particles are manipulated according to the equations

$$v_{in} = w\, v_{in} + c_1 r_1 (p_{in} - x_{in}) + c_2 r_2 (p_{gn} - x_{in}), \qquad (1.17)$$

$$x_{in} = x_{in} + v_{in}, \qquad (1.18)$$

where $n = 1, 2, \ldots, N$; $i = 1, 2, \ldots, NP$ and NP is the size of population; w is the inertia weight; c_1 and c_2 are two positive constants; r_1 and r_2 are two random values in the range $[0, 1]$.

The first equation is used to calculate i-th particle's new velocity by taking into consideration three terms: the particle's previous velocity, the distance between the particle's best previous and current position, and, finally, the distance between swarm's best experience (the position of the best particle in the swarm) and i-th particle's current position. Then, following the second equation, the i-th particle flies toward a new position. In general, the performance of each particle is measured according to a predefined fitness function, which is problem–dependent.

The role of the inertia weight w is considered very important in PSO convergence behavior. The inertia weight is employed to control the impact of the previous history of velocities on the current velocity. In this way, the parameter w regulates the trade–off between the global (wide–ranging) and local (nearby) exploration abilities of the swarm. A large inertia weight facilitates global exploration (searching new areas), while a small one tends to facilitate local exploration, i.e. fine–tuning the current search area. A suitable value for the inertia weight w usually provides balance between global and local exploration abilities and consequently a reduction on the number of iterations required to locate the optimum solution. A general rule of thumb suggests that it is better to initially set the inertia to a large value, in order to make better global exploration of the search space, and gradually decrease it to get more refined solutions, thus a time decreasing inertia weight value is used.

From the above discussion it is obvious that PSO, to some extent, resembles EAs. However, in PSO, instead of using genetic operators, each individual (particle) updates its own position based on its own search experience and other individuals (companions) experience and discoveries. Adding the velocity term to the current position, in order to generate the next position, resembles the mutation operation in evolutionary programming. Note that in PSO, however, the "mutation" operator is guided by particle's own "flying" experience and benefits by the swarm's "flying" experience. In another words, PSO is considered as performing mutation with a "conscience", as pointed out by Eberhart and Shi [12].

In general, PSO has been proved very efficient in a plethora of application in science and engineering [23, 38–40]

Transforming the objective function

Let a point $\bar{w}$ such that there exists a neighborhood $\mathcal{B}$ of $\bar{w}$ with

$$E(\bar{w}) \leqslant E(w), \quad \forall\, w \in \mathcal{B}. \tag{1.19}$$

This point is a local minimizer of the error function and, as already mentioned above, many methods get stuck in such undesired local minima. The main idea of applying a transformation to the error function is to make some undesired local minima disappear, while keeping the location of the global minimizer unchanged. The techniques that will be described below aim at transforming the error function in such a way that convergence to a global minimizer is enhanced for *any* learning algorithm that is equipped with them. Two methods are described: the *deflection procedure* and the *function stretching* technique.

The deflection procedure. Following the *deflection procedure* proposed in [30], when the sequence of weight vectors $\{w^k\}_0^\infty$ converges to a local minimum $\bar{w} \in \mathbb{R}^N$ the error function $E(w)$ is reformulated as follows:

$$F(w) = S(w; \bar{w}, \lambda)^{-1} E(w),$$

where $S(w; \bar{w}, \lambda)$ is a function depending on a weight vector w and on the local minimizer $\bar{w}$ of E; λ is a relaxation parameter. In case there exist m local minima $\bar{w}_1, \ldots, \bar{w}_m \in \mathbb{R}^N$, the above relation is reformulated as:

$$F(w) = S(w; \bar{w}_1, \lambda_1)^{-1} \cdots S(w; \bar{w}_m, \lambda_m)^{-1} E(w).$$

The deflection procedure suggests to find a "proper" $S(\cdot)$ such that $F(w)$ will not have a minimum at $\bar{w}_i, i = 1, \ldots, m$, while keeping all other minima of E locally "unchanged". In other words, we have to construct functions S that provide F with the property that any sequence of weights converging to $\bar{w}_i$ (a local minimizer of E) will not produce a minimum of F at $w = \bar{w}_i$. In addition, this function F will retain all other minima of E. This is *the deflection property* [30]. For example, the function:

$$S(w; \bar{w}_i, \lambda_i) = \tanh\left(\lambda_i \|w - \bar{w}_i\|\right),$$

provides F with this property, as it will be explained below.

Let us assume that a local minimum $\bar{w}_i$ has been determined, then

$$\lim_{w \to \bar{w}_i} \frac{E(w)}{\tanh\left(\lambda \|w - \bar{w}_i\|\right)} = +\infty,$$

which means that $\bar{w}_i$ is no longer a local minimizer of F. Moreover, it is easily verified that for $\|w - \bar{w}_i\| \geqslant \varepsilon$, where ε is a small positive constant, it holds that:

$$\lim_{\lambda \to +\infty} F(w) = \lim_{\lambda \to +\infty} \frac{E(w)}{\tanh\left(\lambda \|w - \bar{w}_i\|\right)} = E(w), \qquad (1.20)$$

since the denominator tends to unity. This means that the error function remains unchanged in the whole weight space.

It is worth noticing that the effect of the deflection procedure is problem–dependent and is related to the value of λ. For an arbitrary value of λ there is a small neighborhood $\mathcal{R}(\bar{w}, \rho)$ with center $\bar{w}$ and radius ρ, with $\rho \propto \lambda^{-1}$, that for any $x \in \mathcal{R}(\bar{w}, \rho)$ it holds that $F(w) > E(w)$. To be more specific, when the value of λ is small (say $\lambda < 1$) the denominator in the above relation becomes one for w "far" from $\bar{w}$. Thus, the deflection procedure affects a large neighborhood around $\bar{w}$ in the weight space. On the other hand, when the value of λ is large, new local

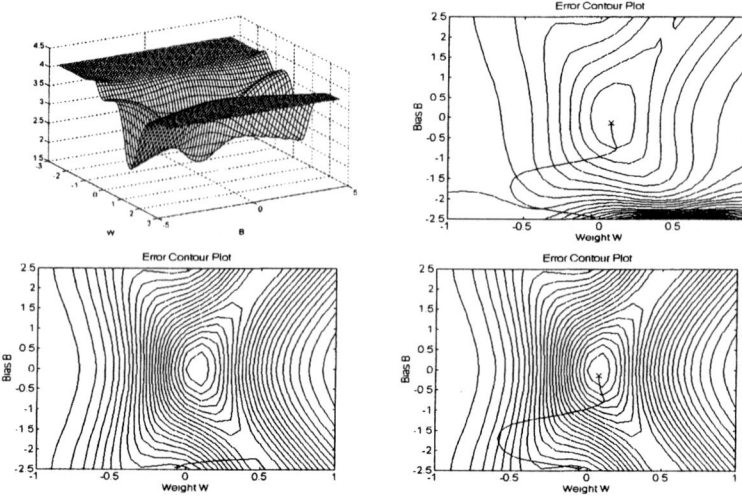

Figure 1.2.   Applying deflection to a simple learning task

minima is possible to be created near the computed minimum $\bar{w}$, like a "Mexican hat". These minima have function values greater than $F(\bar{w})$ and can be easily avoided by taking a proper stepsize or by changing the value of λ.

To better visualize the effect of the deflection procedure, we provide an application example. It concerns training a single neuron using the BP algorithm to associate 8 input–output pairs. The error surface of the problem is shown in Fig. 1.2 (top–left). The desired minimum is located at the center and there are two valleys that lead to undesired local minima. In Fig. 1.2 (bottom–left) we illustrate the weight trajectory when the initial conditions lead the learning algorithm to converge to an undesired local minimum. In Fig. 1.2 (top–right) and in Fig. 1.2 (bottom–right) we present the deflected trajectory of weights drawn on the contour lines of the original and the error function subject to deflection, respectively.

Notice that the deflection procedure can be incorporated in any learning algorithm to help escaping the influence of local minima. In the experiments reported below, the classical BP method has been equipped with the deflection procedure. The resulting scheme is named BP with deflection (BPD).

The function "stretching" technique. The *function "stretching" technique* [37] consists of a two–stage transformation in the form of the original error function $E(w)$ and can be applied soon after a local minimum $\bar{w}$ of the function E has been detected:

$$G(w) = E(w) + \frac{\gamma_1}{2} \|w - \bar{w}\| \left(\text{sign}(E(w) - E(\bar{w})) + 1\right), \quad (1.21)$$

$$H(w) = G(w) + \gamma_2 \frac{\text{sign}\left(E(w) - E(\bar{w})\right) + 1}{2\tanh\left(\mu(G(w) - G(\bar{w}))\right)}, \quad (1.22)$$

where γ_1, γ_2 and μ are arbitrary chosen positive constants, and $\text{sign}(\cdot)$ defines the well known three valued sign function. Note that the sign function can be approximated by the well known logistic function:

$$\text{sign}(w) \approx \text{logsig}(w) = \frac{2}{1 + \exp(-\nu w)} - 1 \equiv \tanh\left(\frac{\nu}{2} w\right),$$

for a large value of ν. This sigmoid function is continuously differentiable and is widely used as a transfer function in artificial neurons.

It is worth noticing that the first transformation stage elevates $E(w)$ and makes disappear all the local minima located above $\bar{w}$. The second stage stretches the neighborhood of $\bar{w}$ upwards, since it assigns higher function values to those points. Both stages do not alter the local minima located below $\bar{w}$; thus, the global minimizer is left unchanged.

At this point it is useful to provide an application example of this technique in order to illustrate its effect. The problem considered is a notorious two dimensional test function, called the *Levy No. 5*:

$$f(x) = \sum_{i=1}^{5} i\cos[(i+1)x_1 + i] \times \sum_{j=1}^{5} j\cos[(j+1)x_2 + j] +$$

$$+(x_1 + 1.42513)^2 + (x_2 + 0.80032)^2, \quad (1.23)$$

where $-10 \leqslant x_i \leqslant 10, i = 1, 2$. There are about 760 local minima and one global minimum with function value $f^* = -176.1375$ located at $x^* = (-1.3068, -1.4248)$. The large number of local optimizers makes extremely difficult for any method to locate the global minimizer. In Fig. 1.3, the original plot of the *Levy No. 5* into the cube $[-2, 2]^2$ is shown.

After applying the transformation of Eq. 1.21 (first stage of function "stretching") to the *Levy No. 5*, the new form of the function is shown in Fig. 1.4 (left). As one can see, local minima with higher functional values than the "stretched" local minimum disappeared, while lower minima as well as the global one have been left unaffected. In Fig. 1.4 (right), the final landscape, derived after applying the second transformation stage to the *Levy No. 5*, is presented. It is clearly shown how the whole neighborhood of the local minimum has been elevated; thus, the former local

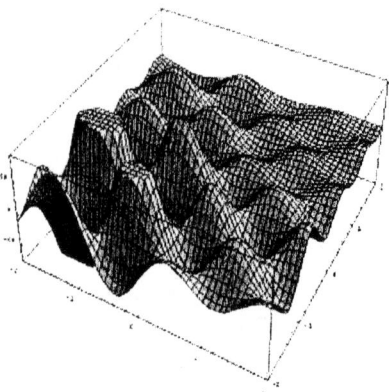

Figure 1.3.   The original plot of the function *Levy No. 5*.

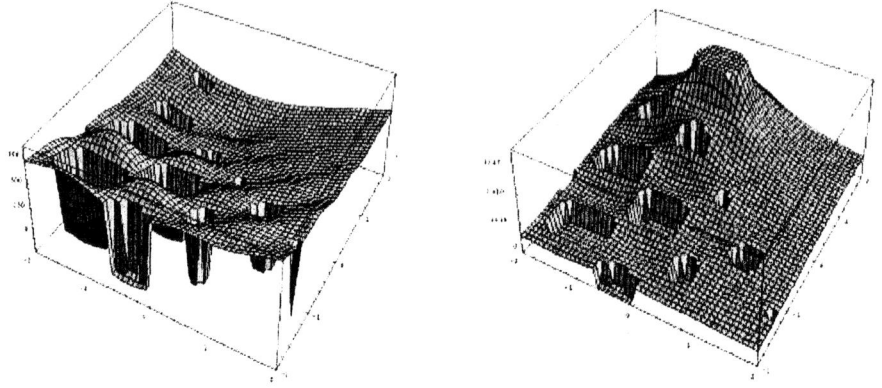

Figure 1.4.   Plot of the *Levy No. 5* after the first stage (left) and after the second stage (right) of the function "stretching" technique.

minimum has now turned to be a local maximum of the function. Details on the performance of the PSO algorithm combined with the function "stretching" technique (SPSO) on two well known test problems, as well as suggestions for selecting parameter values, are presented in the next section.

4.   Experiments and discussion

Experiments have been performed to evaluate the learning methods mentioned in the previous sections and compare their performance. Below, we exhibit results on two notorious for their local minima problems. The algorithms have been tested using initial weights chosen from the uniform distribution in the interval $(-1, 1)$. Note that BPSA and BPD update the weights using BP until convergence to a global or local minimum is obtained: the weight vector w^k is considered as a global minimizer when $E(w^k) \leqslant 0.04$. Convergence to a local minimizer is related to the magnitude of the gradient vector, i.e. when the stopping condition $\|\nabla E(w^k)\| \leqslant 10^{-3}$ is met, w^k is taken as a local minimizer $\bar{w}_i$ of the error function E.

No effort has been made to tune the mutation and crossover parameters, ξ and ρ respectively. We have used the fixed values $\xi = 0.5$ and $\rho = 0.7$, instead. The weight population size NP has been chosen to be twice the dimension of the problem, i.e. $NP = 2N$, for all the simulations considered. Some experimental results have shown that a good choice for NP is $2N \leqslant NP \leqslant 4N$. It is obvious that the exploitation of the weight space is more effective for large values of NP, but sometimes more error function evaluations are required. On the other hand, small values of NP make the algorithm inefficient and more generations are required in order to converge to the minimum.

In all the PSO simulations reported, the values of γ_1, γ_2 and μ were fixed: $\gamma_1 = 10000, \gamma_2 = 1$ and $\mu = 10^{-10}$. The balance between the global and local exploration abilities of the SPSO is mainly controlled by the inertia weights, since the particles' positions are updated according to the classical PSO strategy. A time decreasing inertia weight value, i.e. start from 1 and gradually decrease towards 0.4, has been found to work better than using a constant value. This is because large inertia weights help to find good seeds at the beginning of the search, while, later, small inertia weights facilitate a finer search.

Notice that for the BP, BPM, BBP, SA, BPSA and BPD methods each iteration corresponds to one gradient and one error function evaluation, differently from the BPVS, NMBPM, NMBBP and NMBPVS where, in general, the number of error Function Evaluations (FE) is larger than the number of Gradient Evaluations (GE), due to the use of the line search. In the table below, there are two rows for these algorithms; the first one indicates the statistics for the FE and the second for the GE. On the other hand, a key feature of GA, DE, PSO and SPSO algorithms is that *only* error function values are needed.

1) The XOR classification problem: classification of the four XOR patterns in one of two classes, $\{0, 1\}$, using a 2–2–1 ANN is a classical test problem [50, 54]. The XOR problem is sensitive to initial weights and presents a multitude of local minima [7]. The stepsize is taken equal to 1.5 and the heuristics for SA, BPSA and PSO are tuned to $n = 0.3$, $d = 0.002$ and $c_1 = c_2 = 0.5$. In all instances, 100 simulations have been run and the results are summarized in Table 1.1.

2) The three bit parity problem [50]: a 3-3-1 ANN receives eight, 3-dimensional binary input patterns and must output an "1" if the inputs have an odd number of ones and "0" if the inputs have an even number of ones. This is a very difficult problem for an ANN because the network must determine the proper parity (the value at the output) for input patterns which differ only by Hamming distance 1. It is well known that the network's weight space contains "bad" local minima. The stepsize has been taken equal to 0.5 and the heuristics for SA, BPSA and PSO have been tuned to $n = 0.1$, $d = 0.00025$, $c_1 = 0.1$ and $c_2 = 1$. In all instances, the results of 100 simulations are summarized in Table 1.1.

The results suggest that combination of local and global search methods like BPSA and BPD provide a better probability of success than the BP. Note that the performance of BPSA is not the appropriate one although derivative related information has been used. On the other hand, BPD escapes local minima and converges to the global minimum in all cases. A consideration that is worth mentioning is that the number of function evaluations in BPSA and BPD contains the additional evaluations required for BP to satisfy the local minima stopping condition. The results also indicate that the GA and the DE are promising and effective, even when compared with other methods that require the gradient of the error function, in addition to the error function values. For example, GAs as well as DE_3 and DE_4 have exhibited very good performance for the test problems considered. On the other hand, there have been cases where a discrepancy has been found in DE's behavior; see for example DE_5 and DE_6. For a discussion on the generalization capabilities of the networks generated by the DE algorithms see [43, 45]. Finally, the PSO algorithm combined with the function "stretching" technique (SPSO) has exhibited improved success rate, although it needed additional iterations to converge.

In conclusion, global search methods provide techniques that alleviate the problem of occasional convergence to local minima in neural network learning. Escaping from local minima is not always possible, however these methods exhibit a better chance in locating appropriate solutions and, in that sense, they improve the efficiency of the learning process. Experiments indicate that learning algorithms equipped with the proposed error function transformation techniques are capable to escape from undesired local minima and locate a desired one effectively. The deflection procedure and the function "stretching" technique provide stable convergence and thus a better probability of success for a learning algorithm. In general, the results exhibited by the proposed methods on two notorious for their local minima problems are promising.

Training		XOR Problem			Parity Problem		
Method		Mean	s.d.	Succ.	Mean	s.d.	Succ.
BP		144.1	112.6	42%	932.0	1320.8	91%
BPM		249.7	322.1	49%	219.9	198.9	93%
BBP		93.3	201.5	71%	150.3	137.3	94%
NMBPM	(FE)	260.4	287.8	68%	244.3	205.9	99%
	(GE)	254.4	287.3		235.1	204.4	
NMBBP	(FE)	191.6	328.9	80%	106.6	123.1	99%
	(GE)	102.1	173.4		99.2	164.5	
BPVS	(FE)	199.1	373.1	78%	105.8	186.9	98%
	(GE)	185.2	343.3		100.4	171.6	
NMBPVS	(FE)	208.4	395.2	80%	102.1	109.9	99%
	(GE)	201.3	378.8		95.3	183.5	
SA		424.2	420.8	43%	805.4	2103.1	22%
BPSA		1661.9	2775.7	65%	2634.0	6866.8	66%
GA		422.3	397.5	95%	1091.5	766.2	73%
DE_1		192.9	124.7	75%	622.6	522.1	91%
DE_2		284.9	216.2	80%	1994.1	657.6	61%
DE_3		583.9	256.3	97%	896.3	450.6	99%
DE_4		706.1	343.7	98%	1060.2	716.6	98%
DE_5		300.5	250.2	85%	2112.0	644.9	26%
DE_6		482.9	264.9	93%	2062.5	794.8	44%
PSO		1459.7	1143.1	77%	6422.4	2992.1	42%
SPSO		7869.6	13905.4	100%	9803.6	5436.6	95%
BPD		575.1	387.3	100%	760.0	696.4	100%

Table 1.1. Comparative results

Acknowledgments

The authors would like to thank the European Social Fund, Operational Program for Educational and Vocational Training II (EPEAEK II), and particularly the Program PYTHAGORAS for funding the above work. Dr V.P. Plagianakos and Prof. M.N. Vrahatis also acknowledge the financial support of the University of Patras Research Committee through a "Karatheodoris" research grant.

References

[1] N. Ampazis and S.J. Perantonis, (2002). Two Highly Efficient Second Order Algorithms for Training Feedforward Networks, *IEEE Transactions on Neural Networks*, **13**, 1064–1074.

[2] L. Armijo, (1966). Minimization of Functions Having Lipschitz–continuous First Partial Derivatives, *Pacific Journal of Mathematics*, **16**, 1–3.

[3] J. Barzilai and J.M. Borwein, (1988). Two Point Step Size Gradient Methods, *IMA Journal of Numerical Analysis*, **8**, 141–148.

[4] R. Battiti, (1989). Accelerated Backpropagation Learning: Two Optimization Methods, *Complex Systems*, **3**, 331–342.

[5] R. Battiti, (1992). First– and Second–order Methods for Learning: Between Steepest Descent and Newton's Method, *Neural Computation*, **4**, 141–166.

[6] D.P. Bertsekas, (1995). Nonlinear Programming, Belmont, MA: Athena Scientific.

[7] E.K. Blum, (1989). Approximation of Boolean Functions by Sigmoidal Networks: Part I: XOR and Other Two Variable Functions, *Neural Computation*, **1**, 532–540.

[8] M. Burton Jr. and G.J. Mpitsos, (1992). Event Dependent Control of Noise Enhances Learning in Neural Networks, *Neural Networks*, **5**, 627–637.

[9] L.W. Chan and F. Fallside, (1987). An Adaptive Training Algorithm for Back–propagation Networks, *Computers Speech and Language*, **2**, 205–218.

[10] A. Corana, M. Marchesi, C. Martini, and S. Ridella, (1987). Minimizing Multimodal Functions of Continuous Variables with the Simulated Annealing Algorithm, *ACM Transactions on Mathematical Software*, **13**, 262–280.

[11] J.E. Dennis and R.B. Schnabel, (1983). *Numerical Methods for Unconstrained Optimization and Nonlinear Equations*, Englewood Cliffs, Prentice–Hall.

[12] R.C. Eberhart and Y.H. Shi, (1998). Evolving Artificial Neural Networks, *Proceedings International Conference on Neural Networks and Brain*, Beijing, P.R. China.

[13] R.C. Eberhart, P.K. Simpson and R.W. Dobbins (1996). *Computational Intelligence PC Tools*, Academic Press Professional, Boston, MA.

[14] S.E. Fahlman (1988). Faster–learning Variations on Back–propagation: An Empirical Study, D.S. Touretzky, G.E. Hinton and T.J. Sejnowski (Eds.), *Proceedings of the 1988 Connectionist Models Summer School*, 38–51, San Mateo, Morgan Koufmann.

[15] A.V. Fiacco and G.P. McCormick (1990). *Nonlinear Programming: Sequential Unconstrained Minimization Techniques*, Philadelphia, SIAM.

[16] M. Gori and A. Tesi, (1992). On the Problem of Local Minima in Backpropagation, *IEEE Transactions on Pattern Analysis and Machine Intelligence*, **14**, 76–85.

[17] L. Grippo, F. Lampariello, and S. Lucidi, (1986). A Nonmonotone Line Search Technique for Newton's Method, *SIAM Journal on Numerical Analysis*, **23**, 707–716.

[18] M.T. Hagan and M. Menhaj, (1994). Training Feedforward Networks with the Marquardt Algorithm, *IEEE Transactions on Neural Networks*, **5**, 989–993.

[19] J.H. Holland, (1975). *Adaptation in Neural and Artificial Systems*, University of Michigan Press.

[20] C. Houck, J. Joines, and M. Kay, (1995). *A Genetic Algorithm for Function Optimization: A Matlab Implementation*, NCSU–IE TR, 95–09.

[21] R.A. Jacobs, (1988). Increased Rates of Convergence Through Learning Rate Adaptation, *Neural Networks*, **1**, 295–307.

[22] J. Kennedy and R.C. Eberhart, (1995). Particle Swarm Optimization, *Proceedings IEEE International Conference on Neural Networks*, Piscataway, NJ, IV:1942–1948.

[23] J. Kennedy and R.C. Eberhart, (2001). *Swarm Intelligence*, Morgan Kaufmann Publishers.

[24] S. Kirkpatrick, C.D. Gelatt Jr., and M.P. Vecchi, (1983). Optimization by Simulated Annealing, *Science*, **220**, 671–680.

[25] S. Kollias and D. Anastassiou, (1989). An Adaptive Least Squares Algorithm for the Efficient Training of Multilayered Networks, *IEEE Transactions on Circuits Systems*, **36**, 1092–1101.

[26] Y. Lee, S.H. Oh, and M. Kim, (1993). An Analysis of Premature Saturation in Backpropagation Learning, *Neural Networks*, **6**, 719–728.

[27] G.D. Magoulas, V.P. Plagianakos, and M.N. Vrahatis, (2002). Globally Convergent Algorithms with Local Learning Rates, *IEEE Transactions Neural Networks*, **13**, 774–779.

[28] G.D. Magoulas, V.P. Plagianakos, and M.N. Vrahatis, (2004). Neural Network-based Colonoscopic Diagnosis Using On-line Learning and Differential Evolution, *Applied Soft Computing*, **4**, 369–379.

[29] G.D. Magoulas, M.N. Vrahatis, and G.S. Androulakis, (1997). Effective Back–propagation with Variable Stepsize, *Neural Networks*, **10**, 69–82.

[30] G.D. Magoulas, M.N. Vrahatis, and G.S. Androulakis, (1997). On the Alleviation of Local Minima in Backpropagation, *Nonlinear Analysis, Theory, Methods and Applications*, **30**, 4545–4550.

[31] G.D. Magoulas, M.N. Vrahatis, and G.S. Androulakis, (1999). Improving the Convergence of the Back–propagation Algorithm Using Learning Rate Adaptation Methods, *Neural Computation*, **11**, 1769–1796.

[32] G.D. Magoulas, M.N. Vrahatis, T.N. Grapsa, and G.S. Androulakis,(1997). *Neural Network Supervised Training Based on a Dimension Reducing Method, Mathematics of Neural Networks, Models, Algorithms and Applications*, S.W. Ellacott, J.C. Mason, I.J. Anderson Eds., Kluwer Academic Publishers, Boston, 245–249.

[33] Z. Michalewicz, (1996). *Genetic Algorithms + Data Structures = Evolution Programs*, Springer.

[34] M.F. Möller, (1993). A Scaled Conjugate Gradient Algorithm for Fast Supervised Learning, *Neural Networks*, **6**, 525–533.

[35] J. Nocedal, (1992). Theory of Algorithms for Unconstrained Optimization, *Acta Numerica*, **1**, 199–242.

[36] J.M. Ortega and W.C. Rheinboldt, (1970). *Iterative Solution of Nonlinear Equations in Several Variables*, Academic Press, New York.

[37] K.E. Parsopoulos, V.P. Plagianakos, G.D. Magoulas and M.N. Vrahatis, (2001). Objective Function "Stretching" to Alleviate Convergence to Local Minima, *Nonlinear Analysis, Theory, Methods and Applications*, **47**, 3419–3424.

[38] K.E. Parsopoulos and M.N. Vrahatis, (2002). Recent Approaches to Global Optimization Problems Through Particle Swarm Optimization, *Natural Computing*, **1**, 235–306.

[39] K.E. Parsopoulos and M.N. Vrahatis, (2004). On the Computation of All Global Minimizers Through Particle Swarm Optimization, *IEEE Transactions on Evolutionary Computation*, **8**, 211–224.

[40] N.G. Pavlidis, K.E. Parsopoulos and M.N. Vrahatis, (2004). Computing Nash Equilibria Through Computational Intelligence Methods, *Journal of Computational and Applied Mathematics*, in press.

[41] V.P. Plagianakos, G.D. Magoulas and M.N. Vrahatis, (2002). Deterministic Nonmonotone Strategies for Effective Training of Multi–Layer Perceptrons, *IEEE Transactions on Neural Networks*, **13**, 1268–1284.

[42] V.P. Plagianakos, D.G. Sotiropoulos, and M.N. Vrahatis, (1998). Automatic Adaptation of Learning Rate for Backpropagation Neural Networks, N.E. Mastorakis, (Ed.), *Recent Advances in Circuits and Systems* 337–341, Singapore, World Scientific.

[43] V.P. Plagianakos and M.N. Vrahatis, (1999). Neural Network Training with Constrained Integer Weights, *Proceedings of Congress on Evolutionary Computation (CEC'99)*, 2007–2013, Washington D.C.

[44] V.P. Plagianakos and M.N. Vrahatis, (2000). Training Neural Networks with Threshold Activation Functions and Constrained Integer Weights, *Proceedings of the IEEE International Joint Conference on Neural Networks (IJCNN'2000)*, Vol. **5**, pp.161–166, Como, Italy.

[45] V.P. Plagianakos and M.N. Vrahatis, (2002). Parallel Evolutionary Training Algorithms for "Hardware–Friendly" Neural Networks, *Natural Computing*, **1**, 307–322.

[46] V.P. Plagianakos, M.N. Vrahatis, and G.D. Magoulas (1999). Nonmonotone Methods for Backpropagation Training with Adaptive Learning Rate, *Proceedings of the IEEE International Joint Conference on Neural Networks (IJCNN'99)*, Vol. **3**, pp.1762–1767, Washington D.C.

[47] E. Polak, (1997). *Optimization: Algorithms and Consistent Approximations*, New York, Springer–Verlag.

[48] M. Raydan, (1997). The Barzilai and Borwein Gradient Method for the Large Scale Unconstrained Minimization Problem, *SIAM Journal on Optimization*, **7**, 26–33.

[49] A.K. Rigler, J.M. Irvine, and T.P. Vogl, (1991). Rescaling of Variables in Backpropagation Learning, *Neural Networks*, **4**, 225–229.

[50] D.E. Rumelhart, G.E. Hinton, and R.J. Williams, (1986). Learning Internal Representations by Error Propagation, *Parallel Distributed Processing: Explorations in the Microstructure of Cognition 1*, D.E. Rumelhart, J.L. McClelland Eds., MIT Press, 318–362.

[51] S. Saarinen, R. Bramley, and G. Cybenko, (1993). Ill-conditioning in Neural Network Training Problems, *SIAM Journal on Scientific Computing*, **14**, 693–714.

[52] F. Silva and L. Almeida, (1990). Acceleration Techniques for the Back–propagation Algorithm, *Lecture Notes in Computer Science*, **412**, 110–119, Berlin, Springer–Verlag.

[53] R. Storn and K. Price, (1997). Differential Evolution – A Simple and Efficient Heuristic for Global Optimization over Continuous Spaces, *Journal of Global Optimization*, **11**, 341–359.

[54] P.P. Van der Smagt, (1994). Minimisation Methods for Training Feedforward Neural Networks, *Neural Networks*, **7**, 1–11.

[55] T.P. Vogl, J.K. Mangis, A.K. Rigler, W.T. Zink, and D.L. Alkon, (1988). Accelerating the Convergence of the Back–propagation Method, *Biological Cybernetics*, **59**, 257–263.

[56] M.N. Vrahatis, G.S. Androulakis, J.N. Lambrinos and G.D. Magoulas, (2000). A Class of Gradient Unconstrained Minimization Algorithms with Adaptive Stepsize, *Journal of Computational and Applied Mathematics*, **114**, 367–386.

[57] M.N. Vrahatis, G.D. Magoulas and V.P. Plagianakos, (2000). Globally Convergent Modification of the Quickprop Method, *Neural Processing Letters*, **12**, 159–169.

[58] M.N. Vrahatis, G.D. Magoulas and V.P. Plagianakos, (2003). From Linear to Nonlinear Iterative Methods, *Applied Numerical Mathematics*, **45**, 59–77.

[59] S.T. Weslstead, (1994). *Neural Network and Fuzzy Logic Applications in C/C++*, Wiley.

[60] X.-H. Yu, G.-A. Chen, (1995). On the Local Minima Free Condition of Backpropagation Learning, *IEEE Transactions on Neural Networks*, **6**, 1300–1303.

Chapter 16

EVOLUTIONARY APPROACH TO DESIGN ASSEMBLY LINES

B. Rekiek, P. De Lit, A. Delchambre
Université Libre de Bruxelles, CAD/CAM Department
50, Av. F. Roosevelt CP. 165/14, B-1050 Brussels, Belgium

Abstract   The purpose of this paper is to describe the main problems concerning the design of assembly lines. It is composed of the following steps: (1) input data preparation (2) elaboration of the logical layout of the line, which consists in the distribution of operations among stations along the line and an assignment of resources to the different stations and (3) finally the mapping phase, allowing to check the results with a commercial simulation package. This work presents a new method to tackle the hybrid assembly lines dealing with multiple objective. The goal is to minimize the total cost of the line by integrating design (congestion, machine cost,. . .) and operation issues (cycle time, precedence constraints, availability,. . .). After an overview of the current work in this area, this paper presents, in detail, a very promising approach to solve multiple objective problems: a multiple objective grouping genetic algorithm (a grouping genetic algorithm hybridized with the multi-criteria decision-aid method PROMETHEE II). An approach to deal with user's preferences in design problems is also introduced. The essential concepts adopted by the method are described. An application of the proposed method to an industrial case study is presented.

Keywords: Assembly lines design, multiple objective problems, grouping genetic algorithm, multi-criteria decision-aid.

1.   Introduction

Assembly lines are found in all types of industries, wherever "products" may be imagined to move along from station to station. Assembly is a process by which subassemblies and components are put together yielding the finished products. The assembled product takes shape gradually, starting with one part called the base part, the remaining parts being attached at the various stations the product visits. Tasks are accomplished by a group of workers, machines or robots. After a lapse of time called the cycle time, the conveyor moves, thus positioning each product in front of the next station in the line. Figure 1.1 illustrates our words.

The design of an efficient assembly line is a problem of considerable industrial importance. Line layout problems are divided into logical and physical layout [8]. The goal of the logical layout is to assign tasks to a set of stations and to decide about their order along the line. The physical layout determines the space requirements taking into account station dimensions,

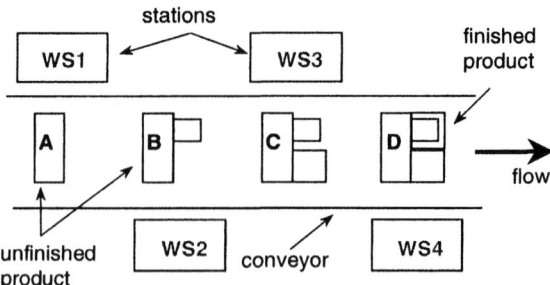

Fig. 1.1   The assembly line concept.

material storage,...In this paper the authors are more concerned with the logical line layout (LLL) of assembly lines.

In the literature, the LLL is divided into assembly line balancing (ALB) and resource planning (RP) problems. The ALB used especially for manual assembly lines (MAL) aims to balance the loads of the stations. This approach is appropriate for such systems, since the global cost of MAL is directly influenced by the number of stations. Thus, the main objective of the ALB problem is either to minimize a number of stations or to distribute a workload among them.

On an hybrid assembly line (HAL) tasks can be executed either manually, by robots or by hard automated equipment. In general, the operating time and the cost depend on the resource used. Given a list of available candidate equipment to complete the operations, the problem in the RP is to decide which resources to use and which tasks to assign to each of them. Here, the main objective is to minimize the total cost of the line by integrating design (congestion, cost,...) and operation issues (cycle time, precedence constraints and availability,...).

Assembly line design (ALD) problems are multi-criteria ones. That is, those problems involve multiple often conflicting objectives to be met (cost, availability, imbalance between station,...) and ask for a compromise among them. Since their early days, the genetic algorithms (GAs) have been viewed to be well suited for multiple objective problems (MOP). Thus, several GA-based techniques have been developed since then.

The paper is organized as follows: in section 2, a more detailed description of the ALD problem is given, discussing its constraints as well as its objectives. Section 3 presents related work concerning ALD techniques. The first phase of the integrated approach which is the "preparation of data" is introduced in section 4. Section 5 is devoted to the optimization phase while the mapping phase is described in section 6. Results of a case study are presented in section 7, and conclusions are drawn in section 8.

2.   Assembly line design

2.1   Task and resource assignment problems

The proposed method is built upon many collaborations with industrials. Its main steps can be summarized as follows (see Figure 1.2):

- Preparation: the designer introduces its input data (tasks, resources, constraints, preferences,...);

- Optimization: the optimization method proposes a line architecture (stations contents, their order,...);

- Mapping: allows the designer to analyze and test the results using a simulation package.

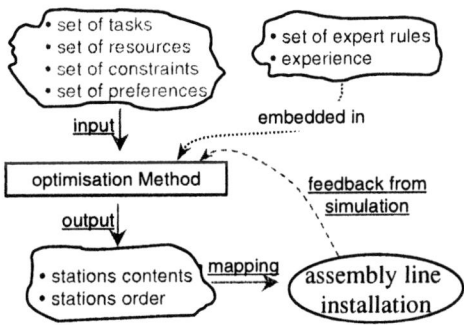

Fig. 1.2 Design method.

2.2   Problem statement

As the reader will remark, the authors will only detail the resource planing problem and not the line balancing. Indeed, if each task has its own equipment and by the way a fixed process time, the problem is transformed to an assembly line balancing one. The same technique is used to solve both of the problems.

A task in the point of view the RP is the combination of feeding, handling and insertion for each assembly operation. A set of possible groups of equipment (feeders, handlers, insertion devices) called "functional groups" (FG) [21] are attributed to each task.

In other words the RP problem can be defined as follows. Given a set of tasks, and for each of them a set of possible resources each characterized by its price, availability and speed in terms of the resulting duration of the task; given a fixed number of stations, a desired cycle time and possible precedence among tasks, find:

- the resources to be allocated to each task, among the possible ones,
- an assignment of the tasks to stations along the line, so that:
  - no precedence constraint is violated;
  - the stations workload is as close as possible to the cycle time.

The following objectives have to be met:

- the total cost of the resources allocated to tasks is as less as possible;
- a maximal availability of the line is attained;
- the workload is as balanced as possible between the stations.

The proposed method is composed of three parts namely the preparation, the optimization and finally the mapping phase. In the next sections the authors will go through these different modules.

3.   Background

Most ALD methods deal with separate objectives namely minimize the number of stations, minimize the idle time, minimize the cost, maximize availability, . . . A detailed survey on the subject can be found in [3] and [26].

Most of the research on balancing deal with the simple assembly line balancing (SALB) problems in which no alternative equipment types are considered. That is, each process time is fixed. The RP problem is proven to be an NP-Hard problem [32]. This state of the art will focus on the review of some meta-heuristic approaches to ALD problem.

On the evolutionary algorithms side, to the best of our knowledge, the first attempt to tackle SALB problems was done by [11]. Falkenauer developed a grouping GA (GGA). The authors generalized a bin packing algorithm to obtain a fast algorithm supplying high-quality approximate solutions. The aim advantage is its ability to handle problems with sparse, even empty precedence constraints.

Sureh and Sahu[28] developed a simulated annealing (SA) heuristic for the SALB problem. They considered the problem of stochastic task durations. Their aim was to minimize the smoothness index which intends to distribute work into the station as evenly as possible.

Leu et al.[17] proposed a sequence-oriented GA to deal with the ALBP. The authors used a set of heuristics to initialize the population. They also proposed several techniques to get valid solutions.

Anderson and Ferris[1] used the station-oriented GA encoding. They used the COMSOAL heuristic [2] to initialize the population. They suggested three approaches to deal with infeasible solutions. The first uses some penalty function to drive the solutions towards feasibility. A second uses a repair routine to force generated solutions to correspond to feasible ones. In the third one, the chromosome is decoded using rules which guarantee a feasible assignment.

Kim et al.[15] developed a GA to solve multiple objective ALBP. They proposed and addressed several types of ALB problems and considered five objectives: (1) minimize the number of stations; (2) minimize the cycle time; (3) maximize workload smoothness; (4) maximize work relatedness; and (5) a multiple objective with (3) and (4). The authors used the sequence-oriented coding and a repair routine to deal with infeasible solutions. Their emphasis is placed on seeking a set of Pareto optimal solutions.

Sureh et al.[29] used a GA to solve the stochastic SALBP. They proposed a GA working with two populations, the first one stores only feasible solutions while the second one deals with infeasible ones. Some solutions are exchanged at regular intervals between the two populations.

Falkenauer proposed a method based on the GGA and the B&B algorithm to deal with the ALB with resource dependant tasks times problem [9]. The GGA distributes the tasks onto stations, while the B&B algorithm selects the optimal resource for each station.

Mînzu and Henrioud[20] proposed a "kangaroo" algorithm to treat the problem of multi-product assembly lines. The method aims to minimize the maximum work content of the stations, which leads to a well balanced line. The computational tests prove that the algorithm supplies good solutions in a small number of iterations.

McMullen and Frazier[19] presented a SA method to address the ALB with multiple objectives. Two kind of objectives were treated: the single and the composite objectives. Three single objectives were considered: (1) minimize the cost of the line, (2) minimize the smoothness, and (3) minimize the probability of lateness due to the stochastic nature of task duration. Also, three composite objectives which are expressed as the weighted sum of the first and the third objective. Several heuristics (called trade, transfer, compression and the expansion) were used as local improvements.

Ponnambalam et al.[22] used the sequence-oriented encoding for a multi-objective GA (MO-GA). They used 14 simple heuristics to initialize the population. The method aims to maximize the line efficiency. During the execution of the MO-GA, a set of Pareto optimal solutions are stored and updated at each generation.

Sabuncuoglu et al.[25] proposed a sequence-oriented GA to deal with the ALBP. They used the two-point crossover and the scramble mutation, that is a random cut-point is selected and the genes after the cut-point are randomly replaced. The authors also proposed a method called "dynamic partitioning" that modifies the chromosome to save computation time. The method modifies the chromosome by allocating tasks to stations (i.e. freezing certain tasks) that satisfy some criteria, and continues with the remaining unfrozen tasks.

Little concern has been given to the physical demands placed on workers when assigning tasks to stations. Carnahan proposed a methodology considering both production objectives (cycle time and number of station) as well as worker physical constraints [6]. The method is based on the sequence-oriented GA and applied to the SALBP-2 [26]. The authors used two search algorithms: (1) a multiple rank heuristic (MRH) and (2) a problem space GA (PSGA). The MRH is a combination of 81 separate heuristics utilizing three search methods, three ranking criteria, three tasks assignment and three weighting factors. The PSGA uses these 81 heuristics to initialize the population.

Lee et al.[16] presented a sequence-oriented GA hybridized with the trade and transfer heuristics. The trade procedure is used to exchange tasks between adjacent stations. The transfer procedure has two varieties: the expansion transfer aims to create additive stations if needed, while the compression transfer aims to decrease the number of stations. The method aims to minimize the weighted sum of the total cost of the line and the lateness across all stations.

The sequence-oriented representation is also used in [18]. The authors put the accent on the initialization phase. The first step is the creation of the precedence table (the tasks having predecessors). Next a list of tasks without predecessors is compiled and finally a task without a preceding task is randomly selected from the list and assigned to the first position of the solution string. The list of tasks without predecessors is updated and another task without preceding is selected and assigned and so on. The mutation operator works as follows: all tasks to the left of the mutation point are deleted from the chromosome and the initialization procedure is called thereafter to reconstruct the solution.

When covering the literature on multiple objective problems (MOP) solving methods, it seems that the main difference among these methods is the way the solutions are ranked. The ranking approach uses methods that can be classified in one of three ways: the aggregating approaches, the non-Pareto approaches and the Pareto approaches. In the late eighties, [13] published his method called non-dominated sorting, and search techniques started to use the concept of Pareto optimality through selection and ranking methods. Since then, numerous approaches to solve MOP have appeared. For a comprehensive review, the reader is suggested to refer to the overviews of the different MO-GA methods presented in [12], [7] and [30].

4.   Data preparation (phase 1)

Once the product and the existing resources of the enterprise has been analyzed a set of assembly plans are proposed as well as their preferable resources. The results of the preparation phase will only be presented. For more details about this phase the reader is suggested to refer to [21]. The method yields the following input for the optimization phase, as illustrated on Figure 1.3:

- the desired number of stations,

- the desired cycle time,

- for each task:

    - the precedence constraints between this task and the other ones,

    - the user's operating mode preferences (manual, automated or robotic),

- an equipment database which yields the features of the different resources (cost, availability, process time, occupied area).

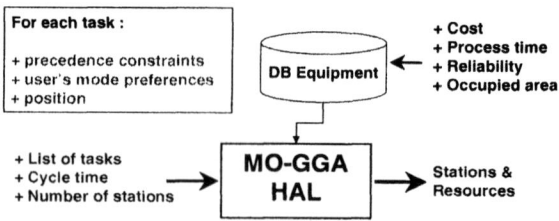

Fig. 1.3   Data flow for the ALD method.

Precedence graph

The precedence graph of a product is a partial ordering in which tasks must be performed. The nodes of the graph represent tasks and the directed arcs (i, j) constitute the precedence relationship as shown in Figure 1.4.

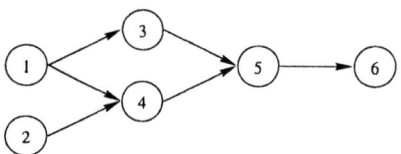

Fig. 1.4   Example of precedence graph of 6 operations.

Cost of a functional group

The cost of a functional group will be given by the sum of the costs of its pieces of equipment. They will include: (1) the purchase cost, (2) the exploitation and the maintenance cost, (3) the manpower cost, including possible training,. . . and (4) the consumption cost [21].

Process time

The estimation of the process time of an elementary task is far from being simple. There is still a lack of reliable tools for the estimation of these times, except for the manual feeding, handling and insertion where the work of [4] has brought a considerable improvement.

Availability

The availability of an equipment is defined as the proportion of total time that it will be available for use. When several FGs are grouped on the same station, the availability of the station is the product of the availabilities of the FGs.

Grouping preferences for HAL

Three possible methods for each operation are considered (manual, robotic and automated). The compatibility between the different modes yields a set of dissociative preferences (manual tasks cannot be grouped with robotic or automated ones). On the other side, one of the industrial preoccupations is the recovery of existing stations (heavy machines or robots). Two types of virtual operations were introduced: (1) fixed operations on stations: some operations have to be fixed on a given station (control station, paint station,. . .) and no additional operation can be added to it, (2) linked operations: a set of operations must be grouped on the same station but additional operations can be added. These preferences are introduced by the designer as hard constraints of the problem.

5. Assembly line design: optimization phase (phase 2)

This part constitutes the evolutionary computation part of the methodology. The approach is based on the genetic algorithm and many industrial designers' ideas, which are embedded in the method as heuristics. Since ALD can be simply described as a problem of assignment of tasks to stations, it can be easily viewed as a grouping problem. In order to deal with multiple objective nature of ALD, a multiple objective grouping genetic algorithm will be presented in section 5.2.

5.1 Grouping genetic algorithm

Falkenauer pointed out the weaknesses of standard GAs [14] when applied to grouping problems, and introduced the grouping genetic algorithm (GGA), which is a GA heavily modified to match the structure of grouping problems. Those are the problems where the aim is to group together members of a set (i.e. find a good partition of the set). The GGA operators (crossover, mutation and inversion) are group-oriented, in order to follow the structure of grouping problems. For more details about the GGA and its applications, the reader will refer to [10].

5.2 Overall architecture of the MO-GGA

Applying GAs to solve MOPs has to deal with the twin issues of searching large and complex solution spaces and dealing with multiple potentially conflicting objectives. Classical methods reduce the problem to a mono-criterion one —such as the popular weighted-sum approach. Many studies adopted the Pareto-based GA search to sample the solution space [7]. Few researchers have suggested ways of integrating multicriteria decision-aid (MCDA) methods and the GA search. The GA iteratively samples the tradeoff surface while the MCDA method narrows the search.

The MOPs involve two "quasi-inseparable" difficulties, namely search and multi-criteria decision making. The space to search inside can be too large to be enumerated, and too complex to be explored by simple search methods. The objectives may be conflicting, this is why a tradeoff has to be made by a rational decision maker (DM).

In MOPs there is no common agreement on what optimum really means. Thus, the MOP is defined as the problem of finding a vector of decision variables which optimizes a vector function whose elements represent the objective functions. The word "optimize" means finding a solution which would give acceptable values for all the involved objectives. It has several interpretations within this context, and it is up to the DM to decide which solution best fits his desiderata.

The two classical (pragmatic) strategies that were applied to the traditional separation of search and multi-criteria decisions can be described as follows.

First, make multi-criteria decisions to aggregate objectives, then apply the search method to optimize the resulting figure of merit. The different objectives are combined to form a scalar objective function, usually through a linear combination (weighted sum) of the attributes. The weights represent the importance of each objective. The main drawback is that it may use the sum of values of two totally different objectives (in the case of ALD, it could be the sum of the cost and the availability values), which makes no sense. In general, the obtained "optimal" solution is a function of the coefficients used to combine the objectives.

Conduct the search using the different objectives at the same level of importance. In general, the objectives compete, in the sense that an improvement of one objective will lead to a degradation of others. The idea is to search not for a single solution but for a set of solutions that represent the "best tradeoffs". This approach yields the Pareto frontier. The search phase is then followed by making multi-criteria decisions to choose among the reduced set. This approach is generally considered to be a "best practice". Nevertheless, the problem is the number of solutions the DM has to choose among which can become unmanageable for a DM.

The approach proposed here is based on a merge of a search and MCDA as illustrated in Figure 1.5. Indeed, in order to come out of the MOP stated by the cost function, the MCDA method PROMETHEE II is used. For more detail about this method, the reader is invited to refer to [5]. It is however important to know that it computes a flow ϕ which is a kind of fitness for each solution. This "fitness" gives a ranking between the different solutions in the population. Note that the weights (associated to the different objectives) used in PROMETHEE II allow an easy matching of the user's wishes.

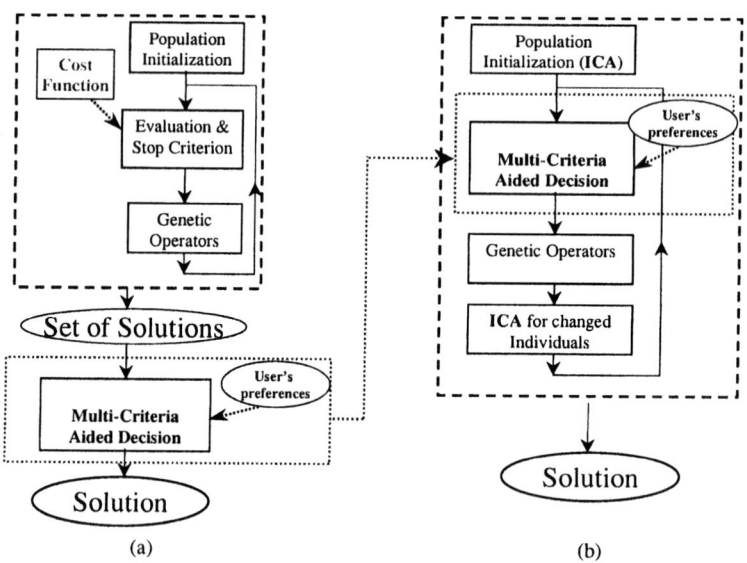

(a)   (b)

Fig. 1.5 Classical GA (a) and the proposed selection approach integrating search and decision making (b).

The choice of one solution over the others requires problem knowledge. It is the DM's task to adjust the weights to help the method to find good solutions. Optimizing a combination of objectives has the advantage of producing a single solution. If the solution proposed cannot be accepted, because of inappropriate setting of the weights, new runs may be required until a suitable solution is found. For a given user's preferences the following multiple objective GA is run:

Generate an initial population with an individual construction algorithm (ICA);
Order individuals using PROMETHEE II;

repeat

    Select parents;

    Recombine best parents from the population;

    Mutate children;

    Reconstruct individuals using the ICA;

    Use PROMETHEE II to order the new population;

    Replace individuals of the population by children;

until a satisfactory solution has been found.

The principal features of the method will be outlined in the next sections.

5.2.1   Individual construction algorithm.

The equal piles for assembly lines (EPAL) method is embedded in the GGA and is used to construct solutions [23]. The essential and distinct concepts adopted by the method will be described below, along with step-by-step execution procedure and an illustrative example.

The Boundary stones algorithm.

In the equal piles problem, the hard constraint is the fixed number of stations (piles). The approach proposed to solve the problem is based on the "boundary-stones". These boundaries will be used as seeds to fill the stations. The algorithm follows the steps described below.

Step 1.   It begins by detecting if the precedence graph is cyclic [27]. This step allows to check the validity of the proposed precedence graph.

Step 2.   It orders the operations using the labels defined below. These labels depend on the number of predecessors and successors. A first formula is given by:

$$label_{1i} = nbpreds_i - nbsuccs_i, \qquad (1.1)$$

where $label_{1i}$ is the ordering criteria of operations i (it heavily depends on the precedence graph), $nbpreds_i$ is the total number of predecessors of operation i, and $nbsuccs_i$ is its total number of successors. The calculation of labels do not take into account the operations duration.

For complex graphs (presenting several sprays), formula (1) falls in a trap (yielding a poor balancing). The more the graph contains sprays the graph contains the more its ordering becomes difficult (for a graph having a diagonal adjacency matrix the labeling is very easy since the graph is "dense" and uniform). Formula (1) was completed to avoid the trap, giving:

$$label_i = nbpreds_i - nbsuccs_i + follows_i, \qquad (1.2)$$

where $follows_i$ is the maximal label value of the direct successors of operation i in the precedence graph.

Table 1.1 gives result of the application of formulas (1) and (2) to the precedence graph illustrated in Figure 1.4. This method permits to order operations, finding the probably first and last operations on the product, and permits to choose the possible seeds of stations.

Step 3.   Boundary stones (or station seeds) are chosen using the sequence obtained at the second step. The number of stones is equal to the number of stations. This step allows to find seeds of piles. Suppose that the aim is an assembly line with three stations (so the number of stones is three). The boundary stones are determined to cluster operations in three clusters corresponding to the three stations. In this example, the first operation in the graph precedence of the product is 1 (it has no predecessor and gets the minimal label). The last operation is

Table 1.1 Example of application of formulas (1) and (2).

Op.	$nbpreds$	$nbsuccs$	$label_1$ (1)	$follows$	$label$ (2)
1	0	4	-4	0	-4
2	0	3	-3	0	-3
3	1	2	-1	3	2
4	2	2	0	3	3
5	4	1	3	5	8
6	5	0	5	6^a	11

[a]There are no successors for the given operation and it is the last operation in the precedence graph, so its label is equal to the total number of operations.

6 (having no successor and corresponding to the maximal label). According to their labels $\{-4, -3, 2, 3, 8, 11\}$, operations are ordered as follows $\{1, 2, 3, 4, 5, 6\}$ (refer to Table 1.1). The first boundary stone is the label corresponding to the first operation:

$$stone_1 = \min(label). \tag{1.3}$$

Boundary stone number i is defined as:

$$stone_{i+1} = stone_i + gap, \tag{1.4}$$

where

$$gap = \frac{\max_i(label_i) - \min_i(label_i)}{N}. \tag{1.5}$$

In this example, $gap = 5$ and the boundary stones are $\{-4, 1, 6\}$.

Step 4.   Once the boundary stones have been fixed, the labels (and consequently the operations) are grouped into as many clusters as stations. The seed (to which corresponds the first operation) of cluster i, $seed_i$ will be a label close to $stone_i$; for the first station, $seed_1$ is set to $label_1$. To this $seed_i$ will also correspond an operation which will be the seed of station i. Note that there can be several possible seeds (operation) for each cluster, which adds randomness to the procedure. Once the seeds have been chosen each cluster i is completed by adding label to it in increasing order, so that

$$\forall cluster_i, \ \forall j \in [1, n_{op}], \ seed_i \leq label_i < seed_{i+1}, \tag{1.6}$$

where n_{op} is the total number of operations. The clustering fixes possible insertion positions (stations) of the remaining unassigned operations. Operations of $cluster_i$ (the one corresponding to the $seed_i$ and the operations in the last cluster excepted) may be assigned to station i or $i + 1$.

For example, suppose the chosen cluster seeds are $\{-4, 2, 8\}$. The corresponding label clusters are $\{-4, -3\}$, $\{2, 3\}$ and $\{8, 11\}$. So operation 1 will be assigned to station 1, operation 3 to station 2 and operation 5 to station 3. Among the remaining operations, forming three clusters $\{2\}$, $\{4\}$, $\{6\}$, operation 2 may be assigned to station 1 or 2, operation 4 to station 2 or 3, and operation 6 to station 3. Figure 1.6 illustrates the authors' words. The operations already assigned are the station seeds. The arrows starting from the clusters (cl_1, cl_2, cl_3) point to the station which the remaining operations can be assigned to.

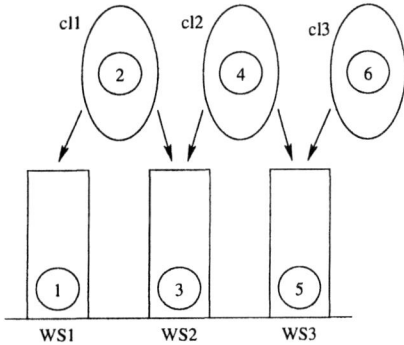

Fig. 1.6 Operation clustering and assignment.

Step 5.   Once the clustering has been done the algorithm assigns the remaining operations to stations according to the rules exposed at step 4, taking into account the precedence constraints and the user's preferences. The operations are randomly extracted from the clusters.

Due to the precedence constraints of the product, most of the time the obtained station loads will exceed the desired cycle time (the maximum stations process time). A local improvement phase attempts to equalize again these loads, by moving operations along the line or exchanging operations between stations.

Steps 4 and 5 are applied each time the GGA is about to construct a new solution (assembly line), e.g. at population initialization, or complete an existing one, e.g. after a crossover or during the mutation.

5.2.2   Dealing with Precedence Constraints.

The EPAL algorithm must yield a solution (set of stations) which respects the precedence constraints of the product. Indeed, if these constraints are violated for a given order of stations along the line the product being assembled has to move against the sense of the line conveyor, thus visiting at least one station several times.

5.2.3   The Heuristics.

Four heuristics are used to improve the solutions obtained by the boundary stones algorithm: the simple wheel and the multiple wheels, the merge and split and the pressure difference heuristics. All of them will be executed on a solution until no improvement is obtained, or a maximum number of trials is reached.

The simple wheel.

This heuristic moves sets of operations along the line. The move will always be accepted if the destination station operating time added to the operating time of the moved operations does not exceed the cycle time. If it exceeds, the move is accepted with some probability. Firstly the heuristic tries to move a set of operations from the first station to the second one. Then it tries to move a set from the new second station to the third and so on until the last station is reached. Next, it begins with moves from the last station to the last but one and so on until the first station is reached as shown in Figure 1.7. This leads to move operations along the line to reduce the imbalance between stations (precedence constraints are always checked).

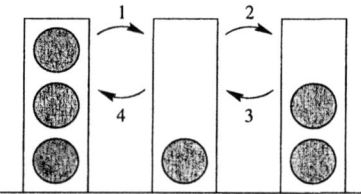

Fig. 1.7   Simple wheel heuristic.

The multiple wheel.

The second idea is to exchange operations between stations. Two adjacent stations are taken at each time (refer to Figure 1.8). All possible exchanges (which do not violate precedence constraints) are executed. The first exchange is made between the first and second station, the second one is performed between the second and third station and so on.

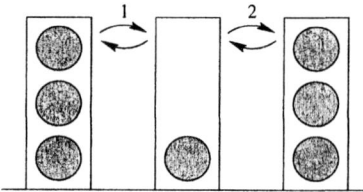

Fig. 1.8   Multiple wheel heuristic.

The first heuristic gives the best results, the second is used only if the algorithm is stuck in a local optimum and fails to improve the solution.

Merge and split.

Figure 1.9 (a) represents a kind of situation one has to face when dealing with the operating mode of hybrid assembly lines. Suppose there are two non-filled adjacent manual stations and an over-filled automated one. In order to find a good balancing, one way is to merge the two manual stations and to split the automated station. Indeed, since the hard constraint is the fixed number of stations, merging two stations obliges to split another one.

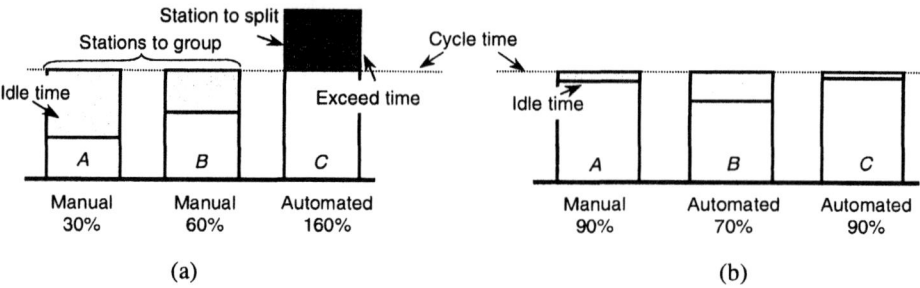

Fig. 1.9   Solution before (a) and after (b) the merge and split heuristic.

Figure 1.9 (b) represents the solution obtained after the merge and split procedure. The result of the heuristic is one manual station and two automated stations. In the balancing point of view, the second solution is better than the first one.

Pressure difference.

The main idea behind this heuristic gave it its name. It begins by finding a station exceeding the cycle time (the high pressure) as well as the station less filled (less pressure). The goal is to move the exceeding process time of the station C in Figure 1.10 (a) to fill the gap (idle time) existing on station A. In this case a task i to move from station C must have all its predecessors in station A (or before). If the move had been from A to C, all the successors of task i would have to be in C or later.

Figure 1.10 (b) represents the solution obtained after executing the procedure. The kind as well as the number of stations obtained is the same before the application of this heuristic. The simple wheel and multiple wheel heuristics cannot improve such a solution, since the two manual stations are separated by an automated one. Note that the operating modes and the precedence constraints of each task has to be verified each time a move is made.

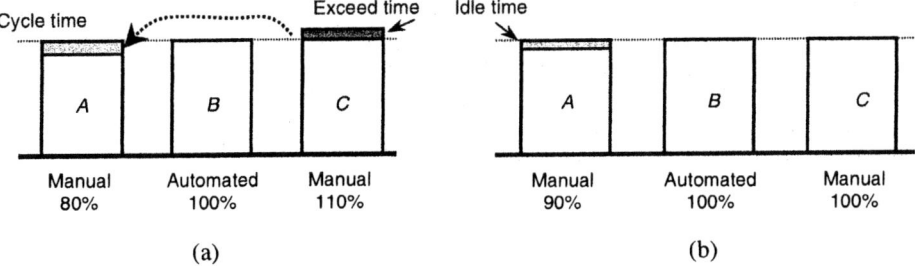

Fig. 1.10 Solution before (a) and after (b) the pressure difference heuristic.

5.2.4   Cost Function.

The objective is to equalize stations loads, under the constraint of a fixed number of stations. The authors propose the following cost function which has to be minimized:

$$f_{\text{EP}} = \sum_{i=1}^{N}(fill_i - cycletime)^2,\qquad(1.7)$$

where N is the number of workstations, $fill_i$ the sum of working times on workstation i, and $cycletime$ the desired cycle time, defined as:

$$cycletime = \frac{\sum_{i=1}^{nop} time_i}{N}\qquad(1.8)$$

5.2.5   Multi-criteria decision-aid methods and genetic algorithms.

The decision (using preferences) and the search are embedded in the algorithm and interact mutually. Evolutionary methods, particularly GAs, possess several characteristics that are desirable for MOPs and make them preferable to classical optimization methods [13].

Selection mechanism.

Selection is a process in which solutions are chosen according to their fitness. In classical GAs, the individual's fitness is computed according to a cost function that leads to a scalar fitness value. The PROMETHEE II net flow (ϕ) is used as an evaluation function for the MO-GGA. The solutions are compared to each other thanks to flows, depending on the current population. Note that the values of the ϕ are context related and have no absolute meaning.

At each generation, a ranking changes the fitness of the individuals according to their environment (the current population). In classical GAs, the fitness of an individual is independent of the other individuals constituting the population. There is no direct feedback from the environment to the individual's fitness; each individual's phenotype remains constant, unaffected by the environment. The authors believe this is an handicap of GA-based methods in the case of MOPs.

The control strategy.

As the fitness of the individuals is contents dependent, solutions have to be compared to the best ever found one. At each generation the best-ever solution takes part in the evaluation of the ϕ flows. The MCDA method ranks an individual by taking into account the presence of the others. This fitness allows the GA to choose the best solution simply by looking for the individual having the maximum value.

6.   The mapping phase (phase 3)

The optimization module yields a logical-layout of the line. The GGA solution contains the following information:

- cycle time,
- the number of stations,
- for each station:
    - the process time,
    - a list of tasks, their mode, order as well as their position,
    - a list of resources.

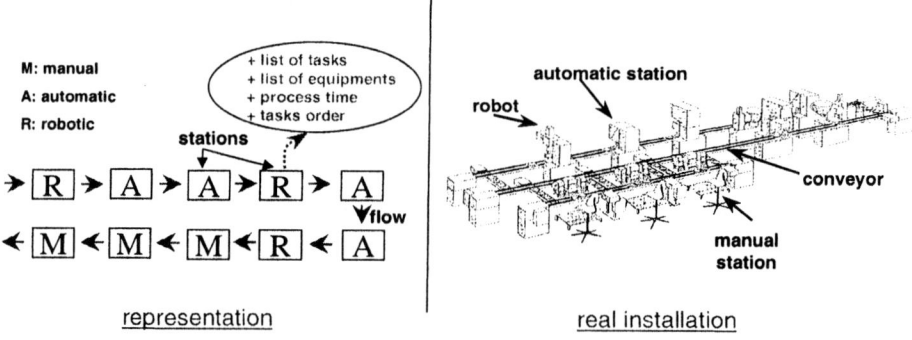

Fig. 1.11   Relationship between the real architecture of the line and its representation.

This information only constitutes the logical-layout of the assembly line presented on the left side of Figure 1.11. The right part shows a real installation of an assembly line and its relative representation which comes from the GGA module. The missing step of the physical layout is replaced by an interactive method. Each station is represented by a object (square) and is defined by a list of tasks, a list of resources, its order among the other stations,... The mapping phase helps the designer to make a first draw of the assembly line.

Figure 1.12 shows the "virtual" representation of an assembly line as it done in AUTOMOD software [31]. It represents four stations connected by a conveyor. Tasks are accomplished by one operator, two dedicated machines and one robot.

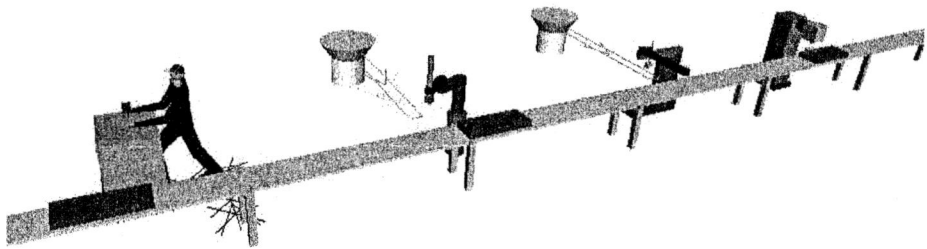

Fig. 1.12   An AUTOMOD representation of an assembly line.

7.   Application

The chosen product is a car's alternator, corresponding to a real industrial case. The desired cycle time of the assembly line is fixed at 15 seconds. A description of the operations performed on product is summarized in tables 1.3, 1.4 and 1.5. Table 1.2 presents the precedence constraints between tasks.

Table 1.2   Precedence constraints of the product.

Op	Preds	Op	Preds	Op	Preds
1	4	17	16	33	32
2	1	18	6	34	31
3	2	19	17	35	33, 34
4	–	20	37	36	35
5	3	21	20	37	33
6	–	22	21	38	31
7	–	23	17	39	36, 38, 46, 47, 48
8	–	24	17	40	39
9	–	25	18	41	40
10	7, 8, 9	26	16	42	41
11	10	27	44	43	42
12	10	28	45	44	29,30
13	11, 12	29	28	45	20, 23, 24, 25
14	13	30	28	46	35
15	14	31	27	47	35
16	15	32	31	48	35

Table 1.3 presents for each task the possible resources to accomplish it and the operating mode (M: manual, R: robotic and A: automated) associated to each equipment. For instance,

Table 1.3 Operating mode and possible resources associated to each task.

TASK	MODE	EQUIP	Task	MODE	EQUIP	TASK	MODE	EQUIP
1	M	0	17	R	25	33	A	49
	A	1		R	26		A	50
	A	2		R	27		A	51
2	M	3	18	M	28	34	A	52
	A	4	19	M	29		A	53
	R	5	20	A	30	35	A	54
3	A	6	21	A	31	36	A	55
	M	7		M	32	37	M	56
4	M	8	22	M	33	38	A	57
5	M	9	23	M	34	39	M	58
6	M	10	24	M	35	40	M	59
7	M	11	25	A	36		A	60
8	M	12		M	37		A	61
9	M	13	26	A	38		A	62
10	A	14		A	39	41	A	63
11	A	15	27	M	41	42	M	65
12	A	16	28	M	42	43	M	66
13	R	17	29	A	43	44	M	67
	R	18		A	44	45	A	68
14	R	19	30	A	45	46	A	69
	R	20		A	46	47	A	70
15	A	21	31	M	47	48	A	71
	A	22	32	M	48			
16	A	23						
		24						

task 1 can use one of the three equipment $\{0, 1, 2\}$, 0 being done manually, whilst 1 and 2 are automated FGs. Table 1.4 shows the process time and the cost of each equipment associated to a given operation. The last column in the table shows for each equipment the number of necessary operators (1 operator for manual tasks and 0 in case of machines or robots). The input data is prepared and structured using the SELEQ software package [21].

Only two criteria are optimized in this example:

■ imbalance of workload: the imbalance between the process time of the stations has to be minimized,

■ cost: the price of the assembly line has to be minimized.

Note that the real number of stations cannot be determined by computing the ratio between the sum of the operating times and the cycle-time. Indeed, that number constitutes the theoretical minimum number of stations without considering the precedence constraints and the operating mode of the operations. The cycle time constraint is complied with by observing that there is a minimal/maximal duration for each task. The theoretical minimal (respectively maximal) number of stations is the sum of the duration of the fastest (respectively slowest) resource of each task over the cycle time. For the case presented here, the theoretical minimum number . of stations is equal to 22, while the maximal number is 25.

In order to generate possible solutions, the following ICA is proposed:

1 assign tasks to the stations according to the equal piles strategy.

2 generate all valid resources combinations for each station.

3 select the best equipment combination for each station using PROMETHEE II.

Table 1.4 Process time, cost (arbitrary units) and number of operators required by each equipment.

EQUIP	TIME	COST	NB_OP	EQUIP	TIME	COST	NB_OP
0	800	1712023	1	35	900	1700000	0
1	700	118396	0	36	400	80687	0
2	800	131218	0	37	500	1835082	0
3	400	1700000	1	38	1500	99613	0
4	200	100484	0	39	1400	476287	0
5	200	344492	0	40	900	1775000	1
6	400	466587	0	41	1500	1700000	1
7	1500	1795355	1	42	600	92387	0
8	300	1700000	1	43	700	468292	0
9	300	1700000	1	44	800	90403	0
10	300	1700000	1	45	800	468292	0
11	300	1700000	1	46	300	1700000	1
12	300	1700000	1	47	300	1700000	1
13	300	1700000	1	48	600	114550	0
14	1500	125000	0	49	600	488751	0
15	0	83931	0	50	700	198341	0
16	0	83931	0	51	400	10424	0
17	600	35915	0	52	400	45570	0
18	700	328029	0	53	1500	75000	0
19	600	18926	0	54	300	70000	0
20	700	471996	0	55	300	1700000	1
21	200	6473	0	56	1400	75000	0
22	300	384361	0	57	300	1700000	1
23	800	77318	0	58	1000	1700000	1
24	900	231324	0	59	500	79298	0
25	500	27659	0	60	500	81960	0
26	700	271667	0	61	600	457187	0
27	600	172932	0	62	1400	25000	0
28	400	1700000	0	63	1500	1700000	1
29	800	1700000	0	64	1500	1700000	1
30	800	45570	0	65	300	1700000	1
31	400	80687	0	66	1400	37500	0
32	500	1835082	0	67	400	70000	0
33	500	1700000	0	68	400	70000	0
34	500	1700000	0	69	400	70000	0

More detail about the method the can be found in [24].

The MO-GGA was applied to this instance for several user's preferences. The results of the method will be examined for the different weights combinations corresponding to the relative importance one might give to each objective.

Table 1.5 summarizes the results, obtained in less than 10 minutes on a Pentium II 333 MHz. It presents the process time on the different stations, the total cost of the line according to the different optimization strategies. The number of stations is given by N, the cost of the line by "COST" and the balancing of the stations by "BALANCE". The columns labeled from 1 to 25 represents the workload of the different stations. Number in bold font represent stations where the cycle time is exceeded. The weight attributed to the balancing is 'B', the one for cost the

Table 1.5 Process time of each station according to the different weights (B, C).

N	B	C	1	2	3	4	5	6	7	8	9	10	11	12	13	14	15	16	17	18	19	20	21	22	23	24	25	BALANCE	COST
22	0	1	23	22	15	12	15	13	13	12	14	15	14	18	14	5	15	15	14	16	15	14	15	15				254	22148352
22	0.5	0.5	15	15	15	15	14	13	14	12	12	14	15	14	18	14	15	15	14	21	15	14	15	15				74	23848352
22	1	0	15	15	15	15	15	15	14	12	13	14	15	15	18	14	15	15	15	21	15	14	15	15				62	29197448
23	0	1	23	22	15	14	13	4	4	8	12	14	14	15	14	23	10	15	15	14	16	15	14	15	15			513	22148352
23	0.5	0.5	10	14	15	12	15	14	13	13	12	14	14	15	14	18	14	10	15	15	16	15	14	15	15			93	24068032
23	1	0	15	15	15	15	15	15	12	16	14	14	15	15	12	15	14	11	15	15	14	13	14	15	15			44	28657204
24	0	1	15	9	18	15	14	13	14	12	12	14	15	12	6	18	10	15	15	14	15	11	5	14	15	15		312	22437168
24	0.5	0.5	10	14	15	12	15	14	13	9	13	12	14	15	14	12	14	10	14	15	15	15	13	14	15	15		132	25768032
24	1	0	15	15	15	15	14	15	14	12	13	14	15	13	7	15	11	11	15	15	14	13	15	14	15	15		122	30675056
25	0	1	12	9	21	15	4	14	13	9	13	12	14	15	14	18	4	5	14	10	15	15	16	15	14	15	15	516	22355208
25	0.5	0.5	14	7	15	9	15	14	13	13	13	12	14	15	14	12	4	14	10	14	15	15	13	15	14	15	15	287	27248352
25	1	0	14	8	15	12	15	14	12	12	9	13	13	14	15	15	11	15	11	15	15	14	13	15	14	15	15	161	33150960

being 'C'. The weights (B, C) represents the relative importance given to each criterion. In this case, three pairs of preferences which are $\{(0, 1), (1, 0), (0.5, 0.5)\}$ were used. The pair $(0, 1)$ means that the cost is the only important objective, no care is given to the imbalance of the line. In contrast, the pair $(1, 0)$ means the opposite. Finally, the pair $(0.5, 0.5)$ means that the same importance is given to the two objectives.

The algorithm was run 12 times using four different N "number of stations" (N varying from the theoretical minimum number of stations to the theoretical maximal number) and three combinations of preferences. For a given number of station the three cases were studied. The results show that the proposed method respects the user's preferences regarding the optimization objective.

Figure 1.13 shows the cost of the line according to the number of stations for several preferences. It demonstrates that the increase of the cost with the number of stations is not a general behavior. For instance the cost of a line with 23 stations is less than with 22 stations (for weights set to $(1, 0)$). For a given number of stations, the cost of the line corresponding to $(1, 0)$ is high in comparison with $(0.5, 0.5)$ which is higher than the cost corresponding to $(0, 1)$.

The results corresponding to the solutions of 24 stations allow to make the following comments:

- the couple ($B = 1$, $C = 0$) yields a minimal process time of 7 (station 13) and the maximal process time of 15 and a cost of 30675056.

- the couple ($B = 0.5$, $C = 0.5$) yields a minimal process time of 9 (station 8) and the maximal process time of 15 and a cost of 25768032.

- the couple ($B = 0$, $C = 1$) yields a minimal process time of 5 (station 21) and the maximal process time of 18 (station 3) and a cost of 22437168.

The preference $(1, 0)$ yields an expensive but well balanced line in comparison to other preferences (see Table 1.5). In contrast, the results obtained using the preference $(0.5, 0.5)$ show clearly that setting an equal weight to the two objectives does not mean that one will obtain the line with the lowest cost and the lowest imbalance simultaneously, but rather the best compromise between the two objective. Finally, the couple $(0, 1)$ leads to a cheapest (minimal cost) and a less balanced line.

Figure 1.14 shows that the preference $(1, 0)$ leads to a good balancing in comparison to the other ones. Since the two objectives (cost and imbalance) are conflicting, improving the quality

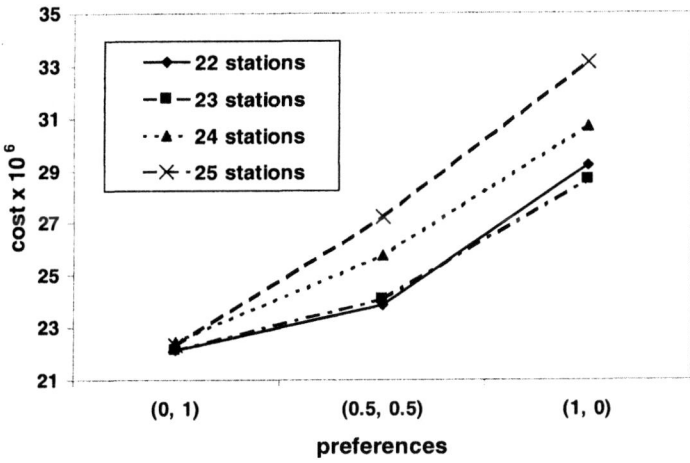

Fig. 1.13   Cost (arbitrary units) of the line according to three preferences.

of one of them decreases the quality of the other. The preference given to the different objectives permits to the algorithm to explore several regions of the search space.

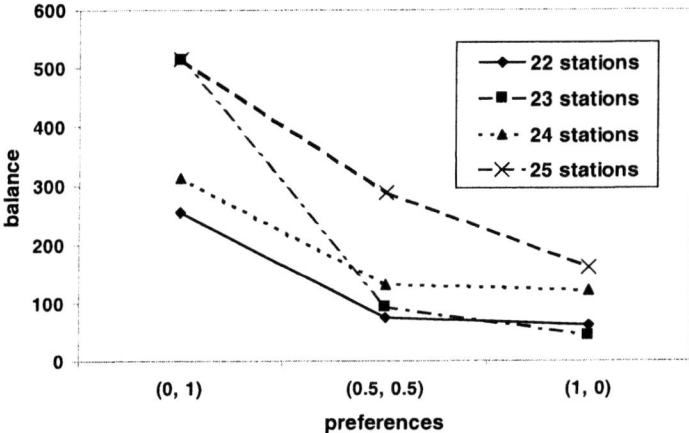

Fig. 1.14   The balancing of the line according to the preferences set for different number of stations.

The station load can exceed the cycle time in some cases, meaning that the desired cycle time cannot be held for the selected number of stations. The line will generally be less expensive as the cycle time constraint is relaxed.

These results show that a solution using 22 stations leads to the cheapest cost if the balancing is not important. Even if the cost of this solution is very small, the process time of some stations exceeds the cycle time (for instance, 23 seconds for station 1) and the quality of the balancing is so poor that this solution will never be accepted in practice. The choice of a solution is user-dependent. A good compromise between the balancing and the cost of the line corresponds to a solution using 24 stations found using the $(0.5, 0.5)$ preference.

The three combinations of weights analyzed show that obtaining a solution having simultaneously the lowest cost and the lowest imbalance is not possible in the proposed instance of the problem. It has also clearly demonstrated that considering each criterion separately leads to very bad results according the other ones. Giving the same preference to the two objectives leads to solutions where the values obtained for each criterion are a good compromise between the two others.

The main advantage of such a computer-aided tool is that it allows to try a lot of different combinations for a lot of different sets of data. This is almost impossible to realize manually due to the very large amount of possible solutions. An important aspect of this approach is that the decision maker stays the master of the optimization process.

8.   Summary and conclusions

This paper presents a new method to design assembly lines. It is based on a multiple objective grouping genetic algorithm. The aim is to assign assembly tasks to station and to select equipment to carry them out.

The accent is put on how to deal with the user's preferences in design problems. The method can deal with the preferences, simply by adjusting the weight of the different objectives. A new paradigm to deal with multiple objectives using genetic algorithms is introduced. The method cannot guarantee that the best absolute solution will be found. But one could ask what the optimal solution to such a problem is. The authors believe the answer is that there is not an optimal solution to such a MOP, because it is first of all a matter of finding the best compromise between different objectives. Note that the choice of one solution over the others requires problem knowledge.

The architecture of the proposed method shows that wishes coming from the industrial world can be taken into account. The authors believe that the method is able to deal with real-world problems, but it still needs more tests and confrontations to industrials' point of view.

Aknowledgements

This paper is based on results of the project "Outils d'aide à la conception interactive des produits et de leur ligne d'assemblage". We particularly thank the "Région Wallonne" which has funded this project.

[1] Anderson, E. J. and M. C. Ferris: 1994. 'Genetic Algorithms for Combinatorial Optimisation: The Assembly Line Balancing Problem'. *ORSA Journal on Computing* **6**, 161–173.

[2] Arcus, A. L.: 1966. 'COMSOAL: A Computer Method of Sequencing Operations for Assembly Lines'. *International Journal of Production Research* **4**, 259–277.

[3] Baybars, I.: 1986. 'A Survey of Exact Algorithms for the Simple Assembly Line Balancing'. *Management Science* **32**, 909–932.

[4] Boothroyd, G.: 1992. 'Assembly Automation and Product Design'. *Marcel Dekker, Inc.*

[5] Brans, J.-P. and B. Mareschal: 1994. 'The PROMCALC & GAIA Decision Support System for Multicriteria Decision Aid'. *Decision Support Systems* **12**, 297–310.

[6] Carnahan, B. J., M. S. Redfen, and B. A. Norman: 1999. 'Incorporating Physical Demand Criteria Into Assembly Line Balancing'. *Internal Report TR99-8, Pittsburgh University, Department of Industrial Engineering*.

[7] Coello, C. A. C.: 1999. 'A Comprehensive Survey of Evolutionary-Based Multiobjective Optimization'. *Knowledge and Information Systems* **1**(3), 129–156.

[8] Delchambre, A.: 1996. 'CAD Method for Industrial Assembly: Concurrent Design of Products, Equipment and Control Systems'. *John Wiley & Sons*.

[9] Falkenauer, E.: 1997. 'A Grouping Genetic Algorithm for Line Balancing with Resource Dependent Task Times'. *In: Proceedings of ICONIP'97. Dunedin, New Zealand, pp . 464–468.*

[10] Falkenauer, E.: 1998. 'Genetic Algorithms and Grouping Problems'. *John Wiley & Sons Inc., First edition.*

[11] Falkenauer, E. and A. Delchambre: 1992. 'A Genetic Algorithm for Bin Packing and Line Balancing'. *In: Proceedings of the 1992 IEEE International Conference on Robotics and Automation. Los Alamitos, CA, pp . 1186–1192.*

[12] Fonseca, C. M. and P. J. Fleming: 1995. 'An Overview of Evolutionary Algorithms in Multiobjective Optimization'. *Evolutionary Computation* **3**(1), 1–16.

[13] Goldberg, D. E.: 1989. 'Genetic Algorithms in Search, Optimization and Machine Learning'. *Addison–Wesley Publishing Company Inc.*

[14] Holland, J. H.: 1975. 'Adaptation in Natural and Artificial Systems'. *Ann Arbor: University of Michigan Press*.

[15] Kim, Y. K., Y. J. Kim, and Y. Kim: 1996. 'Genetic Algorithms for Assembly Line Balancing with Various Objectives'. *Computers & Industrial Engineering* **30**(3), 397–409.

[16] Lee, C. Y., M. Gen, and Y. Tsujimura: 1999. 'Multicriteria Assembly Line Balancing Problem with Parallel Workstations Using Hybrid GAs'. *In: Proceedings of the 3rd International Conference on Design and Automation (EDA'99). Vancouver, Canada, pp* . 115–122.

[17] Leu, Y. Y., L. A. Matheson, and L. P. Rees: 1994. 'Assembly Line Balancing Using Genetic Algorithms with Heuristic-Generated Initial Population and Multiple Evaluation Criteria'. *Decision Sciences* **25**(4), 581–606.

[18] Mapfaira, H. and M. Byrne: 1999. 'A Genetic Algorithms Approach for Assembly Line Balancing'. *In: Proceedings of the 12th International Conference on Engineering Design (ICED99). Munich, Germany, pp* . 957–960.

[19] McMullen, P. R. and G. V. Frazier: 1998. 'Using Simulated Annealing to Solve a Multiobjective Assembly Line Balancing Problem with Parallel Workstations'. *International Journal of Production Research* **36**(10), 2717–2741.

[20] Mînzu, V. and J.-M. Henrioud: 1997. 'Assignment Stochastic Algorithm inMulti-Product Assembly Lines'. *In: Proceedings of the 1997 International Symposium on Assembly and Task Planning (ISATP97). pp* . 109–114.

[21] Pellichero, F.: 1999. 'Computer-Aided Choice of Assembly Methods and Selection of Equipment in Production Line Design'. *Mémoire présenté en vue de l'obtention du grade de docteur en sciences appliquées, Department of Applied Mechanics, Université Libre de Bruxelles, Belgium*.

[22] Ponnambalam, S. G., P. Aravindan, and G. M. Naidu: 1998. 'Assembly Line Balancing Using Multi-Objective Genetic Algorithm'. *In: Proceedings of CARS&FOF'98. Coimbatore, India, pp* . 222–230.

[23] Rekiek, B., P. D. Lit, F. Pellichero, E. Falkenauer, and A. Delchambre: 1999a. 'Applying the Equal Piles Problem to Balance Assembly Lines'. *In: Proceedings of the 1999 IEEE International Symposium on Assembly and Task Planning (ISATP99). Porto, Portugal, pp* . 399–404.

[24] Rekiek, B., F. Pellichero, P. D. Lit, E. Falkenauer, and A. Delchambre: 1999b. 'A Resource Planner for Hybrid Assembly Lines'. *In: Proceedings of the 15th International Conference on CAD/CAM Robotics & Factories of the Future CAR & FOF'99. Aguas de Lindoia, Brazil, pp* . MW6–18–MW6–23.

[25] Sabuncuoglu, I., E. Erel, and M. Tanyer: 1998. 'Assembly Line Balancing Using Genetic Algorithms'. *Internal Report IEOR-9807, Department of Industrial Engineering, Bilkent University.*

[26] Scholl, A.: 1999. 'Balancing and Sequencing of Assembly Lines'. *Heidelberg: Physica, Second edition.*

[27] Sedgewick, R.: 1984. 'Algorithms'. *Addison-Wesley Publishing Company.*

[28] Sureh, G. and S. Sahu: 1994. 'Stochastic Assembly Line Balancing Using Simulated Annealing'. *International Journal of Production Research* **32**(8), 1801–1810.

[29] Sureh, G., V. V. Vinod, and S. Sahu: 1996. 'A Genetic Algorithm for Assembly Line Balancing'. *Production Planning and Control* **7**, 38–46.

[30] Van Veldhuizen, D. A.: 1999. 'Multiobjective Evolutionary Algorithms: Classifications, Analyses, and New Innovations'. *Ph.D. thesis, Graduate School of Engineering. Air Force Institute of Technology, Ohio.*

[31] Wanet, V.: 1999. 'Développement D'une Bibliothèque de Simulations D'éléments D'une Ligne D'assemblage'. *Travail de fin d'études présenté en vue de l'obtention du grade d'ingénieur civil mécanicien, Université Libre de Bruxelles, Brussels, Belgium.*

[32] Wee, T. S. and M. J. Magazine: 1982. 'Assembly Line Balancing as Generalized Bin Packing'. *Operations Research letters* **1**, 56–58.

Chapter 17

AGROECOSYSTEM MANAGEMENT

Ralf Seppelt

Centre for Environmental Research Leipzig – Halle,
Department for Applied Landscape Ecology,
Leipzig, Germany
ralf.seppelt@ufz.de

Ja mach nur einen Plan
Sei nur ein großes Licht!
Und mach dann noch 'nen zweiten Plan
Gehn tun sie beide nicht.

—Bertholt Brecht, 1921

Abstract

This chapter presents applications of agroecological models in the framework of optimum control theory. The question of regional agroecosystem managemt can be answered with this approach. The focus is laid on the estimation of optimum fertiliser input and crop rotation schemes as a dynamic control problem with different time scales. In this context mathematical properties of ecological models are discussed. Ecological models are heterogeneous in mathematical structure and incorporate different characteristic time scales.

Several solutions are presented for a German investigation site. Different assessment scenarios of production schemes are compared with the tool of optimisation. An innovative topic is the estimation of regionalised optimum management strategies, depending on site properties. The proposed methodology supports the step of decision support in precision farming.

Keywords: Agroecosystem, Ecological Modelling, Management Optimization, GIS, Hierarchy, Scales

1.   INTRODUCTION

1.1   EARLIER RELATED WORK

Modeling agroecological processes is a well known branch of environmental modeling. Simulation models offer a deep insight into the ongoing processes of an agricultural site. Second, models allow the prognosis of management consequences. Because of this, simulation models for agricultural sites and regions exist in abundance, see overview and evaluation in Diekkrüger et al. (1995).

Management of agricultural regions is a very good exercise of environmental management. Farmers are directly influencing ecological systems and are mutually dependent on these systems. Therefore the question of optimizing management strategies arises. Optimum control theory offers the connection between simulation models, evaluation of the environmental system states and anthropogenic management.

Clark laid the foundation of investigating ecological optimum control problems (Clark, 1976; Clark et al., 1979). Applications in optimum harvesting of renewable resources, for instance fishing and forest management were presented. This initiated a large number of publications on agricultural models (Cohen, 1987; Falkovitz & Feinerman, 1994; Velten & Richter, 1993).

Costanza introduced the terms *aggregated* and *complex* models to characterize the properties of complexity, scale and hierarchy (Costanza et al., 1993). The more stress is laid on realism and accuracy in model development, the more processes have to be considered. On the other hand, the more complex a model system becomes, the more challenging is its application in optimum control theory. Common in all cited contributions is, that very aggregated models are considered which allow the estimation of an analytical solution of the problem.

More complex simulation models are studied in terms of scenario analysis and retrospective simulation. Because of the complexity and heterogeneity of the underlying simulation models applications in numeric optimum control is rare. For instance applications in ecological economics and agroecosystem management are presented by Doherty et al. (1999) and Duffy et al. (1993). Dynamic programming procedures are used to solve the problem in these contributions.

Additionally, environmental processes show spatial properties. Regional optimization of management strategies requires a regionalised simulation of agroecological processes. Several solutions can be found in recent literature (Han et al., 1995). They study potato yield and nitrogen leaching distribution resulting from site–specific potato management. Voinov et al. offered a regional agroecological simulation model for the Patuxent watershed in Maryland, USA (Voinov et al., 1999). Applications for two agricultural sites in Germany can be found in (Svendsen *et al.*, 1995). In terms of optimization first applications in landscape design are presented by Assfalg and Werner (1993).

1.2   THE TASK

The estimation of optimum management strategies of agroecosystem in terms of optimum control theory requires the following

- formulation of an appropriate and probably complex spatial explicit simulation model for an agricultural region with the incorporation of farmers management

- definition of a performance criterion (and — as needed — constraints) which assess the observed variables and assigns a set of state variables to values identified with "good" or "bad" environmental states, whatever this means in the considered context.

Both items above are difficult to achieve in environmental modeling. The first, because modeling of environmental processes leads to complex and non–linear equation systems, which cannot be treated analytically. The second, because it is not appropriate to set up a performance criterion by a single–disciplinary approach. It requires a multidisciplinary approach comprising social, economic and ecologic issues.

In this chapter several applications of optimum control theory to agroecosystem models are presented. The overall goal is the estimation of optimum management schemes with respect to ecology and economy.

2.   AGROECOSYSTEM MODELING

2.1   SYSTEM ANALYSIS

The first step of the analysis of an agroecosystem is a collection of state variables. In the second step the interaction of these variables are identified. Table 1.1 summarizes a list of most important variables of an agroecosystem and Figure 1.1 shows a conceptual network of processes. This conceptual network focuses on the processes, which are studied in this particular contribution. More general food–webs and conceptual systems of agricultural ecosystems can be found (Begon *et al.*, 1986; Bick, 1993).

Some general remarks are necessary before going into detail of model development. With a look at Table 1.1 and Figure 1.1 one can identify the following properties of agroecosystem models.

1. Modeling the entire system in an integrative way requires input from modeling approaches of different disciplines, such as soil science, biology, chemistry, physics.

2. "The are no Newtonian laws in ecology" Beddington *et al.* (1981). Of course the basic laws of thermodynamics and physics are valid in environmental systems. However, equal processes can be described by different modeling approaches, while looking on the phenomena on different levels of aggregation.

3. The consequence is, that environmental simulation models and especially agroecosystem models are set up by multidisciplinary approaches. For instance, population dynamics are described by algebraic difference or matrix equations; process models are expressed by systems of ordinary differential equations. If spatial processes are to be considered, partial differential equations have to be included.

Table 1.1 Agroecological processes on different time scales and the modeling approaches. Abbrev.: DAE — Difference Algebraic Equation, ODE — Ordinary Differential Equation, PDE — Partial Differential Equation.

Process	Variables	Characteristic Time	Mathematical Model
Growth of microbial Populations	Biomass, Nitrogen Content, Activity	30 minutes	ODE
Nitrification, denitrification	NH_4^+, NO_3^-, N_2O, N_2, microbial activity	1 day to 1 week	System of ODE
Degradation, Volatilation of Agro–Chemicals	Concentration in liquid and solute phase	Minutes to weeks	System of ODE
Pest populations dynamics	Density of eggs, juveniles, larvae, adults	Weeks to vegetation periods	Matrix–Equations, DAE
Crop growth	Organ biomass, Nitrogen content, Leaf Area Index	Month	(Systems of) ODE
Population dynamics of weed	Seed dispersal, Coverage level	Vegetation period	DAE
Water transport in unsaturated soil zone	Water content, -pressure	1 hour	PDE Equation
Solute transport in unsaturated soil zone	Concentration in liquid and solute phase	Large spectrum	PDE coupled with ODE systems
Solute transport in aquifer	Concentration in liquid and solute phase	up to several years	PDE coupled with ODE systems

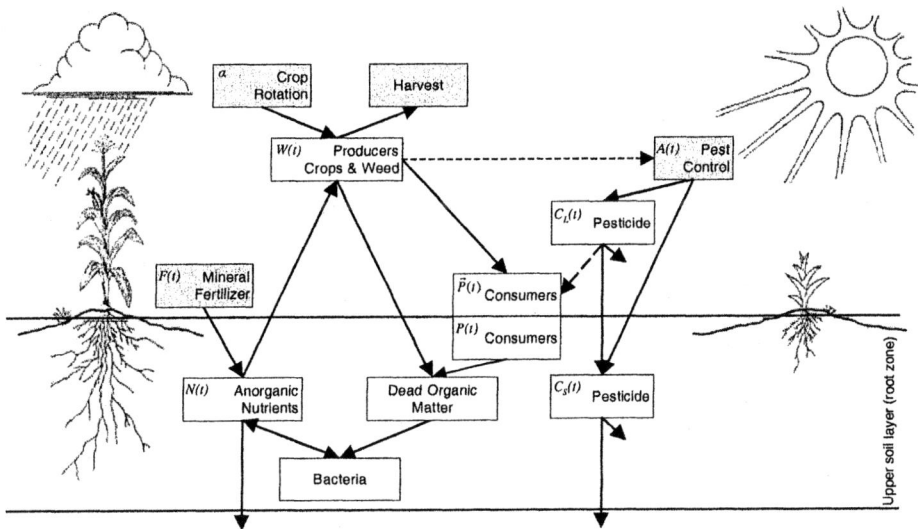

Figure 1.1   Conceptual model of an agricultural ecosystem. White boxes denote compartments, gray boxes denote management variables. Arrows denote general transport of mass, energy. Dashed arrows denote dependencies or flow of information. State and control variables are noted at the upper left corner of the compartment boxes.

4. Processes show a broad spectrum of time scales, starting from very fast processes like pesticide degradation, volatilisation to very slow process like nutrient fate in groundwater.

Environmental modeling is characterized by a large number of different approaches. Depending on the scale of interest, on the accessibility and usability of data sets, and of the aim and scope of the problem to be solved, different simulations models are developed and used. Some authors complain about an enormous redundancy (Müller, 1997).

In terms of application optimum control theory item 3 faces us with the problem, which can be summarized as *mathematical heterogeneity* (Seppelt, 1999; Seppelt, 2003). It is not feasible to model integrated systems in the framework of one mathematical theory like ordinary differential equations. Numerical algorithms of optimum control have to take these mathematical properties into account. The lesser prerequisites an algorithm of numerical control has, the broader is the spectrum of possible application in environmental modeling.

2.2   MODELING THE PROCESSES

Crop growth — Macrophytes. Let $W(t)$ denote the weight of the above ground biomass (in kg/ha) of a specific crop, a primary producer (see Fig. 1.1). There exists an abundant literature for modeling crop growth. The approach used here simulates crop growth in a single differential equation and covers the processes of growth (parameter r in 1/d), mortality (μ in 1/d) and senescence (function $f_s(t)$), c.f. (Richter *et al.*, 1991). f_s is monotonically decreasing and equals unity at $t = 0$. It incorporates the reduction in dry matter biomass during maturite stages of crops into the model.

$$\dot{W} = \Big(r f_s(t) - \mu\Big) W \tag{1.1}$$

The growth rate r is defined by a maximum growth rate and reducing factors, which incorporate reactions of plant growth on nutrient availability and pest infestation

$$r = r_{max}\, r_N(N)\, r_P(P_4, t) \tag{1.2}$$

r_{max} (1/d) denotes the maximum growth rate of a specific crop (1/d). The reductions functions $r_N(N), r_P(P_4, t) \in [0, 1]$ describe the dependence of growth on soil nitrogen and a pest population P_4 (see below). A Michaelis–Menten kinetic is used for r_N, a Weibull function is used for r_P, c.f. (Richter *et al.*, 1991; Schmidt *et al.*, 1993).

Nutrient cycle — Detritus. The plant–accessible pool of mineral nitrogen in soil is denoted by $N(t)$ in kg/ha. Plant growth depends on this pool. The simulation model couples the processes of nitrogen uptake by plants, nitrogen leaching out of the root zone, decomposition, mineralization, NO_2 fixation and fertilization.

$$\dot{N} = -d(W, t) - k_l N + k_m \mu W + r_f \frac{W}{W + k_f} + F(t) \tag{1.3}$$

Leaching and mineralization are modeled by linear flows with the rates k_l and k_m (1/d). Fixation of NO_2 is modeled by a Michaelis–Menten function with the parameters r_f (1/d) and k_f (kg/ha). The demand of nitrogen taken up by the crop $d(W, t)$ is calculated using the reference content of nitrogen in the crop biomass $k_{cn}(t)$ (%) (Schröder & Richter, 1993) and is defined by:

$$d(W, t) = \frac{d}{dt}\Big(k_{cn}(t) W(t)\Big) = k_{cn}\dot{W} + \dot{k}_{cn}W$$

Decreasing biomass leads to negative values of $d(W, t)$. This may be interpreted as mineralization of dead biomass. This completely different process cannot be set up by the same parameter of nitrogen uptake. Therefore d is restricted to be positive or zero.

Pest infestation — Consumers. Pest population modeling is a very good example of mathematical heterogeneity in ecological modeling. Two types of pest populations are considered here. The first has a slow population dynamic, about one to three generations per vegetation period. A population like this has to be take into account in crop rotation design. The second population, with a very fast exponential growth in one year has to be controlled by pesticide applications.

An example for the first type of pest population is the sugar beet cyst nematode *Heterodera schachtii* population $\vec{P}(t_i)$. This population develops in distinct stages. For *H. schachtii* the stages eggs and juveniles (P_1), hatched larvae (P_2), penetrated larvae (P_3) and adults (P_4) are distinguished. From an agricultural point of view P_1, the number of eggs and juveniles in a unit soil in spring is decisive in determining potential crop damage. The adults P_4 stress crop growth of sugar beets, see Eqn. (1.2). Transition probabilities of development to the succeeding stage p_h, p_p and p_d, the fertility f and the probability of survival over winter p_s can be identified in terms of parameter estimation. This approach leads to a difference equation based on a Leslie matrix (Richter *et al.*, 1991; Schmidt *et al.*, 1993):

$$\vec{P}(t_{i+1}) = \begin{pmatrix} p_s & 0 & 0 & f \\ p_h & 0 & 0 & 0 \\ 0 & p_p & 0 & 0 \\ 0 & 0 & p_d & 0 \end{pmatrix} \vec{P}(t_i) \tag{1.4}$$

The transition probabilities depend on climatic situations and on the hosted crop.

Additionally to the modification of the growth rate, shown in Equation (1.2), the population P derives its energy from the host biomass. This is modeled by the extension of Equation (1.1) by another sink

$$\dot{W} = \cdots - \gamma \frac{W}{W + k_p} P$$

An example for the second pest type is powdery mildew (*Erysiphe graminis*). This population also develops in distinct stages. For this model we assume that one has no knowledge for a separation of stages and assume a model for the entire population P in terms of a differential equation. Population growth depends of the ability of extraction of biomass from the host.

$$\dot{P} = r_P \frac{W}{W + k_P} P - \mu_P P - \mu_C r_C \left(C_L\right) C_L P \tag{1.5}$$

For parameterization of the model a worst case scenario is assumed. A fast run off from the crop leaf, a high degredation rate together with a high exponential growth rate of the pest makes the time of pesticide applications an important control variable. The nonlinear dose–response function

$$r_C(C_L) = 1 - e^{-\left(\frac{C_L}{c_{crit}}\right)^2}$$

and the consideration of leached pesticide out of the upper soil layer in the performance criterion will be decisive for the amount of pesticide applied.

Xenobiotica cycle — Agrochemicals. The differential equations for the fate of agrochemicals are derived from the compartment scheme in Figure 1.1 with the assumption of linear fluxes. Let C_L denote the concentration (mg/l) of a pesticide on the crops leaf and C_S the concentration (mg/l) in the upper soil layer. Precipitation is the driving force for transport from leaf surface to soil surface (k_w) and for leaching out of the upper soil horizon (k_L). Degradation of the chemicals is assumed as linear first order process (k_d).

$$\dot{C}_L = -k_d C_L - k_w C_L + \nu(W)A(t) \qquad (1.6)$$
$$\dot{C}_S = -k_d C_S + k_w C_L - k_l C_S + (1 - \nu(W))A(t) \qquad (1.7)$$

The distribution of the amount of applied pesticide $A(t)$ (mg/l) depends of the current state of crop, on the leaf area index. The more leaf area is present, the more pesticide reaches the plant leaf. This is expressed by the function, c.f. (Schröder et al., 1995)

$$\nu(W) = \frac{c_{LAI}W}{c_{LAI}W + 1} \qquad (1.8)$$

Generic Agroecosystem Model. The entire agroecosystem model is set up by the following equation system. This model couples all processes described above.

$$\dot{W} = \left(r_{max}r_N(N)r_P(P_4,t)f_s(t) - \mu\right)W - \gamma\frac{W}{W + k_P}P$$

$$\dot{N} = -k_l N - d(W,t) + k_m W_\mu + r_f\frac{W}{W + k_f} + F(t)$$

$$\dot{C}_L = -k_d C_L - k_w C_L + \nu(W)A(t)$$

$$\dot{C}_S = -k_d C_S + k_w C_L - k_l C_S + (1 - \nu(W))A(t)$$

$$\vec{P}(t_{i+1}) = \begin{pmatrix} p_s & 0 & 0 & F \\ p_h & 0 & 0 & 0 \\ 0 & p_p & 0 & 0 \\ 0 & 0 & p_d & 0 \end{pmatrix} \vec{P}(t_i)$$

$$\dot{P} = r_P\frac{W}{W + k_P}P - \mu_P P - \mu_C r_C(C_L)C_L P$$

In terms of a summary, here are the used parameter functions:

$$r_N = \frac{N}{N + k_N}$$

$$r_P = r_0 \left(e^{-\gamma_1 (P_4 - P_r)^2} - e^{-\gamma_1 P_r^2} \right) + \frac{(1 + r_1)e^{-\gamma_2 P_4}}{1 + r_1 e^{-\gamma_2 P_4}}$$

$$f_s(t) = \begin{cases} 1 & \text{if } t < t_d \\ \frac{(1+\rho_1)e^{-\rho_2 t}}{1+\rho_1 e^{-\rho_2 t}} & \text{else} \end{cases}$$

$$k_{cn}(t) = (k_{max} - k_0)e^{-\frac{t-t_d}{\tau_2}^{\gamma_c}} + k_0$$

$$d(W, t) = \max \left(k_{cn}(t)\dot{W} + \dot{k}_{cn}(t)W, 0 \right)$$

$$r_C(C_L) = 1 - e^{-\left(\frac{C_L}{C_{crit}}\right)^2}$$

$$F(t) = \sum_{i=0}^{q-1} F_i \delta(t - t_i)$$

$$A(t) = \sum_{i=0}^{q-1} A_i \delta(t - t_i)$$

The system is highly nonlinear, an open system and mathematical heterogeneous. The system is solved using high order embedded Runge–Kutta formulae (Prince & Dormand, 1981). Parameters are derived from literature and field experiments. Most of the introduced parameters are itself functions. They depend either

- on the specific properties of the investigation site, like soil parameters, hydrological parameters,

- on climatic conditions, like precipitation, humidity, temperature, or

- on the planted crop on the field, for instance: $r_{max}, k_N, \mu, \rho_1, \rho_2, t_d$ are estimated using parameter estimation procedures (Richter *et al.*, 1991) using data sets from field experiments (McVoy *et al.*, 1995) for the following crops: sugar beet (abbreviation "sub"), winter wheat ("ww"), winter barley ("wb"), oats ("oa"), spring barley ("spb") and the fallow seeds: oil raddish ("or") and field beans ("fb") and fallow, for instance no crop on the field.

From this follows, that the control variable "crop planted" $\alpha(t)$ modifies the model parameters. Additionally, for special cases, such as "fallow", the structure of the model is modified also. The dependence from site specific parameters introduces spatial aspects into the model, which require regional simulations.

Model Regionalisation and Spatial Database. Agroecosystem management is a spatial problem. Spatial dependencies are introduced to the model with the spatial dependence of state variables $W(t, \vec{x})$, $N(t, \vec{x})$, ..., the control variables $F(t, \vec{x})$, $A(t, \vec{x})$ and the model parameters. The model paramters, which show spatial dependencies, such as k_l, k_w, r_{max} are denoted by the vector $\vec{\theta}(\vec{x})$. The spatial localization in the observed region G is denoted by $\vec{x} \in G$.

Spatial information has to be provided, which either specifies spatial distribution of parameters $\vec{\theta}(\vec{x})$ and initial conditions $W(0, \vec{x})$, $N(0, \vec{x})$, ... as a function of location $\vec{x}$, or, which defines the spatial range of validity of control variables, like fertilizer application $F(t, \vec{x})$, pesticide application $A(t, \vec{x})$ or planted crop $\alpha(t, \vec{x})$.

For the described model two layers of spatial information are used:

- The digital soil map (1:5.000) S_{soil} with the attributes "field capacity" and "rooting depth" from which the parameter $k_l(\vec{x})$, the leaching rate, is derived.

- The land use map S_{field} with the field borders identifies every single field by a unique identifier, an integer number.

These two layers of information determine different aspects of an agroecological simulation. The relevant information layer for simulation and the model parameter specification is the soil–map with its pedological attributes. Management variables $F(t, \vec{x})$ and $\alpha(t, \vec{x})$ depend on the observed field S_{field}. In terms of precision farming (Lu *et al.*, 1997) fertilization ought to depend on soil properties S_{soil}.

A well known approach for regional agroecological modeling is the calculation of the intersection–map $S_{field} \cap S_{soil}$ (Breunig, 1996, p. 77). This generates a new map with homogeneous attributes with respect to all underlying layers of information, here: soil properties and field identifiers. All further steps refer to these smallest homogeneous units in the observed region, which are called *ecotopes* (Naveh & Lieberman, 1984, p. 6).

3.   MANAGEMENT IN TERMS OF CONTROL THEORY

The chosen example of an agroecosystem management model is typical for environmental modeling. It shows the solution of two important problems in ecosystem management: the determination of long term strategies with the use of the temporarily structured model and regionalised management optimization.

3.1   ASSESSING THE PROCESS

Integration of Economy and Ecology.   Performance criteria have to integrate economic and ecological issues. Economic issues are for instance prices for yield, farmers income, and prices for fertilizer, farmers expense — the first to be maximized, the latter to be minimized. Further economic issues may focus on taxation of fertilizer, to reduce fertilizer input, or on a limitation of fertilizer input.

The nutrient content in soil or the infestation of pests are examples for the ecological part of the assessment. Whereas the former economic assessment could be performed within a monetary unit, it is difficult to identify units for ecological variables. Ecologic and economic issues are difficult to compare.

Open Systems. Ecological systems are open systems. In a performance criterion only variables can be used, which are represented in the simulation model. For instance, an assessment of possible groundwater contamination with nitrate can only be assessed by the possible outflow of nitrogen out of the plant accessible root zone. In economic terms this means, that *external cost have to internalized.*

Model Variables vs. Measurement Variables. Ecological variables and more often variables of a simulation model are difficult to measure in reality. Yield, and nutrient content in soil are easy to measure. Nutrient outflow into groundwater or the population of a pest like the considered sugar beet cyst nematode are very difficult to measure. Only effects can be observed. Therefore these variables cannot be used in a framework of practical farm management system. One has to identify different variables, so called indicators.

General Performance Criterion. A general notation of a performance criterion which maps state and policy space for the simulation time 0 to T and the entire region G of the given agroecosystem model to a scalar value is

$$
\begin{aligned}
J[W, N, C_S, A, F, \alpha] \ = \ & \lambda_W(\alpha(t, \vec{x})) \int_G W(T, \alpha(T, \vec{x}), \vec{x}) \, d\vec{x} \\
& - \int_G \int_0^T \lambda_F F(t, \vec{x}) + \lambda_A A(t, \vec{x}) \, dt \, d\vec{x} \qquad (1.9) \\
& - \int_G \int_0^T \lambda_N k_l N(t, \vec{x}) + \lambda_C k_l C_S(t) \, dt \, d\vec{x}
\end{aligned}
$$

With the weights λ_i the different state and control variables are aggregated to a scalar performance criterion. Setting up values for the weights requires answers to the above stated problems of comparing economic and ecologic variables, openness and the use of measurable and non–measurable variables. Different sets of λ–values define different assessment scenarios and with this different perspectives to optimality.

λ–values for the different criteria

- can be derived from market prices ("economic" criterion),

Table 1.2 Overview of the considered scenarios and their performance criteria with the assessed variables and the weights λ_i. Weights are set to zero in eqn. (1.9) for variables which are not assessed in a performance criterion, denoted by the missing •. Constraints are understood as constraint for the entire region, e.g. for all $\vec{x} \in G$. All goal functions are quantified by monetary unit per area (DM/ha).

scenario		assessed variables in Eqn. (1.9)					additional constraints
		W	N	C_S	F	A	
J_1	"economic"	•			•	•	
J_2	"taxes"	•			•		
J_3	"ecologic"	•	•	•	•	•	
J_4	"N–limit"	•	•				$N(T) < N_{max}$
J_5	"F–limit"	•			•		$\sum F(t_i) < F_{max}$
weight		λ_W	λ_N	λ_C	λ_F	λ_A	

- may be modified by taxation of fertilizer ("taxes" criterion),

- are estimated by internalization of external effects. This makes use of the approach of an assessment relative to non–disturbance (Nilsson & Bergström, 1995) ("ecologic" criterion),

- can be supported by constraints like limitation of total fertilizer input ("F-limit" criterion) or of nutrient content in soil at harvest time ("N-limit" criterion).

Table 1.2 summarizes these performance criteria. Obviously not all criteria are based on measurable variables. It is necessary to study the results of these different optimization criteria.

3.2   CONTROLLING THE PROCESS

General Task. Based upon the knowledge of the agricultural process summarized in a simulation model we are now able to formulate a general optimum control problem: *Estimate a function of optimum fertilizer input $F^*(t, \vec{x})$, of pesticide application $A^*(t, \vec{x})$ and a sequence $\alpha^*(t_i, \vec{x})$ of planted crop, so that a performance criterion $J[\vec{X}, \vec{U}]$ is maximized.*
    Note that

- Fertilization and pesticide application are discrete events. From this follows that not only the amounts but also the time of optimum application are to be estimated.

- The spatial dependency of fertilizer, pesticide application and model parameters complicate the problem tremendously.

- With the specification of the control variable planted crop α, a set of model parameters and also the model structure is changed during simulation.

Algorithm. Additionally to the characterization of the model system (see end of Section 2.2) one can state that the derived optimization problem is complex and the solution recommends a clear structure and necessitates a simplification of the problem.

The underlying procedures for solving the numerical optimization problems are derived from the dynamic programming approach, introduced by Bellman and Dreyfus (1962). Advantages of this approach are the use of a general dynamic system with lesser prerequisites to the mathematical structure of the problem and the ability of dealing with discrete and continuous control variables. Applications can be found in recent literature in agro–ecosystem management (Duffy & Taylor, 1993; Seppelt, 1999).

The disadvantage of this algorithm is that the computational effort increases polynomially with the increasing number of state and policy variables. Bellman characterizes this property as "curse of dimensionality" (Bellman & Dreyfus, 1962). For an application to environmental system, one can make use of properties of environmental systems, namely the given hierarchies in time, see Table 1.1.

Hierarchy in Time. The optimum control problem is at first structured using the hierarchy in time, which is defined by the process dynamics. One can distinguish between fast processes like crop growth, pesticide dynamics, fertilization and pesticide application and slow processes like population dynamics and crop rotation design. Additionally control variables may be continuous and discrete.

The problem can be structured in the framework of a hierarchical control model. In a first step a *local* optimum control problem is solved, with the optimization of fertilizer input in a vegetation period for each crop. These solutions are stored as a function of initial values and crop identifiers. In the second step, the *global* or entire optimum control problem is solved and an optimum crop rotation is estimated where every crop receives its optimum fertilizing scheme. In this step the performance criterion makes use of the performance criterion of the local problem.

$$\widehat{J}_n[N, J] = \sum_{i=1}^{Q} J_n[W, N, F, A, C_S] - \Lambda_N \int_G \int_0^T k_l N(t, \vec{x}) \, dt \, d\vec{x} \quad (1.10)$$

The solution of this global task makes intensive use of the previously stored local solutions which reduces computational effort and couples a discrete and continuous optimum control solutions.

Spatial Control Problem. Solving the spatial optimization task requires a significant reduction of computational effort, which is induced by the spatial dependence of state and control functions. The important step is the identification of homogeneous areas in the region. A vector–oriented database, which represents the investigated area by a set of irregular polygons, is used instead of a grid–based map. This decomposition of the plain into irregular cells is called *tesselation* (Breunig, 1996, p. 15). Tesselations may be derived from grid based data sets by aggregating grid points with similar attributes using classification algorithms, c.f. (Sadler *et al.*, 1998; Weibel, 1997). The result is a map S of homogeneous areas $s \in S$.

Based on this data–structure the soil–map, the field–map and the ecotope–map can be accessed by unique identifiers stored in the GIS–database. Figure 1.2 shows this technique: Input data consists of the geometry information and the

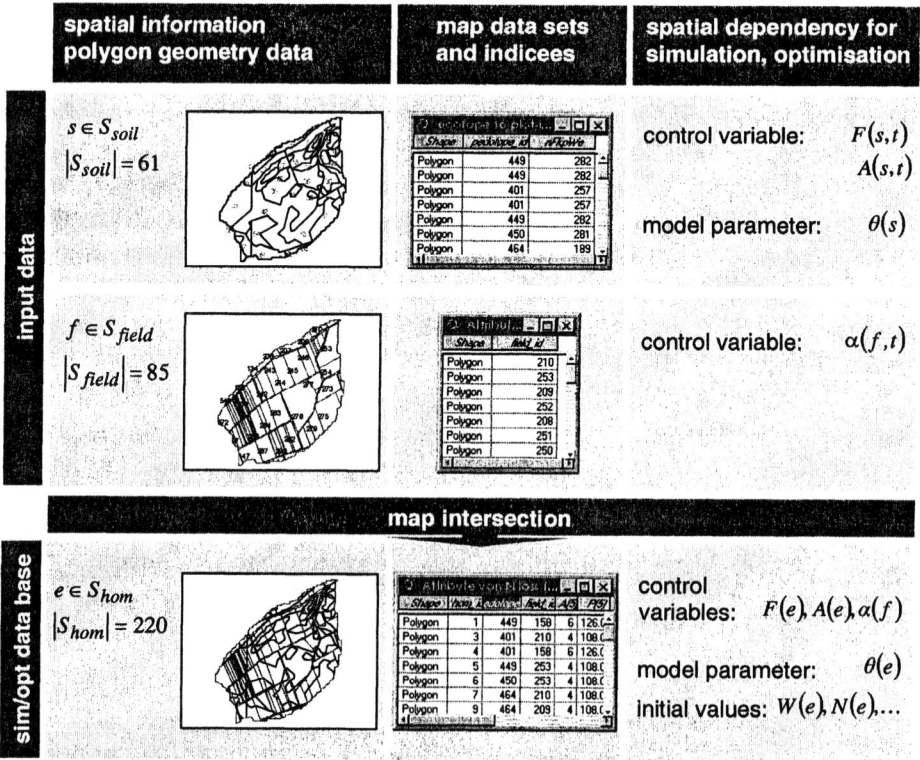

Figure 1.2 Database concept in GIS: The underlying data sets consist of geometry information, a decomposition of the region by irregular polygons (left column) and the associated attribute tables for soil properties and field identifiers (middle column). Dependencies to the simulation model and the optimum control problems are noted in the right column. Associations between units of the input maps are derived from the intersection-map, shown in the last row.

associated database with soil properties and field identifiers. The right column summarizes the model variables which have spatial dependencies. Simulation and optimization is based on the intersected maps shown in the last row. In the observed region the intersection map consists of $N_2 = |S_{hom}| = 220$ ecotopes. For each of these units a simulation is performed and all results are aggregated and visualized by the GIS.

If the ecotope map is used for the solution of the spatial optimization problem $N_2 = 220$ optimization runs are necessary for the complete regional solution. The basic idea for a regional optimization is to estimate a set of optimum control solutions depending on initial conditions *and* soil properties. This reduces the numerical effort to the number of distinct pedological areas $N_3 = |S_{soil}| = 61$.

A regionalisation is performed based on this data set of optimum solutions and the identifier given in the ecotope map S_{hom}. Efficiency increases with an increase of access to pre–calculated results with equal pedological properties.

The basic innovation of this approach is a careful separation between spatial areas with distinct properties. A simulation and optimization is only performed for regions with new and distinct properties. This reduces the computational effort by orders of magnitude. This distinguishes the solution from all grid based modeling approaches.

4. CASE STUDIES

4.1 VEGETATION PERIOD

Fertilizing Schemes. The estimation of optimum fertilizing schemes to different performance criteria (see Table 1.2) are derived first. Table 1.3 contains the results of total fertilizer input, harvest biomass and leached nitrogen for different crops. Literature values are added for comparison (Niesel-Lessenthin, 1988). Optimum fertilizing schemes from "economic" lead to a maximum consumption of fertilizer with a high amount of nitrogen loss. A reduction of fertilizer input can be achieved with governmental restrictions. With the use of "taxed" fertilizer (J_2) or the limitation of harvest nitrogen pool $N(t) < N_{min} = 45$ kg/ha (J_4), total fertilizer input is reduced by 30%. This leads to a reduction of nitrogen loss in the same range, while yield reduction is less than 15%.

Introducing external costs into assessment (J_3) results in the lowest values of leached nitrogen after vegetation period: less than 60% of the results of J_1. This fertilizing scheme incorporates a notable reduction of yield: 25% less than the results of "economic" assessment.

These calculations use a non–measurable variable for assessment, due to the nonzero weight λ_N. One can ask for an assessment based on these results using measurable state variables. This problem is solved using J_5 for assessment. This limits the maximum amount of applied fertilizer to $F_{max} = \sum F_i^*$ using

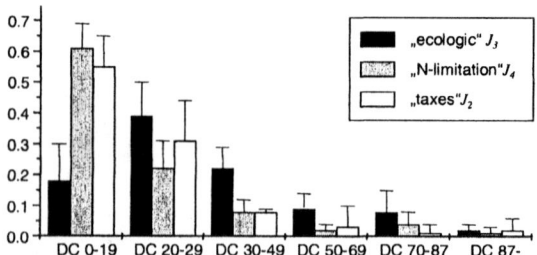

Figure 1.3 Distribution of optimum fertilizer amounts of grains in the main development stages for different performance criteria. Standard deviation is calculated based on fertilization schemes applied to different grains (ww, wb, oa, spb).

the optimum fertilizing schemes F^* calculated on the basis of the "ecological" assessment with J_3.

Table 1.3 Comparison of optimal fertilizing strategies from different performance criteria and crops. The last column shows literature values of expected yield and recommended fertilizer (Niesel–Lessenthin, 1988). Total amounts of applied fertilizer from J_3 printed in italics are used as F_{max} values for J_5. All values in kg/ha. Symbols are: $W(T)$ harvest biomass, N_{tot} total amount of nitrogen leached from root zone during vegetation perod, F_{tot} total amount of fertilizer applied (optimized).

		J_1 "econ."	J_2 "taxes"	J_3 "ecolog."	J_4 "$N(T)$-limit."	J_5 "F_{max}-limit."	reference values
sugar beet	$W(T)$	18000	13700	13000	17400	12100	13000–26000
	N_{tot}	62	47	21	50	26	
	F_{tot}	305	228	*136*	236	128	140–210
winter wheat	$W(T)$	10400	9600	9000	10100	8400	7600–14400
	N_{tot}	67	28	23	35	24	
	F_{tot}	208	138	*121*	160	113	100–210
winter barley	$W(T)$	13600	12800	11200	12300	11300	7600–14400
	N_{tot}	67	53	39	48	44	
	F_{tot}	290	192	*158*	179	147	100–210
spring barley	$W(T)$	15000	14000	10000	13000	9600	6400–11200
	N_{tot}	36	25	11	22	13	
	F_{tot}	294	202	*124*	178	118	100–170
oats	$W(T)$	10300	9400	5600	9000	6100	6400–12400
	N_{tot}	33	21	8	19	9	
	F_{tot}	241	142	*81*	130	75	80–170

On a closer examination the optimum function $F^*(t)$ explains the time dependence of the fertilizer application. For a comparison of the fertilizing schemes the results of the grain–models are taken for detailed study. A comparable time scale can be defined using the stages of development (Zadoks *et al.*, 1974). Figure 1.3 summarizes the distribution of fertilizer (results of optimum control assessed by J_1, J_3, J_4 in the development stages normalized by the total amount). This figure explains why the solution of J_3 attains a 60% reduction of fertilizer and nitrogen loss with a considerable smaller reduction of yield. Fertilizer is applied only in the main stages of growth which are DC 30 to DC 49, see (Zadoks *et al.*, 1974). The solutions derived by other scenarios (J_2, J_4) of assessment lead to optimum fertilizing schemes, which set up a sufficient amount of fertilizer at the beginning of the vegetation period. Obviously, this increases the amount of nitrogen loss.

Pesticide Application.   Figure 1.4 shows the results of optimum pesticide application schemes to the performance criterion of Eqn. (1.9). The parameter of variation is the initial pest infestation $P(0)$ and the time, denoted by main development stages. All figures show the relative distribution of application

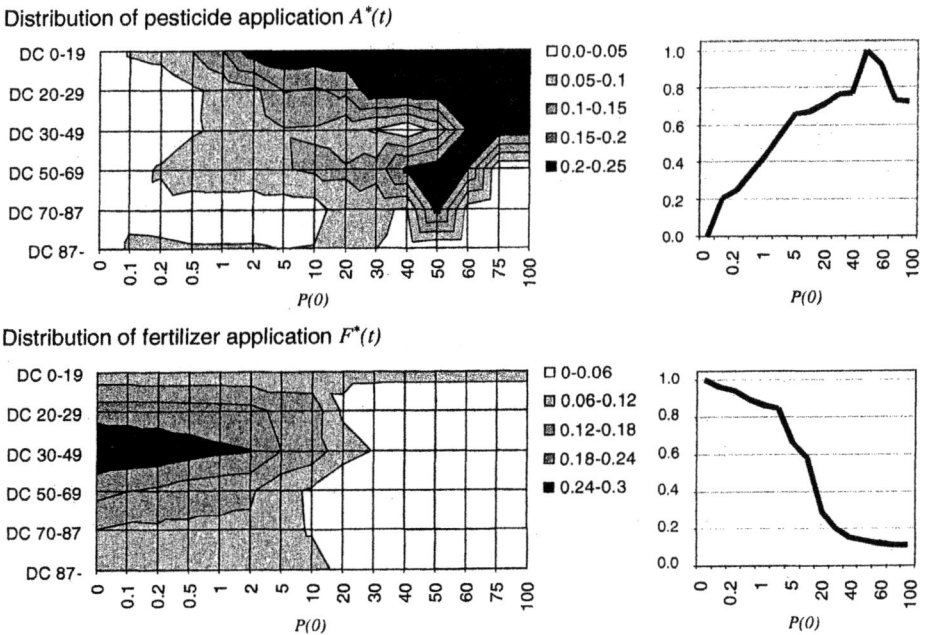

Figure 1.4   Distribution of pesticide and fertilizer application for control of pest population P in a vegetation period using criterion J_3 ("ecologic") for assessment.

amounts $A^*(t)$ with respect to the maximum application amount of a whole vegetation period and the entire spectrum of initial pest investations. The left figures show the distribution function as a density plot and allows an analysis of time dependence. The right figures show the accumulated application amount of fertilizer and pesticide as a function of initial pest population relative to the maximum amount applied.

The distribution of the fertilizer schemes is shown in the lower figures. For low initial pest infestation one can identify the results described in the former paragraph. Above a critical level of investigation the fertilization scheme changes completely. No fertilizer is applied at the stages after DC 20 and the total amount of applied fertilizer is less than 20% of the normal amount.

The reason for this optimum fertilizing scheme is, that fertilization is no more decisive for the outcome of yield. Yield is controlled only by pest control throughout pesticide application.

Optimum pesticide application schemes look as follows:

- The pest population is controlled at the earliest stages with an optimum amount, slightly above the critical dose to get response in the pest population. The pest population shows a high growth rate. To achieve a maximum effect in pest control with a minimum amount of pesticide getting washed into the upper soil layer, an early application date is important.

- For moderate pest populations below the critical level two or three application events can control the pest population. If more application days are necessary, continuous but small applications of pesticide are optimal. This reduces pesticide run off into the upper soil layer.

- Above the critical level of initial pest infestation, the pesticide application is decisive for the yield amount. The amount of applied pesticide reaches a maximum. For the control of the pest population an application during the entire vegetation period becomes necessary.

- With a further increase of the initial pest population it becomes impossible to maintain a sufficient amount of yield. Optimum application amounts are reduced for a limitation of pesticide run off.

4.2   LONG–TERM STRATEGIES

Fertilization and crop rotation design are the selected variables for the estimation of long–term strategies. Figure 1.5 shows an example of a complete solution of the optimum control problem: An optimum crop rotation (based on $\widehat{J_3}$) with locally optimum fertilized crops (continuation of J_3). The figure shows a typical crop rotation of the farming systems in the investigation site:

A sequence of sugar beet and winter wheat/barley with a period length of two to three years.

For detailed analysis the continuation of the local performance criteria J_1 "economic", J_3 "ecologic" and J_4 "N–limitation" are chosen.

Nutrient balance. An important question in the assessment of the nutrient circulation is, if the nutrient balance can be equalized observing a vegetation period or a crop rotation. The weight Λ_N in equation (1.10) incorporates the amount of nitrogen loss into the assessment of the crop rotation. One can include or exclude the local assessment of nitrogen loss with the choice of the local performance criterion. Table 1.4 summarizes the results of a seven year crop rotation with all possible scenarios.

Only if the local assessment does not restrict the application of fertilizer (like J_1), a global assessment of $N(t)$ reduces the amount of nitrogen loss. If a considerable reduction of fertilizer is achieved in the vegetation period, the

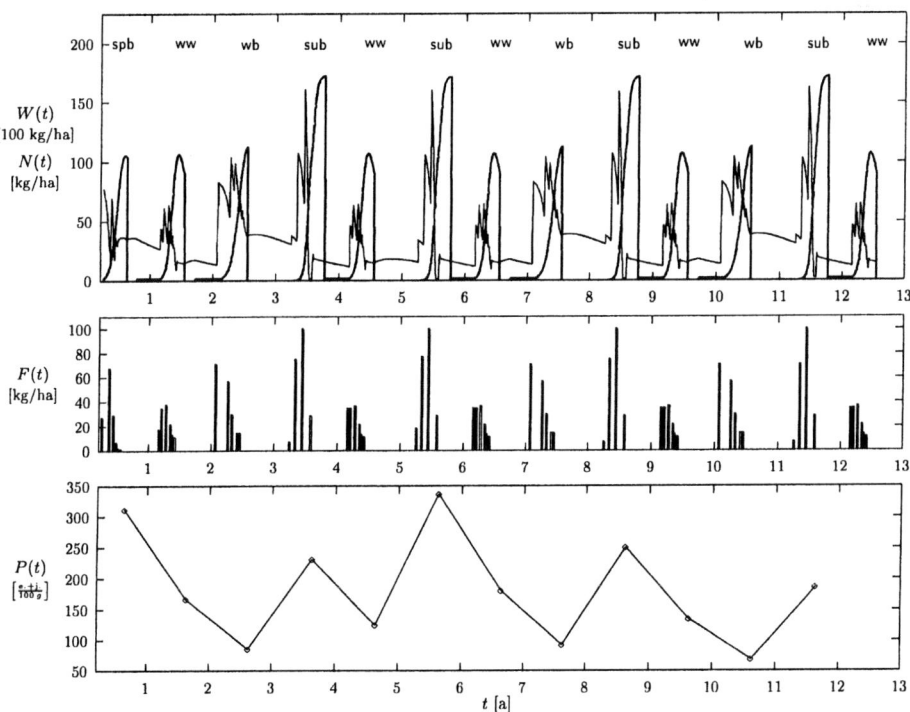

Figure 1.5 Optimum crop rotation of $T = 13$ years with optimum fertilizing schemes (horizontal axis shows t in years). State variables: biomass $W(t)$ (thick line) Nitrogen content in soil $N(t)$ (thin line upper plot), and Population of *H. schachtii* $P(t)$. Control function: fertilizer $F(t_i)$ (center plot) and planted crop ($\alpha(t_i)$) notation in upper plot. Initial values: $N_0 = 50$ kg/ha, $P_0 = 500$ eggs and juveniles per 100g soil.

Table 1.4   Total amount of nitrogen leached after a 7 year crop rotation for long–term performance criteria and literature values for comparison (Niesel–Lessenthin, 1988). The calculation of the average and standard deviation values was carried out with respect to different initial populations $P_1(t_0)$.

	$\widehat{J_1}$	$\widehat{J_3}$	$\widehat{J_4}$	
$\Lambda_N = 0$	650($\pm$40)	190($\pm$45)	275($\pm$35)	(kg/ha)
$\Lambda_N > 0$	450($\pm$45)	180($\pm$10)	275($\pm$32)	(kg/ha)

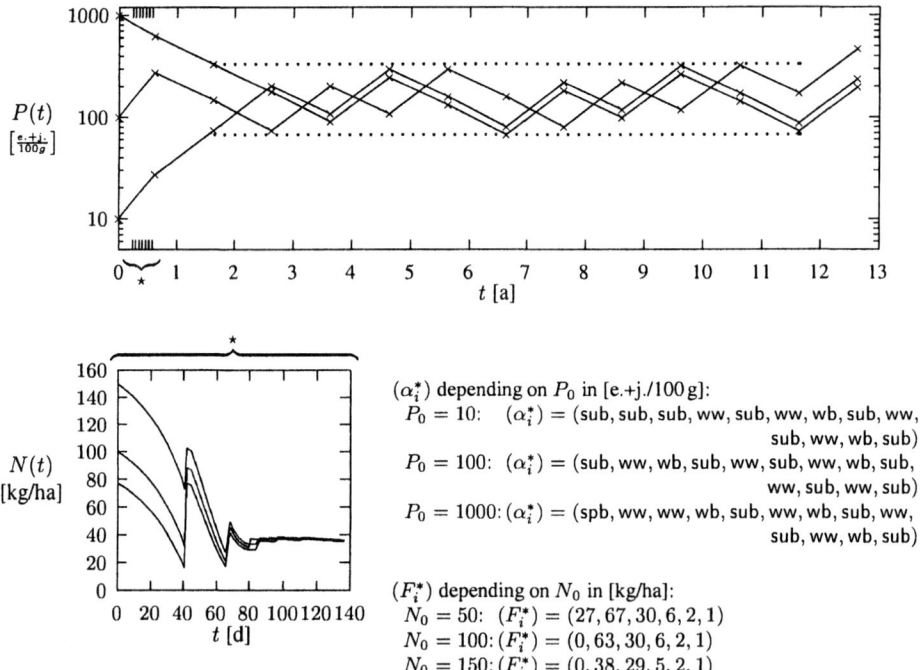

Figure 1.6   Quantitative characterization of the most rapid iterated approach to an optimum path of soil nitrogen content (lower left, first vegetation period) and population of nematodes (upper). assessment.

resulting amount of leaching nitrogen cannot be reduced any more even if $N(t)$ is used in the global performance criterion.

Weed control.   Weed control can be carried out by designing the crop rotation scheme or by pesticide application. Optimum biological weed control requires an optimum crop rotation. In the given example it is the population of *H. schachtii* which effects the yield of sugar beet, the most important, most valuable crop. After plantation of sugar beets the plantation of a non–host crop may reduce the population of nematodes in a crop rotation. For this reason, the optimum crop rotation solution in figure 1.5 consists of a 2 to 3 year crop rotation of sugar beet, with intermediate plantation of wheat. Crops like oil radish or field beans may decrease the population of nematodes more efficiently — these crops are catch crops, which enable *H. schachtii* to hatch but disable the larvae for becoming fertile (see (Schmidt *et al.*, 1993)). On the other hand, these crops do not have a positive effect on farmers income, see Section 4.3.

A general property of all these solutions of the optimum control problem is displayed in figure 1.6. The optimum solution consists of two most rapid approaches to a local and a global optimum path. The local optimum state of nitrogen content in soil is a content of 50 kg/ha or less. This is reached within the first vegetation period with its optimum fertilizing scheme. All further fertilizing schemes start and end with an average nitrogen content of approximately 50 kg/ha. This demonstrates, that the assumption for assessment of Section 3.1 is plausible. The global optimum control path is reached after three vegetation periods. This path is characterized by an interval of 80 to 200 eggs and juveniles in 100 g soil of *H. schachtii*. Figure 1.6 shows three different initial conditions ($P_0 = 10, 100, 1000$ e. + j./100g). Note, the optimal solution leaves this path, if the assessment stops at the end of the simulation interval.

4.3   REGIONAL OPTIMUM MANAGEMENT

The integration of the spatial explicit model into optimum control procedures extends the questions of "What to plant when?" to "What to plant when *and where*?".

The observed region, in which the field experiments for model calibrations are carried out, is the catchment site "Ohebach" which is part of the investigation area "Neuenkirchen" of the CRP 179, c.f. (McVoy *et al.*, 1995). The investigation area is located in the northern forelands of the Harz mountains in Lower Saxony (Niedersachsen), Germany, with an area of 16 km². A local 2 km² "Ohebach" catchment is chosen for detailed study. Soil is set up by unconsolidated quaternary sediments and covered by a $1 - 2$ m thick layer of loess, and topped finally by $0.2 - 1$ m of colluvial sediments, depending on slope. In the loess areas soils are Orthic to Gleyic Luvisols (FAO classification), see McVoy *at al.* (1995) for details. The average field size is 30 ha.

Precision Farming Fertilizing Schemes.   For an analysis of nitrogen applications we focus on a single vegetation period of one crop (winter wheat) on a given field. Figure 1.7 shows the intensive investigation "Field 278" of the "Ohebach" region. Soil properties of the field are characterized by seven homogeneous pedological units with an effective field capacity of 127 up to 268 mm.

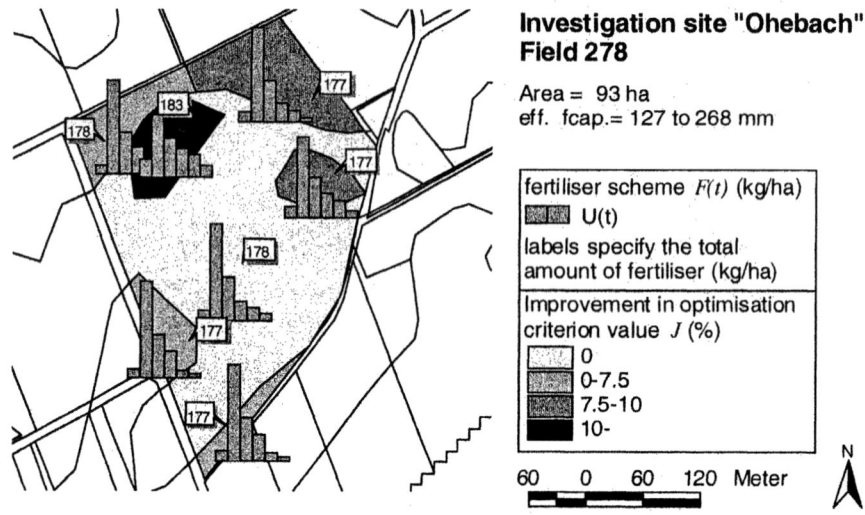

Investigation site "Ohebach"
Field 278

Area = 93 ha
eff. fcap.= 127 to 268 mm

fertiliser scheme $F(t)$ (kg/ha)
▨▨ U(t)
labels specify the total
amount of fertiliser (kg/ha)

Improvement in optimisation
criterion value J (%)
☐ 0
☐ 0-7.5
▨ 7.5-10
■ 10-

60   0   60   120 Meter

N

Figure 1.7 Zoom into "Field 278" of investigation site "Ohebach". The 93 ha field is set up by 7 pedological units. The amount of total fertilizer applied is labelled beside.

The grey shading shows, that regional management can be improved by up to 10% in comparison to an equally applied fertilizer, when applying precision farming approaches (as expressed by the performance criterion value). This is performed by an optimum allocation of fertilizer to the stages of crop development, qualitatively displayed with the bar charts.

Regionalised Crop Rotations. Figure 1.8 completes these results. The whole observation region "Ohebach" is shown, with the results of year three to seven of the optimum crop rotation from the precision farming solution. The lower part of the figure shows the population density of *H. schachtii* and the upper part shows the planted crop. Note that different optimum crop rotations are estimated for minimization of nitrogen loss and pest control. Exemplary two crop rotations are noted for field A and B below the maps. On the less fertile sites also fallow crops are used as catch crops for *H. schachtii*, see Section 4.2. Moreover, less nutrient demanding crops are planted in the more permeable regions of the investigation site (north–western part).

5.   DISCUSSION

It is well known that ecological systems are complex and hierarchical systems. As a consequence ecological simulation models tend to be complex.

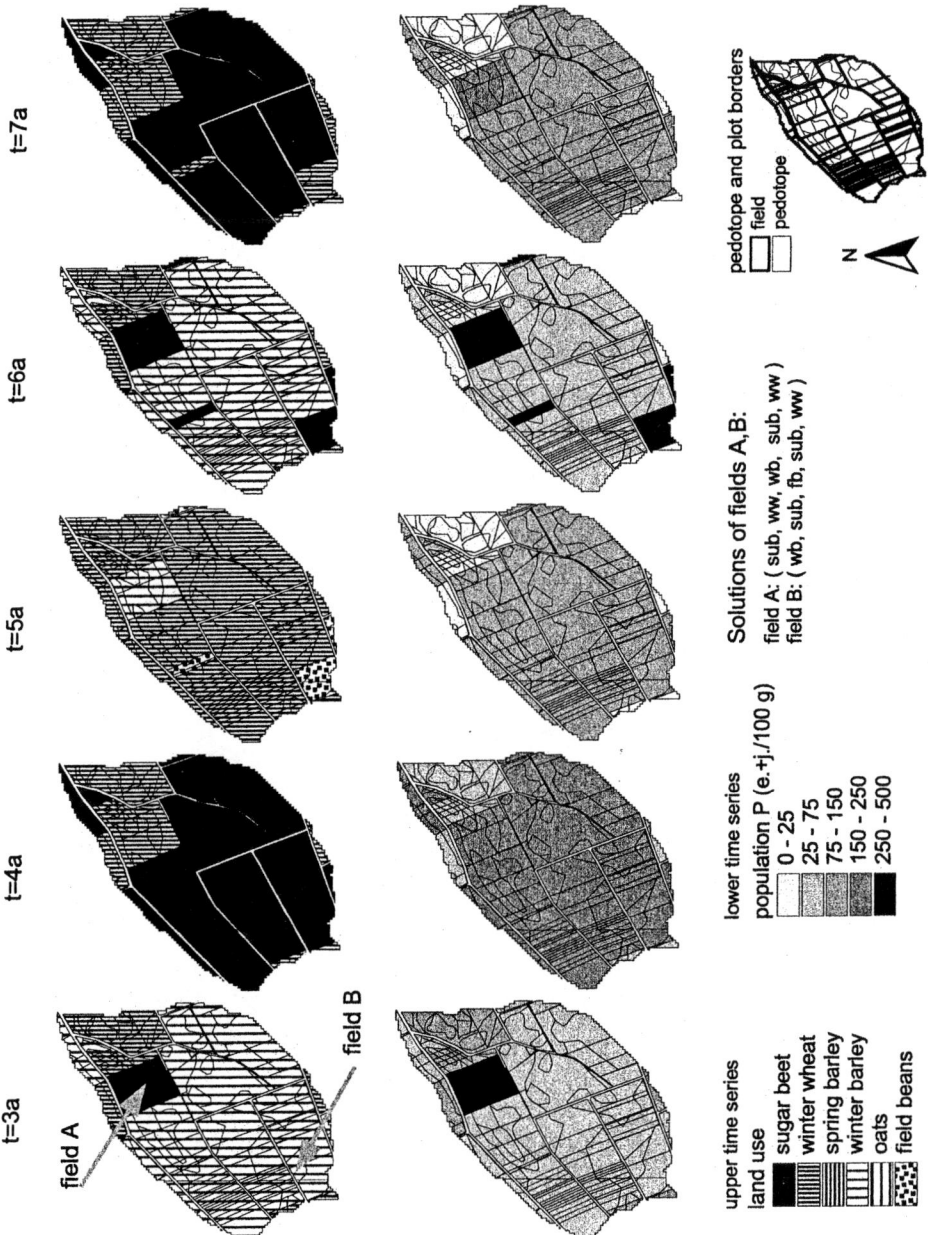

Figure 1.8 Regional optimum crop rotation in investigation site "Ohebach". Optimum crop rotation including precision farming solution of optimum fertilizer input. For detailed study, two crop rotation solutions are noted below the maps.

Because of this, the analysis of human impact to ecosystems using simulation models in terms of scenario analysis is limited.

This chapter showed that a systematic search throughout the policy space of environmental impact can be provided by the application of numerical optimum control theory to environmental models. The approach links ecological process models with the human impact in a clear manner, distinguishing between state, control and assessment variables and models.

Problems and perspectives of this approach in ecological system science are presented. Main difficulties of this approach, which give hints to further research needs, are

- to model a performance criterion or indicator, which takes all important disciplinary aspects of environmental processes into evaluation,

- to find an appropriate level of model aggregation, and

- to apply a suitable procedure of numerical optimum control.

Comparable to the standard approach of scenario analysis is the analysis of different performance criteria and the related optimum solutions. A very interesting outcome of the case studies is to study and to compare different views to optimality by modification of the performance criterion. It allows the comparison of different policy strategies of farm management. The proposed framework allows the analysis of different indicators of environmental assessment and can answer the question wheather environmental variables are aggregated in a suitable way. Facing this, it may also give an answer to the question of how to incorporate externalities into the process assessment.

Overall, the choice of a simulation model on an appropriate level of aggregation is decisive for the success of the approach. A more complex model may be more realistic. On the other hand, this may lead to problems of computational effort or of assessing the process in an appropriate way. Further research is necessary.

The proposed framework for optimum control and the dynamic programming procedure with a hierarchical structure of different time scales comes out as a general concept applicable to a large class of environmental models. Mathematically heterogeneous systems like ecosystem models can be treated. On the other hand, the "dilemma of dimensionality" is an intrinsic property of the underlying procedure. Approaches to face this problem with special respect to environmental models are presented. However, this problem cannot be eliminated completely. Further research should focus on robust procedures of optimum control of mathematically heterogeneous models with lesser prerequisites on the underlying model.

GIS comes out as a framework, which enables the coupling of agroecological simulation models and spatial databases. GIS–functions and robust optimization procedures decrease the numerical effort. Common precision farming approaches focus on the identification of spatially explicit management strategies in terms of pesticide or fertilizer application. Knowing this the most interesting results in the context of regional optimization are that not only the fertilizer schemes should be estimated and applied for each pedological unit. There is also a spatial dependence in the allocation of the optimum crop rotation. It is shown that agroecological process models with regionalised parameter fields integrated in numerical optimum control procedures may support the decision process in precision farming systems.

Overall, applications of environmental models in terms of optimum control theory are a very promising branch of ecological system theory. Optimum control of environmental model should find its way into an application of decision support system for environmental management.

Acknowledgments

Thanks are due to Otto Richter for valuable suggestions and remarks and to Dagmar Söndgerath for carefully reading the manuscript. Parts of the work were founded by the Germany Research Foundation (DFG) in the Special Collaborative Porject 179 "Water- and Matterdynamics in Agroecosystems".

Figures 1.5, 1.6 as well as 1.7, 1.8 reprinted from Seppelt (1999; 2000) with permission of Elsevier Science.

References

Assfalg, W., & Werner, R. 1993. The optimal use of agricultural landscapes. *Applied Geograhy and development*, **42**, 60–97.

Beddington, J., Botkin, D., & Levin, S.A. 1981. Mathematical models and resource management. *Lecture Notes in Biomathematics*, 1–5.

Begon, M., Harper, J.L., & Townsend, C.R. 1986. *Ecology*. Oxford, London, Edinburgh: Blackwell Scientific Publications.

Bellman, R.E., & Dreyfus, S.E. 1962. *Applied Dynamic Programming*. Princeton University Press, Princeton.

Bick, H. 1993. *Ökologie*. 2. edn. Gustav Fischer.

Breunig, M. 1996. *Integration of Spatial Information for Geo-Information Systems*. Lecture Notes in Earth Sciences, vol. 61. Springer, Berlin.

Clark, C.W. 1976. *Mathematical Bioeconomics*. John Wiley & Sons, New York.

Clark, C.W., Clarke, F.H., & Munro, G.R. 1979. The optimal exploitation of renewable resource stocks: problems of irreversible investment. *Econometrica*, **47**(1), 23–47.

Cohen, Y. 1987. Application of optimal impulse control to optimal foraging problems. *Lecture Notes in Biomathematics*, **73**, 39–52.

Costanza, R., Wainger, L., Folke, C., & Mäler, K.-G. 1993. Modeling complex ecological economic systems. *Bioscience*, **43**(8), 545–555.

Diekkrüger, B., Söndgerath, D., Kersebaum, K.C., & McVoy, C.W. 1995. Validity of agroecosystem models. *Ecological Modelling*, **81**, 3–29.

Doherty Jr., P.F., Marschall, E.A., & Grubb Jr., T.G. 1999. Balancing conservation and economic gain: a dynamic programming approach. *Ecological Economics*, **29**(3), 349–358.

Duffy, P.A., & Taylor, C.R. 1993. Long–Term Planning on a Corn–Soybean Farm: A Dynamic Programming Analysis. *Agricultural Systems*, **42**, 57–71.

Falkovitz, M.S., & Feinerman, E. 1994. Minimum leaching scheduling of nitrogen fertilization and irrigation. *Bulletin of Mathematical Biology*, **56**(4), 665–686.

Han, S., Evens, R.G., Hodges, T., & Rawlins, S.L. 1995. Linking a geographic information system with potato simulation model for site-specific crop management. *Journal of Environmental Quality*, **24**, 772–777.

Lu, Y.-C., Daughtry, C., Hart, G., & Watkins, B. 1997. The current state of precision farming. *Food Review International*, **13**(2), 141–162.

McVoy, C.W., Kersebaum, K.C., Arning, M., Kleeberg, P., Othmer, H., & Schröder, U. 1995. A data set from north Germany for the validation of agroecosystem models: documentation and evaluation. *Ecological Modelling*, **81**, 265–300.

Müller, F. 1997. State-of-the-art in ecosystem theory. *Ecological Modelling*, **100**, 135–161.

Naveh, Z., & Lieberman, A.S. 1984. *Landscape Ecology*. Springer, New York.

Niesel-Lessenthin, B. 1988. *Faustzahlen für Landwirtschaft und Gartenbau*. 11. edn. Landwirtschaftsverlag, Münster.

Nilsson, J., & Bergström, S. 1995. Indicators for the assessment of ecological and economic consequences of municipal policies for resource use. *Journal of Ecological Economics*, **14**, 175–184.

Prince, P.J., & Dormand, J.R. 1981. High order embedded Runge–Kutta formulae. *Journal of Computational Applied Mathematics*, **7**, 67–75.

Richter, O., Spickermann, U., & Lenz, F. 1991. A new model for plant growth. *Gartenbauwissenschaft*, **56**(3), 99–106.

Sadler, E.J., Busscher, W.J., Bauer, P.J., & Karlen, D.L. 1998. Agronomic models. Spatial scale requirements for presision farming: a case study in the southeastern USA. *Agronomy Journal*, **90**, 191–197.

Schmidt, K., Sikora, R.A., & Richter, O. 1993. Modelling the population dynamics of the sugar beet cyst nematode *Heterodera schachtii*. *Crop Protection*, **12**(7), 490–496.

Schröder, U., & Richter, O. 1993. Parameter estimation in plant growth models at different levels of aggregation. *Modelling Geo-Biosphere Processes*, **2**, 211–226.

Schröder, U., Richter, O., & Velten, K. 1995. Performance of the plant growth models of Special Collaborative Project 179 with respect to winter wheat. *Ecological Modelling*, **81**, 243–250.

Seppelt, R. 1999. Applications of optimum control theory to agroecosystem modelling. *Ecological Modelling*, **121**(2–3), 161–183.

Seppelt, R. 2000. Regionalised optimum control problems for agroecosystem management. *Ecological Modelling*, **131**(2–3), 221–232.

Seppelt, R. 2003. *Computer-Based Environmental Management*. New York, Weinheim: Wiley-VCH.

Svendsen, H., Hansen, S., & Jensen, H.E. 1995. Simulation of crop production, water and nitrogen balances in two German agro–ecosystems using DAISY model. *Ecological Modelling*.

van Kreveld, M., Nievergelt, J., Roos, Th., & Widmayer, P. (eds). 1997. *Algorithmic Foundations of Geographic Information Systems*. Lecture Notes in Computer Science, vol. 1340. Springer.

Velten, K., & Richter, O. 1993. Optimal maintenance investment of plants and its dependence on environmental conditions. *Bulletin of Mathematical Biology*, **55**(5), 953–971.

Voinov, A., Voinov, H., & Costanza, R. 1999. Surface water flow in landscape models: 2. Patuxent watershed case study. *Ecological Modelling*, **119**(2–3), 211–230.

Weibel, R. 1997. *Generalization of spatial data: principles and selected algorithms*. Vol. 1340 of (van Kreveld *et al.*, 1997). Chap. 5, pages 99–152.

Zadoks, J.C., Chang, T.T., & Konzak, C.F. 1974. A decimal code for the growth of cereals. *Journal of Weed Research*, **14**, 415–421.

Chapter 18

FINDING THE MINIMAL ROOT OF AN EQUATION

Applications and algorithms based on Lipschitz condition

Yaroslav D. Sergeyev

DEIS, University of Calabria, Via P. Bucci, Cubo 41-C, Rende (CS), Italy and Nizhni Novgorod State University, pr. Gagarina 23, Nizhni Novgorod, Russia
email: yaro@si.deis.unical.it

Abstract   In this survey, the problem of finding the minimal root to an equation is discussed. It is supposed that the equation under consideration can have many roots. In the case when the Lipschitz constant for the objective function or its first derivative is known a priori, two methods based on global optimization ideas are presented. The algorithms either find the minimal root or determine the global minimizers (in the case when the objective function has no roots). If the Lipschitz constants are unknown, there are introduced two methods adaptively estimating local Lipschitz constants during the search. This approach allows us to accelerate the search in comparison with the case with known a priori Lipschitz constants. Sufficient conditions for convergence of the new methods to the desired solution are established.

Keywords: Minimal root, multiextremal functions, global optimization.

Introduction

Let us consider the problem of finding the minimal root of an equation $f(x) = 0$ where $x \in [a, b]$, $f(x)$ is multiextremal, and $f(a) > 0$. This problem arises in many applications, such as computer graphics (see [6], [20], [26]), time domain analysis (see [1], [10]), filter theory (see [16], [22]), wavelet theory (see [24]), and phase detection (see [11], [25], [41]). A few of them are presented in the next section of this survey.

In the case when the formula of the objective function is known, interval analysis methods (see [7], [8], [17], [20], [21]) can be applied to solve this problem. In this survey, it is supposed that the formula is unknown and the function $f(x)$ satisfies the Lipschitz condition with the constant L, $0 < L < \infty$, i.e.,

$$|f(x) - f(y)| \leq L|x - y|, \quad x, y \in [a, b]. \tag{1.1}$$

A more special case is also discussed: the function $f(x)$ is Lipschitzian with an unknown constant L and has the Lipschitzian first derivative, i.e., $f'(x)$ satisfies the following condition

$$|f'(x) - f'(y)| \leq K|x - y|, \quad x, y \in [a, b], \tag{1.2}$$

where the constant K, $0 < K < \infty$, is the Lipschitz constant for $f'(x)$.

In many applications, the problem under consideration may be interpreted in the following way. It is necessary to know the behaviour of a device over a time interval $[a, b]$. The device works correctly while a function $f(x) > 0$. Of course, at the initial moment, $x = a$, the device works correctly and $f(a) > 0$. We must either find an interval $[a, x^*)$ such that

$$f(x^*) = 0, \quad f(x) > 0, \quad x \in [a, x^*), \quad x^* \in (a, b], \tag{1.3}$$

or prove that x^* satisfying (1.3) does not exist in $[a, b]$. It is necessary not only to solve the equation $f(x) = 0$ but also to prove that the found root is the minimal over $[a, b]$.

If the point x^* does not exist, then in many applications it is useful to find a measure of the robustness of the device, i.e., a global minimizer x' and the value $f(x')$ such that

$$f(x') = min\{f(x) : x \in [a, b]\}. \tag{1.4}$$

In general, it is difficult to determine the point x^* (or x') in an analytical way and numerical methods are used to find a σ-approximation x^*_σ of the point x^* such that

$$0 \leq f(x^*_\sigma), \quad |x^*_\sigma - x^*| \leq \sigma. \tag{1.5}$$

Analogously for the point x' in the case

$$f(x) > 0, \quad x \in [a, b], \tag{1.6}$$

a point x'_σ such that $|x' - x'_\sigma| \leq \sigma$ is determined.

Two approaches are currently used by engineers for solving these problems. The first one uses standard local techniques for finding equation roots in order to achieve a rapid convergence to the point x^*. The drawback of these methods is that convergence is not assured since $f(x)$ is a multiextremal function on $[a, b]$ and the methods may diverge or converge to a local minimum greater than zero (see [30]). Moreover, if the objective function $f(x)$ has more than one root (and this is usually the case), different choices of the initial conditions can produce different solutions of the equation $f(x) = 0$.

The second approach is based on the use of any simple grid technique which produces a dense mesh starting from the point a and going on by the step σ till the value $f(x)$ becomes less than zero. This approach is very reliable but the number of evaluations of $f(x)$ is too high.

In this survey a new approach (see [12], [27], [38]) is described. Numerical algorithms able either to find a point x_σ^* or determine x_σ' for the problems (1.3), (1.6) are proposed. Four methods based on geometric ideas of the global optimization techniques [28], [33], [34], [36] are described.

Two of them use the exact a priori given Lipschitz constants of the objective function or its first derivative. When these constants are not known a priori, two other methods solving the problem by using adaptive estimates of the local Lipschitz constant during the search are introduced. As it has been shown in the recent literature (see [3], [29], [34], [35], [38]), adaptive balancing local and global information can accelerate the search significantly.

1. ELECTRICAL ENGINEERING APPLICATIONS

Problem 1. The first example deals with a neural Analog to Digital (A/D) converter [9], i.e. an electronic device which transforms the electrical signals from the analog form to the digital one. Let us consider an example of A/D with eight similar blocks linked in series. Each block is built using only an input comparator and a subtractor amplifier. The comparator compares the input signal V_{IN} with a reference signal equal to half of the A/D converter range ($V_{FS}/2$), where V_{FS} is the full-scale voltage. The comparator output signal is subtracted from the input of the block and is amplified two times. The resulting signal being the error signal corresponding to the unconverted part of the input signal becomes the input of the successive conversion stage. Therefore, the input signal shape of the $i - th$ block corresponds to the part of the V_{IN} signal for $x > x_{i-1}$, where x_{i-1} is the time instant in which the input signal of the $(i - 1) - th$ block reaches the value $V_{FS}/2$. Consequently, the A/D converter operates on error signal propagation in cascade blocks. To avoid possible commutation errors due to the simultaneous commutations of the comparators of different blocks, a correction circuit is also applied at the digital output V_{OUT}. The time instants x_i, where $1 \leq i \leq 7$, are unknown but each of them can be found as the minimal root of seven equations $f_i(x) = 0$, where $f_i(x)$ is the input signal to the $i - th$ block, $1 \leq i \leq 7$.

Problem 2. The second example deals with electrical filters (see [16], [22]). *Filters* are basic electronic components used in many fields such as power conversion circuits, electronic measurement instruments and communications systems. Particularly, *electrical filters* can be found in the telephone, television, radio, radar, and sonar. A filter is a device that modifies in a predetermined way the input signal that passes through it. Electrical filters may be classified as: *analog* filters, used to process analog or continuous-time signals; and *digital* filters, used to process digital signals (discrete-time signals).

Let us consider a signal $s(x)$, where x is time. If a signal $s(x)$, composed of a sum of signals $s_1(x), s_2(x), .., s_n(x)$ so that

$$s(x) = s_1(x) + s_2(x) + .. + s_n(x),$$

is the input of an analog filter, the output signal is obtained from the input one by suppressing certain components $s_k(x), k \in \{1, .., n\}$. Let us define for the signal $s(x)$ its *frequency* θ as the number of times that the signal repeats itself in unit time and the *pulse* $\omega = 2\pi\theta$. Below we refer to θ or ω simply as frequency.

As an example, let us consider a radio or a television receiver. The transmission station is assigned an interval of frequencies called the *band of frequencies* or *channel frequencies*, in which it must transmit its signal. Ideally, the receiver should accept and process any signal in the assigned channel and completely exclude signals at all other frequencies so that

$$\omega_{c1} \leq \omega \leq \omega_{c2} \tag{1.7}$$

is the channel of the signal to be received. However, no circuits can produce such a transfer function exactly. In practice, filters are not required to meet the extremely stringent requirements such as those of (1.7) and some filters with a channel approximating (1.7) have been found to be consistently satisfactory. In this example the cutoff frequency can be found as the root.

Problem 3. The last example is related to the problem of measuring the phase angle between two functions with the same frequency. This problem very often arises in electronics and electrical engeneering and some instruments were created for measuring it. Traditional techniques work by converting the functions into two square waves and then measuring the difference either between the roots or between the pulse centers of the square waves. These techniques introduce error in phase measurement when the input functions are distorted by harmonics.

Another way of measuring the phase difference between two functions $f_1(x)$ and $f_2(x)$ may be obtained by considering the difference between the time instants in which the functions cross zero. Thus, the process to obtain the phase difference consists of three steps:

i. The functions $f_1(x)$ and $f_2(x)$ are sampled and digitised by an A/D converter (see Problem 1);

ii. An algorithm is applied twice to find the minimal roots of $f_1(x)$ and $f_2(x)$;

iii. The time difference between the roots for both functions is proportional to phase difference.

2. ALGORITHMS FOR PROBLEMS WHERE THE FIRST DERIVATIVES ARE NOT AVAILABLE

In this section algorithms for solving the problems (1.1), (1.3) and (1.1), (1.4), (1.6) are presented. The case where the function $f(x)$ is non-differentiable or it is differentiable but the first derivatives can not be evaluated is considered. Let us describe the main idea of the methods.

Suppose that the objective function $f(x)$ has been already evaluated at n trial points $x_i, 1 \le i \le n$, and $z_i = f(x_i)$. For every interval $[x_{i-1}, x_i]$, $1 < i \le n$, we construct an auxiliary function $\phi_i(x)$ in such a way that hopefully $\phi_i(x) \le f(x)$, $x \in [x_{i-1}, x_i]$. Adaptively improving the set of functions $\phi_i(x)$, $1 < i \le n$, by adding new points we improve our lower approximation of $f(x)$. This approach is widely used in global optimization (see [13], [19], [29], [39], [40]) applying functions $\phi_i(x)$ with different structures. The papers [14], [28], [29], [34] address methods using only the values of objective functions and the papers [2], [5], [33], [15], [36] describe algorithms, where the first derivatives are also taken into consideration. In this section we propose the methods using support functions from [28], [34]. The next section deals with the methods based on ideas from [33], [36], [40].

In two algorithms presented here the following ideas are used to provide a fast localization of the points x_σ^* from (1.5) or the point x_σ' in the case (1.6):

(A) constructing piece-wise linear auxiliary functions from [28], [34] where they have demonstrated a good performance in the global optimization;

(B) constructing auxiliary functions only for intervals $[x_{i-1}, x_i]$, $1 < i \leq k$, where

$$k = min\{\{n\} \cup \{i : f(x_i) \leq 0, 1 < i \leq n\}\}. \tag{1.8}$$

(C) adaptive estimating the local Lipschitz constant L_i for every interval $[x_{i-1}, x_i]$ (in the second algorithm).

Let us discuss these ideas one after another. (A) Suppose that we have an estimate m_i of the constant L_i such that:

$$m_i > L_i. \tag{1.9}$$

In this case it is possible to construct a piece-wise linear support function $\phi_i(x)$ for $f(x)$ over $[x_{i-1}, x_i]$ (see [28], [34]) as follows:

$$\phi_i(x) = max\{z_{i-1} - m_i(x - x_{i-1}), z_i - m_i(x_i - x)\}. \tag{1.10}$$

In (1.10) z_{i-1} and z_i are the values of the function $f(x)$ at the points x_{i-1} and x_i respectively. It is easy to compute the value

$$R_i = \phi_i(y_i) = min\{\phi_i(x) : x \in [x_{i-1}, x_i]\}.$$

as

$$R_i = 0.5[z_i + z_{i-1} - m_i(x_i - x_{i-1})], \tag{1.11}$$

The minimum takes place at the point

$$y_i = 0.5[x_i + x_{i-1} - (z_i - z_{i-1})/(m_i)]. \tag{1.12}$$

We shall call R_i the *characteristic* of the interval $[x_{i-1}, x_i]$.

Let us discuss the item (B). The new algorithms construct the function $\phi_i(x)$ from (1.10) from left to right. If in a step the characteristic $R_j \leq 0$ has been found, this means that there exists a point $x_\tau \in [x_{j-1}, x_j]$ such that $\phi_j(x_\tau) = 0$.

In this case we determine the point x_τ, set the new trial point $x^{n+1} = x_\tau$ and evaluate $f(x^{n+1})$. If in this new point $f(x^{n+1}) < 0$ then there is no need to consider the interval $(x^{n+1}, b]$ because the solution x_σ^* is in $(a, x^{n+1}]$ (here a, b are from (1.3)). Then we set $n = n + 1$ and restart the procedure. If (1.6) takes place, then the algorithm finds an approximation x_σ' of the point x' from (1.4).

Lastly (C). The values m_i from (1.9) have a decisive influence on the convergence rate (and correctness) of the algorithm. Estimation of the global Lipschitz constant L is a global optimization problem itself. Underestimates of L (or, more generally, lack of the global information about the objective function) can lead to loss of the global solution in both cases when methods use a priori given or adaptive estimates of L (see [3], [18], [37], [39], [40], [42]).

Since in real problems often it is difficult to know the exact value of L from (1.1) a priori, a fixed estimate $H > L$ of L is used during the search in many global optimization methods for different auxiliary functions $\phi_i(x)$ (see, for example, [5], [28]). Our first algorithm A1 uses this approach too and takes $m_i = H > L$.

The main drawback of this approach is that the global Lipschitz constant L gives a very poor information about the behaviour of the objective function over every small interval $[x_{i-1}, x_i]$. In order to avoid this drawback in algorithm A2 we estimate local Lipschitz constants L_i for every interval $[x_{i-1}, x_i]$, $1 < i \leq k$. The problem of estimating Lipschitz constants is under an intensive investigation (see [31], [42]) and many global optimization algorithms do it in different manners (see [14], [31], [29], [39], [42]). In this paper an approach successfully applied in many global optimization techniques (see [34], [35], [36], [40]) for different classes of problems will be used.

Let us now describe the methods. We introduce the general scheme and only **Step 2** will be different for algorithms A1 and A2.

Step 0. Set $x^1 = a, x^2 = b, z^i = f(x^i), i = 1, 2, b^2 = b, k = 2$. Suppose now that $n \geq 2$ trials of the algorithm have already been carried out at points $x^1, .., x^n$. The $(n+1)$th trial point x^{n+1} is chosen according to the following procedure.

Step 1. The points x^j, $1 \leq j \leq n$, $x^j \leq b^n$, where $b^n = x_k$ (k is from (1.8)), of the previous n trials are ordered according to the increase of their coordinates. Below we denote by superscripts, iterations numbers, and by subscripts, the trial points ordered in the course of iterations. Thus, we order the points

$$x^j, 1 \leq j \leq n, x^j \leq b^n$$

as follows:

$$a = x_1 < x_2 < ... < x_i < ... < x_k = b^n \leq b. \tag{1.13}$$

Set $i = 2$.

Step 2. Calculate the estimate m_i of the local Lipschitz constants L_i of the interval $[x_{i-1}, x_i]$ as follows:

For the method A1 (*a priori given Lipschitz constant*): Set

$$m_i = H, \tag{1.14}$$

where the constant H is such that $L < H < \infty$.

For the method A2 (*adaptive estimating the local Lipschitz constants during the search*):

$$m_i = r \cdot max\{\lambda_i, \gamma_i, \xi\}, \tag{1.15}$$

where the parameter $r > 1$ is a reliability parameter and $\xi > 0$ is a small number - the second parameter of the method.

The values λ_i and γ_i from (1.15) relate to changes of local and global information, respectively, obtained during the search. The value

$$\lambda_i = max\{\frac{|z_j - z_{j-1}|}{x_j - x_{j-1}} : j = i - 1, i, i + 1, 1 < j, \le k\} \tag{1.16}$$

where $z_i = f(x_i), 1 \le i \le k$, relates to a local estimate and looks only at adjacent intervals. In contrast, γ_i gives a global estimate, where

$$\gamma_i = \frac{x_i - x_{i-1}}{X^{max}} \lambda^{max}. \tag{1.17}$$

The value X^{max} is the length of the widest interval, i.e.

$$X^{max} = max\{x_i - x_{i-1} : 2 \le i \le k\},$$

and the value λ^{max} is a global estimate of the global Lipschitz constant, i.e.

$$\lambda^{max} = max\{\frac{|z_i - z_{i-1}|}{x_i - x_{i-1}}, 1 < i \le k\}. \tag{1.18}$$

Thus, the formula (1.15 banlances local and global information represented by λ_i and γ_i, respectively.

Step 3. Evaluate the characteristic value R_i from (1.11) and the point $y(i)$ from (1.12). If $R_i \le 0$ then go to **Step 5** otherwise set $i = i + 1$. If $i \le k$ then go to **Step 2** otherwise go to **Step 4**.

Step 4. Find an interval i with the minimal characteristic, i.e.

$$i = argmin\{R_j : 1 < j \le k\} \tag{1.19}$$

and define the new trial point x^{n+1} as follows

$$x^{n+1} = y_i,$$

where y_i is determined following (1.12), then go to **Step 6**.

Step 5. Calculate

$$x^{n+1} = x_{i-1} + z_{i-1}/m_i,$$

i.e. the left root of the equation $\phi_i(x) = 0$ over the interval $[x_{i-1}, x_i]$.

Step 6. If the stopping rule $x^{n+1} - x_{i-1} \leq \sigma$, where σ is from (1.5), is fulfilled then **Stop**. Otherwise calculate the value $f(x^{n+1})$ and go to **Step 1** setting $b^{n+1} = x^{n+1}$ if $f(x^{n+1}) \leq 0$ and $b^{n+1} = b^n$ otherwise.

Convergence conditions of the algorithms proposed are described by the following two theorems (see [27]).

Theorem 1 *Let the situation (1.6) take place, i.e. there is no root in $[a, b]$, and L_t be the local Lipschitz constant of $f(x)$ over the subinterval $[x_{t-1}, x_t], t = t(n)$, a global minimizer x' belongs to during the n-th iteration of A1 or A2. If there exists an iteration number n' such that for all $n > n'$ the inequality*

$$m_t > L_t \qquad\qquad (1.20)$$

holds then, the global minimizer x' will be a limit point of the sequence $\{x^n\}$ generated by A1 or A2 and only global minimizers can be limit points of $\{x^n\}$.

Theorem 2 *Let there exist a point x^* from (1.3), i.e. there exist at least one root x^* in $[a, b]$ and $x^* \in [x_{t-1}, x_t], t = t(n)$, during the n-th iteration of A1 or A2 and L_t be the local Lipschitz constant of $f(x)$ over the interval $[x_{t-1}, x_t]$. If there exists a number n^* such that for all $n > n^*$ the inequality (1.20) holds, then the point x^* will be the unique limit point of the sequence $\{x^n\}$ generated by A1 or A2.*

Thus, if the convergence conditions of the methods are satisfied, after fulfillment of the stopping rule the following situations can take place:

i. $b^{n+1} \neq b$. This means that we can take $x_\sigma^* = x_k$ if $f(x_k) = 0$ or $x_\sigma^* = x_{k-1}$ if $f(x_k) < 0$ because $x_i, i = k - 1$, is the maximal trial point such that $f(x_i) > 0$.

ii. $b^{n+1} = b$ and $R_i > 0$, for all $i, 1 < i \leq k$. This means that no root has been found in the interval $[a, b]$. The point

$$x_\sigma^n = argmin\{f(x_j) : 1 \leq j \leq n\}$$

can be taken as a σ–approximation of the global minimizer x' over $[a, b]$ and the value $f(x_\sigma^n)$ can be used as an estimate of reliability of our device over the interval $[a, b]$.

iii. $b^{n+1} = b$ and there exists an interval j such that its characteristic $R_j \leq 0$. This situation means that it is necessary to take new $\sigma^1 < \sigma$ because the algorithm stops within the interval $[x_{j-1}, x_j]$ with properties $z_{j-1} > 0$, $z_j > 0$, $R_j \leq 0$ and cannot proceed because $|x_{j-1} - x_j| \leq \sigma$.

3. ALGORITHMS FOR PROBLEMS WHERE THE FIRST DERIVATIVES ARE AVAILABLE

In problems where the first derivatives are available, it is possible to use them to construct better support functions. Simple non-smooth (see [12]) and a little bit more complex smooth (see [38], [40]) support functions can be used in order to solve the problems (1.2), (1.3) and (1.2), (1.4), (1.6). Since the smooth ones are closer to the objective function, let us present here two methods based on these structures.

In order to introduce the methods we suppose that the objective function $f(x)$ and its first derivative $f'(x)$ have been already calculated at n trial points $x^i, 1 \leq i \leq n$. We can reorder these points by subscripts in such a way that

$$a = x_1 < x_2 < ... < x_i < ... < x_n = b.$$

We designate the results of trials as

$$z_i = f(x_i), \quad z'_i = f'(x_i), \quad 1 \leq i \leq n,$$

and suppose that we have an estimate m_i of the constant K_i such that:

$$m_i \geq K_i. \tag{1.21}$$

In this case it is possible to construct a support function $\phi_i(x)$ for $f(x)$ over $[x_{i-1}, x_i]$ (see [36]) as follows:

$$\phi_i(x) = \begin{cases} z_{i-1} + z'_{i-1}(x - x_{i-1}) - 0.5m_i(x - x_{i-1})^2, & x \in [x_{i-1}, y'_i] \\ 0.5m_i x^2 + b_i x + c_i, & x \in (y'_i, y_i] \\ z_i - z'_i(x_i - x) - 0.5m_i(x_i - x)^2, & x \in (y_i, x_i] \end{cases} \tag{1.22}$$

where

$$y_i = \frac{x_i - x_{i-1}}{4} + \frac{z'_i - z'_{i-1}}{4m_i} + g_i, \tag{1.23}$$

$$y'_i = -\frac{x_i - x_{i-1}}{4} - \frac{z'_i - z'_{i-1}}{4m_i} + g_i, \tag{1.24}$$

$$g_i = \frac{z_{i-1} - z_i + z'_i x_i - z'_{i-1} x_{i-1} + 0.5m_i(x_i^2 - x_{i-1}^2)}{m_i(x_i - x_{i-1}) + z'_i - z'_{i-1}},$$

$$b_i = z'_i - 2m_i y_i + m_i x_i, \tag{1.25}$$

$$c_i = z_i - z'_i x_i - 0.5m_i x_i^2 + m_i y_i^2. \tag{1.26}$$

The function $\phi_i(x)$ has been constructed (see [36]) by using the Taylor formula for the point x_{i-1} (see the first line in (1.22)) and the point x_i (see the third line in (1.22)). The second line of (1.22) has been obtained using boundness of the $f(x)$ curvature which follows from (1.2). Note that the first derivative $\phi_i'(x)$, for all $x \in (x_{i-1}, x_i)$ exists.

Let us find the point

$$h_i = argmin\{\phi_i(x) : x \in [x_{i-1}, x_i]\} \tag{1.27}$$

and the corresponding value (*characteristic* of the interval $[x_{i-1}, x_i]$)

$$R_i = \phi_i(h_i) = min\{\phi_i(x) : x \in [x_{i-1}, x_i]\}. \tag{1.28}$$

Let us consider two cases. If $\phi_i'(y_i') < 0$ and $\phi_i'(y_i) > 0$, then

$$h_i = argmin\{f(x_{i-1}), \phi_i(\widehat{x}_i), f(x_i)\} \tag{1.29}$$

where:

$$\widehat{x}_i = 2y_i - z_i'm_i^{-1} - x_{i-1}. \tag{1.30}$$

The point $\widehat{x}_i$ is determined from the equation $\phi_i'(x) = 0$, $x \in [y_i', y_i]$. It follows from (1.22) that

$$\phi_i(\widehat{x}_i) = c_i - 0.5m_i\widehat{x}_i^2. \tag{1.31}$$

In the second case there is no point $\widehat{x}_i \in [y_i', y_i]$ such that $\phi_i'(\widehat{x}_i) = 0$ and

$$h_i = argmin\{f(x_{i-1}), f(x_i)\}. \tag{1.32}$$

The algorithms A3 and A4 work similarly to A1 and A2 by constructing the function $\phi_i(x)$ from (1.22) from left to right taking the intervals one after another and calculating their characteristics. If in a step $R_j \leq 0$ has been found, this means that there exists a point $\tilde{x} \in [x_{j-1}, x_j]$ such that $\phi_j(\tilde{x}) = 0$.

In this case we determine the new trial point $x^{n+1} = \tilde{x}$ and evaluate $f(x^{n+1})$ and $f'(x^{n+1})$. If in this new point $f(x^{n+1}) < 0$ then there is no need to consider the interval $(x^{n+1}, b]$ because the solution x_σ is in $(a, x^{n+1}]$. Then we set $n = n + 1$ and restart the procedure.

Let us suppose that n trials, with $n \geq 2$, of the algorithm have already been carried out at points $x^1, .., x^n$. The $(n + 1)$-th trial point x^{n+1} is chosen according to the following procedure.

Step 1. Among the trial points $x^1, .., x^n$ of the previous n iterations form the subset $X^{k(n)}$ such that

$$X^{k(n)} = \{x^j : 1 \leq j \leq n, x^j \leq b^n, b^n = x_k\},$$

where k is from (1.13). Thus, b^n is determined by (1.13) and is the right margin of the search interval during the nth iteration. Reorder the elements of the set $X^{k(n)}$ by subscripts in ascending order as in (1.13).

Step 2. For the first derivative $f'(x)$ calculate the estimate m_i of the local Lipschitz constants K_i of the interval $[x_{i-1}, x_i]$ as follows:

For the method A3 (*a priori given Lipschitz constant*): Set

$$m_i = H, \tag{1.33}$$

where the constant H is such that $L < H < \infty$.

For the method A4 (*adaptive estimating the local Lipschitz constants during the search*): Calculate adaptive estimates m_i for the local Lipschitz constants K_i for the intervals

$$[x_{i-1}, x_i], 1 < i \leq k,$$

as follows:

$$m_i = r \cdot max\{\lambda_i, \gamma_i, \xi\}, \tag{1.34}$$

where $\xi > 0$ and $r > 1$ are parameters of the method.

The values λ_i and γ_i reflect changes of local and global information obtained during the search. The value λ_i is calculated as

$$\lambda_i = max\{v_j : 1 < j \leq k, i - 1 \leq j \leq i + 1\}, \tag{1.35}$$

where

$$v_j = \frac{|2(z_{j-1} - z_j) + (z'_j + z'_{j-1})(x_j - x_{j-1})| + d_j}{(x_j - x_{j-1})^2}, \tag{1.36}$$

and

$$d_j = ([2(z_{j-1} - z_j) + (z'_j - z'_{j-1})(x_j - x_{j-1})]^2 + (z'_j - z'_{j-1})^2(x_j - x_{j-1})^2)^{\frac{1}{2}}.$$

The second component γ_i from (1.34) is calculated as

$$\gamma_i = m(x_i - x_{i-1})/X^{max} \tag{1.37}$$

where m estimates the global Lipschitz constant K from (1.2)

$$m = max\{v_i : 1 < i \leq k\} \tag{1.38}$$

and

$$X^{max} = max\{x_i - x_{i-1} : 2 \leq i \leq k\}.$$

Step 3. Initially set the index sets $I = \emptyset$; $Y = \emptyset$; $Y' = \emptyset$. Set the index of the current interval $i = 2$.

Step 3.0. If $i > k$ then go to **Step 4**, otherwise compute the values y_i, y_i' according to (1.23) and (1.24). If $\phi_i'(y_i') \cdot \phi_i'(y_i) < 0$ then go to **Step 3.2**, otherwise go to **Step 3.1**.

Step 3.1. Calculate $R_i = \phi_i(h_i)$, where h_i is from (1.32). If $h_i = x_i$ then include i in Y else include i in Y'. Go to **Step 3.3**.

Step 3.2. Calculate $R_i = \phi_i(h_i)$, where h_i is from (1.29). Include i in I. Go to **Step 3.3**.

Step 3.3. If $R_i \leq 0$ then go to **Step 5** otherwise set $i = i + 1$ and go to **Step 3.0**.

Step 4. Find an interval i with the minimal characteristic, i.e.

$$i = argmin\{R_j : 1 < j \leq k\} \tag{1.39}$$

and define the new trial at the point x^{n+1} as follows

$$x^{n+1} = \begin{cases} y_i' & \text{if } i \in Y' \\ \widehat{x}_i & \text{if } i \in I \\ y_i & \text{if } i \in Y \end{cases} \tag{1.40}$$

Go to **Step 6**.

Step 5. If $\phi_i(y_i') \leq 0$ then go to **Step 5.1**. Otherwise go to **Step 5.2**.

Step 5.1. Calculate

$$x^{n+1} = x_{i-1} + \frac{1}{m_i}(z_{i-1}' + \sqrt{z_{i-1}'^2 + 2m_i z_{i-1}}), \tag{1.41}$$

i.e. the right root of the equation

$$z_{i-1} + z_{i-1}'(x - x_{i-1}) - 0.5m_i(x - x_{i-1})^2 = 0$$

obtained from the first line of (1.22) and go to **Step 6**.

Step 5.2. If $\phi_i'(y_i') \cdot \phi_i'(y_i) \geq 0$ then go to **Step 5.3**. Otherwise if $\phi_i(\widehat{x}_i) > 0$ then compute

$$x^{n+1} = x_i + \frac{1}{m_i}(z_i' + \sqrt{z_i'^2 + 2m_i z_i}) \tag{1.42}$$

i.e. the right root of the equation

$$z_i - z_i'(x_i - x) - 0.5m_i(x_i - x)^2 = 0$$

obtained from the third line of (1.22) and go to **Step 6**.

If $\phi_i(\widehat{x}_i) \leq 0$ then x^{n+1} is calculated following the formula

$$x^{n+1} = \frac{-b_i - \sqrt{b_i^2 - 2m_i c_i}}{m_i} \tag{1.43}$$

obtained from the second line of (1.22) as the left root of the equation

$$0.5m_i x^2 + b_i x + c_i = 0$$

then go to **Step 6**.

Step 5.3. If $\phi_i(y_i) > 0$ then calculate x^{n+1} using (1.42) and go to **Step 6**. Otherwise use (1.43) for calculating x^{n+1}.

Step 6. If the stopping rule $|x_i - x_{i-1}| \leq \sigma$, where σ is from (1.5), is fulfilled then **Stop**. Otherwise calculate the value $f(x^{n+1})$ and go to **Step 7** setting $b^{n+1} = x^{n+1}$ if $f(x^{n+1}) < 0$.

Step 7. Calculate the value $f'(x^{n+1})$. Set $n = n+1$ and go to **Step 1**.

The results obtained after satisfaction of the stopping rule are considered similarly to the results for the algorithms A1 and A2. The following convergence conditions (see [38]) hold for the methods A3 and A4.

Theorem 3 *Let K_t be the local Lipschitz constant of $f'(x)$ over the interval $[x_{t-1}, x_t] \ni x^*$, $t = t(n)$, during the n-th iteration of A3 or A4. If there exists an iteration number n^* such that for all $n > n^*$ the inequality*

$$m_t \geq K_t \qquad (1.44)$$

holds then the point x^* will be the unique limit point of the trial sequence $\{x^n\}$ generated by A3 or A4.

Theorem 4 *Let the situation (1.6) take place, i.e. there is no root in $[a, b]$, and K_t is the local Lipschitz constant of $f'(x)$ over the interval $[x_{t-1}, x_t] \ni x'$, where x' is a global minimizer and there exists a number n' such that (1.44) takes place. Then x' will be a limit point of the trial sequence $\{x^n\}$ generated by A3 or A4 and only global minimizers can be limit points of $\{x^n\}$.*

It should be notice that in all the four methods presented here to have convergence to the desired solution it is not necessary to estimate correctly the global Lipschitz constant K (or L in the case when the first derivatives are not available) over the whole region $[a, b]$. It is enough to do it only for the local constant K_i (or L_i for the methods A1 and A2) for the subinterval $[x_{t-1}, x_t]$. This condition is significantly weaker than the corresponding convergence results for the methods using estimates of Lipschitz constants (see [14], [15], [18], [39]).

4.   NUMERICAL EXAMPLES

In this section three numerical experiments are considered. In the first of them four algorithms and a grid method solve a test problem.

The second and third experiments deal with real applications. Let us start with the following test function

$$f(x) = \cos(x) - \sin(5x) + 1.$$

It has six roots over the interval $[0, 7]$ and the minimal root is $x^* = 1.57079$.

The parameters of the algorithms have been chosen as follows: $r = 1.1$ for the algorithm A2 and $r = 1.2$ for A4, $\epsilon = 10^{-6}$ for both of them. The value $\sigma = 10^{-4}(b - a)$ has been used in the stopping rule for all algorithms. The exact Lipschitz constants have been applied in A1 and A3 in all experiments. The methods A1–A4 have found the desired solution with the required accuracy in 53, 36, 11, and 10 iterations, correspondingly. The grid method with the step $\sigma = 10^{-4}(b - a)$ has solved this problem in 2016 iterations.

The second and third experiments are related to practical electrical engineering problems of finding the cutoff frequency for the filters problem presented in Section 1. Let us consider a Chebyshev filter (see [16]) with parameters: $R = 1\Omega$; $L = 2H$; $C = 4F$. The cutoff frequency can be found as the minimal root of the function

$$f(\omega) = F(\omega)^2 - 1/2F_{max}^2,$$

$$F(\omega) \mid \frac{V_{out}(\omega)}{V_{in}(\omega)} \mid= \frac{1}{\sqrt{1 + R^2C^2\omega^2}} \cdot \frac{1}{\sqrt{(2 - \omega^2LC)^2 + \omega^2L^2/R^2}}.$$

The solution has been found at the point $\omega = 0.8459 rad/s$. This result was obtained in 2745 iterations by the grid method, in 11 iterations by the algorithm A3 and in 10 iterations by the algorithm A4.

The last example considers a passband filter (see [16]). The transfer function of this filter is given by

$$F(\omega) =\mid \frac{V_{out}(\omega)}{I_{in}(\omega)} \mid= \frac{\omega L_1 R_1}{\sqrt{(Z_1^2 + Z_2^2)^2 \cdot Z_3}},$$

where

$$Z_1 = -\omega^3 R_1 L_1 L_2 + \omega R_1 L_2 + \omega R_1 L_1 C_1/C_2 - R_1/(\omega C_2) + 2\omega L_1 R_1 + \omega L_1 R_2,$$

$$Z_2 = \omega^2 L_1 L_2 + \omega^2 R_1 R_2 L_1 C_1 - R_1 R_2 - L_1/C_2,$$

$$Z_3 = (\omega L_1)^2 + (\omega^2 R_1 L_1 C_1 - R_1)^2.$$

The parameters for this filter have been chosen as follows

$$R_1 = 3108\Omega, \quad L_1 = 40e^{-3}H, \quad C_1 = 1e^{-6}F,$$

$$R_2 = 477\Omega, \quad L_2 = 350e^{-2}H, \quad C_2 = 0.1e^{-6}F.$$

The cutoff frequency was found as the minimal root of the function

$$F(\omega) = -(f(\omega)^2 - 1/2F_{max}^2).$$

The solution has been found at the point $\omega = 4824.43 rad/s$. This result was obtained in 4474 iterations by the grid method, in 44 iterations by the algorithm A3 and in 27 iterations by the algorithm A4.

5.   CONCLUDING REMARKS

In this survey, we have considered a problem very often arising in engineering applications, namely, the problem of finding the minimal root of an equation $f(x) = 0$, where $f(x)$ is a multiextremal black-box function which either satisfies the Lipschitz condition (1.1) with a constant L, $0 < L < \infty$, or $f(x)$ is such that its first derivative $f'(x)$ satisfies the corresponding Lipschitz condition(1.2) with a constant K, $0 < K < \infty$. Since the objective function, $f(x)$, is multiextremal, local search techniques cannot be used to solve the problem and, therefore, methods based on ideas of global optimization have been considered. On the other hand, the fact that the function is given in a black-box form means that its formula is not available. This means that interval analysis techniques cannot be used and the Lipschitz information becomes extremely important.

Four methods proposed for solving the problem have been described in this survey. All of them construct auxiliary support functions during their work in order to locate the desired solution. The first and the second methods can be applied when the first derivatives are not available. The first algorithm uses the exact a priori given Lipschitz constant L of the objective function. When L is not known a priori, the second method can be used successfully. It solves the problem using adaptive estimation of the local Lipschitz constant during the search. It should be also mentioned that it uses the obtained estimates very efficiently in order to accelerate the search of the minimal root.

When the first derivatives are available, it is possible to use the third and the fourth methods. The third algorithm uses the exact a priori given Lipschitz constant K of the first derivative. When K is not known a priori the fourth method solves the problem by using adaptive estimates of the local Lipschitz constant during the search. Again, the usage of the local Lipschitz estimates allows us to accelerate the search.

It is worthwhile to notice that all the introduced methods start the search of the minimal root without any knowledge about existence of at least one root over the search interval. During the search the algorithms either determine an approximation of the minimal root or an approximation of a global minimizer (in case where there are no roots).

[1] G. ANTONELLI, F. BINASCO, G. DANESE, D. DOTTI, *Virtually Zero Cross-Talk Dual Frequency Eddy Current Analyzer Based on Personal Computer*, IEEE Transaction on Instrumentation and Measurement, 43 (1994), pp. 463–468.

[2] W.P. BARITOMPA, *Accelerations for a Variety of Global Optimization Methods*, J. of Global Optimization, 4(1) (1994), pp. 37–45.

[3] W.P. BARITOMPA, C.P. STEPHENS, *Global optimization requires global information*, J. Optim. Theory Appl., 96(3) (1996), pp. 575–588.

[4] D. BEDROSIAN, J. VLACH, *Time Domain Analysis of Network with Internally Controlled Switches*, IEEE Trans. Circuit Syst., CAS-39(3) (1992), pp. 192–212.

[5] L. BREIMAN, A. CUTLER, *A Deterministic Algorithm for Global Optimization*, Math. Programming, 58 (1993), pp. 179–199.

[6] O. CAPRANI, L. HVIDEGAARD, M. MORTENSEN, T. SCHNEIDER, *Robust and efficient ray intersection of implicit surfaces*, Reliable Computing, 1 (2000), pp. 9–21.

[7] L. CASADO, I. GARCIA, Ya.D. SERGEYEV, *Interval branch and bound algorithm for finding the First-Zero-Crossing-Point in one-dimensional functions*, Reliable Computing, 2 (2000), pp. 179–191.

[8] L. CASADO, I. GARCIA, Ya.D. SERGEYEV, *Interval algorithms for finding the minimal root in a set of multiextremal non-differentiable one-dimensional functions*, SIAM Journal on Scientific Computing, 24(2) (2002), pp. 359–376.

[9] F. CENNAMO, P. DAPONTE, D. GRIMALDI, E. LOIZZO, *An improved neural based A/D converter*, IEEE 36th Midwest Simposium on Circuit and System, Detroit, Michigan (USA), (1993), pp. 430–433.

[10] L. O. CHUA, Charles A. DESOER, Ernest S. KUH, *Linear and Non linear Circuits*, MacGraw Hill, Singapore, 1987.

[11] L. D. COSART, L. PEREGRINO, A. TAMBE, *Time Domain Analysis and its Practical Application to the Measurement of Phase Noise and Jitter*, IEEE Instrumentation and Measurement Technology Conference, Brussels, Belgium, (1996), pp. 430–1435.

[12] P. DAPONTE, D. GRIMALDI, A. MOLINARO, Ya.D. SERGEYEV, *An Algorithm for Finding the Zero-Crossing of Time Signals with Lipschitzian Derivatives*, Measurement, 16 (1995), pp. 37-49.

[13] C.A. FLOUDAS, P.M. PARDALOS, *State of the Art in Global Optimization*, Kluwer Academic Publishers, Dordrecht, 1996.

[14] E.A. GALPERIN, *The Alpha Algorithm and the Application of the Cubic Algorithm in Case of Unknown Lipschitz Constant*, Computers Math. Applic., 25(11–12) (1993), pp. 71–78.

[15] V.P. GERGEL, *A Global Search Algorithm Using Derivatives*, Systems Dynamics and Optimization, N. Novgorod University Press, 1992, pp.161–178.

[16] D. E. JOHNSON, *Introduction to Filter Theory*, Prentice Hall Inc., New Jersey, 1976.

[17] E. HANSEN, *Global Optimization Using Interval Analysis*, vol. 165 of Pure and applied mathematics, Marcel Dekker, Inc, New York, 1992.

[18] P. HANSEN, B. JAUMARD, and S.-H. LU, *On the use of estimation of the Lipschitz constant in global optimization*, J. Optim. Theory Appl., 75 (1992), pp. 195–200.

[19] R. HORST, P.M. PARDALOS, *Handbook of Global Optimization*, Kluwer Academic Publishers, Dordrecht, 1995.

[20] D. KALRA, A.H. BARR, *Guaranteed Ray Intersections with Implicit Surface*, Computer Graphics, 23(3) (1989), pp. 297–306.

[21] R. B. KEARFOTT, *Rigorous Global Search: Continuous Problems*, Kluwer Academic Publishers, 1996.

[22] H.Y-F. LAM, *Analog and Digital Filters-Design and Realization*, Prentice Hall Inc., New Jersey, 1979.

[23] D. MacLAGAN, T. STURGE, W. BARITOMPA, *Equivalent Methods for Global Optimization*, State of Art in Global Optimization, eds. C.A. Floudas, P.M. Pardalos, (1996), pp. 201–212.

[24] S. MALLAT, *Zero-Crossing of a Wavelet Transform*, IEEE Trans. on Inf. Theory, 37(4) (1991), pp. 1019–1033

[25] R. MICHELETTI, *Differenza di Fase tra Due Grandezze Sinusoidali*, Elettronica Oggi, 99 (1990), pp. 121–127

[26] D. MITCHELL, *Robust Ray Intersection with Interval Arithmetic*, in Graphics Interface'90, 1990, pp. 68–74.

[27] A. MOLINARO, Ya.D. SERGEYEV, *Studying the minimal root problem for the multiextremal and non-differentiable functions*, Report N 5, ISI-CNR, Rende-Cosenza, 1998.

[28] S.A. PIJAVSKII, *An Algorithm for Finding the Absolute Extremum of a Function*, USSR Math. Math. Physics, 12, 1972, pp.57–67.

[29] J. PINTÉR, *Global Optimization in Action*, Kluwer Academic Publishers, 1996.

[30] W.H. PRESS, Y.B.P. FLANNER, S.A. TEUKOLSKY, W.T. VETTERLING, *Numerical Recipes: the Art of Scientific Computing*, Cambridge University Press, Cambridge, 1986.

[31] J.B. OLIVEIRA, *Evaluating Lipschitz Constants for Functions Given by Algorithms*, Computational Optimization and Applications., 16(3) (2000), pp. 215- 229.

[32] Ya.D. SERGEYEV, *A Global Optimization Algorithm Using Derivatives and Local Tuning*, ISI–CNR Report, 1 (1994), Rende, Italy.

[33] Ya.D. SERGEYEV, *A Method Using Local Tuning for Minimizing Functions with Lipschitz Derivatives*, Developments in Global Optimization, eds. E. Bomze, T. Csendes, R. Horst and P.M. Pardalos, Kluwer Academic Publishers, (1994), pp. 199–216.

[34] Ya.D. SERGEYEV, *A One-Dimensional Deterministic Global Minimization Algorithm*, Comput. Maths. Math. Phys, 35(5) (1995), pp. 705–717.

[35] Ya.D. SERGEYEV, *An Information Global Optimization Algorithm with Local Tuning*, SIAM J. Opt., 5(4) (1995), pp. 858–870.

[36] Ya.D. SERGEYEV, *Global One-Dimensional Optimization Using Smooth Auxiliary Functions*, Mathematical Programming, 81(1) (1998), pp.127-146.

[37] Ya.D. SERGEYEV, *On convergence of "Divide the Best" global optimization algorithms*, Optimization, 44(3) (1999), pp.303–325.

[38] Ya.D. SERGEYEV, P. DAPONTE, D. GRIMALDI, A. MOLINARO, *Two Methods for Solving Optimization Problems Arising in*

Electronic Measurements and Electrical Engineering, SIAM Journal on Optimization, 10(1) (1999), pp. 1-21.

[39] R.G. STRONGIN, *Numerical Methods on Multiextremal Problems*, Nauka, Moscow, 1978.

[40] R.G. STRONGIN, Ya.D. SERGEYEV, *Global optimization with non-convex constraints: Sequential and parallel algorithms*, Kluwer Academic Publishers, 2000.

[41] P. TURCZA, R. SROKA, T. ZIELINSKI, *Implementation of an Analytic Signal Method of Instantaneous Phase Detection in Real-Time on Digital Signal Processor*, Proc. of TC-4 IMEKO Modern Electrical and Magnetic Measurement, Prague, (1995), pp. 482–486.

[42] G.R. WOOD, B.P. ZHANG, *Estimation of the Lipschitz Constant of a Function*, J. of Global Optimization, 8 (1996), pp. 91–103.

Chapter 19

OPTIMIZATION OF RADIATION THERAPY DOSE DELIVERY WITH MULTIPLE STATIC COLLIMATION

Tervo J.
Department of Mathematics and Statistics, University of Kuopio, Kuopio, Finland

Kolmonen P.
Department of Applied Physics, University of Kuopio, Kuopio, Finland

Pintér J.D.
Pintér Consulting Services and Dalhousie University, Halifax, Nova Scotia, Canada

Lyyra-Laitinen T.
Department of Clinical Physiology, Kuopio University Hospital, Kuopio, Finland

Abstract
In radiation therapy new delivery techniques have been recently developed. Especially the multileaf collimator (MLC) has provided better facilities to deliver the dose for a cancer patient. The MLC based techniques allow the construction of 3–dimensional and conformal dose distributions. The succesful use of MLC delivery method requires the global optimization of the treatment plan. The paper gives one potential approach to optimize the treatment plan applying the so called multiple static MLC technique. For numerical optimization, the LGO global optimization software system is used. For the comparison of the numerical results, simulated annealing algorithm was used.

Keywords: Radiation therapy, inverse treatment planning, multileaf collimator, global optimization in high dimensional problems.

Introduction

Advanced technical and computational resources have provided new techniques for radiation therapy treatment. Two recent techniques of high interest are *intensity modulated* radiation therapy (IMRT) and *inverse* radiation therapy treatment *planning* (IRTTP). These two new techniques together provide more conformal dose distribution to the target volume, as well as saving the other tissues. IMRT and IRTTP are now becoming routinely used techniques in leading clinics, and they are subject of intensive research.

External radiation therapy is usually delivered by a linear accelerator (Figure 1.1, left). A relatively new accessory of linear accelerator, necessary for IMRT, is the multileaf collimator (MLC) (Figure 1.1, right). It is a collimator where a number of opposite segments (the leaves of the collimator) are able to move in parallel. So the open field can be changed as a function of time. In planning the collimator plane, the treatment space is a 2-dimensional rectangle divided into smaller rectangles (2.5 mm × 2.5 mm), called bixels. Since the duration of time which the bixel is open (beam–on–time of the bixel) is directly correlated to the intensity weight of the bixel, intensity modulated fields can be created by moving the leaves of the MLC. The locations or velocities of heads (edges) of MLC leaves (*the multileaf parameters*) must be determined in the course of treatment planning. The dose can be delivered by *dynamical collimation* or *multiple static collimation*. Applying dynamical collimation, the radiation is on during the movement of leaves. Using multiple static collimation, the radiation is interrupted when the leaves are moving. In the multiple static collimation, the time lengths (beam–on–times) of the subsequent subfields can be chosen as planning parameters. Finally, the collimator can also rotate around its transversal axis. The *collimator angle* gives an additional parameter for the multiple static collimation. For reviews of intensity modulated treatment planning we refer to Börgers, 1997; Brahme, 1995; Shepard et al., 1999; Webb, 1993; Webb, 1997. In the literature rotational tomotherapy technique is also reported Oelfke and Bortfeld, 1999; Shepard et al., 1999; Yang et al., 1997, but in the following we shall not consider rotational therapy. The successful choice of the treatment parameters demands the application of an *inverse treatment planning algorithm* which is based on a suitable *optimization strategy*.

In the IRTTP the desired dose is determined in the planning target volume (macroscopic and microscopic tumor, possible lymphnode regions, margins) and appropriate limitations are set for the dose in certain critical organs and, also, in the healthy normal tissue. Then, the treatment plan parameters are optimized to meat (approximate) the given dose distribution. This is just the opposite of routine forward planning, in which the parameters are set by the planner and the dose distribution is then calculated and analyzed, whether it fulfills the requirements or not. Especially, when the shape of the target volume is complex, this forward planning is time consuming and the results may be far from optimal.

Conventionally the IRTTP problem is solved in two steps. First, one tries to optimize the intensity distributions (weights or beam–on–times of bixels) of fields in such a way that the corresponding fluxes incident on the patient body surface produce the desired dose distribution in the patient space. Second, one optimizes the MLC parameters to generate the obtained intensity profiles. In our optimization strategy we link the dose to the MLC parameters. The objective functions used are expressed *directly* with the help of these parameters.

The optimization criteria which guarantee an acceptable plan are not uniquely determined. However, the most common approach is to select the treatment parameters (directions, multileaf parameters and time lengths) in such the way that the tumor receives the *prescribed dose* and that the doses in the critical organs and in the normal tissue are *below some prescribed values*. In addition, one often describes *(partial) dose volume constraints*. The dose volume constraint is associated with a certain structure of the patient space, e.g. with the critical organ. Typically, it is required that the volume fraction that receives a dose greater than a given threshold must be under a prescribed value Börgers, 1997; Webb, 1993; Bortfeld, 1999. The above optimization criteria are called *physical criteria*. In addition to developing optimization algorithms for IRTTP in the above sense, one often tries to find only *feasible solutions* Censor et al., 1988; Censor and Zenios, 1997; Kolmonen et al., 1998. In the feasible solution approach the aim is to find a solution which guarantees that the dose in the tumor is between the prescribed limits, and that the doses in organs at risk and normal tissue are under the prescribed limits. Dose volume constraints can be added also in this approach.

Instead of the physical criteria, there exist also *biological criteria* to optimize the treatment plan Brahme, 1999. The objective function is constructed using probabilities to destroy the tumor and, at the same time, to save the critical organs and normal tissue. In the literature well known related concepts are the tumor control probability (TCP) and the normal tissue complication probability (NTCP) Börgers, 1997; Webb, 1993; Webb, 1997. However, such biological objective functions require some statistical model parameters which are not reliably known. For this reason, optimization based on such foundations is not commonly accepted. In this work we shall use only physical objective functions.

It is easy to check numerically that the objective function types derived in this model are highly *multiextremal*. Hence, it is necessary to apply a global optimization approach to find as good solution as possible for the constrained extremum problem. Especially in the past decade, the advances in global optimization algorithms, convergence analysis and increasing computational power have given real possibilities to handle large dimensional global optimization problems in practice. However, numerical complexity increases exponentially with the number of decision variables and constraints. Therefore, without any a priori information the search strategies applied in global optimization software systems may still be too time–consuming for truly real world problems. We hope that this contribution gives some inspiration and insight for the optimization experts to improve the solution methods for the optimization problems presented.

The notion of IRTTP problem always requires a *dose calculation model* (the solution for the forward problem). A commonly used approach is based upon the *pencil beam model*. In this approach, one divides the penetrating beam into beamlets, so called pencil beams. The effect of radiation of each pencil beam emerging from the field S_l to a point of the patient space is covered by the "dose deposition kernel" h_l. One of the disadvantages of this method is that the kernel function h_l is not known analytically: however, various discrete or semianalytical expressions have been developed for it which are based on the classical Bathon law, Gaussian type kernels or on more sophisticated foundations Janssen et al., 1997; Johns and Cunningham, 1983; Mackie et al., 1988; Tervo and Kolmonen, 1998; Ulmer and Harder, 1996; Wang and Jette, 1999; Webb, 1993. The advantage of the pencil beam modelling approach is simple implementation of dose calculations. The most rigorous dose calculation models are based on the *Boltzmann transport equation* Cercignani, 1988; Dautray and Lions, 1993; Jette, 1995; Larsen, 1997; Tervo et al., 1999a. The difficulty of transport equation based models is the partial lack of knowledge of differential cross sections of different particles and interactions. In addition, the large dimensionality of numerical schemes, e.g. the schemes for grid based finite element or collocation approximations, causes further difficulties. However, advances in the knowledge of cross section data and the increase of computing power will diminish these disadvantages in the near future. For simplicity, in this contribution we shall use the pencil beam model in dose calculation. In some previous studies we have already considered IRTTP problem using MLC locations as decision parameters. In Tervo and Kolmonen, 2000 we sought a feasible solution applying the Cimmino algorithm together with initialization. The paper Tervo et al., 2003b considered the problem for dynamical MLC technique. The optimization was based on simulated annealing (SA) algorithm. The contribution Tervo and Kolmonen, 2002 linked the model to the Boltzmann transport based dose calculation. The numerical test applied the SA algorithm.

In this paper we use two different global optimization algorithms. One of them is the LGO (Lipschitz Global Optimizer) software system. In Tervo et al., 2003a we preliminarely used LGO software. Here we enlarge considerations of global optimization for the IRTTP problem. LGO is an integrated system of global and local solvers in which a (theoretically convergent) global search phase is followed by local search procedures. In the local phase the gradient information is also used. For theoretical background of the LGO system we refer to Pintér, 1996. Current LGO features are described in Pintér, 1998; Pintér, 2000. The other method which we use for the comparison is the well-known simulated annealing algorithm as described e.g. in Press et al., 1986; Corana et al., 1987.

1. Inverse Planning with Multileaf Parameters

1.1 Dose Calculation Based on the Pencil Beam Model

Sections 1 and 2 review the mathematical model for leaf control, as presented in Tervo et al., 2003a; Tervo et al., 1999a. See also Tervo and Kolmonen, 2002; Tervo et al., 2003b. We adopt the notation and formalism of these references. As mentioned above, our method does not require the determination of the intensity distribution as an intermediate step. Specifically, we compute the dose $D = D(x)$ in the patient space as a function of the decision variables. Then we are able to optimize or to find feasible solutions directly by using MLC operations based decision variables.

Assume that the collimator plane (treatment head) is a rectangle $U = [-a, a] \times [-b, b] \subset \mathbf{R}^2$. Denote an arbitrary point of U by $u = (u_1, u_2)$. The collimator leaves are orthogonal to the u_2-axis. Suppose that the leaves have a positive width d and that there are N leaf pairs (B_i, A_i), $i = 1, ..., N$. Let $U_i := [-a, a] \times]u_{2,i-1}, u_{2,i}[$, $i = 1, ..., N$ be the strips (so called channels) along the u_1-axis determined by the leaf pairs (B_i, A_i). Figure 1.1 illustrates the concepts and the notation.

Assume that we have L fields S_l, $l = 1, ..., L$. Let $\Psi_l = \Psi_l(u) = \Psi_l(u_1, u_2)$ be the flux density per unit area of the considered field S_l. We assume that Ψ_l is a piecewise continuous function $U \mapsto \mathbf{R}$ such that $\Psi_l(u) \equiv \Psi_{li}(u_1)$ for $u \in U_i$. Clearly, only such fluxes can be generated with the MLC technique corresponding to an individual field S_l (in the case where the collimator angle is fixed). Let Ψ_0 be the uniform flux den-

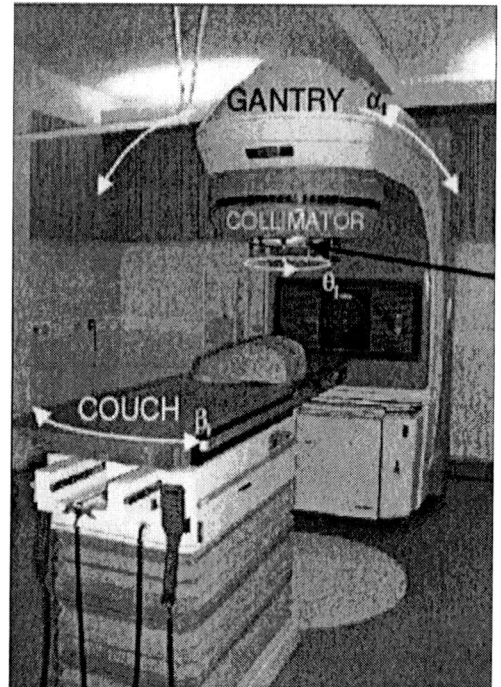
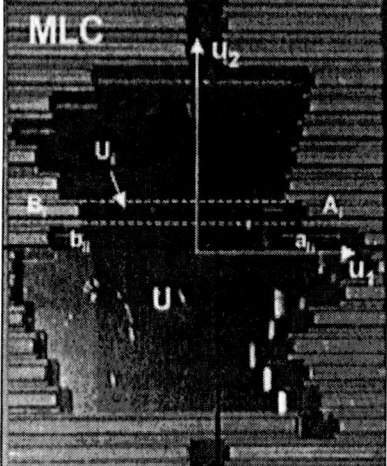

Figure 1.1. Left: The linear accelerator, couch and MLC. The angles α_l, β_l and θ_l describe the gantry, couch and collimator angles, respectively. Right: The MLC seen from the lower side. In the photograph the number of leaf pairs $N = 17$.

sity per unit time and unit area on the MLC and let T_l be the treatment time corresponding to S_l. Denote the right and left hand location of the head (edge) of leaf A_i and B_i by $a_{li}(t) \in [-a, a]$ and by $b_{li}(t) \in [-a, a]$ at time moment $t \in [0, T_l]$, respectively.

In Tervo et al., 2003a we have shown that, in the case where head scattering, leakage and the so called tongue and groove effects are omitted, we have

$$\Psi_{li}(u_1) = \Psi_0 \int_0^{T_l} (H(a_{li}(t) - u_1) - H(b_i(t) - u_1))dt, \ u \in U_i \qquad (1.1)$$

or equivalently

$$\Psi_{li}(u_1) = \Psi_0 \int_0^{T_l} H(a_{li}(t) - u_1)H(u_1 - b_i(t))dt, \ u \in U_i \qquad (1.2)$$

where H is the Heaviside function

$$H(x) = \begin{cases} 1, & x \geq 0 \\ 0, & x < 0 \end{cases}.$$

In the following model formulation we shall use (1.1).

In the mathematical modelling of radiation response using the pencil beam model we apply the integral equation Gustafsson et al., 1994; Kolmonen et al., 1998

$$D(x) = \sum_{l=1}^{L} \int_{U} h_l(x, u) \Psi_l(u) du, \ x \in V. \tag{1.3}$$

Here $h_l(x, u)$ is the so called *dose deposition kernel*. It tells how much energy is deposited at the point x of patient space V from the point u of the treatment space U. As mentioned the exact analytical expression of $h_l(x, u)$ is unknown, but it can be approximately or numerically calculated with various methods. Here we use a rather simple approximation in the calculation of $h_l(x, u)$ based on the Fermi equation. The strips U_i are mutually disjoint and $U = \cup_{i=1}^{N} U_i$. Thus we have

$$D(x) = \sum_{l=1}^{L} \int_{U} h_l(x, u) \Psi_l(u) du = \sum_{l=1}^{L} \sum_{i=1}^{N} \int_{U_i} h_l(x, u) \Psi_{li}(u_1) du. \tag{1.4}$$

Substituting (1.1) into (1.4) we obtain

$$D(x) = \Psi_0 \sum_{l=1}^{L} \sum_{i=1}^{N} \int_{U_i} \int_{0}^{T_l} h_l(x, u) \left(H(a_{li}(t) - u_1) - H(b_{li}(t) - u_1) \right) dt du. \tag{1.5}$$

1.2   Leaf Constraints

Below we shall describe the essential physical contraints which the leaves must satisfy. The symbols γ, $\overline{\gamma}$, κ will denote positive model parameters.

1 The leaves can not overlap:

$$b_{li}(t) \leq a_{li}(t), \ l = 1, ..., L, \ i = 1, ..., N, \ t \in [0, T_l] \tag{1.6}$$

2 The position of the leaf has to be in the interval $[-a, a]$:

$$a_{li}(t) \leq a, \ b_{li}(t) \geq -a, \ l = 1, ..., L, \ i = 1, ..., N, \ t \in [0, T_l] \tag{1.7}$$

3 The movement of the leaves over the central axis of the treatment space U is restricted:

$$a_{li}(t) \geq -\kappa, \ b_{li}(t) \leq \kappa, \ l = 1, ..., L, \ i = 1, ..., N, \ t \in [0, T_l] \quad (1.8)$$

4 The opposite leaves are not too near each other or they are fully closed:

$$a_i(t) - b_i(t) \geq \overline{\gamma} \text{ or } a_{li}(t) - b_{li}(t) = 0,$$
$$l = 1, ..., L, \ i = 1, ..., N, t \in [0, T_l]$$

which in view of (1.6) is equivalent to

$$a_{li}(t) - b_{li}(t) \geq 0$$

and

$$(a_{li}(t) - b_{li}(t))(a_{li}(t) - b_{li}(t) - \overline{\gamma}) \geq 0,$$
$$l = 1, ..., L, \ i = 1, ..., N, t \in [0, T_l] \quad (1.9)$$

5 The leaf movements satisfy the so called inter-digitation condition:

$$b_{li}(t) \leq a_{l(i+1)}(t), \ b_{l(i+1)}(t) \leq a_{li}(t),$$
$$l = 1, ..., L, \ i = 1, ..., N - 1, t \in [0, T_l] \quad (1.10)$$

6 The distance between the extreme A_i and correspondingly B_i leaves is bounded:

$$a_{li}(t) - a_{lj}(t) \leq \gamma, \ b_{li}(t) - b_{lj}(t) \leq \gamma,$$
$$l = 1, ..., L, \ i, j = 1, ..., N, t \in [0, T_l] \quad (1.11)$$

1.3   The Inverse Problem

As mentioned in the Introduction, the patient space is modelled as a 3-dimensional region V: it typically contains a given planning target volume (PTV) **T**, organs at risk (OAR) **C**, and normal (or complementary) tissue **N**. $V = \mathbf{T} \cup \mathbf{C} \cup \mathbf{N}$, and **T**, **C**, **N** are mutually disjoint volumes.

In practice the dose calculations are done with respect to some fixed coordinate systems. For the origin O of the patient coordinate system (x_1, x_2, x_3) one often chooses the *isocentre (point)*. The isocentre is the point in the patient space around which the different parts (gantry, collimator, couch) of the treatment unit are rotating (see Figure

1.2). For a more detailed explanation of the terminology see e.g. Webb, 1993; Webb, 1997). The treatment coordinate system (u_1, u_2) is orthogonally transversal to the line OS_l which connects the isocentre O and the (field) *focus* S_l. The axis OS_l is called *field central axis*. *Gantry and couch angles* α_l, β_l are the rotation angles with respect to the x_3 and x_2 axes, respectively. The *collimator angle* θ_l is the rotation angle with respect to the OS_l axis.

Using the above definitions, the basic problem of radiation therapy for leaf trajectories can be stated as follows:

Suppose that D_0 is the prescribed (uniform) dose in **T**. *Furthermore, suppose that the upper bounds of dose in critical organ(s) and normal tissue are D_C and D_N, respectively. Let T_l be positive numbers.*

Find the number L of fields S_l, gantry, couch and collimator angles α_l, β_l, θ_l of the fields S_l, respectively. For any $i = 1, ..., N$, $l = 1, ..., L$ determine under the above described constraints (1.6-1.10) the leaf trajectories $a_{li} : [0, T_l] \mapsto \mathbf{R}$, $b_{li} : [0, T_l] \mapsto \mathbf{R}$ of the fields S_l such that the overall dose distribution computed by the integral (1.5) satisfies

$$D(x) = D_0, \quad x \in \mathbf{T}, \tag{1.12}$$

$$D(x) \leq D_C, \quad x \in \mathbf{C}, \tag{1.13}$$

$$D(x) \leq D_N, \quad x \in \mathbf{N}. \tag{1.14}$$

Figure 1.2 illustrates the coordinate systems and the problem setting. We suppose that the parameters L, α_l, β_l are given. Instead of the requirement $D(x) = D_0$, $x \in \mathbf{T}$ one may only demand that

$$d_T \leq D(x) \leq D_T, \quad x \in \mathbf{T} \tag{1.15}$$

where d_T (D_T) is the lower (upper) bound of the dose in PTV. Then the problem is called *a feasibility problem*.

In addition to the requirements (1.12-1.14) one often requires that the so called *dose volume constraints* Börgers, 1997; Webb, 1993 are fulfilled. The dose volume contraints are expressed with respect to a certain structure. For example, for the critical organs **C** these conditions can be described as follows. Let $v(D)$ be the volume fraction of **C** that receives a dose greater than D. We can then prescribe that

$$v(D) \leq v_0, \text{ for all } D \geq d_{0,C} \tag{1.16}$$

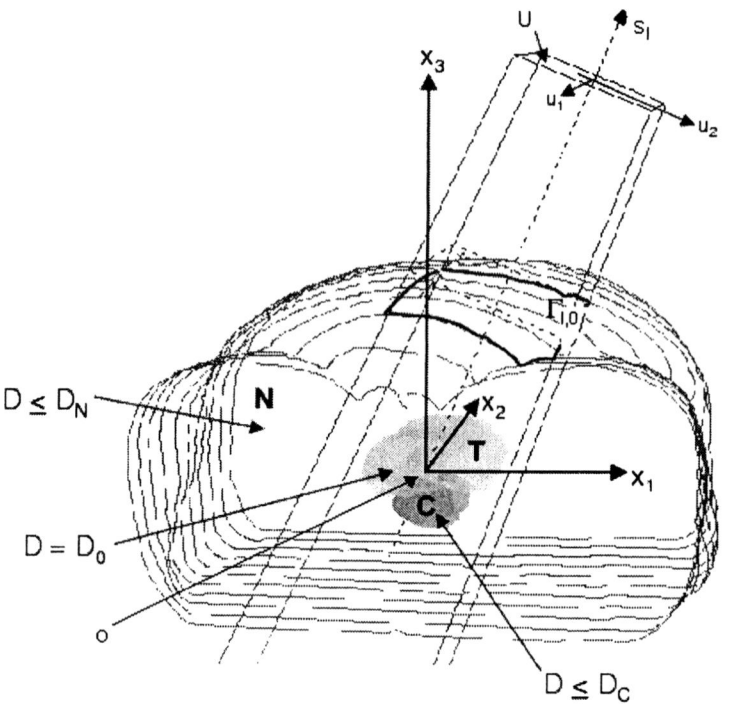

Figure 1.2. The problem setting. The patient space coordinate system (x_1, x_2, x_3) and the treatment space coordinate system (u_1, u_2).

where v_0 is a given volume fraction and $d_{0,C}$ is a given dose. The function $v = v(D)$ is the so called *(differential) dose volume histogram*. Because the dose volume histogram is a decreasing function of D the condition (1.16) is equivalent to

$$v(d_{0,C}) \leq v_0. \tag{1.17}$$

In practice the condition (1.17) is handled as follows. We define a partition $\{V_m | m = 1, ..., M\}$ of $\mathbf{C}$ and choose a point in each partition $x_m \in V_m$. We demand that

$$\frac{|\{m = 1, ..., M | D(x_m) \geq d_{0,C}\}|}{M} \leq v_0 \tag{1.18}$$

where $|\{\ \}|$ denotes the numer of elements of a set. The condition (1.18) can be expressed by the approximation (cf. Shepard et al., 1999)

$$\frac{1}{M} \sum_{m=1}^{M} \mathrm{erf}_\epsilon(D(x_m) - d_{0,C}) \leq v_0 \tag{1.19}$$

where erf_ϵ is the error function given below, see (2.4). The positive number ϵ is so small that the function erf_ϵ approximately equals the Heaviside function. In the case, where dose volume contraints are taken into account, conditions of the form (1.18) are added to the requirements (1.12-1.14).

2. Optimization of IRTTP with Multiple Static Collimation

2.1 Multiple Static Collimation

Multiple static collimation means that corresponding to each field S_l the time interval $[0, T_l]$ is divided into n_l subintervals $[t_{l0}, t_{l1}], ..., [t_{l(n_l-1)}, t_{ln_l}]$ and within each subinterval the MLC configuration is fixed. The radiation is interrupted, when the subsequent leaf configurations are changed. In the following we consider the computation of dose in the case of multiple static collimation. We also compute the leaf constraints.

Let $\{t_{l0}, ..., t_{ln_l}\}$ be a partition of the interval $[0, T_l]$. Furthermore, let χ_{lk} be the characteristic functions

$$\chi_{lk}(t) = \left\{ \begin{array}{ll} 1, & t \in [t_{l(k-1)}, t_{lk}] \\ 0, & \text{otherwise} \end{array} \right. .$$

In our mathematical model of multiple static collimation, the leaf trajectories a_{li}, b_{li} are of the form

$$a_{li}(t) = \sum_{k=1}^{n_l} a_{lik} \chi_{lk}(t), \; b_{li}(t) = \sum_{k=1}^{n_l} b_{lik} \chi_{lk}(t). \qquad (2.1)$$

The leaf configurations corresponding to the subintervals $[t_{l(k-1)}, t_{lk}]$, $k = 1, ..., n_l$ are called *subfields of the fields* S_l, $l = 1, ..., L$. The practical interpretation of parameters a_{lik}, b_{lik} is that the edges of the ith leaf pair (B_i, A_i) corresponding to the kth subfield of the lth field are at points b_{lik} and a_{lik}. The choice of parameters a_{lik}, b_{lik} must be optimized. In the following we find that our modelling enables to set also the time lengths of subfields $t_{lk} - t_{l(k-1)}$ as optimization parameters.

The exact Heaviside function over the interval $[-a, a]$ is given in Section 1. In practice it is reasonable to replace the Heaviside function with a continuous approximation: examples are

$$\tilde{H}(x) = C_1 + C_2 \overline{\text{arc}} \tan(C_3 x) \qquad (2.2)$$

(here $\overline{\text{arc}} \tan$ denotes the principal branch of arc tan function)

$$\tilde{H}(x) = C_1 + C_2(1 - \tanh(C_3 x)) \qquad (2.3)$$

or

$$\tilde{H}(x) = \mathrm{erf}_\tau(x) = \frac{1}{\sqrt{\pi}\tau} \int_{-\infty}^{x} e^{-s^2/\tau^2}\,\mathrm{d}s. \tag{2.4}$$

The unknown parameters d_j, C_1, C_2, C_3, τ for (2.2-2.4) can be obtained with data fitting techniques by using measurement or Monte Carlo data. The replacement of Heaviside function with a smooth function is necessary also because of head scatter and leakage Tervo et al., 2003a. We shall not consider the corrections in this contribution. In the following we express the formulations with $\tilde{H} = H$.

The coefficients $(a_{li1}, ..., a_{lin_l})$, $(b_{li1}, ..., b_{lin_l}) \in \mathbf{R}^{n_l}$ are called *multileaf parameters*. The multileaf parameters and time lengths $\delta t_{lk} := t_{lk} - t_{l(k-1)}$ are *decision variables* in the multiple static collimation. Denote $\bar{a}_{li} = (a_{li1}, ..., a_{lin_l})$, $\bar{b}_{li} = (b_{li1}, ..., b_{lin_l}) \in \mathbf{R}^{n_l}$ and

$$\begin{aligned}
(\mathbf{a}, \mathbf{b}) &= (\bar{a}_{11}, ..., \bar{a}_{1N}, ..., \bar{a}_{L1}, ..., \bar{a}_{1N}, \bar{b}_{11}, ..., \bar{b}_{1N}, ..., \\
&\quad \bar{b}_{L1}, ..., \bar{b}_{LN}) \in \mathbf{R}^{2N(n_1 + ... + n_L)}.
\end{aligned}$$

From (1.5) we obtain the dose expressed with the decision variables, that is $D(x) = D(x, \mathbf{a}, \mathbf{b}, \delta t)$ where

$$\begin{aligned}
D(x, \mathbf{a}, \mathbf{b}, \delta t) &= \Psi_0 \sum_{l=1}^{L} \sum_{i=1}^{N} \int_{U_i} \int_{0}^{T_l} h_l(x, u) \\
&\quad \cdot \left(H\left(\sum_{k=1}^{n_l} a_{lik}\chi_{lk}(t) - u_1 \right) \right. \\
&\quad \left. - H\left(\sum_{k=1}^{n_l} b_{lik}\chi_{lk}(t) - u_1 \right) \right) dt\,du \\
&= \Psi_0 \sum_{l=1}^{L} \sum_{i=1}^{N} \sum_{k=1}^{n_l} \int_{t_{l(k-1)}}^{t_{lk}} \int_{U_i} h_l(x, u) \\
&\quad \cdot \left(H\left(\sum_{k=1}^{n_l} a_{lik}\chi_{lk}(t) - u_1 \right) \right. \\
&\quad \left. - H\left(\sum_{k=1}^{n_l} b_{lik}\chi_{lk}(t) - u_1 \right) \right) du\,dt \\
&= \Psi_0 \sum_{l=1}^{L} \sum_{i=1}^{N} \sum_{k=1}^{n_l} \left[\int_{U_i} h_l(x, u)(H(a_{lik} - u_1) \right. \\
&\quad \left. - H(b_{lik} - u_1))du \right] \delta t_{lk}
\end{aligned} \tag{2.5}$$

where we denoted $\delta t = (\delta t_1, ..., \delta t_L) \in \mathbf{R}^{n_1 + ... + n_L}$, $\delta t_l = (\delta t_{l1}, ..., \delta t_{ln_l}) \in \mathbf{R}^{n_l}$. We observe that the time lengths δt_{lk} of the subintervals $[t_{l(k-1)}, t_{lk}]$

can also be chosen as parameters of the optimization processes. In the sequel we shall assume that the decision variable is $(\mathbf{a}, \mathbf{b}, \delta t) \in \mathbf{R}^{(2N+1)(n_1 + \ldots + n_L)}$.

The time lengths δt_{lk} must satisfy the constraint

$$0 \le \delta t_{lk} \le T_l, \; l = 1, \ldots, L, \; k = 1, \ldots, n_l. \tag{2.6}$$

Instead of (2.6) we demand that for some $c_0 > 0$

$$c_0 \le \delta t_{lk} \le T_l \text{ or } \delta_{lk} = 0 \tag{2.7}$$

which is equivalent to

$$0 \le \delta t_{lk} \le T_l, \;\; \delta t_{lk}(\delta t_{lk} - c_0) \ge 0, \; l = 1, \ldots, L, \; k = 1, \ldots, n_l \tag{2.8}$$

This restriction will exclude subfields that are too short for clinical use.

The leaf constraints described in Section 1.2 can easily be computed. Since $a_{li}(t) = a_{lik}$, $b_{li}(t) = b_{lik}$ for $t \in [t_{l(k-1)}, t_{lk}]$, the constraints (1.6-1.10) are equivalent to the following conditions:

$$b_{lik} \le a_{lik}, \; l = 1, \ldots, L, \; i = 1, \ldots, N, \; k = 1, \ldots, n_l \tag{2.9}$$

$$a_{lik} \le a, \; b_{lik} \ge -a, \; l = 1, \ldots, L, \; i = 1, \ldots, N, \; k = 1, \ldots, n_l \tag{2.10}$$

$$a_{lik} \ge -\kappa \text{ and } b_{lik} \le \kappa, \; l = 1, \ldots, L, \; i = 1, \ldots, N, \; k = 1, \ldots, n_l \tag{2.11}$$

$$(a_{lik} - b_{lik})(a_{lik} - b_{lik} - \overline{\gamma}) \ge 0, \; l = 1, \ldots, L, \; i = 1, \ldots, N, \; k = 1, \ldots, n_l \tag{2.12}$$

$$b_{lik} \le a_{l(i+1)k}, \; b_{l(i+1)k} \le a_{lik}, \; l = 1, \ldots, L, \; i = 1, \ldots, N-1, \; k = 1, \ldots, n_l \tag{2.13}$$

$$a_{lik} - a_{ljk} \le \gamma, \; b_{lik} - b_{ljk} \le \gamma, \; l = 1, \ldots, L, \; i, j = 1, \ldots, N, \; k = 1, \ldots, n_l. \tag{2.14}$$

As mentioned above, in multiple static collimation we add the constraint

$$0 \le \delta t_{lk} \le T_l, \;\; \delta t_{lk}(\delta t_{lk} - c_0) \ge 0, \; l = 1, \ldots, L, \; k = 1, \ldots, n_l \tag{2.15}$$

which is sometimes replaced only by the more simple (linear inequality) constraint

$$0 \le \delta t_{lk} \le T_l, \; l = 1, \ldots, L, \; k = 1, \ldots, n_l. \tag{2.16}$$

REMARK 1 *The constraints (2.12) and (2.15) are nonlinear and nonconvex. They make the numerical solution more difficult because one must find global minima on nonconvex feasible sets.*

2.2   Optimal Solution

Applying the above concepts, we are now able to formulate different kinds of optimality and feasibility problems with the help of the decision variables introduced. The dose $D(x, \mathbf{a}, \mathbf{b}, \delta t)$ is obtained from (2.5). First we formulate the *feasibility problem*. The patient space V is divided into subregions (so called voxels) V_p, $p = 1, ..., P$. Suppose that $x_p \in V_p$. Then we have

$$D(x_p) \;=\; D(x_p, \mathbf{a}, \mathbf{b}, \delta t) \tag{2.17}$$

$$=\; \Psi_0 \sum_{l=1}^{L} \sum_{i=1}^{N} \sum_{k=1}^{n_l} \left[\int_{U_i} h_l(x_p, u)(H(a_{lik} - u_1) \right.$$
$$\left. - H(b_{lik} - u_1)) du \right] \delta t_{lk}.$$

Divide the index set $J = \{1, ..., P\}$ into three disjoint sets $J = J_T \cup J_C \cup J_N$ where

$$J_T = \{p \in J | x_p \in \mathbf{T}\},$$
$$J_C = \{p \in J | x_p \in \mathbf{C}\},$$
$$J_N = \{p \in J | x_p \in \mathbf{N}\}.$$

Let D_N, D_C, D_T be the prescribed upper limit of dose in normal, critical and tumor tissue, respectively and let d_T be the lower limit of dose in tumor tissue. We can state the following feasibility problem.

Find $(\mathbf{a}, \mathbf{b}, \delta t) \in \mathbf{R}^{(2N+1)(n_1 + ... + n_L)}$ *for which the inequalities*

$$d_T \leq D(x_p, \mathbf{a}, \mathbf{b}, \delta t) \leq D_T, \; p \in J_T, \tag{2.18}$$

$$D(x_p, \mathbf{a}, \mathbf{b}, \delta t) \leq D_C, \; p \in J_C, \tag{2.19}$$

$$D(x_p, \mathbf{a}, \mathbf{b}, \delta t) \leq D_N, \; p \in J_N, \tag{2.20}$$

are satisfied under the additional constraints (2.9-2.15).

In addition to the requirements (2.18-2.20), one can consider the dose volume constraint for the critical organs

$$\frac{1}{|J_C|} \sum_{p \in J_C} \mathrm{erf}_\epsilon(D(x_p, \mathbf{a}, \mathbf{b}, \delta t) - d_{0,C}) \leq v_0 \tag{2.21}$$

where (as above) $|J_C|$ denotes the number of elements in J_C. Analogous requirement can be added for the case of normal tissue.

Second, we state the following *optimization problem*. Define the objective function

$$F(\mathbf{a}, \mathbf{b}, \delta t) \;=\; c_1 \sum_{p \in J_T} |D_0 - D(x_p, \mathbf{a}, \mathbf{b}, \delta t)|^2$$

$$+ \quad c_2 \sum_{p \in J_C} |(D_C - D(x_p, \mathbf{a}, \mathbf{b}, \delta t))_-|^2$$

$$+ \quad c_3 \sum_{p \in J_N} |(D_N - D(x_p, \mathbf{a}, \mathbf{b}, \delta t))_-|^2$$

$$+ \quad c_4 \left| \left(v_0 - \frac{1}{|J_C|} \sum_{p \in J_C} \mathrm{erf}_\epsilon(D(x_p, \mathbf{a}, \mathbf{b}, \delta t) - d_{0,C}) \right)_- \right|^2$$

$$+ \quad c_5 \sum_{p \in J_T} \|\nabla_x D(x_p, \mathbf{a}, \mathbf{b}, \delta t)\|^2, \tag{2.22}$$

where the subscript $_-$ refers to the negative part of a function. ∇_x is the gradient of the dose distribution with respect to variable x; the coefficients c_1, c_2, c_3, c_4 and c_5 are suitably chosen positive weights. The last term of the objective function increases the smoothness of the dose distribution in PTV by minimizing its gradient. We have also inserted the penalty for the dose volume constraint of the critical organ. We can now state the optimization problem:

Find the global minimum

$$\min_{(\mathbf{a}, \mathbf{b}, \delta t) \in \mathbf{R}^{(2N+1)(n_1 + \ldots + n_L)}} F(\mathbf{a}, \mathbf{b}, \delta t) \tag{2.23}$$

under the constraints (2.9-2.15).

REMARK 2 *A. We remark that the feasible problem can be converted to the global optimization problem by replacing the term* $c_1 \sum_{p \in J_T} |D_0 - D(x_p, \mathbf{a}, \mathbf{b}, \delta t)|^2$ *by*

$$c_1 \sum_{p \in J_T} |(D_T - D(x_p, \mathbf{a}, \mathbf{b}, \delta t))_-|^2 + c_2 \sum_{p \in J_T} |(D(x_p, \mathbf{a}, \mathbf{b}, \delta t) - d_T)_-|^2. \tag{2.24}$$

Consult Horst and Tuy, 1996 for more details.

B. One can also consider the optimization problem

$$\min \sum_{p \in J_T} \|\nabla_x D(x_p, \mathbf{a}, \mathbf{b}, \delta t)\|^2 \tag{2.25}$$

under the constraints (2.18)-(2.21). In this case one can omit the issue of finding suitable weights $c_1 - c_5$.

3.   Optimization Algorithms and Simulations

We demonstrate the calculation of the treatment scheme based on the determination of parameters in the inverse problem. We use discrete square norm based objective functions that is, the optimization problem of Section 3.2. Only the first three terms of the objective function (2.21) are included and the first term is substituted by (2.24). In the optimization both the leaf positions and time intervals were set as decision parameters leading to objective functions of type $F(\mathbf{a}, \mathbf{b}, \delta t)$. To model the intensity response of the MLC we used the simple Heaviside function. The dose calculation was done, for simplicity, applying the solution to the Fermi equation Börgers and Larsen, 1996.

3.1   Algorithms

In realistic multiple static collimation problems the number N of leaf pairs is typically $4 - 25$. Assuming that the number of fields (L) is between 2 and 10 and that the number n_l of subfields, is 3 to 10, we find that $(2N + 1)(n_1 + ... + n_L) = 36 - 5100$. Hence, the above model leads to a large scale global optimization problem. The constraints are mainly linear inequality constraints; added are the nonconvex constraints (2.12) and (2.15) unless those are omitted. The objective function F is defined in a subset of $\mathbf{R}^{(2N+1)(n_1+...+n_L)}$ and it is highly nonlinear. Therefore the optimization algorithm must be chosen so that it is suitable to handle nonlinear optimization problems with (linear inequality) constraints. In addition, the large dimensionality of the problem must be taken into account and the algorithm must be carefully initialized or it must have a capability to search the global extremum. Here we solve the problem by applying the latter alternative. In global search we have utilized the LGO software Pintér, 1996; Pintér, 1998; Pintér, 2000 and simulated annealing Press et al., 1986; Corana et al., 1987. The current LGO system is described in this monograph Pintér, 2000.

The simulated annealing (SA) method generates random trial values F of the objective function to the problem in the neighborhood of the current best solution F_{best} Corana et al., 1987. The constraints are handled as penalty terms included in the object function. If the solution is better than the current best $(\Delta F = F - F_{\text{best}} \leq 0)$ it is accepted. If, however, the solution is worse than the current best solution it can still be accepted with probability $p(\Delta F) = \exp(-\Delta F/T)$.

The fact that worse solutions can be accepted is the key feature of the method. It enables SA to climb out of a local minimum. An important parameter in the algorithm is the temperature T. It governs the acceptance probability p. During iteration the temperature is slowly cooled.

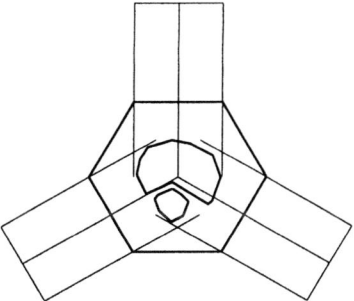

Figure 1.3. The phantom that was used in simulations. A large PTV is shown with a smaller OAR next to it. Three treatment fields are also shown.

In the beginning the acceptance level is "high" and many points in the search space are evaluated. In the end only the solutions that are better than the current best solution tend to be accepted. Heuristically, this should ensure that in the end the global minimum is found with a desired accuracy.

3.2   Phantom and treatment fields

In the numerical example discussed here, we considered only a 2-dimensional phantom (a slice of a 3-dimensional phantom) to diminish the computational complexity. Hence, we need only one leaf pair ($N = 1$). The phantom together with treatment field setting is shown in Figure 3. The phantom was hexagonal shaped because the treatment fields could be placed so that the central axes of the fields were perpendicular to the phantom surface. Thus complications arising from oblique incidence were avoided.

The relative dose constraints together with relative weights for different areas of the phantom are shown in table 1.1.

Table 1.1. Relative dose constraints and weighting of the constraints for the phantom simulation. PTV is the target with lower and higher dose constraints. OAR is the organ at risk having a higher constraint. Compl. is the complementary tissue belonging neither to the PTV nor OAR. This tissue had a higher limit for dose

Area	Dose	Weight
PTV low	0.95	2.0
PTV high	1.05	0.5
OAR	0.6	1.0
Compl.	0.8	2.0

Three fields were used ($L = 3$). Each field had three multiple static subfields ($n_1 = n_2 = n_3 = 3$). As one subfield of a field had two leaves (left and right) the total number of leaf parameters was 18 and weight parameters 9. Hence the total number of decision parameters was 27.

3.3 Results

The optimization was at first carried out using LGO software package. For global optimization, an adaptive random search was used with 90000 function evaluations. After that a constrained local search with 10000 function evaluations was used.

For reference the simulated annealing algorithm was used. A considerably large number (33750) of function evaluations was permitted together with a slow cooling scheme to ensure that the final solution would be near the global minimum of the objective function. In the end the Nelder-Meade Simplex method of Matlab was used as a local optimizer to obtain the accurate (possibly local) minimum for the simulated annealing method.

In Figure 1.4. are shown the dose distributions after global optimization. In Figure 1.5 are the cumulative dose-volume histograms determined from the dose distributions. Table 1.2 shows information about the dose distributions.

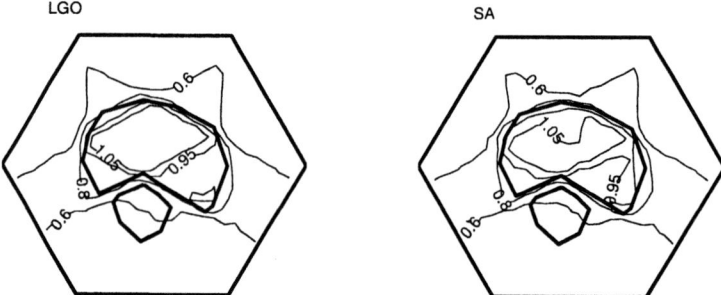

Figure 1.4. The resulting dose distributions after the MLC parameters were globally optimized. Isodose curves represent the dose. Left: LGO, right: Simulated annealing.

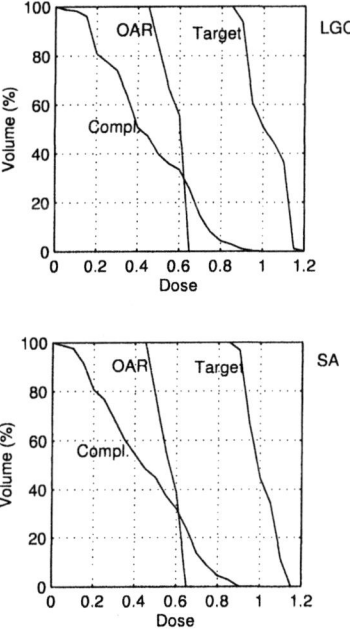

Figure 1.5. The resulting dose distributions of figure 1.4 represented using cumulative dose-volume histograms. Upper: LGO, lower: Simulated annealing.

Table 1.2. Information about the solutions when the two global algorithms were used: LGO and Simulated Annealing (SA). Contents are for each region (PTV, OAR and Compl.): the minimum dose (D_{min}), maximum dose (D_{max}), mean dose (D_{mean}) and the standard deviation (D_{std}) of dose.

Method	Region	D_{min}	D_{max}	D_{mean}	D_{std}
LGO	PTV	0.861	1.158	1.016	0.093
	OAR	0.464	0.615	0.569	0.055
	Compl.	0.019	0.980	0.451	0.217
SA	PTV	0.875	1.139	1.004	0.073
	OAR	0.462	0.615	0.557	0.055
	Compl.	0.016	0.962	0.449	0.222

4.   Discussion

In the present study, we optimized a radiation therapy treatment plan for a phantom model using a novel multiple static collimation algorithm and two global optimization strategies. Both strategies give comparable results in the simple phantom geometry with limited dimensions. This suggests that both strategies are able to find the global minimum. However, calculation times for solving the problems (typically 1 - 5 hours with Pentium 400 MHz PC) are long, as compared to the problem dimensions.

IMRT is a topic of increasing interest today because it offers possibilities to conform the dose strictly to the target region Nutting et al., 2000; Eisbruch et al., 1999; Dawson et al., 2000. This, naturally, enables sparing of normal tissue and critical organs. In these studies doses for critical organs have been reduced as compared to a conventional treatment plan. On the other hand, it provides a way for dose escalation in the target. These studies have also shown the therapeutic benefit of dose escalation, because the local regional recurrences after treatment were found in the areas of high risk in planning, i.e. areas included into the tumor.

The usual way in the intensity modulated IRTTP is to optimize the intensity distributions for each field to fullfil the desired dose distribution in the patient domain. After this intermediate step, the MLC parameters are optimized to generate the required intensity distribution. We have developed an IRTTP algorithm for the optimization of the dose distribution using the MLC parameters directly, as decision variables. Using this strategy, optimization of the intensity distribution is not needed as an intermediate step.We have studied the direct optimization model earlier in Kolmonen et al., 2000; Tervo and Kolmonen, 2000; Tervo et al., 2003a; Tervo et al., 2003b; Tervo and Kolmonen, 2002. This approach has some obvious benefits such as more easy and reliable way to take into account the (physical) restrictions of the movements of MLC leaves. MLC constraints are properly taken into account in the inverse problem. Also, head scattering, leakage and "groove and tongue effects" can be modelled more accurately. Dose volume constraints are also more natural to implement by using this approach. The overall treatment planning is in principle more simple, because no intermediate (two step) optimization is needed. The 'disadvantage' of the method is that it leads to large dimensional, highly nonlinear (multiextremal) problems. Hence, careful expert initialization or the use of proper global optimization strategies Horst and Tuy, 1996; Pintér, 1996 is necessary.

In our IRTTP algorithm, we use a physical objective function in the optimization. In the objective function, dose limits for target and upper dose limits for OAR and normal tissue were given. Use of the physical objective function has an advantage of exact determination of the dose Bortfeld, 1999. Moreover, an extensive literature is devoted to the target dose vs. tumor control relationship, for different types of cancers. Using such statistical data, we are able to construct so called biological objective functions. Of course, using a biological objective function in principle gives a benefit of direct determination of the dose to reach the tumor control and possibly avoid the complications Brahme, 1999. However, the radiobiological approach includes many uncertainties due to the limited knowledge of complicated biology in tissue irradiation. This, at least currently, does not strongly support the use of radiobiological objective function.

In IRTTP, various optimization methods have been applied including e.g. simulating annealing, genetic optimization, tomographic image reconstruction, active set methods, (alternating) projection methods, generalized gradient methods. However, very little attention has been paid to more sophisticated global optimization algorithms. In this study, we used two different global optimization strategies, LGO and simulated annealing plus local search. Both optimization strategies had significant calculation times (several hours). The important finding is, however, that the dose distribution given by the both global optimization strategies is very similar. This proposes that the strategies approximate well the global minimum.

In simulations, we used a hexagonal phantom with concave target partly surrounding the critical organ. Three portals with three subfields were used. This is a low amount of portals. Sauer et al Sauer et al., 1999 have proposed a suitable amount of portals to be five to seven, in order to achieve homogeneous dose distribution. Also, with three subfields, the step-and-shoot method is capable to provide only a coarse, skyscaper like intensity distribution. The intensity distributions with this dose delivery technique will become smoother only by adding subfields. Our aim in limiting the number of portals and subfields was, however, to keep the number of decision parameters low in the illustrative optimization problem. For this reason, the optimal distribution in our solution does not provide a very homogeneous dose distribution for the target. Raising the number of portals and subfields (i.e. the decision variables) will certainly improve the distribution, but the optimization time will also increase.

In conclusion, we have developed a new IRTTP model and solved it applying two global optimization strategies. The optimal treatment plans given by the both strategies are very similar, which proposes the usefulness of the both optimization strategies in this kind of IRTTP problems. The applicability of our IRTTP method in a simple case, demonstrated in this study, also makes possible to apply it to more complex problems in future.

Acknowledgments

This work was supported by TEKES (Technology Development Centre Finland) and Varian Medical Systems Finland, Espoo, Finland.

References

Bortfeld T. (1999). Optimized planning using physical objectives and constraints. *Semin. Radiat. Oncol.*, 9:20–34.

Brahme A. (1995). Treatment optimization using physical and radiological objective functions. In *Radiation Therapy Physics* (Ed. Smith A.). Springer, Berlin.

Brahme A. (1999). Optimized radiation therapy based on radiobiological objectives. *Semin. Radiat. Oncol.*, 9,35–48.

Börgers C. and Larsen E.W. (1996). On the accuracy of the Fokker–Planck and Fermi pencil beam equations for charged particle transport. *Med. Phys.*, 23:1749–1759.

Börgers C. (1997). The radiation therapy planning problem. In *Computational radiology and imaging: Therapy and diagnostic* (Ed. Börgers C. and Natterer F.). Springer, Berlin.

Censor Y., Altschuler M.D. and Powlis W.D. (1988). On the use of Cimmino's simultaneous projections method for computing a solution of the inverse problem in radiation therapy treatment planning. *Inv. Prob.*, 4:607–623.

Censor Y. and Zenios S.A. (1997). *Parallel Optimization: Theory, Algorithms and Applications*. Oxford University Press, New York.

Cercignani C. (1988) *The Boltzmann Equation and Its Applications*. Springer, Heidelberg.

Corana A., Marchesi M., Martini C. and Ridella S. (1987). Minimizing multimodal functions of continuous variables with the "simulated annealing" algorithm. *ACM Transactions on Mathematical Software*, 13:262–280.

Dautray R. and Lions J.-L. (1993). *Mathematical Analysis and Numerical Methods for Science and Technology*, Vol. 6. Springer, Berlin.

Dawson L.A., Anzai Y., Marsh L., Martel M.K., Paulino A., Ship J.A. and Eisbruch A. (2000). Patterns of local-regional recurrence following parotid-sparing conformal and segmental intensity-modulated radiotherapy for head and neck cancer. *Int. J. Radiat. Oncol. Biol. Phys.*, 46: 1117–1126.

Eisbruch A., Dawson L.A., Kim H.M., Bradford C.R., Terrell J.E., Chepeha D.B., Teknos T.N., Anzai Y., Marsh L.H., Martel M.K., Ten Haken R.K., Wolf G.T. and Ship J.A. (1999). Conformal and intensity modulated irradiation of head and neck cancer: the potential for improved target irradiation, salivary gland function, and quality of life. *Acta Otorhinolaryngol. Belg.*, 53: 271–275.

Gustafsson A., Lind B.K., Svensson R. and Brahme A. (1994). Simultaneous optimization of dynamic multileaf collimation and scanning patterns of compensation filters using a generalized pencil beam algorithm. *Med. Phys.*, 22:1141–1156.

Horst R. and Tuy H. (1996). *Global Optimization - Deterministic Approaches*. Springer, Berlin.

Janssen J.J., Riedeman D.E.J., Morawska-Kaczyńska M, Storchi P.R.M. and Huizenga H (1997). Numerical calculation of energy desposition by high-energy electron beams: III. Three-dimensional heterogeneous media. *Phys. Med. Biol.*, 39:1351–1366.

Jette D. (1995). Electron beam dose calculation. In *Radiation Therapy Physics* (Ed. Smith A.R.). Springer, Berlin.

Johns H.E. and Cunningham J.R. (1983). *The Physics of Radiology*, 4th ed. Thomas Springfield, IL.

Kolmonen P., Tervo J. and Lahtinen T. (1998). The use of Cimmino algorithm and continuous approximation for dose deposition kernel in the inverse problem of radiation treatment planning. *Phys. Med. Biol.*, 43:2539–2554.

Kolmonen P., Tervo J., Jaatinen K. and Lahtinen T. (2000). Direct computation of the "step-and-shoot" IMRT plan. In *The Use of Computers in Radiation Therapy* (Ed. Schegel W. and Bortfeld T.). Springer, Berlin.

Larsen E.W. (1997). The nature of transport calculations used in radiation oncology. *Trans. Theory Stat. Phys.*, 26:739–751.

Mackie T.R., Bielajew A., Rogers D.W.O. and Battista J.J. (1988). Generation of photon energy kernels using EGS Monte Carlo code. *Phys. Med. Biol.*, 33:1–20.

MATLAB Optimization Toolbox User's Guide. MathWorks Inc. Natick, Massachusetts, USA.

Nutting C.M., Convery D.J., Cosgrove V.P., Rowbottom C., Padhani A.R., Webb S. and Dearnaley D.P. (2000). Reduction of small and large bowel irradiation using an optimized intensity-modulated pelvic radiotherapy technique in patients with prostate cancer. *Int. J. Radiat. Oncol. Biol. Phys.*, 48:649–656.

Oelfke U. and Bortfeld T. (1999) Inverse planning for x-ray rotation therapy: a general solution of the inverse problem. *Phys. Med.*, 44:1089–1104.

Pintér J.D. (1996). *Global Optimization in Action*. Kluwer Academic Publishers, Dordrecht.

Pintér J.D. (1998). A model development system for global optimization. In *High Performance Algorithms and Software in Nonlinear Optimization* (Ed. DeLeone, Murli, Pardalos, Toroldo). Kluwer Academic Publishers, Dordrecht.

Pintér J.D. (2000). *LGO-A Model Development System for Continuous Global Optimization. User's Guide*. Pintér Consulting Services, Halifax, NS.

Press W.H., Flannery B.P., Teukolsky S.A. and Wetterling W.T. (1986). *Numerical Recipes: The Art of Scientific Computing*. Cambridge University Press, Cambridge.

Sauer O.A., Shepard D.M. and Mackie T.R. (1999). Application of constrained optimization to radiotherapy planning. *Med. Phys.*, 26: 2359–2366.

Shepard D.M., Ferris M.C., Olivera G.H. and Mackie T.R. (1999). Optimizing the delivery of radiation therapy to cancer patients. *SIAM Review*, 41:721-744.

Tervo J. and Kolmonen P. (1998). Data fitting model for the kernel of integral operator from radiation therapy. *Math. Comput. Modelling*, 28:59–77.

Tervo J., Kolmonen P., Vauhkonen M., Heikkinen L.M. and Kaipio J.P. (1999). A finite element model of electron transport in radiation therapy and related inverse problem. *Inv. Probl.*, 15:1345–1361.

Tervo J. and Kolmonen P. (2000). A model for the control of multileaf collimator in radiation therapy treatment planning. *Inv. Probl.*, 16:1–21.

Tervo J. and Kolmonen P. (2002). Inverse radiotherapy treatment planning model applying Boltzmann transport equation. *Math. Models Methods Appl. Sci.*, 12:109–141.

Tervo J., Kolmonen P., Lyyra-Laitinen T., Pintér J.D. and Lahtinen T. (2003). An optimization-based approach to the multiple static technique in radiation therapy. *Annals Oper. Research*, 119:205–227.

Tervo J., Lyyra-Laitinen T. , Kolmonen P. and Boman E. (2003). An inverse treatment planning model for intensity modulated radiation therapy with dynamic MLC. *Appl. Math. Comput.*, 135:227–250.

Ulmer W. and Harder D. (1996). Application of a triple Gaussian pencil beam model for photon beam treatment planning. *Z. Med. Phys.*, 6:68–74.

Wang L. and Jette D. (1999). Photon dose calculation based on electron multiple-scattering theory: Primary dose desposition kernels. *Med. Phys.*, 26:1454–1465.

Webb S. (1993). *The Physics of Three-Dimensional Radiation Therapy*. Institute Of Physics Publishing, Bristol.

Webb S. (1997). *The Physics of Conformal Radiotherapy: Advances in Technology.* Institute Of Physics Publishing, Bristol.

Yang J.N., Mackie T.R., Reckwerdt P., Deasy J.O. and Thomadsen B.R. (1997). An investigation of tomotherapy beam delivery. *Med. Phys.,* 24:425–436.

Chapter 20

PARALLEL TRIANGULATED PARTITIONING FOR BLACK BOX OPTIMIZATION

Yong Wu
wuyong@pmail.ntu.edu.sg

Linet Özdamar*
lozdamar@hotmail.com, mlozdamar@ntu.edu.sg

Arun Kumar
makumar@ntu.edu.sg
Division of Systems and Engineering Management
School of Mechanical & Production Engineering
Nanyang Technological University
50, Nanyang Avenue, 639798, Singapore

Abstract   We propose a parallel triangulation based partitioning algorithm (TRIOPT) for solving low dimensional bound-constrained black box global optimization problems. Black box optimization problems are important in engineering design where restricted numbers of input-output pairs are provided as data. Optimization is carried out over sparse data in the absence of a formal mathematical relationship among inputs and outputs. In such settings, function evaluations become expensive, because system performance assessment might be conducted via simulation studies or physical experiments. Thus, the optimal solution should be found in a minimal number of function evaluations. In TRIOPT, input-output pairs are treated as samples located in the search domain and search space coverage is obtained over these samples by triangulation. This produces an initial partition of the domain. Thereafter, each simplex is assessed for re-partitioning in parallel. In this assessment, performance values at the vertices are transformed and mapped to [0,1] interval using a non-linear transformation function with dynamic parameters. Transformed values are then aggregated into a group measure

*Corresponding author

upon which the decision for re-partitioning is taken. Simplices whose group measures overcome a given threshold value are re-partitioned in parallel. Here, the performance of TRIOPT is measured on several applications from different fields and compared with powerful partitioning techniques such as LGO, DIRECT, and MCS.

Keywords: black box optimization problems, partitioning, adaptive parallel search

1.   Introduction

Black box optimization problems such as optimal system design are often found in engineering practice and in other fields. In such applications, the function defining system performance may not have a precise mathematical expression and a function evaluation may imply a system simulation or a physical experiment. In these situations, an optimization method should focus on minimizing the number of function evaluations while searching for the global optimum.

We consider the bound constrained global optimization problem expressed below.

$$\text{find } \mathbf{x}^* \in \mathcal{D} \text{ such that } f(\mathbf{x}^*) \geqslant f(\mathbf{x}), \ \forall \mathbf{x} \in \mathcal{D} \qquad (1.1)$$

where $\mathcal{D} \subset \mathbb{R}^n$ is the feasible domain, and $f : \mathcal{D} \to \mathbb{R}$ is the objective function. We assume that the form of f is unknown or that only sample information is available for the solver.

Partitioning approaches are global optimization techniques that divide given domains into smaller sub-spaces whose potential of holding the global optimum is determined either reliably (deterministic approaches, e.g. Hansen and Jaumard, 1995; Pintér, 1996) or in a heuristic manner (stochastic approaches, e.g. Özdamar and Demirhan, 2000; Özdamar and Demirhan, 2001). Deterministic partitioning algorithms can be classified under two major categories: algorithms based on interval methods (e.g., Horst and Tuy, 1996; Kearfott, 1996; Moore and Ratschek, 1988) and algorithms based on certain a priori assumptions on functions, such as Lipschitz methods (e.g., Gourdin et al., 1994; Hansen and Jaumard, 1995; Pintér, 1988). The first category of algorithms use inclusion functions where the range of f within a specific partition is estimated by an inclusion function that always covers actual function range, hence, leading to reliable box disposal. On the other hand, Lipschitzian algorithms generate a partition of the domain based on an assumed rate of change, that is, the Lipschitz constant. If the latter constant is known with accuracy, these methods are also reliable. There exist Lipshitzian approaches that eliminate the necessity of specifying the Liptshitz constant, e.g., DIRECT (Jones et al., 1993) where the Lipshitz constant is taken as a weighting parameter that balances global and local features of the search. DIRECT conducts dynamic parallel partitioning on boxes that are non-dominated with respect to two criteria: box size and box value. A powerful commercial software that integrates deterministic and probabilistic global and local search within a global partitioning framework is the Lipschitz(-Continuous) Global Optimizer (LGO) that has been developed by Pintér (Pintér, 1996; Pintér, 1997). LGO includes random search components such as multi-start random search and local optimization components such as generalized reduced gradient method. An efficient black box partitioning approach is the Multilevel Coordinate Search (MCS) (Huyer and Neumaier, 1999) where non-uniform partitioning is performed by introducing a partition bias that divides boxes in the vicinity of samples having better function values. Samples are collected from box boundaries leading to sample sharing by neighbor boxes. Apart from interval and Lipschitz methods, there exists yet a third category of partitioning approaches that are based on fuzzy box assessment where evidence within each partition is collected by using random search techniques, such as Simulated Annealing, hill climbing methods, etc. (Demirhan and Özdamar, 1999; Özdamar and Demirhan, 2000).

All of the approaches above involve rectangular partitioning. There are also simplical partitioning algorithms that are at times more efficient in terms of the number of new partitions obtained per additional sample taken. As indicated in Clausen and Žilinskas, 2002, Wood, 1991, and Zhang et al., 1993, simplical partitioning might be comparatively more flexible because a wide range of partitioning schemes can be used to result in more regular geometrical shapes that are proportional to the topology of the domain. Similar to rectangular partitioning, simplical partitioning is also convergent when nested partitioning is applied, because edge length is reduced to zero exponentially.

In simplical partitioning, each re-partitioning iteration on a given simplex can be interpreted as a simplical direct search move in the category of contraction. Direct search methods are derivative free local search techniques that explore the domain in a prescribed manner, moving from one solution to a hopefully better one. An well-known example is the pattern search that moves according to a given design of exploratory moves (Lewis and Torczon, 1999). For a recent review of direct search methods, the reader is referred to Kolda et al., 2003. Simplical direct search methods, such as Nelder-Mead (Nelder and Mead, 1965), are quite popular and they are widely used by researchers and practitioners (Walters et al., 1991). Lewis et al., 2000 discuss the advantages of simplical direct search methods in their extensive review of direct search, however noting Nelder-Mead's instability in convergence (McKinnon, 1998).

Here, a dynamic, parallel and global simplical partitioning approach is introduced. We call the approach TRIOPT (Triangulated Optimization). The novelty in TRIOPT is that simplices are assessed for re-partitioning after vertex function values are transformed and mapped onto the [0,1] interval. During function transformation, adaptive parameters are utilized. For instance, such a parameter is the function threshold whose value can increase or decrease during the search. The threshold acts as a virtual line where larger function values are treated as peaks and those below as valleys. Hence, an adaptive threshold re-defines peaks and valleys leading to dynamic re-direction of the search and a changing degree of parallelism where all such peaks are looked into. The adaptive threshold scheme is enabled by the removal of small simplices and their vertices from the available sample set as well as the acquisition of new sample data. Unlike other partitioning approaches where the current upper bound always improves during the search, the removal of simplices leads to a decrease in the threshold value rather than an increase. Similar to DIRECT and other partitioning algorithms, TRIOPT is a convergent algorithm.

2.   TRIOPT Algorithm

Generic algorithm. Given a search domain $\mathcal{D}$, an initial triangulation is formed, and a working simplex list $\mathcal{W}$ that includes all simplices is generated. Then, a simplex $\mathcal{S}_c$ is selected from $\mathcal{W}$ and if its size is lower than δ, it is added to $\mathcal{P}$, that is the set of simplices to be reported. Vertices in $\mathcal{S}_c$ that are not shared with other simplices are removed from the set of available samples, $\mathcal{V}$, leading to a possible change in the maximum and minimum function values (denoted as $f_{\max}$, $f_{\min}$, respectively) in $\mathcal{V}$. If the size of $\mathcal{S}_c$ is greater than δ, then its entropy, $E_{\mathcal{S}_c}$, is checked. If $E_{\mathcal{S}_c}$ is greater than the entropy threshold β, then it is re-partitioned, removed from $\mathcal{W}$, and its child simplices are added to a temporary list, $\mathcal{T}$. The new sample taken during re-partitioning is added to $\mathcal{V}$. This step is repeated until all simplices $\mathcal{S}_j \in \mathcal{W}$ are evaluated. After this scan is completed, child simplices in $\mathcal{T}$ are added to $\mathcal{W}$ and $\mathcal{T}$ is flushed. In the next iteration, all simplex entropies are re-calculated using updated values for $f_{\max}$, $f_{\min}$ and the function threshold, t.

The following notation is provided for presenting the pseudocode of TRIOPT.

$\mathcal{W}$: List of all pending simplices that require a decision for re-partitioning;

$\mathcal{P}$: Final list of simplices reported to have a potential of enclosing $\mathbf{x}^*$;

$\mathcal{S}_c$: Currently processed simplex;

$size(\mathcal{S}_c)$: Size of currently processed simplex;

$\mathcal{T}$: Temporary list of child simplices;

$\mathcal{V}$: Set of currently available samples;

E_j: Entropy of simplex j;

δ: Tolerance area for removing a simplex from $\mathcal{W}$;

β: Entropy threshold for re-partitioning a candidate simplex;

The pseudocode is given below.

Step 0. Construct a triangulation on an initial sample $\mathcal{V} \in \mathcal{D}$, and initialize $\mathcal{W} = \{all\ simplices\}$. For each simplex $\mathcal{S}_j$, calculate E_j.

Step 1. Select a simplex $\mathcal{S}_c \in \mathcal{W}$.

Step 2. If $size(\mathcal{S}_c) < \delta$, then, remove $\mathcal{S}_c$ from $\mathcal{W}$, insert it in $\mathcal{P}$, remove unshared vertices of $\mathcal{S}_c$ from $\mathcal{V}$ and go to Step 3. Otherwise, go to Step 6.

Step 3. If $\mathcal{T} \bigcup \mathcal{W} = \emptyset$, or termination criterion is satisfied, then stop, report $\mathcal{P}$. Otherwise, continue.

Step 4. If re-partitioning decision of all $\mathcal{S}_j \in \mathcal{W}$ is not completed, then go to Step 1. Otherwise, continue.

Step 5. Add simplices in $\mathcal{T}$ to the set $\mathcal{W}$, update E_j, empty $\mathcal{T}$, and go to Step 1.

Step 6. If $E_{\mathcal{S}_c} \geqslant \beta$, split $\mathcal{S}_c$, add new vertex to $\mathcal{V}$, remove $\mathcal{S}_c$ from $\mathcal{W}$, add child simplices to $\mathcal{T}$.

Step 7. Go to Step 4.

The termination criterion mentioned in Step 3 might be a maximum number of function evaluations (number of physical experiments conducted) or a minimum percentage of improvement in the best function value obtained at the end of the last assessment cycle.

If, in any iteration, none of the simplices in $\mathcal{W}$ overcome the entropy threshold β, then, TRIOPT can stop prematurely. In this case, the simplex with the maximum $E_{\mathcal{S}_j}$ is re-partitioned. Otherwise, Step 4 enables the assessment all successful candidates in parallel.

Function transformation and simplex entropy. Function values of samples $\mathbf{x} \in \mathcal{V}$ are sorted in descending order of $f(\mathbf{x})$ and a threshold value t is selected from the sorted list. t is equal to a sample function value that is near $f_{\max}$. The transformation function $\ddot{f}$ classifies function values according to parameter t where $f(\mathbf{x}) \geqslant t$ are treated as peaks and $f(\mathbf{x}) < t$ as valleys. $\ddot{f}$ is calculated by using dynamic parameters $f_{\max}$, $f_{\min}$ and t. All $f(\mathbf{x})$ are transformed and mapped onto the $[0.1, 1.0]$ interval using the following formula.

$$\ddot{f} = \ddot{f}(f(\mathbf{x}), t) = \begin{cases} 0.1 + 0.4 \left(1.0 - \frac{t - f(\mathbf{x})}{t - f_{\min}}\right)^p, & \text{if } f(\mathbf{x}) < t; \\ 0.5 + 0.5 \left(\frac{f(\mathbf{x}) - t}{f_{\max} - t}\right)^q, & \text{if } f(\mathbf{x}) \geqslant t. \end{cases} \tag{1.2}$$

Here, the deflation and inflation parameters, $p > 1.0$ and $q < 1.0$, are both positive real numbers. p and q reduce function values below t and increase the ones above, respectively, leading to a more discriminating classification of simplices. The slope of $\ddot{f}$ changes at t and the lower bound specified in (1.2) is a precaution against possible splitting degeneracies.

Let us define $\dot{f}$ as a piecewise linear transformation of $f(\mathbf{x})$ where $\ddot{f}$ in (1.2) is applied with both parameters p and q set to *one*.

The linear transformation function $\dot{f}$ is monotone increasing in f. Further, $\ddot{f}$ includes $\dot{f}$, that is, $\dot{f} \subseteq \ddot{f}$ for any $\dot{f}(\mathbf{x})$, $\mathbf{x} \in \mathcal{V}$. This holds because for any peak, $\ddot{f} > \dot{f}$ and for any valley, $\ddot{f} < \dot{f}$. Furthermore, when $f = t$, $\ddot{f} = \dot{f}$.

REMARK 1 *Due to the relationship explained above, $\ddot{f}$ does not change the topology of f constructed from $\mathcal{V}$. The use of $\ddot{f}$ increases the attraction power of a peak $\mathbf{x}$ neighboring a valley $\mathbf{y}$ by the order of $\mathbf{x}^{1-q}\mathbf{y}^{p-1}$ as compared to $\dot{f}$.*

Any other transformation function satisfying this property can be used to transform f. After calculating transformed function values $\ddot{f}$ for all $\mathbf{x} \in \mathcal{V}$, an aggregate entropy measure, E_j, is computed for each simplex $\mathcal{S}_j$ by considering the contribution of its vertices. E_j is calculated follows.

$$
E_j = \begin{cases} \dfrac{1}{k} \displaystyle\sum_{\mathbf{x} \in \mathcal{S}_j : f(\mathbf{x}) \geqslant \alpha} \ddot{f}(\mathbf{x}) \exp(1.0 - \ddot{f}(\mathbf{x})), & \text{if } \exists\, \ddot{f}(\mathbf{x}) \geqslant \alpha; \\ 0.0, & \text{otherwise.} \end{cases} \tag{1.3}
$$

Here, each vertex $\mathbf{x}$ contributes to the entropy if $\ddot{f}(\mathbf{x}) \geqslant \alpha$. k is the number of vertices satisfying this condition. The cut α on $\ddot{f}$ prevents inferior vertices in a simplex from affecting the potential reflected by superior ones. The expression given for E_j is a multiplicative entropy form and it is adapted from Pal and Pal, 1989.

Parallelism. Since *all* simplices with $E_j \geqslant \beta$ are selected for re-partitioning in parallel, the number of simplices re-partitioned in parallel is dynamic and depends on global information ($f_{\max}$, $f_{\min}$, t) updated during the execution of the search. There are two occasions where these parameters might change. The first one takes place when a simplex size reaches the tolerance level, δ and all vertices of the simplex that are not shared by others are discarded from $\mathcal{V}$. Hence, $f_{\max}$, $f_{\min}$ and t might be modified to result in readjusted $\ddot{f}$ values. The second occasion is encountered regularly when new sample, $\mathbf{x}_{new}$ changes global information. Both occasions lead to the re-definition of peaks and valleys, however, the first occasion is important for convergence. In particular, if $f_{\max}$ is reduced with the removal of a small simplex, previously inferior simplices might now become peaks to be re-partitioned. In the limit, the algorithm converges to the simplex that contains the global optimum even if its vertices do not initially indicate a promise of holding it.

Initial partition and simplex re-partitioning strategies. In Step 0, the initial partition over $\mathcal{D}$ is constructed by triangulating (by using Delaunay triangulation on $\mathbb{R}^n$ by Quickhull algorithm, Barber et al., 1996) an initial set of available samples $\mathcal{V}$, if such a set exists. Otherwise, a sample is collected from each vertex of $\mathcal{D}$ and one from its mid-point. The flexibility of partitioning on an arbitrary set of samples becomes an advantage for TRIOPT in black box optimization problems where a set of physical experiments are already available. However, it should be noted that if the initial sample does not contain all vertices of the hypercube $\mathcal{D}$, the whole search domain cannot be covered. The latter necessity restricts the method to low dimensional problems.

In Step 6, when a simplex is re-partitioned, either *interior simplex splitting* or *longest-edge simplex splitting* is conducted. These two partitioning schemes are utilized because interior simplex splitting may result in very irregular simplices if applied repetitively. The following rule is used for making a decision on the type of splitting.

Simplex Partitioning Rule: *If a simplex $\mathcal{S}_j$ is regular, re-partitioning is carried out by interior-simplex splitting, otherwise longest-edge splitting is applied.*

A simplex is considered to be *regular* if the length of each edge is between the shortest edge and twice the shortest edge.

For both partitioning schemes, the location of the new vertex is selected by taking the weighted average over transformed function values of the relevant vertices (all vertices are relevant in interior-splitting and 2 vertices are relevant in longest-edge splitting). Below we provide the formula for locating the new vertex, $\mathbf{x}_{new}$, in interior splitting.

$$\mathbf{x}_{new} = \sum_{\mathbf{x} \in \mathcal{S}_j} \left[\mathbf{x} \cdot \frac{\ddot{f}(\mathbf{x})^r}{\sum_{\mathbf{x} \in \mathcal{S}_j} \ddot{f}(\mathbf{x})^r} \right]. \tag{1.4}$$

Here, r is a non-negative constant that regulates the bias towards good vertices.

In longest-edge splitting, $\mathbf{x}_{new}$ is located by taking the weighted combination of the two vertex coordinates forming the edge.

3.   Numerical Studies

Testing Applications

Several applications from different fields compose the test bed. These are described below.

QFT Bounds Calculation Problem The Quantitative Feedback Theory (QFT) is an engineering design technique for uncertain feedback systems where closed-loop frequency-domain specifications are translated into Nichols chart domains specifying the allowable range of the nominal open-loop response. In the maximization problem considered below, an upper bound for an active noise and vibration control system (ANVCS) (Nataraj and Sheela, 2002) is seeked.

$$\max f(\mathbf{x}) = -\frac{180}{\pi} \arctan\left[\left(2x_2\frac{\omega}{x_1}\right) \Big/ \left(1 - \left(\frac{\omega}{x_1}\right)^2\right)\right]$$
$$x_1 = [0.75, 1.25], \quad x_2 = [0.02, 0.06], \quad \omega = 1.0$$

Optimal Shape Design Optimizing shape design (structural optimization, or redesign) aims at reducing new product development cycle length in industries such as automotive, marine, and aerospace. Here, a truss structure design system is given in Figure 1.1 as a sample shape design problem. The objective is to minimize the volume or weight of the truss structure depending on the location of node D, i.e.:

$$\min f(\mathbf{x}) = \sum_{i=1}^{4} A_i \cdot L_i = \frac{1}{\sigma_0} \sum_{i=1}^{4} |N_i(\mathbf{x})| \cdot L_i.$$

where i is the index of rods, $\mathbf{x} = (x_w, y_w)$, $N_i(\mathbf{x})$ are axial forces, A_i is cross sectional area of rods, L_i is rod length and σ_0 is an allowable stress. Parameters are assumed to be as follows (Pownuk, 2000): $L = 1\,\mathrm{m}$, $P = 10\,\mathrm{kN}$, allowable stress $\sigma_0 = 190\,\mathrm{MPa}$.

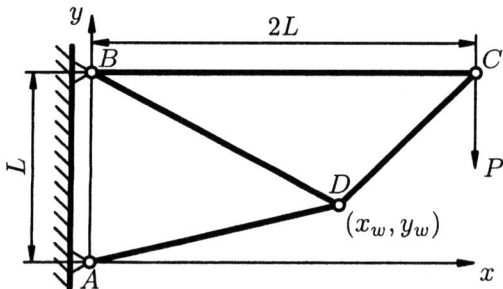

Figure 1.1.   Optimal shape design of a truss.

Weeds and Palmer 4 (CUTEr) Weeds and Palmer 4 are data fitting problems where a given data set is fit into a nonlinear function. In Weeds, a logistic growth function is used, and the aim is to identify the optimal function parameters $\mathbf{x} = \{a, b, c\}$ that provide the best data fit. Palmer 4 is a 4D data fitting application and the objective is to find the optimal coefficients $\mathbf{x} = \{a, b, c, d\}$.
Weeds:

$$\min f(\mathbf{x}) = \sum_{i} \left(y_i - \frac{a}{1 + \exp(ci - b)}\right)^2$$

where $a = [-100, 100]$, $b = [-100, 100]$ and $c = [1.0, 3.0]$, $\mathbf{y}$ is the vector of input data to be fitted.

Palmer 4:

$$\min f(\mathbf{x}) = \sum_i \left(y_i - a z_i^2 - \frac{b}{c + z_i^2/d} \right)^2$$

where $a = [0, 20]$, $b = c = d = [0, 1]$, $\mathbf{y}$ and $\mathbf{z}$ are known output and input data to be fitted, respectively.

Pressure vessel design (Krikanov, 2000) Cylindrical pressure vessels are widely used for commercial and aerospace applications. The use of composite materials improves the performance of vessels and leads to a significant amount of material and weight savings.

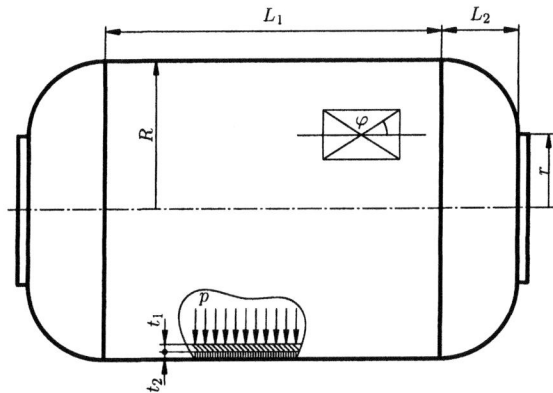

Figure 1.2. Cylindrical pressure vessel.

Consider a laminated cylindrical pressure vessel loaded with internal pressure p as shown in Figure 1.2. We assume the vessel consists of two angle-ply layers. The first layer has thickness t_1 and is formed by symmetric filament winding under angle $\pm\varphi_1$ which is known, because the preassigned radius of the polar opening should satisfy geodetic winding conditions.

The second layer is an arbitrary angle-ply laminate characterized with thickness t_2 and angle $\pm\varphi_2$. The design parameters for the vessel shown in Figure 1.2 are t_1, t_2 and φ_2. The objective function is to minimize the mass of the cylindrical part of the vessel, G, while satisfying the stress and strain constraints. Assume that $g_i = t_i \rho_i$, $a_i = \sigma_i/\rho_i$, $K_i = E_i/\rho_i$; here subscript i indicates the ith layer, g is the mass per unit area, ρ the density, σ the stress, a and K are normalized stress and modulus. We also assume that $\bar{a}$ is the ultimate normalized stress, $\bar{\varepsilon}$ the ultimate strain. Then, the corresponding minimization problem can be written as:

$$\min g_1 + g_2$$

subject to

$$
\begin{aligned}
pR/2 &= a_1 g_1 \cos^2 \varphi_1 + a_2 g_2 \cos^2 \varphi_2 \\
pR &= a_1 g_1 \sin^2 \varphi_1 + a_2 g_2 \sin^2 \varphi_2 \\
a_1/K_1 &= \varepsilon_x \cos^2 \varphi_1 + \varepsilon_y \sin^2 \varphi_1 \\
a_2/K_2 &= \varepsilon_x \cos^2 \varphi_2 + \varepsilon_y \sin^2 \varphi_2 \\
a_1 &\leqslant a_1^*, \quad 0 < a_2 \leqslant \bar{a}_2 \\
\varepsilon_x &\leqslant \bar{\varepsilon}_x, \quad \varepsilon_y \leqslant \bar{\varepsilon}_y
\end{aligned}
$$

where

$$a_1^* = \bar{a}_1 (3 \sin^2 \varphi_2 - 2) \cos \varphi_1 / (\sin^3 \varphi_2 \sqrt{\sin \varphi_2^2 - \sin \varphi_1^2})$$

From the first four equations, a_1, a_2, ε_x and ε_y can be represented in terms of g_1, g_2 and φ_2, and substituted into the objective, leaving the last four stress and strain bound constraints to be incorporated into the objective with a penalty factor as indicated below.

$$\min g_1 + g_2 + \sum r \max(v_i - \bar{v}_i, 0)$$

where, v_i represent a_i and ε_i, and $\bar{v}_i$ are the corresponding ultimate normalized stress and strain, r is the penalty coefficient.

Protein folding (Stillinger et al., 1993) This problem involves two amino acids "A" and "B" that are linked together by unit length bonds forming two dimensional linear polymers. $n - 2$ angles of bend are to be optimized at each of the nonterminal residues. Here, $n = 5$ and we solve $3D$ protein folding problems. In this model, it is assumed that two kinds of interactions compose the intramolecular potential energy, V_1, backbone bend potentials, and V_2, non-bonded interactions with species dependent Lennard-Jones (12,6) form. Residue species along the backbone are encoded by a set of binary variables, $\xi_1, \ldots, \xi_n$. If $\xi_i = 1$, i^{th} residue is A, and if it is -1, residue is B. Below the intramolecular potential-energy function is expressed generally for an n-mer. Here, r_{ij} is distance, and coefficient $C(\xi_i, \xi_j)$ is 1 for an AA pair, 1/2 for a BB pair and $-1/2$ for an AB pair. In the experiments, all possible sequences of the $n = 5$ problem are solved, but only the ones found to produce different results under the portfolio of tested methods are reported (8 problems).

$$f(\theta) = \sum_{i=2}^{n-1} V_1(\theta_i) + \sum_{i=1}^{n-2} \sum_{j=i+2}^{n} V_2(r_{ij}, \xi_i, \xi_j)$$

where,

$$-\pi \leqslant \theta_i \leqslant \pi$$

$$r_{ij} = \left\{ \left[\sum_{k=i+1}^{j-1} \cos \left(\sum_{l=i+1}^{k} \theta_l \right) \right]^2 + \left[\sum_{k=i+1}^{j-1} \sin \left(\sum_{l=i+1}^{k} \theta_l \right) \right]^2 \right\}^{1/2}$$

$$V_1(\theta_i) = (1 - \cos \theta_i)/4$$

$$V_2(r_{ij}, \xi_i, \xi_j) = 4 \left[r_{ij}^{-12} - C(\xi_i, \xi_j) r_{ij}^{-6} \right]$$

$$C(\xi_i, \xi_j) = (1 + \xi_i + \xi_j + 5\xi_i\xi_j)/8$$

Summary of Applications

All applications are summarized in Table 1.1 together with their global optima and related references. We now briefly describe some features of these applications. ANVCS is a smooth function but it is discontinuous in the vicinity of the global optimal solution. The optimal solution lies on the edge of the discontinuity. The truss structure has a very flat surface near the global optimum that might lead to wasted function evaluations. Palmer 4 fitting problem involves even degree polynomials, hence its surface is expected to be more irregular as compared to the truss structure and ANVCS. Protein folding problems are highly multi-modal and therefore quite difficult to solve, so is the pressure vessel design problem.

Table 1.1. Summary of applications.

Function	Dimension	Optimum	Reference
ANVCS	2	89.9999	Nataraj and Sheela, 2002
Truss structure	2	0.4102	Pownuk, 2000
Weeds	3	1046.42[a]	CUTEr
Palmer 4	4	2285.38[a]	CUTEr
Pressure vessel	3	12.7	Krikanov, 2000
Protein folding	3	N.A.	Stillinger et al., 1993

[a]: Reported minimum value of the objective function.

4. Numerical Results

The experimental and parametric settings for TRIOPT are as follows. The initial cover is obtained with samples collected from each vertex of the hypercube $\mathcal{D}$ and one additional sample at its mid-point. The size tolerance level, δ, is set to $(\prod_{i=1}^{n} l_i)/8000$ where l_i is bound length of the ith coordinate. The parameters (p, q) used in Equation 1.2 are set to $(1.5, 0.5)$ for non-smooth surfaces (data fitting and protein folding problems), and $(3.0, 0.5)$ for the remaining application problems. The α and β cuts are fixed to 0.7 and 0.9, respectively to enable fast convergence. With such a high value set for the α cut, the parameter p has no effect on entropies but it does affect the location of the new sample when a simplex is re-partitioned.

Performance assessment is measured by the number of function evaluations (*fe*)that each method takes to converge to a near optimal solution that is within a small percentage below/above the global optimum solution. This relative error ε is set to 1.0% and 0.01% in these sets of experiments. An exception is made for the protein folding problem where the deviation from the best function value obtained over all methods is reported within given numbers of *fe*, 100, 300, 500, 1000, 2000, respectively. This exception is due to the significantly large numbers of *fe* reported for convergence (Stillinger et al., 1993).

Empirical analysis of threshold-based transformation and simplex removal

The experiments described in this section aim at illustrating the impacts of two features on performance: the existence of a dynamic threshold in the transformation function and the removal of vertices from $\mathcal{V}$. The latter two features actively contribute to dynamic concurrency in re-partitioning.

Table 1.2. Number of parallel simplices during TRIOPT runs, rounded up to next integer

No. of parallel nodes	P(min)	P(avg)	P(max)
1.0% accuracy	4	7	28
0.01% accuracy	4	10	28
Protein folding 500 evaluations	4	22	124
Protein folding 1000 evaluations	3	17	166
Protein folding 2000 evaluations	2	16	164

First, we illustrate the scale of parallelism. Table 1.2 provides a summary of the minimum (P(min)), maximum (P(max)), and average numbers (P(avg)) of simplices partitioned in parallel for all test problems. For brevity, these three numbers are *averaged* over the first five applications and separately for the 8 protein folding problems. It is observed that simpler applications with smoother surfaces result in lower degrees of parallelism (maximum number of parallel nodes is in the range of [4-6]) whereas more difficult ones such as Weeds and Palmer 4 have a maximum of 33 and 63 nodes in parallel activated at one point during the search. The protein folding problems have their maximum degree of concurrency in the range of [100-200]. In Figure 1.3 and Figure 1.4, the number of simplices re-partitioned in parallel at each cycle is illustrated for the applications ANVCS and Palmer 4 (0.01% accuracy level run). The former has a smooth surface and the degree of concurrency has a narrower range as compared to the latter.

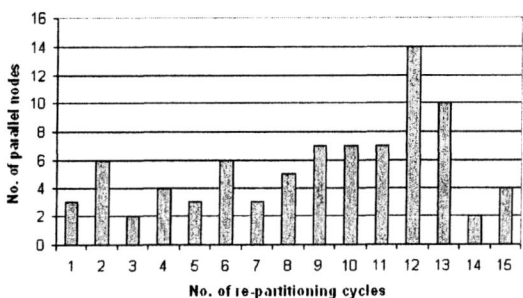

Figure 1.3. Number of parallel simplices in each cycle for ANVCS.

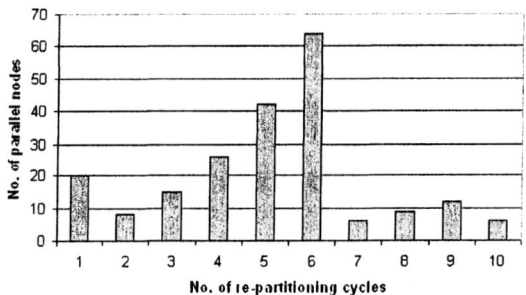

Figure 1.4. Number of parallel simplices in each cycle for Palmer 4.

We now design the two tests that illustrate the effects of the two features mentioned above: the removal of unshared vertices belonging to extinct simplices and the dynamic threshold defined in the transformation function. The first test, N_Rem, executes TRIOPT without removing such vertices, and the second test, N_t, assumes that there is no threshold over which $\ddot{f}$ is defined and all f are transformed using the formula:

$$\ddot{f} = \ddot{f}(f(\mathbf{x})) = 0.1 + 0.9 \left(1.0 - \frac{f(\mathbf{x}) - f_{\min}}{f_{\max} - f_{\min}} \right)^q. \tag{1.5}$$

The test N_t illustrates the impact of t on classification power as well as its contribution to the dynamic re-definition of peaks and valleys, and, N_Rem shows the impact of vertex removal when a dynamic threshold exists.

Table 1.3. Results illustrating impacts of TRIOPT's features

Function	TRIOPT	N_Rem	N_t
1.0% accuracy			
ANVCS	33	33	34
Truss structure	6	6	6
Weeds	8387	—	—
Palmer 4	30	30	19
Pressure vessel	113	113	—
0.01% accuracy			
ANVCS	80	80	314
Truss structure	52	53	790
Weeds	8387	—	—
Palmer 4	139	139	—
Pressure vessel	113	113	—

Table 1.3 summarizes the impacts of vertex removal and thresholding strategy on TRIOPT's performance for all applications except protein folding. Results are given as the number of function evaluations (*fe*) needed to converge to 1.0% and 0.01% tolerance levels. In the first column, we provide TRIOPT's results with all features included. These can be compared to *fe* under N_Rem and N_t columns. In all tables, '—' indicates that the corresponding method cannot solve the problem at the given accuracy level within 20000, and 50000 *fe*.

It is observed that for applications with comparatively smooth surfaces, vertex removal has no effect on $f_{\max}$, however, in a more difficult problem such as Weeds, the simplex removal enables convergence in a finite number of *fe*. The dynamic threshold scheme is much more effective at both accuracy levels leading to convergence.

Table 1.4. Results illustrating TRIOPT's features in 8 protein folding problems with 500,1000, 2000 function evaluations.

	TRIOPT	*N_Rem*	*N_t*
500 evaluations			
Average % dev.	0.032	0.207	10.860
Maximum % dev.	0.111	1.093	53.607
No. of best	8	2	0
1000 evaluations			
Average % dev.	0.011	0.075	10.808
Maximum % dev.	0.035	0.231	53.619
No. of best	8	2	0
2000 evaluations			
Average % dev.	0.008	0.044	8.468
Maximum % dev.	0.035	0.179	38.183
No. of best	8	2	0

Table 1.4 summarizes the results obtained for 8 protein folding sequences AAAAA, AAAAB, AAABA, AABAA, AABBA, ABAAB, ABBAB. Here, there are 3 sets of experiments: the number of *fe* is limited to 500, 1000, and 2000, respectively, and the deviation from best function value obtained over all methods is reported. (We omit 100 and 300 *fe* for brevity.) For each method, the average and maximum percentage deviations from the best result, and the number of best solutions found in 8 problems are reported. In these problems both features are observed to play an important part in improving convergence.

Obviously, re-definition of peaks and valleys through a dynamic threshold is one of the primary factors impacting TRIOPT's performance. The concept of producing artificial peaks and valleys by using a threshold and changing global information by removing samples work together in a coordinated manner to result in an effective concurrency control strategy.

Comparison with other methods

TRIOPT is compared with three global partitioning methods. The first one is DIRECT (Jones et al., 1993) and the corresponding module *glbSolve* of the commercial software package TOMLAB™ (*http://www.tomlab.biz*) is utilized here. Similar to TRIOPT, DIRECT has an adaptive concurrent partitioning scheme. DIRECT is implemented with its default settings in *glbSolve*. The second one is MCS (Huyer and Neumaier, 1999) and it is implemented here by using the Matlab code provided in *http://www.mat.univie.ae.at/~neum/software/mcs/*. MCS has an important parameter that affects results. This parameter is the maximum depth of the tree, s_{max}. Here, it is set to two values, $5n + 10$ (recommended one) and $15n + 10$, and these are labelled as MCS1 and MCS2, respectively, in the summary of results. It is necessary to increase this parameter when MCS cannot converge to the global optimum. MCS has a local search option in its code, we also include it in the test runs, as MCS2L, i.e., we fix s_{max} to $15n+10$. The last partitioning algorithm is LGO (Pintér, 1997) that incorporates derivative free random sampling techniques within a global partitioning algorithm. The commercial software for LGO is utilized here (Pintér, 2003) with the two random search options Global Adaptive Random Search (GARS) and Multi Start Global Random Search (MS). However, in these applications both give the same results.

We summarize the results of the first five applications in Table 1.5 and protein folding problem results in Table 1.6. In the last rows of 1% and 0.01% accuracy results, the number of best and second best solutions found are indicated. In the second column of both tables a second set of results are given for TRIOPT where the local search method Nelder-Mead (NM) is appended to the global partitioning approach. NM is applied when a simplex becomes small enough to be removed from the list. In that case, the simplex becomes a source simplex for NM, if its best vertex is not in the list of previous source vertices for NM. NM stops when the largest edge of the working simplex is less than 0.00001 or when there is no improvement 25 times consecutively.

Table 1.5.   Number of function evaluations required for 1.0% and 0.01% accuracy levels.

Function	TRIOPT	TRI-NM	DIRECT	MCS1	MCS2	MCS2L	LGO
1.0% accuracy level							
ANVCS	33	32	—	—	—	69	4
Truss Structure	6	6	5	17	26	34	12
Weeds	8387	143	15335	—	36879	93	188
Palmer 4	30	19	57	2324	8624	75	93
Pressure vessel	113	48	10087	12199	19293	19293	337
Best/second best	3	5	1	0	0	1	1
0.01% accuracy level							
ANVCS	80	64	—	—	—	—	4
Truss Structure	52	52	79	70	123	49	21
Weeds	8387	151	15533	—	36879	93	278
Palmer 4	139	73	905	7759	9000	89	120
Pressure vessel	113	67	10131	12235	19471	19471	353
Best/second best	1	4	0	0	0	3	2

Table 1.6. Results for 8 protein folding problems for different *fe*.

	TRIOPT	TRI-NM	DIRECT	MCS1	MCS2	MCS2L	LGO
100 evaluations							
Average % dev.	4.133	3.595	21.750	13.800	21.750	0.0	5.424
Maximum % dev.	19.408	13.628	92.331	37.614	93.331	0.0	14.825
Best/second best	4	6	0	0	0	8	0
300 evaluations							
Average % dev.	0.041	0.005	3.461	4.906	12.197	0.0	4.448
Maximum % dev.	0.102	0.025	10.319	16.217	31.132	0.0	14.848
Best/second best	0	8	0	0	0	8	0
500 evaluations							
Average % dev.	0.057	0.0	3.202	3.730	9.250	0.0	0.0
Maximum % dev.	0.195	0.0	10.151	15.707	31.132	0.0	0.0
Best/second best	3	8	2	0	0	8	8
1000 evaluations							
Average % dev.	0.034	0.0	1.522	3.723	5.300	0.0	0.0
Maximum % dev.	0.100	0.0	4.567	15.707	14.530	0.0	0.0
Best/second best	4	8	2	0	3	8	8
2000 evaluations							
Average % dev.	0.029	0.0	1.512	3.729	2.986	0.0	0.0
Maximum % dev.	0.100	0.0	4.531	15.707	11.781	0.0	0.0
Best/second best	6	8	2	0	4	8	8

An overview of the results shows that when the assumption of continuity does not hold (ANVCS), methods that do not incorporate local search do not converge except for TRIOPT. Among methods with local search, LGO converges fastest in this problem. NM usually improves the performance of TRIOPT and this improvement is drastic especially in Weeds. In Weeds, all methods require significant numbers of function evaluations if they are not supported by local search. On the other hand, in Palmer 4, TRIOPT's performance is satisfactory both with and without NM whereas MCS and DIRECT need local search support at the higher accuracy level. In the Truss Structure, where the global optimum lies on a very flat surface, LGO converges faster than others.

Among different s_{max} levels, that of MCS2 seems to be appropriate for convergence. For MCS, it is advised to use the default level since it results in the least number of *fe* required to converge, however, the risk of non-convergence is higher, because ending boxes of the tree might be too large to satisfy the required accuracy level. Again, the efficient local search code embedded in MCS improves performance significantly except in Pressure Vessel. Unfortunately, such a local search module could not be appended to DIRECT implementation of TOMLAB. However, in a difficult problem such as the Pressure Vessel, local search does not help MCS and one cannot predict the possible effect of local search on DIRECT, if it were available. However, TRIOPT converges with or without NM. The next best performance is LGO's. An observation on LGO's performance over all problems is that its convergence is quite reliable although it is not the fastest method to converge.

In the 8 protein folding problems that are highly multi-modal, MCS2L is clearly the best performing method. TRIOPT and TRI-NM find better results at the level of 300 *fe*, in fact quite close to those of MCS2L, and from 500 *fe* onwards, TRI-NM's performance is as good as MCS2L. LGO also catches up with the latter two from this point onwards. Although the support of NM is required to obtain best results, TRIOPT also produces solutions of acceptable quality without NM. To summarize, in the light of these applications, the proposed triangulated partitioning method seems to be as reliable as other well-established methods in the literature for solving black box problems from different fields.

5.   Conclusion

Black box optimization techniques have an important role in real world applications, especially when a function is expensive to evaluate, or, additional information on the function is difficult to obtain.

A triangulated partitioning approach (TRIOPT) that manages the degree of search concurrency is proposed here. TRIOPT involves a flexible parallel search strategy based on dynamic threshold-based function transformation. The partition assessment and selection methodology in TRIOPT is based on a group entropy measure that reflects a potential for containing the optimum. During the search, the same simplex may be viewed both as promising and non-promising at different times, depending on re-defined peaks and valleys designated by the adaptive threshold and overall function range. A feature such as simplex/vertex removal, enables the search to re-direct itself to less promising simplices that might contain the global optimum. This, and the threshold-based non-linear transformation function strategy lead to a faster convergence rate.

One of the advantages of TRIOPT is related to its initial partitioning strategy. If some experimental results (function evaluations) are already obtained from a feasible domain without a specific pattern, an initial partition is easily generated by Delaunay triangulation. Such cases might be more difficult to handle by other partitioning methods that require a careful design of initial samples minimizing the number of function evaluations in further partitioning.

Current work is under way to implement TRIOPT on a grid-computing platform where each simplex may be assessed, re-partitioned and stored by separate processors. This should enable the solution of higher dimensional challenging problems.

References

Barber, C. B., Dobkin, D. P., and Huhdanpaa, H. (1996). The Quickhull algorithm for convex hulls. *ACM Transactions on Mathematical Software*, 22(4):469–483.

Clausen, J. and Žilinskas, A. (2002). Subdivision, sampling, and initialization strategies for simplical branch and bound in global optimization. *Computers and Mathematics with Applications*, 44:943–955.

CUTEr. CUTEr: A constrained and unconstrained testing environment, revisited. http://cuter.rl.ac.uk/cuter-www/problems.html.

Demirhan, M. and Özdamar, L. (1999). A note on the use of a fuzzy approach in adaptive partitioning algorithms for global optimization. *IEEE Transactions on Fuzzy Systems*, 7:468–475.

Gourdin, E., Hansen, P., and Jaumard, B. (1994). *Global optimization of multivariate Lipschitz functions: a survey and computational comparison*. Les Cahiers du GERAD. McGill University, Montreal.

Hansen, P. and Jaumard, B. (1995). Lipschitz optimization. In Horst, R. and Pardalos, P. M., editors, *Handbook of Global Optimization*, pages 407–493. Kluwer Academic Pulishers, Dordrecht.

Horst, R. and Tuy, H. (1996). *Global Optimization. Deterministic Approaches*. Springer Verlag, Berlin.

Huyer, W. and Neumaier, A. (1999). Global optimization by multilevel coordinate search. *Journal of Global Optimization*, 14:331–355.

Jones, D. R., Perttunen, C. D., and Stuckman, B. E. (1993). Lipschitzian optimization without the Lipschitz constant. *Journal of Optimization Theory and Applications*, 79:157–181.

Kearfott, R. R. (1996). *Rigorous Global Search: Continuous Problems*. Kluwer, Dordrecht.

Kolda, T. G., Lewis, R. M., and Torczon, V. (2003). Optimization by direct search: new perspectives on some classical and modern methods. *SIAM Review*, 45(3):385–482.

Krikanov, A. A. (2000). Composite pressure vessels with higher stiffness. *Composite Structures*, 48:119–127.

Lewis, R. M. and Torczon, V. (1999). Pattern search methods for bound constrained minimization. *SIAM Journal on Optimization*, 9(4):1082–1099.

Lewis, R. M., Torczon, V., and Trosset, M. W. (2000). Direct search methods: Then and now. *Journal of Computational and Applied Mathematics*, 124:191–207.

McKinnon, K. I. M. (1998). Convergence of the Nelder-Mead simplex method to a nonstationary point. *SIAM Journal on Optimization*, 9(1):148–158.

Moore, R. E. and Ratschek, H. (1988). Inclusion functions and global optimization II. *Mathematical Programming*, 41:341–356.

Nataraj, P. S. V. and Sheela, S. (2002). A new subdivision strategy for range computations. *Reliable Computing*, 8:83–92.

Nelder, J. A. and Mead, R. (1965). A simplex method for function minimization. *Computer Journal*, 7(4):308–313.

Özdamar, L. and Demirhan, M. B. (2000). Experiments with new stochastic global optimization search techniques. *Computers & Operations Research*, 27:841–865.

Özdamar, L. and Demirhan, M. B. (2001). Comparison of partition evaluation measures in an adaptive partitioning algorithm for global optimization. *Fuzzy Sets and Systems*, 117(1):47–60.

Pal, N. R. and Pal, S. K. (1989). Object background segmentation using new definitions of entropy. *IEE Proc.*, 136:284–295.

Pintér, J. (1988). Branch and bound algorithms for solving global optimization problems with lipschitzian structure. *Optimization*, 19:101–110.

Pintér, J. (1996). *Global Optimization in Action*. Kluwer, Dordrecht.

Pintér, J. (1997). LGO—A program system for continuous and Lipschitz global optimization. In Bomze, I. M., Csendes, T., Horst, R., and Pardalos, P. M., editors, *Developments in Global Optimization*, pages 183–197. Kluwer Academic Publishers, Dordrecht / Boston / London.

Pintér, J. (2003). *LGO: A Model Development and Solver System for Continuous Global Optimization User Guide*. Pinter Consulting Services, Halifax, Nova Scotia, Canada.

Pownuk, A. (2000). Optimization of mechanical structures using interval analysis. *Computer Assisted Mechanics and Engineering Science*, 7(4):699–705.

Stillinger, F. H., Head-Gordon, T., and Hirshfeld, C. L. (1993). Toy model for protein folding. *Physical Review E*, 48(2):1469–1477.

Walters, F. H., Parker, Jr., L. R., Morgan, S. L., and Deming, S. N. (1991). *Sequential Simplex Optimization*. Chemometrics Series. CRC Press, Inc., Boca Raton, Florida.

Wood, G. (1991). Multidimensional bisection applied to global optimization. *Computers and Mathematics with Applications*, 21:161–172.

Zhang, B., Wood, G., and Baritompa, W. (1993). Multi-dimensional bisection: the performance and the context. *Journal of Global Optimization*, 3:337–358.

Chapter 21

A CASE STUDY: COMPOSITE STRUCTURE DESIGN OPTIMIZATION

Zelda B. Zabinsky,[1] Mark E. Tuttle,[2] and Charoenchai Khompatraporn[1,3]

[1] *Industrial Engineering Program, Box 352650*
University of Washington, Seattle, Washington, 98195 USA
zelda@u.washington.edu, ckhom@u.washington.edu

[2] *Mechanical Engineering Department, Box 352600*
University of Washington, Seattle, Washington, 98195 USA
tuttle@u.washington.edu

[3] *Department of Production Engineering, Faculty of Engineering*
King Mongkut's University of Technology Thonburi
91 Pracha-Utit Rd., Thungkru, Bangkok, 10140 Thailand

Abstract   This chapter presents an engineering application of global optimization with a case study on designing large composite structures. The problem of designing large composite structures often becomes a global optimization problem when practical issues such as performance and manufacturability of the structures are considered. We demonstrate how problems on this type can be formulated and solved using two design examples. A stochastic optimization approach is used to obtain the solutions in both examples.

Keywords: Composite materials, composite manufacturing, global optimization, large structural design optimization.

Introduction

The demand for composite structures is rapidly growing as the need for lighter yet stronger and stiffer structures increases. Weight saving of a composite structure in comparison to the same structure made of a conventional material counterpart such as aluminum or steel alloys is made possible through the use of new fiber-reinforced composite materials. This weight saving advantage becomes greatly beneficial, particularly in weight-sensitive applications such as in aerospace structures, because it enhances the fuel efficiency and the thrust-to-weight ratio.

In addition to the high strength-to-weight and stiffness-to-weight ratios, composite materials also allow the engineer to *design the material* as well as the structure by *tailoring* the directional stiffness and strength as required for the structure [6, 7, 27]. On one hand, this can be viewed as an advantage because it offers added flexibility in the design and the ability to precisely meet the requirements. On the other hand, designing the material and the structure simultaneously increases the burden to the design engineer because the number of design variables increases dramatically. The interactions between the design variables also amplify the complexity of the design process by requiring additional computational effort. Fortunately, the capability of today's computer technology enables the design engineer to utilize computationally intensive optimization algorithms to explore the interaction amongst the design variables, while evaluating overall system performance and objectives.

The goal of this research has been to develop an optimization approach suitable for designing large composite structures. The optimization approach developed allows the design engineer to explore the design space and identify robust values for the design variables. It also allows the engineer to evaluate tradeoffs between various objectives, including weight and performance.

This chapter highlights some key considerations when designing large composite structures, and it is organized as follows. The next section briefly addresses some background in designing large composite structures. Sections 2 and 3 respectively discuss optimization problem formulation, and issues relating to selection of optimization algorithms. Two composite structure design examples are illustrated in Section 4. The chapter is concluded in Section 5, with the engineering specifications and the material properties of the two examples detailed in the Appendix.

1.　　Background

Large composite structures (such as aircraft) are typically composed of a collection of composite panels. These panels are commonly designed based on lamination theory [22]. Unlike conventional engineering materials which are often assumed homogeneous, each of these panels are composed of individual *laminae* or *plies*, stacked on top of each other and bonded together. The bonded assembly is called *laminate*. For example, a graphite epoxy ply consists of graphite fibers embedded in an

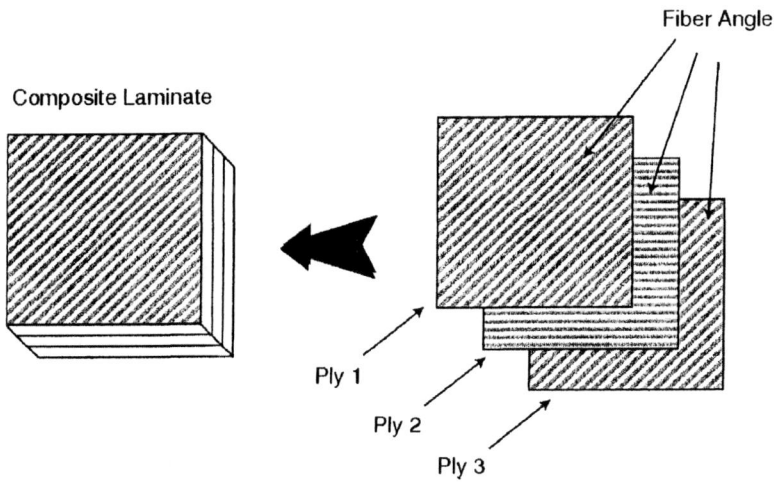

Figure 1.1. A 3-ply composite laminate with variation in fiber angle from one ply to the next

epoxy resin matrix. Each ply contains either unidirectional or woven fibers. To simplify our discussion, we concentrate only on unidirectional fiber plies throughout this chapter. The specific order and fiber orientations of the plies in a laminate is referred to as the *stacking sequence*. Lamination theory relates the complex interactions between these plies to predict mechanical properties such as stiffness and strength of the overall laminate.

Figure 1.1 shows a simple composite laminate consisting of three unidirectional fiber plies. The laminate in Figure 1.1 also illustrates that the orientation of the fibers in one ply may differ from the orientation of another ply. By convention, the fiber orientation θ associated with a given ply i is identified using a subscript. Thus, the fiber angle in ply 1 is denoted by θ_1, the fiber angle in ply 2 is denoted by θ_2, etc. It is interesting to note that for conventional engineering materials (such as steel) a thicker structure naturally implies a stronger structure. This implication, however, is not appropriate for a composite structure because of the strong influence of fiber directions. In fact, the mechanical performance of any composite structure depends so heavily on the fiber directions of the laminate that a thinner laminate with appropriate fiber angles can easily be superior to a thicker one with fiber angles in suboptimal directions.

There are many pertinent issues to consider in designing large composite structures. The next three subsections address some of these issues.

1.1　Performance Issues

As with any other structure, composite structures are designed to perform under different criteria depending upon the functional tasks. The most common requirements are load requirements. In order to meet these load requirements, the overall composite structures rely upon the properties of the individual plies.

Strength, the ability to withstand loading without fracture, is a common requirement imposed in composite structural design. The strength of a composite laminate is predicted based on the *strain* induced in individual plies. Strain is a measure of the deformations induced by a given loading. Hence, the overall *strength* of a composite laminate must be such that the loads applied do not induce ply *strains* above an allowable limit. A second requirement imposed during design is overall laminate *stiffness*. Stiffness is a measure of the deformations caused by per unit loading. Once again, the overall stiffness of a composite structure is directly related to the strains induced in individual plies by a given loading. Note that laminate stiffness and strength are two different requirements although they are closely related. For instance, for a given application a composite laminate may exhibit sufficient strength but insufficient stiffness, and vice versa for another application.

Conceptually, any laminate stacking sequence can be evaluated during the design process. However, *symmetric* laminates are almost always used in real applications. Non-symmetric laminates are very difficult to produce without warping and so they are rarely used in practical designs. In a symmetric laminate, ply fiber angles are arranged symmetrically about the geometric mid-plane of the laminate. For example, if a symmetric laminate contains p even number of plies, then $\theta_1 = \theta_p, \theta_2 = \theta_{p-1}, \theta_3 = \theta_{p-2}$, etc.

A favorable characteristic of composite is their high stiffness-to-weight ratios. Another method of enhancing the stiffness-to-weight ratio is through the use of *sandwich panels* [3, 8]. In any sandwich panel, a very light weight (and typically low strength and/or stiffness) *core material* is sandwiched between two *facesheets* which possess high stiffness and strength. This separation of the facesheets results in an enormous increase in *bending* stiffness with very little increase in weight. In the case of a composite sandwich structure, the two facesheets are composite laminates. Several different core materials may be used, including poly-

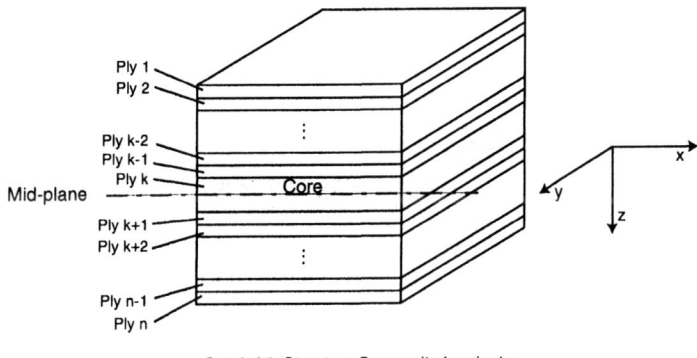

Sandwich Structure Composite Laminate

Figure 1.2. Cross-sectional view of the sandwich structure composite laminate with n total number of plies

meric foams or aluminum honeycomb. To specify the stacking sequence of this panel, one must specify the number of plies and the corresponding fiber angles in both facesheets, as well as the thickness of the core.

An example of a composite sandwich panel with convention notations describing the panel is shown in Figure 1.2. An x-y-z coordinate system is used, where the x-y plane coincides with the geometric mid-plane of the laminate. There are a total of n layers (or plies) shown. The core is denoted as ply k. The plies within the first composite facesheet are numbered as ply 1 through ply $k-1$, and the plies within the second composite facesheet are numbered as ply $k+1$ through ply n. Two "levels" of symmetry may be defined for a composite sandwich panel like this. In a *singly symmetric* sandwich panel both facesheets are themselves symmetric, but fiber angles within the two facesheets are not otherwise related. This implies that for a singly symmetric sandwich panel:

$$\theta_1 = \theta_{k-1} \qquad\qquad \theta_{k+1} = \theta_n$$
$$\theta_2 = \theta_{k-2} \qquad\qquad \theta_{k+2} = \theta_{n-1}$$
$$\theta_3 = \theta_{k-3} \qquad and \qquad \theta_{k+3} = \theta_{n-2}$$
$$\vdots \qquad\qquad\qquad\qquad \vdots$$

and so on   and so on

The second level of symmetry is termed a *doubly symmetric* sandwich panel. In this case, both facesheets are symmetric, and furthermore the

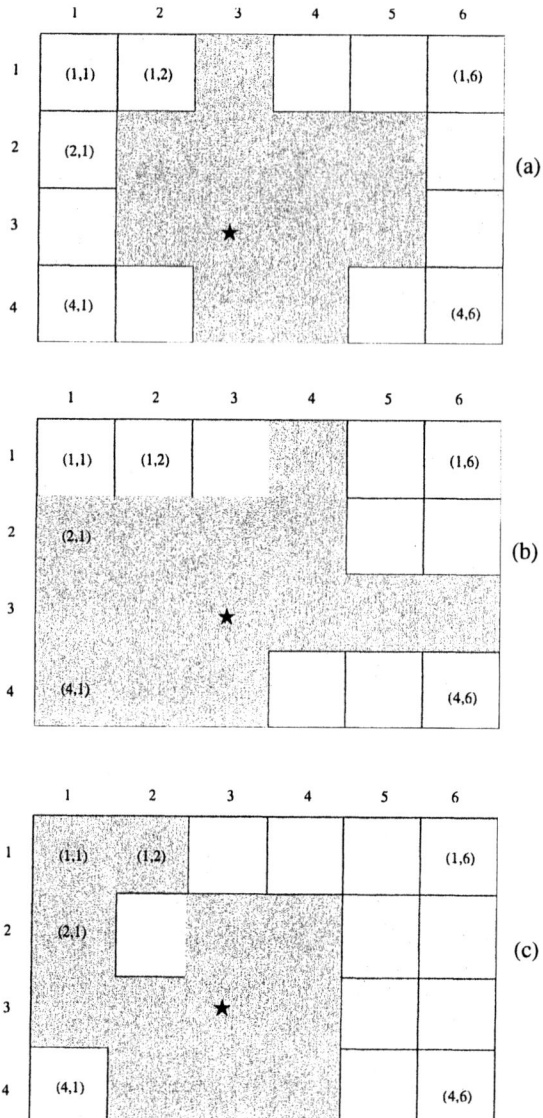

Figure 1.5. Example ply configurations on a 4 × 6 composite panel. Configuration (a) complies with the "greater-than-or-equal-to" blending rule but configurations (b) and (c) violate the rule [13]

as well as various optimization techniques based on local optimization schemes are addressed in some of the aforementioned list [24].

stacking sequence present in both facesheets is identical. For a doubly symmetric sandwich panel, we have:

$$\theta_1 = \theta_{k-1} = \theta_{k+1} = \theta_n$$
$$\theta_2 = \theta_{k-2} = \theta_{k+2} = \theta_{n-1}$$
$$\vdots$$

and so on

Doubly symmetric sandwich laminates are commonly used in practice. The examples presented later in this chapter assume that the panels are doubly symmetric.

An optimization approach that is useful to a design engineer must be able to satisfy performance constraints, including strain and stiffness requirements, as well as a type of symmetry property such as a singly or doubly symmetric structure laminate. It must also satisfy manufacturing issues which are discussed next.

1.2   Manufacturing Issues

In real-life applications, a large composite panel is rarely subjected to a uniformly distributed load. Actual loading usually varies over the length and width of the panel, with some areas experiencing relatively high loads and others experiencing relatively low loads. An example is the keel panels in a commercial aircraft (i.e. panels used in the underbelly of an aircraft fuselage). Typically, there is an opening in the keel panel to accommodate the landing gear. The region of the keel panel just aft of the landing gear normally experiences relatively high loads, while the magnitude of loading gradually decreases as regions are further away from the gear. The design engineer is typically provided with a "map" of predicted load contours over the length and width of the keel panel, as shown schematically in Figure 1.3(a). The smooth lines shown at internal regions of the panel represent lines of constant load intensity. These load contours are predicted during preliminary analysis of the overall aircraft. The job of the engineer is to design a panel that can safely sustain these loads. In the case of a composite sandwich keel panel, this means that the engineer must specify the stacking sequence of both facesheets as well as the core material and thickness.

Of course, one solution would be to identify a stacking sequence that can support the highest concentration of load, and then to use this stacking sequence throughout the entire length and width of the panel. While such a design would be mechanically feasible, it is certain to be heavier than is needed because the regions of the panel that experience relatively low loads would be overdesigned. It is possible to change the stacking sequence used in a composite panel by "dropping" plies (i.e. discontinuing a ply) in regions of low loads. To do so, the panel is divided into regions, as shown in Figure 1.3(b), during the design process. The magnitudes of the loads applied to each region are assumed to be constant, and are inferred from the load contours shown in Figure 1.3(a). The star symbol ★ denotes the region that experiences the highest loading, and we call this region the *key* region.

In order to avoid overdesigning the panel, an optimization approach must be able to tailor the design of specific regions to the anticipated loads. However, the number of plies and fiber orientations cannot vary too dramatically or else the panel would be impractical to manufacture. From the manufacturing view point, it is preferred to have a common stacking sequence over as many regions as possible, and then reduce the number of plies gradually between adjacent regions. The design engineer must avoid abrupt changes in fiber orientation within a ply to eliminate *seams* that are mechanically weak. A mechanism of allowing a ply to continue through adjacent regions until it is no longer required is called *blending*. The optimization approach presented here uses blending rules to achieve a blended panel that is not only manufactureable, but also tailored to allow load concentrations. The next paragraph explains how the blending rules are applied to a composite panel design.

Given the load conditions, the key region of the panel can be identified. Because the key region undergoes the largest load concentration, it is the region that requires the most number of plies. Starting from the key region, the number of plies in adjacent regions may be dropped (i.e. some plies can be removed from the stack) if the required stiffness and strengths of these regions are satisfied. Once a ply is dropped, it cannot be added back to the stacking sequence in later regions. Thus, as we move further away from the key region, the stacking sequence will, in general, have fewer and fewer plies. This rule of consistently dropping plies from the key region is called the "greater-than-or-equal-to" blending rule [13, 32]. Figure 1.4 demonstrates how the blending rules are applied to a 4×6 panel [13]. The key region (3,3) is marked by a star. The "$\geq$" symbol indicates the direction in which the ply can be dropped. For instance, if a ply exists in region (3,3), it may or may not exist in region (2,3). But if a ply exists in region (2,3), the same ply must exist in region (3,3). Figure 1.5 shows three different ply configurations. Configuration (a) complies with the blending rules, but configurations (b) and (c) do not. In configuration (b), only region (1,4) violates the

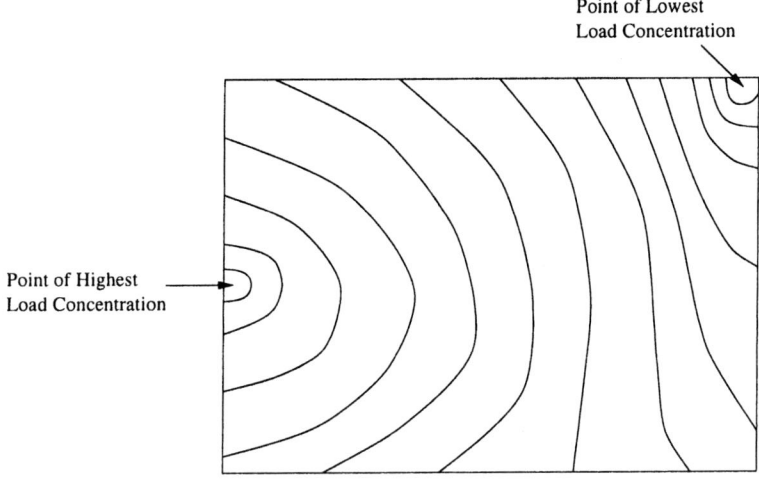

(a) Contours of constant load in a large composite panel

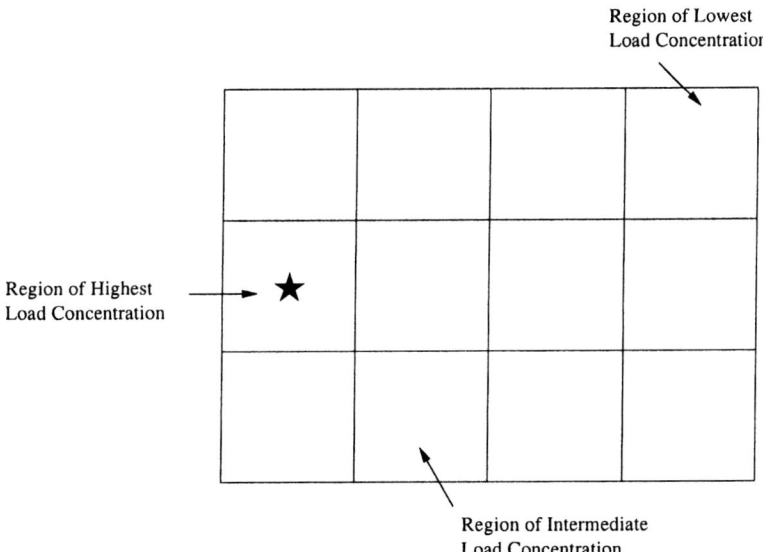

(b) Discrete representation of load contours in a composite panel

Figure 1.3.   Non-uniform loading of a large composite panel

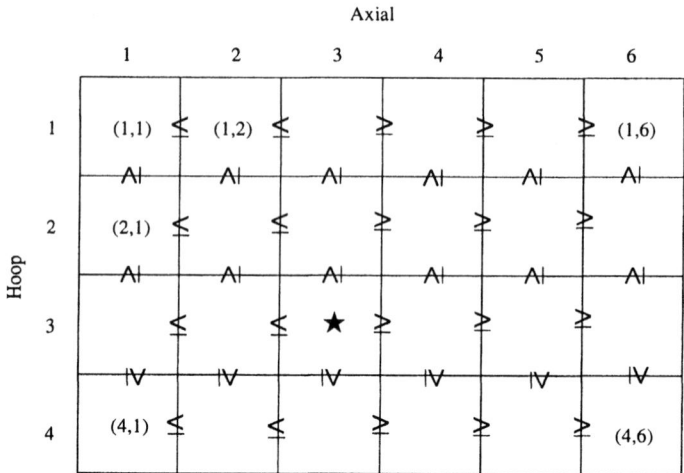

Figure 1.4.   The "greater-than-or-equal-to" blending rule applied to a 4×6 composite panel [13]

"greater-than-or-equal-to" blending rules, whereas regions (1,1), (1,2), and (2,1) in configuration (c) all violate the blending rules.

1.3   Optimization Issues

A design problem using an optimization scheme requires selections of design variables, the type (continuous or discrete) of these variables and their ranges when applicable, the proper objective function to minimize, and a set of constraints placed on the behavior of the structure [16]. For a given set of loads over a composite panel, the design variables must be able to represent multiple material systems, the number of plies, ply orientations, and the stacking sequence. With the number of design variables involved, one can easily imagine that solving even a simple panel design problem to the optimum soon exceeds a manageable level. A design procedure based on numerical optimization algorithms is imperative to fully exploit the promise of advanced composite materials.

There are a number of books and papers discussing basic concepts of structural design optimization of composite structures. The following is a short list of references found in the literature [9, 10, 15, 17, 19, 20, 23, 24]. A myriad of objective functions are explored in these references. Some objective functions minimize weight [9, 10, 19, 20, 23, 24], while others maximize strength or stiffness [15, 17]. Multi-criteria optimization formulation, i.e. many objectives must be optimized simultaneously,

The optimum design of laminated composite structures subjected to multiple loading conditions is truly a global problem in which the associated design space includes many local minima [25]. A series of studies [6, 7, 12, 13, 16, 28, 30, 32] have taken this global issue into account. In [6, 7, 16], the objective function was to maximize the performance of the composite structures, whereas the objective function in [12, 13, 28, 30, 32] was to minimize the weight of the composite structures. This chapter is based mainly on this series of studies, and we focus here on optimization problems whose objective is to minimize the weight of the structure in the later examples. Before we describe these examples, let us first discuss a general problem formulation of the composite structure design and issues relating to optimization algorithm selection in the next two sections.

2.   General Problem Formulation

A mathematical statement of the composite structure optimization problem can be written as follows. Find a set of design variables x that will:

$$
\begin{aligned}
&minimize &&f(x) \\
&subject\ to &&g_j(x) \geq g_{min} && j \in \{1, 2, \ldots, m\} \\
& &&x_i^l \leq x_i \leq x_i^u && i \in \{1, 2, \ldots, n\}
\end{aligned}
$$

where n is the number of variables and m is the number of constraint equations. The objective function $f(x)$ is a measure of a characteristic of interest (such as the total weight of the composite structure) pertaining to the problem. The design variables x_i whose respective values are to be determined during the design process are contained in a design vector x. The objective function may consist of any combination (linear or nonlinear) of the design variables. The inequality constraints $g_j(x) \geq g_{min}$ impose limits on the feasible search region of the design space. The upper and lower bounds $x_i^l \leq x_i \leq x_i^u$ are side constraints on the design variables.

As mentioned earlier, weight saving is one of the important concerns when designing composite structures. Hence, a typical objective function in composite structure design problems is to minimize weight while satisfying constraints on strain, stiffness, hygrothermal expansion (expansion due to moisture and temperature variations within the laminate), and thickness [3]. To simplify our discussion, we only consider the relationship between the weight and the strain, stiffness, and thickness constraints. The details of the effects of hygrothermal expansion to a composite laminate can be found in [9].

3.   Optimization Algorithm Consideration

Optimization algorithms play an important role in solving optimization problems. Since composite structure design is a global problem, only global optimization algorithms are relevant to this discussion. Arora [1, 2] suggested that a good optimization algorithm should have several attributes including:

- *Robustness.* The algorithm must not be problem dependent. It must be able to handle any optimization problem regardless of problem structure. The algorithm should be general and does not oppose any further restriction on the constraint functions, such as linear versus nonlinear, or equality versus inequality. Theoretically, it must be accurate in a sense that it guarantees the convergence to the solution even when it starts from a poor initial design estimate or initial seed.

- *Ease of Use.* Experts or nonexperts alike must be able to grasp the fundamental philosophy of the algorithm. Algorithms that are difficult to use in practice are usually those that are problem dependent.

- *Efficiency.* The algorithm must be efficient in an engineering sense. This means that it should (i) have a relatively fast convergence rate and require few iterations, and (ii) keep the analysis and calculation minimal so computation time is realistic.

There are always tradeoffs between these attributes. Typically, efficiency can be increased only by decreasing robustness and ease of use. A computational penalty, e.g. the CPU time, is often paid in order to achieve a robust and easy to use algorithm [21].

The global optimization algorithms available today are based on one of two approaches, namely *deterministic* and *stochastic* approaches [5]. An important tradeoff of the deterministic approach when handling global optimization problems is that it usually requires some knowledge about the structure of the objective and/or constraint functions. The stochastic approach is more robust and general than the deterministic approach, but it can more computationally intensive.

The stochastic approach was selected as the method to solve the example problems in the next section primarily because it is capable of handling a mixture of continuous and discrete variables with ill-structured, nonconvex functions. Stochastic approaches, such as simulated annealing, genetic algorithms, and evolutionary programming, are useful for optimizing engineering problems. The reader may consult [2, 5, 8, 9, 11, 29, 31, 33] for more details of various techniques on the stochastic approaches.

It is important to realize that a stochastic search approach does not guarantee the optimal solution in all the searches. The probability of obtaining the global optimum is unity only in the limit, but it is never possible to implement an infinite number of iterations in practice. In a lot of cases, however, we do not need the absolute optimal solution. A Nobel laureate economist, Herbert Simon, once mentioned that people do not need to optimize, but rather to "satisfice" by setting goals for various objectives [4]. This is also applicable for engineering design problems. The price we pay to obtain the absolute optimum is the time spent on searching through the whole feasible design space. Since global optimization problems are known to be NP-hard [26], the computation needed to solve a large problem with hundreds of design variables and constraints to optimality can be impractical. A sub-optimal solution that can be obtained quickly may be more valuable from a practical point of view. Each example presented in the following section found a sub-optimal but useful solution in four to five hours on a SUN SPARC 10 workstation.

4.   Example Problems

Two optimization examples of composite design problems are presented. The first example is a point design problem. The objective is to find a feasible design with the least weight in each of the regions in the panel. The feasibility implies that the panel must comply with all the constraints. There are three loads, namely load F_x in the x direction, load F_y in the y direction, and load F_{xy} in the xy or shear direction, in each of the regions. The magnitude of the loads and the dimensions of the panel are given in Figure 1.6.

The second example deals with the same overall composite panel but the blending rules are applied. In other words, the stacking sequence in the second example must abide by the blending rules. The objective of this second example is also to minimize the total weight of the whole panel. The two examples provide us a tradeoff comparison. The point design has a minimal weight but may be impractical to manufacture, while the blended design may be slightly heavier (with more plies) but the blending rules ensure a practical design. The two examples are formulated according to the doubly symmetric sandwich structure and

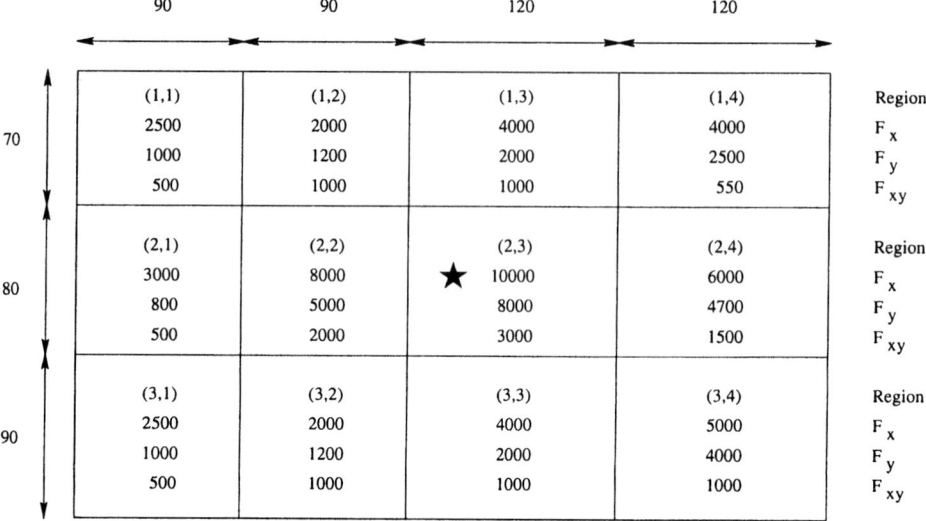

Figure 1.6. Loads in lbfs and dimensions in inches of the example composite panel

classical lamination theory [22]. A software package called COSTADE [14] was used to complete the computations described.

The mathematical formulation of the two examples can be summarized as follows.

$$minimize \quad f(\Theta, T) = weight$$
$$subject \ to$$

$$\text{for each ply in each region} \begin{cases} g_x(\theta_i, t_i) \geq \epsilon_x^{allow} \\ g_y(\theta_i, t_i) \geq \epsilon_y^{allow} \\ g_{xy}(\theta_i, t_i) \geq \epsilon_{xy}^{allow} \end{cases}$$

$$\text{for the core in each region} \begin{cases} g_x(t_k) \geq \epsilon_{x,core}^{allow} \\ g_y(t_k) \geq \epsilon_{y,core}^{allow} \\ g_{xy}(t_k) \geq \epsilon_{xy,core}^{allow} \end{cases}$$

$$\text{for each region} \begin{cases} G_x(\Theta, T) \geq S_x^{req} \\ G_y(\Theta, T) \geq S_y^{req} \\ G_{xy}(\Theta, T) \geq S_{xy}^{req} \end{cases}$$

$$and \quad -90 \leq \theta_i \leq 90 \qquad i \in \{1, 2, \ldots, n\}$$
$$\Theta = (\theta_1, \theta_2, \ldots, \theta_n)$$
$$t_i = \{1, 0\} \qquad i \in \{1, 2, \ldots, k-1, k+1, \ldots, n\}$$
$$t_k \geq 0$$
$$T = (t_1, t_2, \ldots, t_k, \ldots, t_n)$$

where θ_i denotes the angle of ply i, Θ is the vector of all ply angles, and n is the total number of plies. The binary variable t_i, for $i \in \{1, 2, \ldots, k-1, k+1, \ldots, n\}$, is the variable indicating whether ply i exists in that region. When there exists ply i in the region, t_i is one, and zero otherwise. The variable t_k is the thickness of the core which situates at ply k. The vector T contains all the indicator variables and the core thickness. There is a set of strain and stiffness constraints, g_x, g_y, g_{xy}, for each region (j, k) in the panel. It is only implicitly described in the formulation because the explicit description is too cumbersome. The strain constraints ϵ_x^{allow}, ϵ_y^{allow}, and ϵ_{xy}^{allow} refer to the allowable strain of each ply in x, y, and shear direction, respectively. The values of these ϵ's can be found in Table 1.A.1. Similar strain constraints are imposed for the core, and the value of these ϵ_{core}'s can be found in Table 1.A.2. Furthermore, each region must have the region's stiffness G of each direction equal to or greater than the corresponding required stiffness S^{req} of each direction, given in Table 1.A.3.

In some cases, only certain discrete values of fiber angles such as -90, -45, 0, 45, and 90 degrees are used to manufacture a composite panel. Incorporating this limit on the values of the fiber angles makes the design problem more realistic, but the new design is likely to be sub-optimal in terms of weight. The only change in the formulation to accommodate this fiber angle limitation is to allow θ_i to take on only the allowable values of the fiber angles. For instance, we can express the fiber angles previously mentioned as:

$$\theta_i = \{-90, -45, 0, 45, 90\} \quad \text{for} \quad i \in \{1, 2, \ldots, n\}.$$

The above expression can be modified to accommodate other sets of fiber angle values such as every 5 degrees or every 10 degrees. For the purpose of the discussion in this chapter, we allow the fiber angles to take on any integer values between -90 and 90 in the two examples. More details on incorporating discrete value and manufacturing tolerances into the composite design optimization problem can be found in [12, 18].

4.1   Point Design/Non-Blended Panel

In this point design problem, each region is experiencing different loading conditions and is optimized separately. The thickness of the core is held constant at 0.25 inch. This implies that the thickness of the entire panel varies as the number of required plies changes in each region. Table 1.1 shows the optimization results of the point or non-blended design.

Table 1.1. Layup result of the regions in the non-blended/point design panel. The total weight is 896.02 lbs

Region	Weight Core Thickness	Strain MS Stiffness MS	Number of Plies Ply Angles
1,1	46.04	1.583	16
	0.25	0.031	$[44, -56, 0, 12]_{\mathbf{s}}$
1,2	46.04	1.314	16
	0.25	0.021	$[23, 10, 64, -30]_{\mathbf{s}}$
1,3	61.39	0.437	16
	0.25	0.016	$[-56, 9, 6, 49]_{\mathbf{s}}$
1,4	96.91	1.179	20
	0.25	0.211	$[42, -82, 2, 29, -19]_{\mathbf{s}}$
2,1	52.62	3.144	16
	0.25	0.084	$[27, 5, 45, -26]_{\mathbf{s}}$
2,2	76.59	0.319	24
	0.25	0.331	$[14, -42, 72, 44, 60, 34]_{\mathbf{s}}$
2,3 ★	118.10	0.115	28
	0.25	0.407	$[59, 67, 38, 45, -48, 3, 4]_{\mathbf{s}}$
2,4	86.14	0.363	20
	0.25	0.018	$[72, 57, 41, -22, -7]_{\mathbf{s}}$
3,1	59.19	1.583	16
	0.25	0.031	$[44, -56, 0, 12]_{\mathbf{s}}$
3,2	59.19	1.314	16
	0.25	0.021	$[23, 10, 64, -30]_{\mathbf{s}}$
3,3	78.93	0.437	16
	0.25	0.016	$[-56, 9, 6, 49]_{\mathbf{s}}$
3,4	114.88	0.903	20
	0.25	0.002	$[86, -75, 20, 45, 30]_{\mathbf{s}}$

The abbreviation MS in the table stands for *margin of safety*. More details on the margin of safety can be found in the appendix. The subscript $\mathbf{s}$ after a set of ply angles denotes the double symmetry of the ply angles in the stack. For instance, the $[44, -56, 0, 12]_{\mathbf{s}}$ ply stack in region (1,1) refers to a laminate with ply angles of $44, -56, 0, 12, 12, 0, -56, 44$, core, $44, -56, 0, 12, 12, 0, -56, 44$ stacking sequence.

From the table, the ply angles of one region are hardly similar to those of the adjacent regions. This abrupt change in fiber angles makes the panel impractical to be manufactured as a whole, so the panel is unlikely to be produced. The total weight of the panel of this point design, however, can be used as a lower bound on the minimum achievable weight. This lower bound on the weight is an excellent benchmark when comparing the current design with any future designs of the panel.

We next consider the same panel under the same loading condition, only this time the blending rules are applied. The blending rules further constrain the feasible design space and limit the design solutions to those that are manufacturable.

4.2   Blended Panel

The thickness of the entire panel is held constant throughout in this example. Instead of designing each region separately, the blended panel relates the different regions so that they must be designed all together. The core thickness of the key region is fixed at 0.25 inch, but the core thickness of other regions is adjusted such that the thickness of all the regions in the panel is the same throughout, which results in a constant thickness panel. In order to do so, after a feasible stacking sequence of the key region is found, the thickness of the panel is determined based on this solution of the key region. The thickness of the key region is the sum of all the thickness of the required facesheet plies plus its 0.25 inch core thickness. The core thickness of the other regions is calculated in a similar fashion, by subtracting the total thickness of the required facesheet plies of that region from the total thickness of the key region.

The optimization results of the blended design is shown in Table 1.2. Notice that the stacking sequence of all the regions is a subset of the stacking sequence of the key region. This is a consequence of the blending rules.

The solution found by the stochastic algorithm in the second example is not intuitive, and is unlikely to be found with a trial and error method. Although the total weight of the blended design is slightly more, 61.94 lbs or 6.91% increase, than that of the point design, this blended design is preferred over the point design due to the reasons previously discussed. The solution of the blended panel can possibly be further improved by repeatedly running the software with different starting points and then select the least weight design among the results obtained, or allowing the software to execute more iterations in each run. We limited the implementation time for both examples to four to five hours for the sake of a fair comparison.

5.   Summary

This chapter demonstrates how a stochastic optimization algorithm can be utilized in designing composite structures. The design of composite structures has been formulated as a mixed continuous/integer global optimization problem that includes performance and manufacturing constraints. The optimization algorithm used in the chapter obtained solutions that would probably not be found through trial and error. Moreover, the stochastic optimization approach implemented is

Table 1.2. Layup result of the regions in the blended panel. The total weight is 957.96 lbs

Region	Weight Core Thickness	Strain MS Stiffness MS	Number of Plies Ply Angles
1,1	47.96 0.367	1.708 0.048	16 $[16, -45, -1, 51]_s$
1,2	47.69 0.367	0.872 0.048	16 $[16, -45, -1, 51]_s$
1,3	77.29 0.338	0.426 0.136	20 $[16, -45, -45, -1, 51]_s$
1,4	63.95 0.367	0.299 0.048	16 $[16, -45, -1, 51]_s$
2,1	66.25 0.338	2.064 0.124	20 $[16, -45, 42, -1, 51]_s$
2,2	77.68 0.308	0.228 0.239	24 $[69, 16, -45, 42, -1, 51]_s$
2,3 ★	134.08 0.250	0.126 0.415	32 $[69, 16, -45, -45, 42, -1, 47, 51]_s$
2,4	103.58 0.308	0.529 0.239	24 $[69, 16, -45, 42, -1, 51]_s$
3,1	61.66 0.367	1.585 0.048	16 $[16, -45, 42, -1]_s$
3,2	61.66 0.367	1.215 0.048	16 $[16, -45, 42, -1]_s$
3,3	116.52 0.308	1.025 0.289	24 $[69, 16, -45, -45, 42, -1]_s$
3,4	99.37 0.338	0.321 0.186	20 $[69, 16, -45, 42, -1]_s$

efficient in the sense that it found solutions, albeit sub-optimal ones, within a reasonable amount of time. The solutions found in the two examples offer the design engineer a systematic way to evaluate tradeoffs between weight and performance, particularly strain and stiffness, of the structure.

The blending rules force the design solution to be practical from a manufacturing point of view when a large panel experiences a range of loads. The formulation of the problems can be easily adjusted to incorporate future modifications in manufacturing specifications. The optimization process established here is useful and versatile to design large composite structures.

6.   Acknowledgement

The authors would like to thanks the Boeing company and their staff, particularly G.E. Mabson, for the generous support and collaboration.

The work of Zelda B. Zabinsky and Charoenchai Khompatraporn has been partially supported by NSF Grant No. DMI-0244286.

References

[1] Arora, J.S., "Computational Design Optimization: A Review and Future Directions," *Structural Safety 7*, 131-148, 1990.

[2] Arora, J.S. (ed.), *Guide to Structural Optimization*, ASCE Manuals and Reports on Engineering Practice, No. 90, American Society of Civil Engineers, New York, 1997.

[3] Baier, H.J., "Composite Laminate and Sandwich Optimization with Application," *Optimization of Large Structural Systems, Volume II*, edited by G.I.N. Rozvany, Kluwer Academic Publishers, the Netherlands, 997-1009, 1993.

[4] Belton, B. (ed.), "Math That Helps Nearly Everyone Make Decisions," *BusinessWeek Online*, at http://www.businessweek.com/bwdaily/dnflash/oct2000/nf20001013_580.htm, October 13, 2000.

[5] Boender, C.G.E., and Romeijn, H.E., "Stochastic Methods," *Handbook of Global Optimization*, edited by Horst, R. and Pardalos, P.M., Kluwer Academic Publishers, the Netherlands, 829-869, 1995.

[6] Graesser, D.L., Zabinsky, Z.B., Tuttle, M.E., and Kim, G.I., "Designing Laminated Composites Using Random Search Techniques," *Composite Structures 18*, 311-325, 1991.

[7] Graesser, D.L., Zabinsky, Z.B., Tuttle, M.E., and Kim, G.I., "Optimal Design of Composite Structures," *Composite Structures 24*, 273-281, 1993.

[8] Gürdal, Z., Haftka, R.T., "Optimization of Composite Laminates," *Optimization of Large Structural Systems, Volume II*, edited by G.I.N. Rozvany, Kluwer Academic Publishers, the Netherlands, 623-648, 1993.

[9] Gürdal, Z., Haftka, R.T., Hajela, P., *Design and Optimization of Laminated Composite Materials*, John Wiley & Sons, Inc., New York, 102-221, 1999.

[10] Haftka, R.T. and Gürdal, Z., *Elements of Structural Optimization*, Third revised and expanded edition, Kluwer Academic Publishers, the Netherlands, 1992.

[11] Khompatraporn, C., Pinter, J.D., Zabinsky, Z.B., "Comparative Assessment of Algorithms and Software for Global Optimization," *Journal of Global Optimization*, forthcoming.

[12] Kristinsdottir, B.P., Zabinsky, Z.B., Tuttle, M.E., Csendes, T., "Incorporating Manufacturing Tolerances in Near-Optimal Design of Composite Structures," *Eng. Opt. 26*, 1-23, 1996.

[13] Kristinsdottir, B.P., Zabinsky, Z.B., Tuttle, M.E., Neogi, S., "Optimal Design of Large Composite Panels with Varying Loads," *Composite Structures 51*, 93-102, 2001.

[14] Mabson, G.E. and Graesser, D.L., *Cost Optimization Software for Transport Aircraft Design Evaluation (COSTADE) - Users Manual*, NASA Contractor Report 4738, August 1996.

[15] Massard, T.N., "Computer Sizing of Composite Laminates for Strength," *J. Reinforced Plastics Comp. 3*, 300-345, 1984.

[16] Neogi, S., Zabinsky, Z.B., Tuttle, M.E., "Optimal Design of Composites Using Mixed Discrete and Continuous Variables," *Proceedings of the ASME Winter Annual Meeting, Symposium on Processing, Design and Performance of Composite Materials, MD-Vol.52*, 91-107, 1994.

[17] Park, W.J., "An Optimal Design of Simple Symmetric Laminates under the First Ply Failure Criteria," *J. Comp. Mater. 16*, 341-355, 1982.

[18] Romeijn, H.E., Zabinsky, Z.B., Graesser, D.L., and Neogi, S., "New Reflection Generator for Simulated Annealing in Mixed-Integer/Continuous Global Optimization," *Journal of Optimization Theory and Applications 101(2)*, 403-427, 1999.

[19] Schmit, L.A., "Structural Synthesis—Its Genesis and Development," *AIAA J. 19*, 1249-1263, 1981.

[20] Schmit, L.A., Farshi, B., "Optimum Laminate Design for Strength and Stiffness," *Int. J. Numer. Meth. Engng. 7*, 519-536, 1973.

[21] Thanedar, P.B., Arora, J.S., Li, G.Y., Lin, T.C., "Robustness, Generality and Efficiency of Optimization Algorithms for Practical Applications," *Structural Optimization 2*, 203-212, 1990.

[22] Tuttle, M.E. *Structural Analysis of Polymeric Composite Materials*, Marcel Dekker, Inc., New York, 2004.

[23] Vanderplaats, G.N., "Structural Optimization—Past, Present, and Future," *AIAA J. 20*, 992-1000, 1982.

[24] Vanderplaats, G.N., *Numerical Optimization Techniques for Engineering Design: With Applications*, McGraw Hill, New York, 250-282, 1984.

[25] Vanderplaats, G.N., Wesshaar, T.A., "Optimum Design of Composite Structures," *Int. J. Numer. Meth. Engng. 27*, 437-448, 1989.

[26] Vavasis, S.A., "Complexity Issues in Global Optimization: A Survey," *Handbook of Global Optimization*, edited by Horst, R. and Pardalos, P.M., Kluwer Academic Publishers, the Netherlands, 27-41, 1995.

[27] Vinson J.R., Chou, T-W. *Composite Materials and Their Use in Structures*, John Wiley & Sons, New York, 28-32, 1975.

[28] Zabinsky, Z.B., "Global Optimization For Composite Structural Design," *Proceedings of the 35th AIAA/ASME/ASCE/AHS/ASC Structure, Structural Dynamics, and Materials Conference*, April 1994.

[29] Zabinsky, Z.B., "Stochastic Methods for Practical Global Optimization," *Journal of Global Optimization 13*, 433-444, 1998.

[30] Zabinsky, Z.B., "Optimal Design of Composite Structures," *Encyclopedia of Optimization*, edited by Floudas, C.A. and Pardalos, P.M., Kluwer Academic Publishers, the Netherlands, Vol. 4, 153-160, 2001.

[31] Zabinsky, Z.B., *Stochastic Adaptive Search for Global Optimization*, Nonconvex Optimization and its Applications Series, Vol. 72, Kluwer Academic Publishers, the Netherlands, 2003.

[32] Zabinsky, Z.B., Seifer, J.D., Tuttle, M.E., Kristinsdottir, B.P., and Neogi, N., "Optimal Design of Composite Panels with Varying Loads," *Proceedings of the ICCE/2 conference*, New Orleans, 1995.

[33] Zabinsky, Z.B., Smith, R.L., McDonald, J.F., Romeijn, H.E., and Kaufman, D.E., "Improving Hit and Run for Global Optimization," *Journal of Global Optimization 3*, 171-192, 1993.

Appendix

Factor of Safety and Margin of Safety

A *factor of safety* (FS) is normally used in any engineering design. It is basically the numerical multiples of the design's requirement. For example, suppose a composite structure is required to withstand an in-plane unidirectional load of 1000 lbs, but the final designed structure can withstand an in-plane unidirectional load up to 2200 lbs. The corresponding factor of safety is $2200/1000 = 2.200$. The factor of safety can be written as:

$$FS = \frac{capability}{requirement}.$$

A second measure of safety that is more commonly used in the aerospace community is the *margin of safety* (MS). The MS measures the numerical marginal amount in multiples of the designed structure capability. Hence, from the previous example the margin of safety is $(2200\text{-}1000)/2200 = 0.545$. If a composite structure is designed with a margin of safety equal to zero, then that structure is designed to exactly meet the requirements. A composite structure design with negative margin of safety indicates that the design does not meet the requirement. The margin of safety can be expressed mathematically as:

$$MS = \frac{capability - requirement}{capability}.$$

Calculations within this chapter are based on margins of safety rather than factors of safety.

Material Properties

The following Tables 1.A.1 and 1.A.2 show the properties of the materials used to obtain the solutions in the example problems. Note that directions used for ϵ_1^{allow}, ϵ_2^{allow}, and ϵ_{12}^{allow} are such that direction 1 represents the direction that aligns with direction of the fibers, direction 2 represents the direction that is normal to the direction of the fibers, and direction 12 represents the shear direction of the ply.

Table 1.A.1. Material properties of the unidirectional graphite epoxy (facesheet material)

E_1 Msi	E_2 Msi	E_{12} Msi	ν_{12}	ρ lb/in^3	Ply thickness in	ϵ_1^{allow}	ϵ_2^{allow}	ϵ_{12}^{allow}
17.4	1.36	0.76	0.32	0.057	0.0073	0.01	0.005	0.007

Table 1.A.2. Material properties of the fiber honeycomb core

E_1 Msi	E_2 Msi	E_{12} Msi	ν_{12}	ρ lb/in^3	Ply thickness in	$\epsilon_{1,core}^{allow}$	$\epsilon_{2,core}^{allow}$	$\epsilon_{12,core}^{allow}$
1E−5	1E−5	1E−7	0.32	0.3	0.25 or more	0.1	0.1	0.1

Minimum Effective Panel Stiffness

The notation S^{req} in the following Table 1.A.3 stands for the minimum stiffness requirement. The subscript x, y, or xy of each S^{req} indicates the associating direction.

Table 1.A.3. Minimum required effective panel stiffness

S_x^{req} lbf/in	S_y^{req} lbf/in	S_{xy}^{req} lbf/in
1.00E6	0.08E6	0.27E6

Chapter 22

Neural Network Enhanced Optimal Self-tuning Controller Design for Induction Motors

Q.M.Zhu[§1], L.Z.Guo[§2], Z.Ma[§3]

§1. Faculty of Computing,Engineering and Mathematical Sciences, University of the West of England, Frenchay Campus, Coldharbour Lane, Bristol, BS16 1QY, UK

§2. Department of Automatica Control and Systems Engineering, Univerisity of Sheffield, Mappin Street, Sheffield S1 3JD, UK

§3. Automation Research Center, Darlian Maritime University, 1 Linghai Road, Darlian, Liaoning, 116026 P.R.China

Key words: Neural networks, Optimal self-tuning controller, Induction motors

Abstract: This is a case study of application of Global Optimisation (GO), in which a discrete optimal self-tuning controller is designed to regulate the speed of rotor and the amplitude of the rotor flux in an induction motor drive system. Firstly the non-linear dynamics of the induction motor is approximated by a linear model, around its operation point, through a recursive least-squares algorithm. Then the errors between the outputs of the identified linear model and actual rotor are used to train a back-propagation neural network for determining the future control output. With this type of structure, the two control goals of regulating rotor speed and rotor flux amplitude are de-coupled in a nature so that power efficiency can be optimised without affecting speed regulation. A simulated case study is presented to demonstrate the effectiveness of the proposed approach.

1.   INTRODUCTION

Over the last three decades, considerable efforts have been devoted to both the theoretical justification and the development of algorithms for global optimisation. (Dixon and Szego 1975). Whenever a problem can be formulated as the determination of a set of parameters such that the values of some objective function have maxima or minima, then an optimisation algorithm might be employed to achieve the solution. Such problems commonly arise in the general field of engineering design.

Optimisation problems also frequently occur in control engineering. In general, these optimisation problems are concerned either with system identification, to find an optimal model to approximate the behaviour of the system, or determining the optimal static and/or dynamical performance in operation. As more advanced control techniques developed, optimisation methodology has played a significant role in the controller design procedure. As a type of optimisation method, self-tuning control (Astrom and Wittenmark 1995) was proposed to improve the robustness of dynamic systems with regard to adapting the change of environment and to the variation of the parameters within system operation. The general self-tuning controller design has been successfully applied to the control of linear dynamic systems. However, the expansion to non-linear systems has not reached maturity, although there have been a large number of studies published.

Mathematically   global   optimisation   is   concerned   with   the characterisation and computation of global minima or maxima of a non-linear function. The general global minimisation problem can be defined as

Given: $K \subseteq R^n$ compact set, $f: K \rightarrow R$ continuous function. Find: $x^* \in K, f^* = f(x^*)$ such that $f^* \leq f(x)$ for all $x \in K$.

There exist a huge number of algorithms in various books and papers concerned with the solution of the global optimisation problems. Interested readers may refer to relevant materials.

As mentioned above, the (global) optimisation problem in control engineering mainly can be divided into two areas: system identification and control performance optimisation, both of which will be involved in this case study.

Consider a discrete-time dynamic system with following general form

$$y(t+1) = f(Y, U) \tag{1.1}$$

where f is a smooth function over $R^l \times R^{m+l}$, $Y = (y(t), y(t-1), ..., y(t-l+1))^T$ and $U = (u(t), u(t-1), ..., u(t-m))^T$, and $y(t)$ and $u(t)$ are the plant output and control input signals respectively, t is the time index.

In general, system identification determines the parameters of a system from a set of measured input and output data. It is closely related to the function estimation. Now the famous, generalised least squares algorithm for linear systems has been widely used as a typical approach in control engineering and other industrial modelling fields.

For optimal control, consider the following generic performance index

$$J(u(t)) = \sum_{t=1}^{N} g(y(t), u(t)) \tag{1.2}$$

where g is a given real-valued function which is often of some norm form. The optimal control problem of the system (1.1), with the performance index (1.2), is to determine the optimal control input $u(t)$ within the class of allowable control input such that the given performance index reaches the minima or maxima values. Depending on the nature of the performance index, the nature of the constraints, and the desired output etc, the optimal control problems are different. For linear time invariant systems with a quadratic performance index, the optimal control theory has been well developed (Ogata 1995) whist the optimal control problem for non-linear dynamic systems remains has not been completely resolved.

There has been a renaissance of the neural network methodology (Rojas 1996) during 1980s, mainly due to the development of back-propagation optimisation techniques. Neural networks have been widely employed in both linear and non-linear control system design (Hunt, et al 1992, Narandra 1996), including self-tuning controller design for non-linear systems. In general, there have been three major streams in neuro-controller design. The first one is that of using neural networks as direct controllers (Narendra and Parthasarathy 1990, Chen 1990). The second is that of predictive control using neural networks as predictors (Liu, Kadirkamanathan and Billings 1998), and the third is that of using neural networks as plant models (Chen and Khalil 1995, Zhu, Ma and Warwick 1999). However, none of the above streams of neural network-based methods has reached a satisfactory stage of development because of the lack of a comprehensive understanding necessary to design neuro-control systems.

The development of non-linear control techniques and the rapid improvements in power devices and microelectronics have made possible induction motor drives for high performance applications. Adaptive control, particularly feedback linearisation and input-output de-coupling techniques have been widely used in this field (Ben-Brahim, Tadakuma and Akdag 1999, Liaw, Chao, and Lin 1992, Marino, Peresada and Valigi 1993). However those methods assumed either the full state measurements are available, or the flux observer can be constructed or the unknown parameters such as load torque are constant. With the neural network enhanced optimal self-tuning control approach described here, it is shown that those assumptions can be easily removed without significant effect on the performance of the controlled system. The preliminary theoretical research results were reported by Zhu, Hayns and Garvey (1999), and Zhu, Ma and Warwick (1999).

In this case study, a novel neural network-based self-tuning control method is used, combining the attributes of neural network learning with a generalised minimum variance self-tuning control strategy. This is applied to control the rotor speed and the rotor flux amplitude of an induction motor.

2.   A NEURO OPTIMAL MIMO SELF-TUNING CONTROLLER

Zhu, Ma, and Warwick (1999) presented a general framework for a neural network enhanced non-linear Single Input and Single Output (SISO) control system, shown in Figure 1.

In this study, the framework for a SISO control system is expanded to a Multi-Input and Multi Output (MIMO) system to control an induction motor. The control system design consists of following three steps:

1. Optimal (minimum mean variance) controller design

2. Linear sub-model identification design

3. Neural network training algorithm design.

Assume a class of MIMO dynamic plants can be described as follows

$$\dot{Y} = f(Y, U) \tag{2.1}$$

where Y and U are the output and the control input vectors of the plant respectively, and f is either a linear or a non-linear vector function over $Y \times U$ representing the dynamic of the plant. Generally this input-output relation can be readily approximated with a (discrete or continuous time) linear model. Here a discrete model is used for the discrete self-tuning controller design, which is described as follows

$$A(z^{-1})y(t+1) = B(z^{-1})u(t) + F_0(.,.) \tag{2.2}$$

where

$$A(z^{-1}) = diag(1 + a_1^i z^{-1} + \cdots + a_l^i z^{-l})$$
$$B(z^{-1}) = diag(b_0 + b_1^i z^{-1} + \cdots + b_m^i z^{-m}) \tag{2.3}$$

and $F_0(.,.)$ is, defined as an error agent, a corrective item representing the error of the linear model. It is actually a function of the input and output variables of the plant.

There are at least two ways to obtain this type of models. In trivial case, $A(z^{-1}) = I$, $B(z^{-1}) = I$ and $F_0 = f$, in which I is an identity matrix. In another way, f can be expanded according to Taylor series, whose linear part as linear model and those derived from second and higher order derivative can be considered as F_0.

Now assume the plant model to be stable and de-coupled.

2.1   Controller design

An obvious advantage of GO concept and algorithms for linear time-varying system design is to avoid the complicated numerical computation for global search, whose computation time is very vital to practical control system implementation on line. To derive, in terms of generalised minimum variance, an optimal controller, the closed loop system performance index can be described as follow

$$J = \left\| S(z^{-1})y(t+1) - R(z^{-1})v(t) - Q(z^{-1})u(t) - H(z^{-1})F_0(.,.) \right\| \tag{2.4}$$

where $\mathbf{v}(t)$ is a bounded reference input vector. $\|\bullet\|$ is an any kind of appropriate norms defined in the suitable space. $S(z^{-1})$, $Q(z^{-1})$, $R(z^{-1})$ and $H(z^{-1})$ are all diagonal weighting polynomial matrices of z^{-1}, which are defined as

$$
\begin{aligned}
S(z^{-1}) &= diag(1 + s_1^i z^{-1} + \cdots + s_{ns}^i z^{-ns}) \\
Q(z^{-1}) &= diag(q_0 + q_1^i z^{-1} + \cdots + q_{nq}^i z^{-nq})
\end{aligned}
\tag{2.5}
$$

Now define an auxiliary output vector

$$
\phi(t+1) = S(z^{-1})\mathbf{y}(t+1)
\tag{2.6}
$$

According to the standard generalised minimum variance controller design procedure, the optimal predictor $\phi^*(t+1/t)$ for the auxiliary output vector $\phi(t+1)$ can be obtained by

$$
\phi^*(t+1/t) = G(z^{-1})\mathbf{y}(t) + C(z^{-1})B(z^{-1})\mathbf{u}(t) + C(z^{-1})\mathbf{F}_0(.,.)
\tag{2.7}
$$

in which $C(z^{-1})$ and $G(z^{-1})$ are diagonal polynomial matrices and defined as

$$
\begin{aligned}
C(z^{-1}) &= diag(1 + c_1^i z^{-1} + \cdots + c_{nc}^i z^{-nc}) \\
G(z^{-1}) &= diag(g_0 + g_1^i z^{-1} + \cdots + g_{ng}^i z^{-ng})
\end{aligned}
\tag{2.8}
$$

and which satisfy

$$
S(z^{-1}) = C(z^{-1})A(z^{-1}) + z^{-1}G(z^{-1})
\tag{2.9}
$$

The solution to minimise the performance of equation (2.4) is then found to be

$$
\phi^*(t+1/t) = R(z^{-1})\mathbf{v}(t) + Q(z^{-1})\mathbf{u}(t) + H(z^{-1})\mathbf{F}_0(.,.)
\tag{2.10}
$$

According to equation (2.7) and equation (2.10), the controller output $\mathbf{u}(t)$ which satisfies the minimisation requirement can be obtained from

$$
\mathbf{u}(t) = [-Q(z^{-1}) + C(z^{-1})B(z^{-1})]^{-1}(R(z^{-1})\mathbf{v}(t) + H(z^{-1})\mathbf{F}_0(.,.) - G(z^{-1})\mathbf{y}(t) - C(z^{-1})\mathbf{F}_0(.,.))
\tag{2.11}
$$

By combining equation (2.11) and equation (2.2), the closed-loop system equation can be expressed as follow

$$B(z^{-1})S(z^{-1}) - A(z^{-1})Q(z^{-1})y(t+1) = B(z^{-1})(R(z^{-1})v(t) +$$
$$H(z^{-1})\mathbf{F}_0(.,.) - C(z^{-1})\mathbf{F}_0(.,.)) + \mathbf{F}(.,.)$$

$$(2.12)$$

where $\mathbf{F}(.,.) = (-Q(z^{-1}) + C(z^{-1})B(z^{-1}))\mathbf{F}_0(z^{-1})$.

To produce satisfactory dynamics, a stable polynomial matrix $T(z^{-1})$ is chosen to specify the desirable closed-loop characteristic equation as follow:

$$T(z^{-1}) = diag(t_0^i + t_1^i z^{-1} + \cdots + t_{nt}^i z^{-n_t})$$

$$(2.13)$$

The following relationship therefore needs to be accommodated in order to achieve the required control action

$$B(z^{-1})S(z^{-1}) - A(z^{-1})Q(z^{-1}) = T(z^{-1})$$

$$(2.14)$$

So $S(z^{-1})$ and $Q(z^{-1})$ can be determined from equation (2.14) in which, at the time of calculation, $A(z^{-1})$, $B(z^{-1})$ and $T(z^{-1})$ are all known.

In order to cancel the static offset, $R(z^{-1})$ can be selected as

$$R(z^{-1}) = diag\left[\frac{T^1(1)}{B^1(1)}\right]$$

$$(2.15)$$

And to eliminate the effect, in the steady state, of the non-linear part, $H(z^{-1})$ can be chosen as

$$H(z^{-1}) = diag\left[\frac{Q^i(1)}{B^i(1)}\right]$$

$$(2.16)$$

Thus, using equations (2.14), (2.15), (2.16) and (2.9), all parameters in the polynomial matrices can be calculated effectively.

2.2   Linear submodel identification design

For the linear sub-model parameter estimation, a standard recursive least squares (RLS) algorithm can be applied directly (Astrom and Wittenmark 1995). Note that least squares algorithm is a GO algorithm for linear systems. Also note that the MIMO RLS algorithm can be used to generate the estimation for all parameters simultaneously, or the SISO RLS algorithm can be used to give the parameters separately for each SISO sub-model, because the coupling effect among the output has been removed. Here the MIMO RLS algorithm is presented to identify the parameters for a MIMO plant as follows

$$\hat{\mathbf{y}}(t+1) = \Phi(t)^T \hat{\theta}(t) + \hat{\mathbf{F}}_0(t+1) \tag{2.17}$$

where is the estimation of the output vector of the plant at the time $t+1$, $\Phi(t)$ is the regressor and θ is the parameter vector. These are defined as follows

$$\Phi(t) = diag\big(\phi_1(t), \phi_2(t), \cdots, \phi_{n_y}(t)\big)$$
$$\hat{\theta}(t) = \big(\hat{\theta}_1(t)^T, \hat{\theta}_2(t)^T, \cdots, \hat{\theta}_{n_y}(t)^T\big)^T$$
$$\phi_i(t) = (y^i(t), \cdots, y^i(t+1-l); u^i(t), \cdots, u^i(t-m))^T$$
$$\hat{\theta}_i(t) = (\hat{a}_1^i, \cdots, \hat{a}_l^i; \hat{b}_0^i, \cdots, \hat{b}_m^i)^T, \ i = 1, \cdots, n_y \tag{2.18}$$

The update procedure is then

$$\hat{\theta}(t) = \hat{\theta}(t-1) + P(t)\Phi(t)\varepsilon(t)$$
$$\varepsilon(t) = \mathbf{y}(t) - \Phi(t)^T \hat{\theta}(t-1)$$
$$P(t) = \frac{1}{\kappa} \times [P(t-1) - P(t-1)\Phi(t)(\kappa\mathbf{I} + \Phi(t)^T P(t-1)\Phi(t))^{-1}\Phi(t)^T P(t-1)] \tag{2.19}$$

where $0 < \kappa < 1$ is a forgetting factor to affect the convergence of the parameter estimates and to cope with slowing time-varying plants.

2.3   Neural network training algorithm design

A three-layer BP neural network with hyperbolic tangent activation functions in hidden layer is employed to detect the error agent $\mathbf{F}_0(.,.)$. Note a neural network is inherently suitable to deal with a MIMO function relationship so that it is natural to accommodate it within a MIMO plant. Currently there exist plenty of GO algorithms for improving the efficiency of neural network training (Hsin, Li, Sun, and Sclabassi 1995 and Moler, 1993). With regard to the computational efficiency in real time control, a standard BP network with momentum is selected in this study whose structure is shown in Figure 1. The objective of the network training is to adjust all variable weights to minimise the error $E(t)$ from real plant output and linear sub-model, where $E(t)$ is defined as

$$E(t) = \frac{1}{2}(\mathbf{F}_0(.,.)^T \, \mathbf{F}_0(.,.)) \tag{2.20}$$

Actually, during each sampling period the parameters of the linear-sub-model are estimated by the RLS algorithm and then the non-linear vector function $\mathbf{F}_0(.,.)$ can therefore be numerically detected by

$$\mathbf{F}_0(.,.) = \mathbf{y}(t+1) - \Phi(t)^T \, \hat{\theta}(t) \tag{2.21}$$

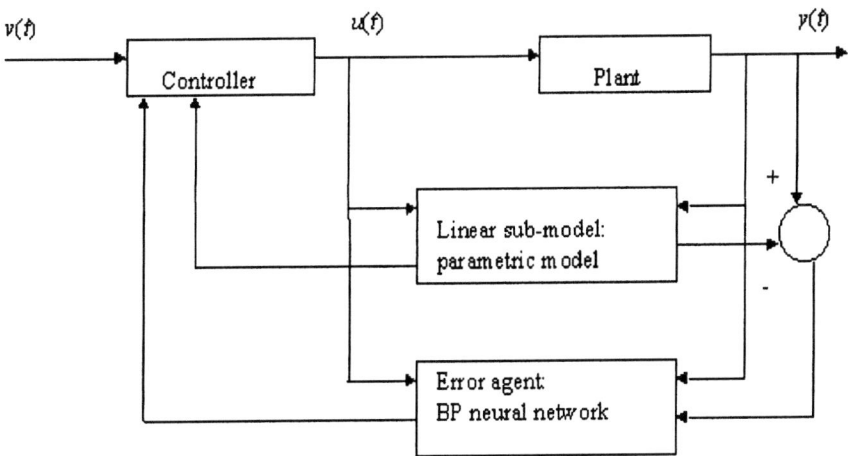

Figure 1. Configuration of the proposed neuro optimal self-tuning control for field-oriented induction motor drive

The error training signal of the neural network may be provided from equation (2.21) whist its input signal vector can be chosen from the delayed controller output (i.e. plant input) and plant output, for example. Thus

$$I(t+1) = (\mathbf{y}(t)^T, ..., \mathbf{y}(t-l+1)^T; \mathbf{u}(t)^T, ..., \mathbf{u}(t-m)^T)^T \tag{2.22}$$

In general, these input variables may be chosen as historical data from the input and output of the controlled plant. Furthermore, some measurable variables related to disturbances can also be used as input element to the network to give some insight into forthcoming disturbances at the current sampling instant. In practice, the input variables of the network depend on the underlying physical background of the problem and must encapsulate the real characteristics of the process.

The training procedure for a neural network can be described by three phases: forward propagation phase, back propagation phase and weights updating phase.

In the forward propagation phase the output of the hidden layer is first calculated as

$$O_j(t) = \frac{1}{1+\exp(-Net_j)}; \ j = 1,..., M$$

$$Net = W^{(1)} \times I(t);$$

(2.23)

The output of the overall network is, in case of linear activation function, then

$$\hat{\mathbf{F}}_0(.,.) = W^{(2)} \times O$$

(2.24)

where matrices $W^{(1)}$ and $W^{(2)}$ are weights between the input and hidden layers, and, the hidden layer and output layer.

In the back propagation phase, so-called "delta-vectors" are computed as follows

$$\Delta^{(2)}(t) = E(t)$$

$$\Delta^{(1)}(t) = O(t)(I - O(t)). \times W^{(2)^T} \Delta^{(2)}(t)$$

(2.25)

Then the weights updating equations can be obtained thus:

$$W^{(1)}(t+1) = W^{(1)}(t) + \alpha[W^{(1)}(t) - W^{(1)}(t-1)] + \eta \Delta^{(1)}(t)^T I(t)$$

$$W^{(2)}(t+1) = W^{(2)}(t) + \alpha[W^{(2)}(t) - W^{(2)}(t-1)] + \eta \Delta^{(2)}(t)^T O(t)$$

(2.26)

where $0 \le \alpha < 1$ and $0 < \eta < 1$ are the momentum constant and learning rate in the computation of the weights respectively.

In summary, the computation for the whole control system at each time instant t can be outlined as follows:

(1) Sample the plant output $\mathbf{y}(t)$ and form the regression vector $\Phi(t-1)$ by means of the plant input $\mathbf{u}(t-1)$ and output $\mathbf{y}(t-1)$ historical data sequences, and then form the neural network input vector $I(t)$ by means of $\mathbf{u}(t-1)$ and $\mathbf{y}(t)$ data sequences,

(2) Predict the next error $\mathbf{F}_0(.,.)$ using the trained neural network,

(3) Estimate the parameters of $A(z^{-1})$ and $B(z^{-1})$ from equations (2.18-2.19),

(4) Calculate the controller parameters from equations (2.14), (2.9),

(5) Generate the controller output $\mathbf{u}(t)$ from equation (2.11),

(6) Produce the next step neural network training input signal vector $I(t+1)$,

(7) Train the neural network for a pre-selected number of epochs using equations (2.22-3.26),

(8) Wait for the sampling clock pulse, then go to step 1.

3. INDUCTION MOTOR MODELS

The voltage equations of induction motors in a synchronously rotating frame can be expressed according to the general theory of electric machines (Krause 1986) as

$$\begin{bmatrix} v_{qs} \\ v_{ds} \\ 0 \\ 0 \end{bmatrix} = \begin{bmatrix} R_s + sL_s & \omega_e L_s & sL_m & \omega_e L_m \\ -\omega_e L_s & R_s + sL_s & -\omega_e L_m & sL_m \\ sL_m & \omega_{sl} L_m & R_r + sL_r & \omega_{sl} L_r \\ -\omega_{sl} L_m & sL_m & -\omega_{sl} L_r & R_r + sL_r \end{bmatrix} \begin{bmatrix} i_{qs} \\ i_{ds} \\ i_{qr} \\ i_{dr} \end{bmatrix} \quad (3.1)$$

where v_{qs} is the quadrature-axis (q-axis) stator voltage, v_{ds} is the direct-axis (d-axis) stator voltage, i_{qs} is the q-axis stator current, i_{ds} is the d-axis stator current, i_{qr} is the q-axis rotor current, i_{dr} is the d-axis rotor current, R_s is the stator resistance, R_r is the rotor resistance, L_s is the stator inductance, L_r is the rotor inductance, L_m is the magnetising inductance, P is the number of poles, ω_e is the electrical angular speed and ω_s is the slip angular speed.

The mechanical equations of the induction motor can be of the following form

$$T_e = \frac{3P}{4} L_m (i_{qs} i_{dr} - i_{ds} i_{qr}) = J \frac{d\omega_r}{dt} + B\omega_r + T_L \qquad (3.2)$$

where T_e is the generated torque, ω_r is the rotor angular speed, T_L is the load torque, J is the inertia constant and B is the damping ratio.

Equations (3.1) and (3.2) can be rearranged to yield the following state equations

$$
\begin{bmatrix} \dot{i}_{ds} \\ \dot{i}_{qs} \\ \lambda_{dr} \\ \lambda_{qr} \end{bmatrix} =
\begin{bmatrix}
-\frac{R_s}{\sigma L_s} - \frac{R_r(1-\sigma)}{\sigma L_r} & \omega_e & \frac{L_m R_r}{\sigma L_s L_r^2} & \frac{P\omega_r L_m}{2\sigma L_s L_r} \\
-\omega_e & -\frac{R_s}{\sigma L_s} - \frac{R_r(1-\sigma)}{\sigma L_r} & -\frac{P\omega_r L_m}{2\sigma L_s L_r} & \frac{L_m R_r}{\sigma L_s L_r^2} \\
\frac{L_m R_r}{L_r} & 0 & -\frac{R_r}{L_r} & \omega_e - \frac{P}{2}\omega_r \\
0 & \frac{L_m R_r}{L_r} & -(\omega_e - \frac{P}{2}\omega_r) & -\frac{R_r}{L_r}
\end{bmatrix}
\begin{bmatrix} i_{ds} \\ i_{qs} \\ \lambda_{dr} \\ \lambda_{qr} \end{bmatrix}
+ \begin{bmatrix} v_{ds} \\ v_{qs} \\ 0 \\ 0 \end{bmatrix} \frac{1}{\sigma L_s}
$$

$$(3.3)$$

and

$$\dot{\omega}_r = -\frac{B}{J}\omega_r + \frac{1}{J}(T_e - T_L) = -\frac{B}{J}\omega_r + \frac{3P}{4}\frac{L_m}{JL_r}(i_{qs}\lambda_{dr} - i_{ds}\lambda_{qr}) - \frac{1}{J}T_L \quad (3.4)$$

where $\sigma = 1 - L_m^2 /(L_s L_r)$, $\lambda_{qr} = L_m i_{qs} + L_r i_{qr}$ and is the q-axis rotor flux, and $\lambda_{dr} = L_m i_{ds} + L_r i_{dr}$ and is the d-axis rotor flux.

Let $\lambda = (\lambda_{qr}^2 + \lambda_{dr}^2)^{1/2}$. Here λ is the rotor flux amplitude. Define λ and ω_r to be the plant outputs and v_{qs}, v_{ds} to be control inputs. This is a typical MIMO non-linear dynamic system.

4.   DESIGN AND SIMULATION OF A FIELD-ORIENTED INDUCTION MOTOR CONTROL SYSTEM

In this study, the neuro optimal self-tuning controller is used to implement the field-oriented control of induction motors. The actual physical dynamic of an induction motor is significantly non-linear and also the physical parameters involved, such as inductance, resistance and load torque etc., are most often not precisely known or are time-varying.

Model structure selection is the first step in designing the control system. A de-coupled MIMO linear model is used to approximate the dynamics of the induction motors and a BP neural network is used to accommodate the errors induced by the MIMO linear model. Therefore the completed model is selected as

$$(1 + a_1^1 z^{-1})\lambda(t+1) = (b_0^1 + b_1^1 z^{-1})v_{ds}(t) + f_\lambda$$
$$(1 + a_1^2 z^{-1})\omega_r(t+1) = (b_0^2 + b_1^2 z^{-1})v_{qs}(t) + f_\omega$$

(4.1)

The forgetting factor for recursively estimating the associated parameters in this linear model was selected to be 0.98. The error agents f_λ and f_ω are obtained by the BP neural network.

The neural network for predicting the error was set-up with $n_i=15$, $n_h=1$ and $n=2$ for the input layer, hidden layer and output layer respectively. The momentum constant is 0.2 and learning rate is 0.05. The network was trained twice in each sampling period. The input signal vector to the BP neural network was chosen as

$$I = [-1, \lambda(t), \lambda(t-1), \lambda_l(t), \lambda_l(t-1), v_{ds}(t), v_{ds}(t-1), \lambda(t) - \lambda_l(t),$$
$$\omega_r(t), \omega_r(t-1), \omega_{rl}(t), \omega_{rl}(t-1), v_{qs}(t), v_{qs}(t-1), \omega_r(t) - \omega_{rl}(t)]$$

(4.2)

where λ_l and ω_{rl} are the linear model outputs.

The second step is to design the controller, following the formulations presented in section 2.1. The desired closed-loop dynamic polynomial matrix $T(z^{-1})$ is assigned as

$$T(z^{-1}) = diag(1 - 0.5z^{-1}, 1 - 0.5z^{-1})$$

(4.3)

As a case study, an induction motor of three phase, delta-connected, four-pole, 1 Hp, 60 Hz, 220 V is chosen with following parameters

$R_s = 3.20\ \Omega$; $R_r = 2.349\ \Omega$; $L_s = 0.1294$ H;
$L_r = 0.1329$ H; $L_m = 0.1267$ H; $J = 0.009$ kg m^2.

A MATLAB script has been developed to test the designed control system. The simulation results are shown in Figures 2 to 5.

Figure 2 shows the simulated responses of the rotor flux amplitude and rotor angular speed to indicate that the flux amplitude is actually affected shortly after the rotor angular speed is changed. Figure 3 shows the estimated parameters of the linear model using a recursive least squares algorithm, which finally converge to the constants. The simulation results show that the entire closed loop system is stable with good transient and static performance. Figure 4 shows the neural network output versus the actual error between linear model and real outputs. Figure 5 shows the control inputs v_{ds} and v_{qs} from which it can be seen that the rotor flux amplitude and rotor speed are de-coupled by the proposed control strategy.

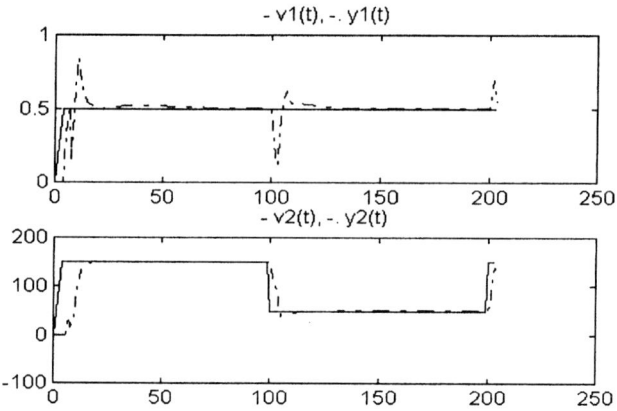

Figure 2. The simulated output response of the induction motor: $v1$, $v2$ are reference inputs and $y1$, $y2$ are rotor flux amplitude and speed

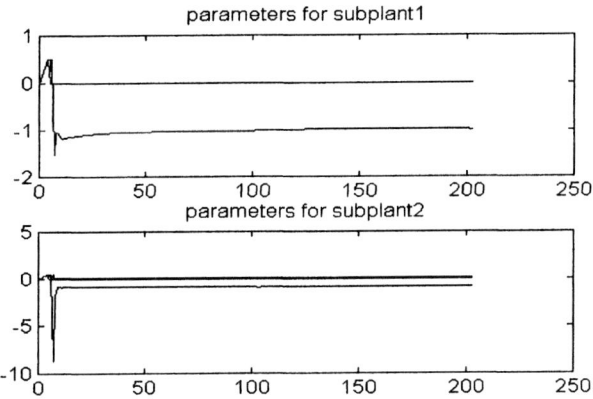

Figure 3. Estimated parameters of linear models by RLS

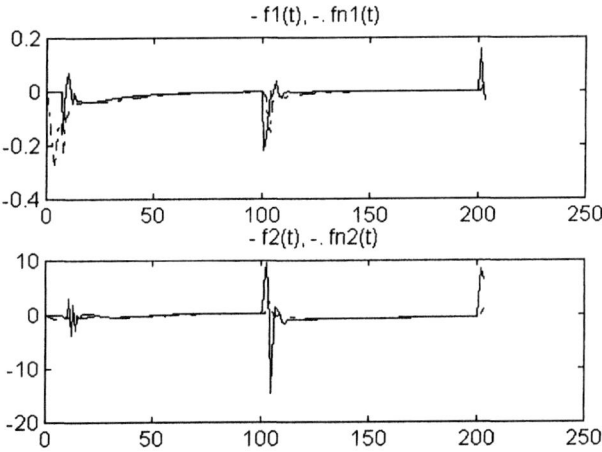

Figure 4. Neural network output and actual error

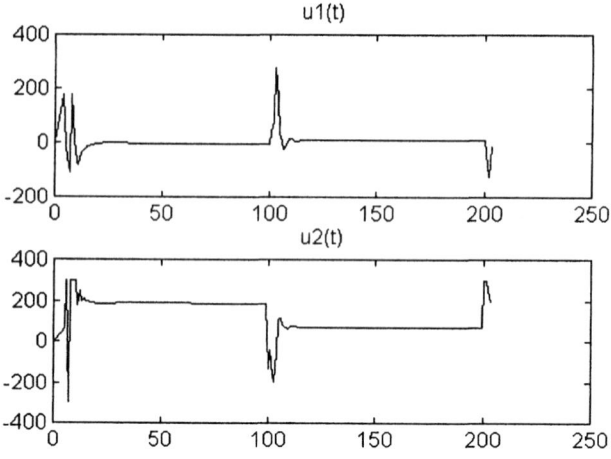

Figure 5. Control input: $u1$, $u2$ are Vds and Vqs

5.   CONCLUSIONS

GO has been a very important concept and the relevant algorithms have been well developed. As a case study, this paper takes advantage of GO concept and computation algorithms to develop an integrated procedure for control system design of induction motor. The methodology shows that (1) exact electrical and mechanical models of the induction motor are not necessary, (2) advanced knowledge of parameters such as resistance, inductance and load torque is not necessary, (3) the correlation among inputs and outputs can be easily de-coupled, (4) the system states are not necessarily required as feedback signals, and (5) the use of linear time-varying model can effectively relieve the complexity of GO search computations, this is a vital consideration in practical real time systems. The simulation results show good performances in both the rotor speed following and rotor flux amplitude regulation have been achieved.

ACKNOWLEDGEMENTS

The first two authors are grateful for the support from the grant (GR/M84305) of EPSRC, UK. We are also appreciated the constructive comments and suggestions from the anonymous reviewers and the editor.

Astrom, K.J. and Wittenmark, B, Adaptive control (second edition), Addison-Wesley Publishing Company, Inc., 1995.

Ben-Brahim, L., Tadakuma, S. and Akdag, A., Speed control of induction motor without rotational transducers, *IEEE Trans. Indust. Appl.* Vol. 35, No. 4, pp. 844-849, 1999.

Chen, F.C., Back-propagation neural networks for non-linear self tuning adaptive control, IEEE Cont. Syst. Mag. (Special Issue on Neural Networks for Control System), Vol. 34, No. 4, pp. 57-80, 1990.

Chen, F.C., and Khalil, H.K., Adaptive control of non-linear systems using neural networks, Int. J. Cont., Vol. 55, pp. 1299-1317, 1992.

Chen, F.C., and Khalil, H.K., Adaptive control of a class of non-linear discrete-time systems using neural networks, IEEE Trans.-AC, Vol. 40, No. 5, pp. 791-801, 1995.

Dixon, L. and Szego, G.(eds.), Towards global optimisation, North-Holland/American Elsevier, 1975.

Hsin, H.C., Li, C. C., Sun, M., and Sclabassi, R.J., An adaptive training algorithm for back-propagation neural networks, IEEE Trans. on System, Man & Cybernetic, Vol. 25, pp. 512-514, 1995.

Hunt, K.J., Sbarbaro, D., Zbikowski, R., and Gawthrop, P. J., Neural networks for Control Systems - A survey, Automatica, Vol. 28, pp. 1083-1112, 1992.

Krause, P.C., Analysis of electric machinery, New York: McGraw-Hill, 1986.

Liaw, C.M., Chao, K.H., and Lin, F.J., A discrete adaptive field-oriented induction motor drive, IEEE Trans. Pow. Electron., Vol. 7, No. 2, pp. 411-419, 1992.

Liu, G.P., Kadirkamanathan, V. and Billings, S.A., Predictive control for nonlinear systems using neural networks, Int.J.Cobtrol, Vol.71, pp. 1119-1132, 1998.

Marino, R., Peresada, S. and Valigi, P., Adaptive input-output linearizing control of induction motors, IEEE Trans. Automat. Contr., Vol.38, No. 2, pp. 208-221, 1993.

Moler, M. F., A scaled conjugate gradient algorithm for fast supervised learning, Neural Networks, Vol. 6, pp. 525-533, 1993.

Narandra, K.S., Neural networks for control: theory and practice, Proceedings of the IEEE, Vol. 84, No. 10, pp. 1385-1406, 1996.

Narendra, K.S. and Parthasarathy, K., Identification and control of dynamic systems using neural networks, IEEE Trans. Neural Networks, Vol. 1, pp.4-27, 1990.

Ogata, K., Discrete-time control systems (Second edition), Prentice Hall, 1995.

Rojas, R., Neural networks -- A systematic introduction, Springer, 1996.

Zhu, Q.M., Hayns, M.R., and Garvey, S.D., Design of a neural network enhanced auto-tuning PID controller, Computational Intelligence for Modelling, Control, and Automation, Vol. 54, pp. 69-74, IOS Press, Oxford, 1999.

Zhu, Q.M., Ma, Z. and Warwick, K., A neural network enhanced generalised minimum variance self tuning control for non-linear discrete-time systems, IEE Proc. -Control Theory & Appl., Vol. 146, No. 4, pp. 319-326, 1999.